Atmospheric Particles

IUPAC Series on Analytical and Physical Chemistry of Environmental Systems

Series Editors

Jacques Buffle, *University of Geneva, Geneva, Switzerland*
Herman P. van Leeuwen, *Agricultural University, Wageningen, The Netherlands*

Series published within the framework of the activities of the IUPAC Commission on Environmental Analytical Chemistry. Analytical Chemistry Division.

Managing Editor, P. D. Gujral, IUPAC Secretariat, Oxford, UK
INTERNATIONAL UNION OF PURE AND APPLIED CHEMISTRY (IUPAC)
Secretariat, Bank Court Chambers, 2–3 Pound Way, Templars Square, Cowley, Oxford, OX4 3YF, UK

Previously published volumes (Lewis Publishers):

Environmental Particles Vol. 1 (1992)   ISBN 0 87371 589 6
Edited by **Jacques Buffle** *and* **Herman P. van Leeuwen**

Environmental Particles Vol. 2 (1993) ISBN 0 87371 895 X
Edited by Jacques Buffle *and* Herman P. van Leeuwen

Previously published volumes (John Wiley & Sons):

Metal Speciation and Bioavailability in Aquatic Systems
Edited by **André Tessier** *and* **David R. Turner**   ISBN 0 471 95830 1

Structure and Surface Reactions of Soil Particles
Edited by **P. M. Huang, N. Senesi** *and* **J. Buffle**   ISBN 0 471 95936 7

Atmospheric Particles
Edited by **R. M. Harrison** *and* **R. van Grieken**   ISBN 0 471 95935 9

Atmospheric Particles

Edited by
Roy M. Harrison
The University of Birmingham, UK

René Van Grieken
University of Antwerp, Belgium

IUPAC SERIES ON ANALYTICAL AND PHYSICAL CHEMISTRY OF ENVIRONMENTAL SYSTEMS VOLUME 5

JOHN WILEY & SONS
Chichester • New York • Brisbane • Toronto • Singapore

Other Wiley Editorial Offices

John Wiley & Sons, Inc., 605 Third Avenue,
New York, NY 10158-0012, USA

WILEY-VCH Verlag GmbH, Pappelallee 3,
D-69469 Weinheim, Germany

Jacaranda Wiley Ltd, 33 Park Road, Milton,
Queensland 4064, Australia

John Wiley & Sons (Asia) Pte Ltd, 2 Clementi Loop #02-01,
Jin Xing Distripark, Singapore 129809

John Wiley & Sons (Canada) Ltd, 22 Worcester Road,
Rexdale, Ontario M9W 1L1, Canada

Library of Congress Cataloging-in-Publication Data

Atmospheric particles / edited by Roy M. Harrison, Rene Van Grieken.
    p.   cm. — (IUPAC series on analytical and physical chemistry
of environmental systems; v. 5)
    Includes bibliographical references and index.
    ISBN 0 471 95935 9 (alk. paper)
    1. Aerosols—Environmental aspects—Technique. 2. Atmospheric
chemistry.   I. Harrison, Roy M., 1948– . II. Grieken. R. van
(René) III. Series.
QC882.46.A86   1998
551.51' 13—dc21
                                                        97-28545
                                                           CIP
British Library Cataloguing in Publication Data

A catalogue record for this book is available from the British Library

ISBN 0 471 95935 9

Typeset in 10/12pt Times by Pure Tech India Ltd, Pondicherry.
Printed and bound in Great Britain by Biddles, Guildford and King's Lynn.
This book is printed on acid-free paper responsibly manufactured from sustainable
reforestation, for which at least two trees are planted for each one used for paper production.

Contents

Series Preface

This fifth volume of the Series on Analytical and Physical Chemistry of Environmental Systems focuses on the particles in a specific environmental compartment, i.e. the air. The chemical and physical diversity of atmospheric particles is almost inconceivably large and therefore it is a tremendous job, as well as a great challenge, to tackle their analysis and physicochemical properties. We express the hope that this book will provide a stimulating basis for further development of our knowledge of environmental processes involving atmospheric particles.

We have been most fortunate to find Professors Roy M. Harrison and René E. Van Grieken prepared to edit this volume. They occupy leading positions in the field of atmospheric particles and were thus able to define a well-balanced selection of front-line topics. We not only thank them for dedicating their professional expertise to the production of this book, but also for their enthusiasm and perseverance.

The preparation of this volume was realized within the frame of our activities for the IUPAC Commission on Fundamental Environmental Chemistry in the division Chemistry and the Environment. We thank the responsible IUPAC officers for their support and assistance. We also thank the International Council of Scientific Unions (ICSU) for financial support of the work of the commission. This enabled us to organize the plenary discussion meeting (Paris, 1996) which formed such an essential step in the preparation and the harmonization of the various chapters of this book.

With the release of this volume on Atmospheric Particles, the Series has covered by and large the subject of Environmental Particles, from aquatic systems to soils and air. Future volumes will deal with other topics in envirochemistry. In different stages of preparation are the following subjects: "*In-situ* Monitoring", "Physical Chemistry of Microbial Behaviour in Environmental Systems" and "Biogeochemistry of Iron". As before, we shall collect critical reviews that characterize the current state of the art and provide guidelines for future research.

Jacques Buffle
Herman P. van Leeuwen,
Series editors

Preface

Atmospheric aerosol particles have been studied for a long time. For two or three decades, aerosols have been of great interest as carriers of heavy metal pollution, and many studies have been devoted to heterogeneous aerosol chemistry, e.g. in the context of acid rain and photochemical smog. Recently, however, there has been a revival of the field, mostly for two reasons. Firstly, it has been shown that aerosol can influence the global climate significantly, both by scattering the incoming sunlight and by enhancing cloud formation, both leading to lower surface temperatures; too many uncertainties still exist, though, to estimate accurately to what extent these effects might compensate for the well-known global warming by gases such as carbon dioxide. Secondly, important new data have been gathered about the direct health effects of aerosol particles; there are now indications that micrometre- and submicrometre-sized particles can cause increases in morbidity and even mortality in urban dwellers, possibly more than any gaseous air pollutant at current concentrations. Additionally, aerosols are attracting renewed attention nowadays because interesting new ways for the synthesis of neoceramic materials pass through an aerosol phase and because it is expected that, in the near future, more and more medicines will be delivered as an aerosol, rather than orally.

The IUPAC Commission on Fundamental Environmental Chemistry is generating a series of broadly based books on environmental particles in surface and ground-waters, air, sediments and soils. In previous volumes of this series, some individual and specific chapters have already been devoted to aerosols, but not in a systematic way; the main thrust of these previous books has been on spectroscopy and colloid chemistry aspects of environmental particles, particularly aqueous ones. Therefore it seemed appropriate to dedicate an entire volume to bring together all relevant information on the most important aspects concerning atmospheric aerosol particles. Classical and fundamental aspects of aerosol science are covered, as well as practical and technological angles, and radically new and concurrent aspects for which reviews are scarce or non-existent hitherto.

For the present volume, we have been extremely fortunate to engage some of the world's leading specialists to cover the particular aerosol chemistry or physics field for which they are renowned authorities; we are truly grateful for these scientists having dedicated so much of their precious time to this volume. We also thank Professors Jacques Buffle, previous chairman of the IUPAC Commission on Environmental Chemistry and now member of the Environmental Chemistry Division Committee, and Herman van Leeuwen, present chairman of the IUPAC Commission on Fundamental Environmental Chemistry, for their interest and continued guidance of this work.

We hope that this book will be useful for the novice in the field as well as for the experienced scientist, since we believe it brings together current state-of-the-art knowledge for the most relevant aspects of aerosol science, a field which is now undoubtedly growing spectacularly in importance once again.

R. M. Harrison
R. Van Grieken

Contributors

H. Cachier
Centre des Faibles Radioactivités, Laboratoire mixte CNRS-CEA, avenue de la Terrasse, 91198 Gif sur Yvette, France

M. Claes
Department of Chemistry, University of Antwerp (UIA), Universiteitsplein 1, B2610 Antwerpen, Belgium

C. I. Davidson
Carnegie Mellon University, Dept. of Civil Engineering, 5000 Forbes Avenue, Pittsburgh, PA 15213, USA

L. De Bock
Department of Chemistry, University of Antwerp (UIA), Universiteitsplein 1, B2610 Antwerpen, Belgium

K. Gysels
Department of Chemistry, University of Antwerp (UIA), Universiteitsplein 1, B2610 Antwerpen, Belgium

S. Harrad
School of Chemistry, University of Birmingham, Edgbaston, Birmingham, B15 2TT, UK

R. M. Harrison
Environmental Health, School of Chemistry, University of Birmingham, Edgbaston, Birmingham, B15 2TT, UK

J. Heintzenberg
Institute für Tropospharenforschung, Permoser Strasse 15, 04303 Leipzig, Germany

C. N. Hewitt
Institute of Environmental & Biological Sciences, University of Lancaster, Lancaster, LA1 4YQ, UK

H. Horvath
Institut für Experimentalphysik, Universität Wien, Boltzmanngasse 5, A-1090 Wein, Austria

J. Injuk
Department of Chemistry, University of Antwerp, (UIA), Universiteitsplein 1, B2610 Antwerpen, Belgium

R. Jaenicke
Institute for Physics of the Atmosphere, Johannes-Gutenberg Universität, Postfach 3980, W-6500 Mainz, Germany

S. G. Jennings
Department of Physics, University College Galway, Ireland

D. Mark
Occupational and Environmental Health, University of Birmingham, Edgbaston, Birmingham B15 2TT, UK

S. Matthias-Maser
Institute for the Physics of the Atmosphere, Johannes-Gutenberg Universität, Postfach 3980, W-6500 Mainz, Germany

J. M. Pacyna
Norwegian Institute for Air Research, PO Box 64, N-2001 Lillestrom, Norway

S. Potukuchi
Department of Mechanical Engineering, University of Delaware, Newark, DE 19716, USA

D. J. T. Smith
Environmental Health, School of Chemistry, University of Birmingham, Edgbaston, Birmingham B15 2TT, UK

R. L. Tanner
Tennessee Valley Authority, Muscle Shoals, AL 35662-1010, USA.

R. Van Grieken
Department of Chemistry, University of Antwerp (UIA), Universiteitsplein 1, B2610 Antwerpen, Belgium

A. S. Wexler
Dept of Mechanical Engineering, University of Delaware, Newark, DE 19716, USA

M. J. Zufall
Carnegie Mellon University, Department of Civil Engineering, 5000 Forbes Avenue, Pittsburgh, PA 15213, USA

1 Atmospheric Aerosol Size Distribution

RUPRECHT JAENICKE

University of Mainz, Germany

1 INTRODUCTION

The atmospheric aerosol [*] has many sources, natural and anthropogenic: gas-to-particle conversion, cloud-to-particle conversion, [1] soil and deserts, oceans, the biosphere, [2] volcanoes, biomass burning, direct injection. In addition, during its lifetime, the aerosol experiences many transformation processes, like coagulation, impaction and sedimentation. Despite (or maybe because of) all these differences, the aerosol exhibits size distributions, with very typical characteristics.

* For unfamiliar terms, See the glossary.

Atmospheric Particles, Edited by R. M. Harrison and R. Van Grieken.
© 1998 John Wiley & Sons Ltd.

The size distribution of any aerosol, especially of the atmospheric aerosol, is one of its core physical parameters. It determines how the various properties like mass and number density, or optical scattering, are distributed over the particle size (or radius). For the atmospheric aerosols this size (or radius) range covers more than five orders of magnitude, from one nanometre to several hundred micrometres. This large size range makes the measurements of aerosol size distribution very difficult, time-consuming, and susceptible to error. Like air masses or atmospheric water and ice clouds, the aerosol size distribution varies from place to place, with altitude, and with time.

Because of this size distribution, many integral parameters of the atmospheric aerosol, like the total mass, the optical extinction coefficient and the radioactive properties, have their centre of gravity focused in a certain size range only. And this size range determines the residence time and the atmospheric transport characteristics of that integral parameter. So some chemical compounds reside on larger particles, thus they have a short residence time and are transported only regionally. Other parameters, like some radioactive loads, are confined to smaller particles, thus have longer residence times and may be transported intercontinentally.

Our knowledge about the atmospheric aerosol size distribution has grown considerably in the past decades, but still remains fractional and limited, especially as a function of altitude and for specific size distributions. We have a rather good idea about uniform air masses and their characteristic aerosol size distribution. Because of the large variability of the aerosol over continents and in the lower parts of the atmosphere, our knowledge in these volumes is limited.

There have been several attempts to speculate why the atmosphere has such characteristic size distributions, despite the large number of production mechanisms. One of the models receiving major attention is the self-preserving size distribution and the measurements to confirm its validity. [3] For most of the atmosphere, however, it means that the time needed to reach equilibrium for self-preservation is too short. So a comprehensive picture is still lacking of why we have such size distributions.

The chapter will summarize the knowledge about the aerosol size distribution today, present some measurements and their limitations, discuss the variability and residence time, the influence of direct measurements on aerosol parameters, like chemistry and other physical properties.

2 THE MATHEMATICAL HANDLING OF THE SIZE DISTRIBUTION

The atmospheric aerosol is a continuum in radius and in its dependent properties, like number density and so on. To start with, one mathematical presentation is

$$N(r) = \int_{x=r}^{x=\infty} n(x)\mathrm{d}x \tag{1}$$

where $N(r)$ is the integral or cumulative size distribution (cm^{-3}), r the particle radius (cm), x the integration variable and $n(r)$ the differential size distribution ($\mathrm{cm}^{-3}\,\mathrm{cm}^{-1}$). This equation connects the *cumulative* $N(r)$ and the *differential* $n(r)$ number size distribution. It becomes clear that the differential number distribution is a number *density*, rather than a number *concentration*. This situation is comparable to that in statistics, with the Gaussian distribution for instance being a *probability density* rather than a probability.

On the other hand then

$$n(r) = \frac{\mathrm{d}N(r)}{\mathrm{d}r} \tag{2}$$

Often, this difference between concentrations and concentration densities has caused confusion. Not discriminating between these quantities and/or mixing them up causes misinterpretations, shift in the maximal and minimal radius of the distributions and so on.

Because of the many orders of magnitude present in atmospheric aerosol concentrations and radii, a logarithmic aerosol size distribution has often been used

$$N(r) = \int_{x=r}^{x=\infty} n^*(x)\mathrm{d}\log x \tag{3}$$

with

$$n^*(r) = \frac{\mathrm{d}N(r)}{\mathrm{d}\log(r)} \quad \text{in } \mathrm{cm}^{-3} \tag{4}$$

and presented on double-logarithmic graphs. Often the arguments have been raised that $\mathrm{d}\log(r)$ is not a proper mathematical expression, because of the logarithm of r having the dimension of a length, whether in cm or µm. However, if that expression is transformed, the meaning becomes clear:

$$\mathrm{dlog}(r) = \log(r + \mathrm{d}r) - \log(r) = \log\left(\frac{r + \mathrm{d}r}{r}\right) \tag{5}$$

The problem with that presentation is that because of its dimensions in cm^{-3}, it seems to mask $\mathrm{d}N(r)/\mathrm{d}\log(r)$ as a number concentration rather than the number concentration density it really is. Often, thus, the face values at the ordinate are misinterpreted as concentrations.

All these differential distributions, of cause, can be converted from one form into the other as Table 1 shows. This table shows clearly that some presentations are equal to others. Some people might be confused by this.

More confusion is produced, if the natural logarithm (abbreviated in this chapter as ln) is written as "log" (used in this chapter for the decadic logarithm).

Figure 1 demonstrates the different appearances for the model rural distribution, discussed later. The presentation with a linear ordinate definitely emphasizes the primary maxima and suppresses secondary maxima.

Table 1. Conversion matrix for different presentations of differential distributions. The upper row is converted into the left-hand column by using the indicated factor (thus for example the following equation develops from the table $\mathrm{d}N(r)/\mathrm{d}\log(r) = \ln(10) \cdot r\frac{\mathrm{d}N(r)}{\mathrm{d}r}$

	$\frac{\mathrm{d}N(r)}{\mathrm{d}r}$	$\frac{\mathrm{d}N(r)}{\mathrm{d}\log(r)}$	$r\frac{\mathrm{d}N(r)}{\mathrm{d}r}$	$\frac{\mathrm{d}N(r)}{\mathrm{d}\ln(r)}$
$\frac{\mathrm{d}N(r)}{\mathrm{d}r} =$	1	$\frac{1}{r\ln(10)}$	r	r
$\frac{\mathrm{d}N(r)}{\mathrm{d}\log(r)} =$	$r\ln(10)$	1	$\ln(10)$	$\ln(10)$
$r\frac{\mathrm{d}N(r)}{\mathrm{d}r} =$	$\frac{1}{r}$	$\frac{1}{\ln(10)}$	1	1
$\frac{\mathrm{d}N(r)}{\mathrm{d}\ln(r)} =$	$\frac{1}{r}$	$\frac{1}{\ln(10)}$	1	1

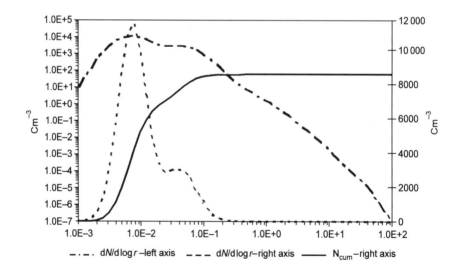

_ . _ . dN/d log r –left axis   _ _ _ dN/d log r–right axis   —— N_{cum}–right axis

Figure 1. Apparently different look and shape of the rural model number distribution depending on linear or logarithmic scale and differential or cumulative presentation

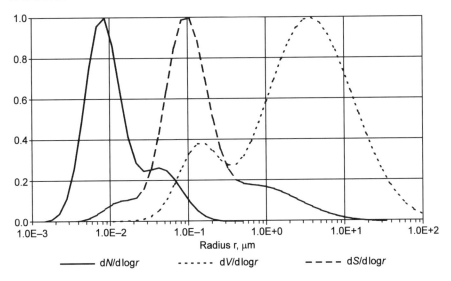

Figure 2. The different distributions number, surface, and volume for the rural (see Table 2) aerosol presented as a normalized linear scale

If only the shape of the distribution is of importance, normalized distributions are used (often termed frequency distributions). Such a presentation might be

$$\frac{1}{N_0}\frac{\mathrm{d}N(r)}{\mathrm{d}\log r} \quad \text{with } N_0 = \int_{r=0}^{r=\infty}\frac{\mathrm{d}N(r)}{\mathrm{d}r}\,\mathrm{d}r \tag{6}$$

the total concentration of all particles. But other normalizing factors are also used, like the maximum of the distribution.

So far only number concentration densities have been discussed. If needed for special questions, volume, mass or surface distributions are presented. The equations are

$$\frac{\mathrm{d}M(r)}{\mathrm{d}r} = \frac{4}{3}\pi r^3 \rho\frac{\mathrm{d}N(r)}{\mathrm{d}r} \tag{7}$$

$$\frac{\mathrm{d}V(r)}{\mathrm{d}r} = \frac{4}{3}\pi r^3\frac{\mathrm{d}N(r)}{\mathrm{d}r} \tag{8}$$

$$\frac{\mathrm{d}S(r)}{\mathrm{d}r} = 4\pi r^2\frac{\mathrm{d}N(r)}{\mathrm{d}r} \tag{9}$$

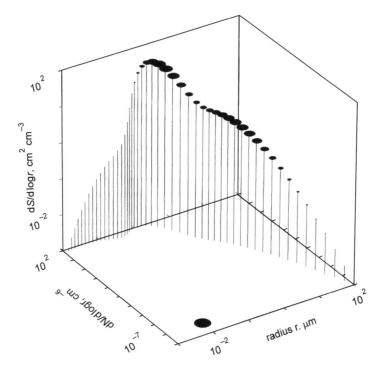

Figure 3. Three-dimensional presentation of the background (see below) model aerosol distribution as number and surface density in logarithmic scale. The volume density is represented by the size of the solid circles. The circle outside of the distribution represents $dV/d\log r = 5 \times 10^{-12}$ cm^3 cm^{-3} in a linear scale. It can be clearly seen that the number as well as the volume distribution show two maxima. The maxima of the number distribution below and around 0.1 μm radius are at different radii from the maxima of the volume distributions above 0.1 μm radius. In contrast, the surface density distribution shows only one maximum around 0.2 μm radius

where $M(r)$ is the mass concentration (g cm^{-3}), $S(r)$ the surface concentration (cm^2 cm^{-3}), $V(r)$ the volume concentration (cm^3 cm^{-3}) and ρ the bulk density of the particles (g cm^{-3}). The distributions are presented on double or semi-logarithmic graphs.

The connection between three of these distributions is shown in Figure 2 for the rural model aerosol size distribution. These distributions are normalized to their maximum value and presented on a semi-logarithmic graph. The maxima of the individual distributions are thus pronounced and shifted to different size ranges. As for most atmospheric aerosols, the number size distribution has its maximum in the Aitken particle size range ($r < 0.1$ μm), the surface distribution around 0.1 μm (large particles: $0.1 < r < 1$ μm), and the volume (mass) distri-

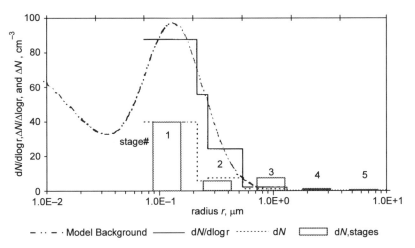

Figure 4. Simulated measurements and presentations for the model background distribution. It can be clearly seen how the distributions are distorted and how, in special cases, artificial secondary extrema are formed, like at 0.3 µm for the differential presentation (whether in impactor stages or histogram) without normalizing to the radius interval (dashed line and columns)

bution in the giant particle size range ($r > 1$ µm). Most probably, this kind of presentation (or weighing) has inspired Whitby (1973) [4] to his classification of the aerosol into modes (nucleation: $d < 0.1$ µm, accumulation: $0.1 < d < 1$ µm, and coarse mode: $d > 1$ µm).

To emphasize this property, Figure 3 shows a three-dimensional presentation of number and volume distribution.

Especially for chemical analysis, particle collectors such as impactors, sampling particles in discrete size ranges, are often used. If the selected radius ranges are not of equal width, it must be avoided presenting the data without normalizing to the radius range. Figure 4 shows what could happen in such a case. The model background distribution has been used and measurements with a typical collector, like an impactor, simulated. The $\Delta N / \Delta \log r$ measured distributions are presented, as well as ΔN only, as is often done. While the simulated distribution reflects rather closely the model distribution, the ΔN produces artefacts, such as a secondary minimum. Often the data are presented as a function of stages, rather than radius ranges. That data presentation is also calculated in Figure 4 and indicates possible misinterpretations.

The discussion shows how important it is to be extremely careful in any comparison with published data and theoretical expectations. It is not sufficient to compare "distributions" *per se* without further specification of the type of presentation.

3 PHYSICAL PRINCIPLES FOR MEASURING THE SIZE DISTRIBUTION

The aerosol size distribution covers about five orders of magnitude in radius and roughly 12 orders of magnitude in the ordinate. Such extended size ranges can hardly be measured with one single instrument, so several have to be used simultaneously. The nature of the particles requires different physical principles to be used for collection and/or identification. Examples might be diffusion batteries, electrical mobility analizers, optical particle counters (white light as well as lasers), impactors, and so on.

In particular, the monochromatic (laser) optical particle counters could be subject to misinterpretations, as Hanusch and Jaenicke [5] showed. The oscillation of the scattering curve might produce artefact minima and/or maxima in the distribution.

Another problem is the loss of particles in non-isokinetic inlets and suction ducts. This applies usually to the giant particles, while the smaller ones are lost due to diffusion in tubes. So often measured distributions are lacking certain particles [6] (for the biological particles).

Finally, the problem of the rather short residence time of the atmospheric aerosol exists [7] and its dependence on the particle size. Such short residence times produce large variations in the particle concentration. [8, 9] If measurements with various instruments are not carried out simultaneously,

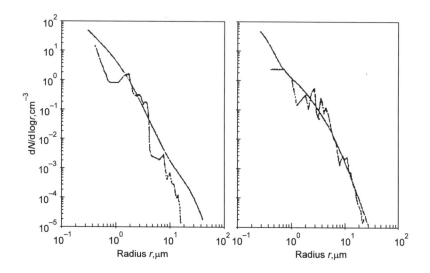

Figure 5. Simulation of published (dashed) size distributions (left, continental model; and right, maritime aerosol) with model input (smooth line) [5]

the next instrument in the suite might measure an aerosol different from the one already drifting away from the measuring location. Combining such measurements into one size distribution might produce an aerosol not existing at all.

4 MODELS OF THE ATMOSPHERIC AEROSOL SIZE DISTRIBUTION

Since Junge [10] proposed the existence of a continuous atmospheric aerosol size distribution, it has been the goal to describe the aerosol with analytical functions as well as producing model aerosol size distributions.

4.1 SIZE DISTRIBUTION EQUATIONS

Junge [10] presented his well-known power function, later modified to

$$n^* = \frac{\mathrm{d}N}{\mathrm{d}\log r} = n^*(r_0) \left(\frac{r}{r_0}\right)^{-v^*} \tag{10}$$

appearing on double logarithmic graphs as a straight line. The apparent simplicity of that equation (only two parameters) has attracted many scientists to such a degree, that they coined the equation Junge *law* and expected the atmospheric aerosol to function that way under all circumstances. However, that distribution does not allow us to show modulations in the shape of the distribution, as they are needed if cloud physical processes are described. In addition to the two parameters (r_0 and v^*) describing the shape of the distribution, a lower radius limit must be introduced, summing the needed parameters up to three.

The gamma distribution has three parameters, for reasons of proper dimensions better written as

$$n(r) = n(r_0) \left(\frac{r}{r_0}\right)^{\gamma} \exp\left(-b\left(\frac{r}{r_0}\right)^{\beta}\right) \tag{11}$$

with b, β and γ being the parameters shaping the distribution. In similar notation, it is occasionally called the modified gamma distributions. [11] Cloud physicists often use the Marshall–Palmer distribution describing cloud droplet distributions. [12] Other considerations, like that of Preining, [13] never became very popular.

For atmospheric aerosols Whitby [13] later proposed three modes classified according to assumed production mechanism as nucleation, accumulation and coarse mode. Davies [14] started using seven superimposed log-normal distribution describing aerosol size distributions

$$\frac{dN}{d\,\ln r} = \frac{N_0}{\sqrt{2\pi}\ln\sigma_g}\exp\left(-\frac{1}{2}\frac{\ln^2\left(\frac{r}{r_M}\right)}{\ln\sigma_g}\right)^2 \tag{12}$$

where $\ln\sigma_g$ is the standard deviation of $\ln r$ and r_M the median number radius. It is one of the confusing facts that the written term "log-normal distribution" is used for both presentations, natural and decadic logarithms.

Equation (12) converts with the same parameters into the presentation with the decadic logarithm according to the above conversion table:

$$\frac{dN}{d\,\log r} = \frac{N_0}{\sqrt{2\pi}\log\sigma_g}\exp\left(-\frac{1}{2}\frac{\log\left(\frac{r}{r_M}\right)}{\log\sigma_g}\right)^2 \tag{13}$$

Today three superimposed logarithmic normal (log-normal) distributions are often used. This permits in many cases testing of Whitby's three modes:

$$\frac{dN}{d\,\log r} = \sum_{i=1}^{i=3}\frac{n_i}{\sqrt{2\pi}\log\sigma_i}\exp\left(-\frac{1}{2}\frac{\log\left(\frac{r}{R_i}\right)}{\log\sigma_i}\right)^2 \tag{14}$$

4.2 GEOGRAPHICALLY SORTED ATMOSPHERIC MODEL DISTRIBUTIONS

In a first attempt to sort into geographically distinct atmospheric aerosols, Junge [15] classified 1963 atmospheric aerosols depending on their location in space into background, maritime, remote continental, and rural. This system later was expanded and quantified. [7] The data for the remote continental aerosol were added from later measurements. [16]

Of course such model distributions only reflect certain average values. Individual distribution vary, depending on local weather and wind, vertical mixing, horizontal transport, gas-to-particle conversion, season, and so on. This is especially well documented for the stratospheric aerosol layer, which depends heavily on the time passed after the outbreak of a major volcano. [17] Examples for the change of the size distribution inside and outside the polar vortex are available. [18] The agreement with recent data [19] for the urban aerosol is quite good, especially in the location of the relative maxima of the volume size distribution.

Table 2. Parameters of model aerosol distribution using equation (14), with additional data. [7] The background aerosol is expected to represent the troposphere above the cloud level. It thus might fill about 80% of the troposphere

Parameter	Polar	Background	Maritime	Remote continental	Desert dust storm	Rural	Urban	Stratosphere
n_1 (cm^{-3})	2.17×10^1	1.29×10^2	1.33×10^2	3.20×10^3	7.26×10^2	6.65×10^3	9.93×10^4	
R_1 (μm)	6.89×10^{-2}	3.60×10^{-3}	3.90×10^{-3}	1.00×10^{-2}	1.00×10^{-3}	7.39×10^{-3}	6.51×10^{-3}	
$\log \sigma_1$	2.45×10^{-1}	6.45×10^{-1}	6.57×10^{-1}	1.61×10^{-1}	2.47×10^{-1}	2.25×10^{-1}	2.45×10^{-1}	
n_2 (cm^{-3})	1.86×10^{-1}	$5.97 \times 10^{+1}$	$6.66 \times 10^{+1}$	2.90×10^3	1.14×10^3	1.47×10^2	1.11×10^3	4.49×10^0
R_2 (μm)	3.75×10^{-1}	1.27×10^{-1}	1.33×10^{-1}	5.80×10^{-2}	1.88×10^{-2}	2.69×10^{-2}	7.14×10^{-3}	2.17×10^{-1}
$\log \sigma_2$	3.00×10^{-1}	2.53×10^{-1}	2.10×10^{-1}	2.17×10^{-1}	7.70×10^{-1}	5.57×10^{-1}	6.66×10^{-1}	2.48×10^{-1}
n_3 (cm^{-3})	3.04×10^{-4}	6.35×10^{-1}	3.06×10^0	3.00×10^{-1}	1.78×10^1	1.99×10^3	3.64×10^4	
R_3 (μm)	4.29×10^0	2.59×10^{-1}	2.90×10^{-1}	9.00×10^{-1}	1.08×10^1	4.19×10^{-2}	2.48×10^{-2}	
$\log \sigma_3$	2.91×10^{-1}	4.25×10^{-1}	3.96×10^{-1}	3.80×10^{-1}	4.38×10^{-1}	2.66×10^{-1}	3.37×10^{-1}	

5 OBSERVATIONS OF DISTRIBUTIONS

In reporting about the size distribution, any information about the shape of the particles is neglected. "Neither mass nor radius describe a particle to any degree of completeness but very often our description of the aerosol amounts to little more than a very coarse, smoothed-out approximation to the distribution of particle 'size' (itself an ambiguous term, but under the circumstances quite adequate, especially since very often mass- to-radius conversions or the reverse are made some value for density as well as assuming a spherical shape)." [20] Photographs [2, 21] show clearly how variable the shape of atmospheric particles is. "It is apparent that neither the submicron nor the micron-range particles are generally spherical and somewhat irregular shapes such as those shown are prevalent—both spherical and regular crystalline forms (cubic, hexagonal and so on) are relatively rare." [20*]

However, the photographs show only a very special state of the particles. Under most atmospheric circumstances, water will be condensed on the particles, giving them a spherical shape and moving the density to that of water: the deliquescence point of major atmospheric aerosol components [22] is in the mid atmospheric relative humidity range and all aerosol particles show a hysteresis [23] in taking up and giving away water. It is only recently [24] that especially desert aerosols are observed to be non-spherical and thus exhibit an increased scattering under 90°.

Typically only certain size distributions of limited size ranges have been reported and the above model distributions have been compiled from such fractions. It is obvious that in this way large limitations are produced. Only recently have measurements of a much wider size range become available. One example is Mirme [25] with measurements at Mace Head, Ireland. Those data cover $10-10^4$ nm in diameter. While their "continental" aerosol is claimed to have one mode only, it agrees quite nicely with the "rural" model of Table 2. Their "marine" aerosol is claimed to have two modes, one definitely around, 25 nm radius. In the light of the remarks below about the variability of the aerosol, Mirmes measurements are performed within 15 minutes of time.

Screening all the possible sources for the atmospheric aerosols and especially the gas-to-particle conversion with its dependence on radiation, temporal variations should be expected. While it was possible to see temporal variations in total mass or number concentrations, [26] daily or annual changes of the size distributions have only been reported for strong influences of the gas-to-particle conversion. [25,27] However, the data are not sufficient to develop a picture of the typical daily behaviour.

---

* Most probably for that uncertainty, Twomey [11] has not included any atmospheric size distribution at all.

5.1 VERTICAL DISTRIBUTIONS

To a certain extent, information about the vertical distribution of the aerosol is included in the models (for instance, the background aerosol is defined for the

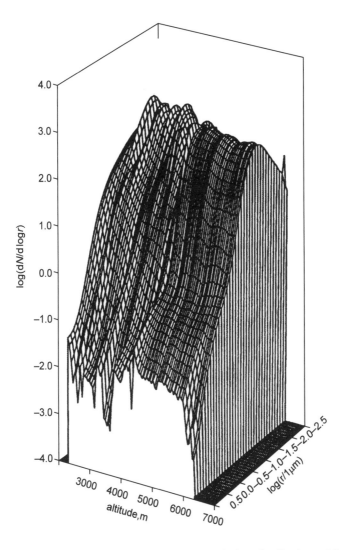

Figure 6. Example for the variation of the aerosol size distribution with altitude. The measurements were obtained over the Arctic region on 6 April 1994 between the surface and 7000 m altitude. It shows the occurrence of deep structures in the size distribution. Most pronounced are the variations for the giant particles and for those smaller than 0.1 µm radius. The range between 0.1 and 1 µm remains rather stable.

region above the clouds in the troposphere and the stratospheric aerosol covers the layer around 20 km altitude).

Some measurements have been published for layers in the atmosphere (Hobbs *et al.* [28] showed distribution bands in order to indicate the variance for particles larger than 0.01 µm), while others show selected size distribution, [29] most of them for the size range of optical instruments. Leaitch and Isaac [30] parametrized the vertical distribution in the classical way of Junge [10] (see equation 11). The large set of recent measurements [31] remain to be treated. Jaenicke [32] has parametrized the vertical tropospheric distribution of the number and mass concentration of atmospheric aerosols. If number and mass concentration do not have the same vertical dependency, it indicates a change of the distribution with altitude.

The few data available show that the distribution varies dramatically with altitude, often within metres and seconds. The general picture of rather uniform bodies of aerosols, like the background aerosol, certainly needs to be modified. Presently no information about the parametrization of such vertical distributions exists.

5.2 HORIZONTAL (TEMPORAL) DISTRIBUTIONS

The variation in the horizontal axis (location and distance) is even more pronounced. It also indicates the rapid changes the aerosol experiences with time (temporal). Figure 7 shows recent results from the Arctic region, obtained with a set of very fast instruments combined and inverted into size distributions. [33]

The rapid temporal variation is reflected even more in Figure 8. Within seconds, the number density of the giant particles varies up to several orders of magnitude. The large variation is best reflected in calculating the relative standard deviation or variation v it from measurements. For

$$v = \frac{\text{standard deviation}}{\text{average}} \tag{15}$$

Figure 9 shows the results from measurements in the Arctic region. The variation of the giant particles exceeds 10. The Aitken particles variation is rather constant with size in this stable Arctic air.

5.3 CONSTITUENT (OR SPECIFIC) DISTRIBUTIONS

Size distributions of constituents would be a logical next step, after total parameters like number and mass distributions have been presented. The problem of getting size distributions of constituents is that, usually, very long collection times are needed in the order of hours or even days. As the measurements of the physical aerosol properties show, considerable changes occur during such times.

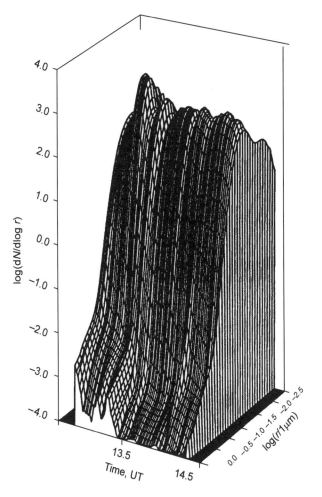

Figure 7. A 2 hour (UT = Universal time) horizontal profile at 4500 m altitude in the Arctic, north of Greenland in spring 1994. Air parcels with production of particles (high number densities at about 0.01 μm and low for particles larger than 0.1 μm) alter rapidly with high concentrations around 0.1 μm

To add to the problem, data obtained for certain compounds are converted into spheres of pure substance, thus distorting the distribution.

5.3.1 Ionic Compounds

Measurements of chemical species carry the problem of modification of the collected chemical species on the substrate, depending on the composition of

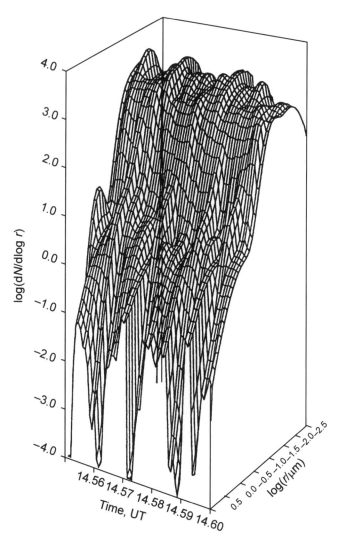

Figure 8. A three-minute time period over Western Germany at 1500 m altitude. The time is given in 1/100 hours. [34] The particle concentration around 10 μm radius varies about three orders of magnitude during this time.

the total chemistry matrix and the content of trace gases. [35] It turns out that most chemical distributions have been obtained only for selected species without the determination of the corresponding other chemical content and/or the total aerosol distribution.

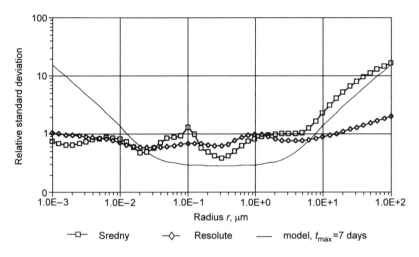

Figure 9. Relative standard deviation or variation *v* as a function of particle size. *v* has been calculated from about 5000 individual number density measurements in the western (Resolute) and eastern (Sredny) Arctic at altitudes between 150 and 8000 m in April 1994. For explanation of the model see section 6, Residence Time and Variability.

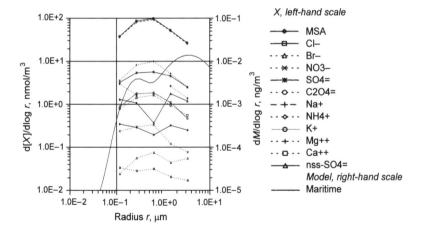

Figure 10. Average size distribution (15 measurements) of different ions at Cape Grim, Tasmania in December 1993. Part of the data analysis procedure is published. [39] The data have been supplied by Dr Bettina Schäfer, Institute for Meteorology and Geophysics, University of Frankfurt/Main, Germany. The data have been supplemented for comparison of the shape with the total mass distribution from the maritime model in Table 2. While the ion concentration distribution is given in $nmol\,m^{-3}$, the model distribution is given in $ng\,m^{-3}$ on the right-hand scale

An early data set [36] most probably includes some collection and size discriminating problems. One of the data sets about the ionic aerosol mass size fractionated composition in surface air for urban, maritime and rural aerosols [35] in the size range of 0.03–8 μm (16 μm) radius suffers from an improper handling of the total and specific mass distribution and an unclear handling of presenting the results as mass concentrations (μg m^{-3}) [37] or mass density distribution (dM/dlog$r$, μg m^{-3}). [35] In addition, the long collection times and the discussion above about variations in the aerosol, rate even this comprehensive data set as a first estimate.

The results have been digested in several, not necessary consistent, publications. [35,37,38] However, this handling problem most probably does not affect the shape of the distribution, because the instrument used had almost logarithmic equally spaced particle size ranges. They show how the ions are distributed across the particle sizes.

The data show the position of the maxima of various ions. [37] The authors report the giant particle nitrate being combined with sodium and the fine particle nitrate with ammonium, just in agreement with earlier data from coastal areas. Even for Deuselbach being located about 400 km from the ocean, Mehlmann *et al* [37] see evidence for the occurrence of sea salt. The size distribution of nitrate is dominated by coarse particles. That portion should represent background conditions. The absence of nitrate in the smaller

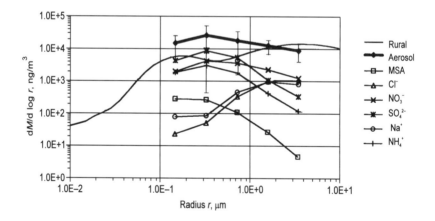

Figure 11. Average size distribution (22 measurements) of various ions at the island of Helgoland, Germany (North Sea) in June 1993. The procedure of data analysis is published. [41] The data have been supplemented with the "rural aerosol" model given in Table 2. Error bars of one standard deviation have been added to the total aerosol giving an impression about the variations observed. Data by Dr Bettina Schäfer, Institute for Meteorology and Geophysics, University of Frankfurt/Main, Germany—private communication

size ranges may have been caused by the acidification of ammonia and/or the thermal dissociation of ammonium nitrate.

More recently ion species have been measured at two locations with different character: Helgoland in the North Sea and Cape Grim at Tasmania. Helgoland is influenced by the United Kingdom and continental Europe, depending on the wind direction. Cape Grim under most circumstances is within marine influence.

The ion size distributions of Cape Grim are difficult to compare with any model total distributions, because the mass density is given in $nmol\,m^{-3}$ rather than ngm^{-3}. Most of the distributions show a monomodal shape, except for Cl^-, Br^- and MSA. The shape of non-sea-salt sulfate agrees nicely with earlier measurements. [40]

For Helgoland, Cl^- and Na^+ (and MSA) seem to follow the shape of the total distribution for particles larger than $1\,\mu m$. Below $1\,\mu m$, NH_4^+ and SO_4^{2-} seem to dominate the shape.

5.3.2 Insoluble Fraction

Clouds evaporate 10 times before they finally precipitate. [1] During their existence as cloud droplets, other particles, including insoluble ones, are incorporated. Therefore a large fraction of the resulting aerosol should not only contain water-soluble material, but also a core of insoluble material. The water-insoluble material is of great importance, because it adds mass (and size)

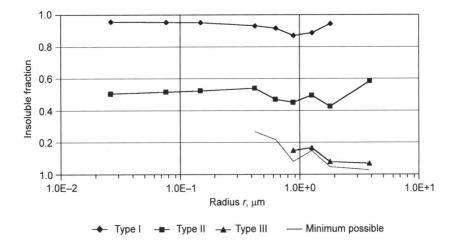

Figure 12. Water-insoluble volume fraction of aerosol particles, measured on the "Kleiner Feldberg", a mountain close to Frankfurt/Main, Germany, in October and November 1990 [43] and November 1993 (The line "minimum possible" indicates the operational lower limit of the measuring method. Thus it might well be that Type III exists below $0.8\,\mu m$ radius.) [42]

to any aerosol particle, thus increasing the surface of the water-soluble compounds with all the consequences for cloud physics and chemistry. [22] Primary biological particles are definitely water-insoluble particles and are covered separately. [2]

In recent studies [42, 43] it turned out that three types of aerosol particles exist, type I with 81%, type II with 50% and type III with 12% volume fraction of insoluble material. That observation holds quite uniformly for particles in the range 0.02–4 μm radius and is shown in Figure 12. The volume size distribution of such insoluble particles thus has the same shape as the total aerosol size distribution.

Type I has been interpreted [42] as biological material, because observations [2] indicate such a percentage. Nothing has been proposed for Type II. Type III could be evaporated cloud droplets, again because of the observed percentage in cloud droplets and because it has not been detected for Aitken particles (evaporated cloud droplets should be larger than Aitken particles).

5.3.3 Radioactivity

During the 1960s, the radioactivity of the aerosol has attracted much attention. Radioactivity on atmospheric aerosols has three major sources: emanations of radon and thoron, radioactive isotopes from cosmic rays, and anthropogenic radioactive isotopes. Most of these processes produce primary aerosol particles,

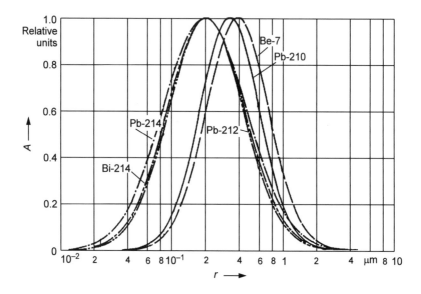

Figure 13. Atmospheric activity A in relative units as a function of particle size for several radioactive isotopes. [7]

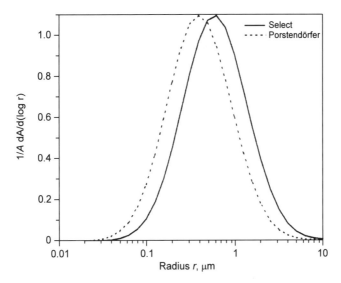

Figure 14. Be-7 activity distribution $(1/A_{max})(\Delta A/\Delta \log r)$ in Mainz, Germany from six measurements in March and April 1995, [45] each collected for about four days, compared to the Be-7 activity A in Figure 13

which later attach to larger particles by thermal diffusion. Thus the majority of radioactivity should reside on the large particles.

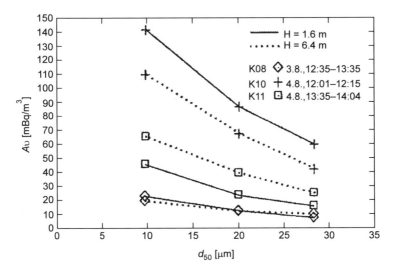

Figure 15. Activity concentration in mBq/m^{-3} as a function of collection height H and cut-off diameter d_{50} during anthropogenic activities in Kopatchi, 3–4 August 1993. [46]

Measurements for Pb-212 and Pb-214 in urban to rural aerosols [44] indicate a maximum of the activity distribution around 0.2 μm, while that of Be-7 and Pb-210 is moved to larger radii. The maximum observed recently is shifted towards smaller particle sizes, if compared to the activity A of Be-7 in Figure 13.

The activity size distribution as an example for resuspension of surface material has been measured near Chernobyl (Chornobyl), Ukraine.

As far as altitude is concerned, we are far from having any substantial knowledge about the chemical, biological and radioactive aerosol size distribution in the vertical. There have been several attempts in the past to reveal the chemistry from measurements of other parameters, [47] but mainly size unresolved. [36]

5.4 COLUMN SIZE DISTRIBUTIONS

Despite what has been said above about sampling and averaging strategies, there is the additional case of "column" or vertically averaged size distributions. These distributions are obtained by using optical instruments and retrieving the size distribution from sky brightness measurements [24, 48] or deriving it from satellites. The retrieved size distributions of course are based only on the optically active particles, that means particles in the range 0.1–10 μm radius. Its use therefore is mainly for climate forcing estimates.

One would expect the results to be mainly influenced by the background and the stratospheric aerosol. The authors claim of seeing "gas-to-liquid conversion" over cities, a dominant coarse particle mode during dust storms, and stratospheric particles after Pinatubo. [48]

6 RESIDENCE TIME AND VARIABILITY

There are several concepts about characterizing the time aerosol particles spend in the atmosphere: relaxation time, residence time, lifetime. [25] Such a time in the atmosphere is influenced by various processes, like wet [50] and dry [51] deposition and transformation in clouds. In the past the model of a first-order removal process has proved to be feasible to describe the residence time:

$$\tau = \frac{M}{S} \qquad (16)$$

where τ is the residence time, M the content of the reservoir, i.e. mass, and S the source strength (or removal strength at equilibrium), i.e. mass per time.

In a more elaborated view, for the atmospheric aerosol such a residence time is not only a measure of the time an aerosol particle spends chemically in that

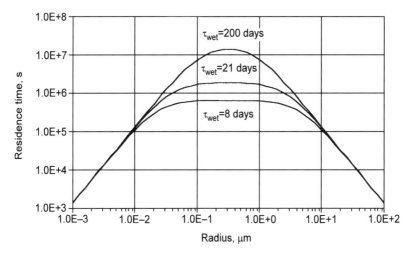

Figure 16. Equation (17) for different wet removal times, reflecting the lower and upper troposphere and the stratosphere

reservoir, it could also serve as a measure of the spatial and temporal variation of that species. In addition, it also reflects the time the aerosol particle spends in a certain size bin. If the aerosol particle is "moved" to another size bin (by coagulation, cloud incorporation and evaporation), its influence on atmospheric processes changes, and is thus terminated in its present form.

From a combination of individual estimates, an empirical size-dependent model [7] has been developed:

$$\frac{1}{\tau} = \frac{1}{1.28 \times 10^8} \left(\frac{r}{0.3}\right)^2 + \frac{1}{1.28 \times 10^8} \left(\frac{r}{0.3}\right)^{-2} + \frac{1}{\tau_{wet}} \tag{17}$$

where τ is the residence time (s), τ_{wet} the wet removal time (s) and r the particle radius (μm).

Figure 16 shows the result for various wet removal times, as they are typical for the lower troposphere and the stratosphere. It can be clearly seen, that the residence time for very large and very small particles is rather short. This is the result of large settling velocities of the giant particles and high mechanical mobility (coagulation) of the Aitken particles. The size range for which wet removal seems to be very effective, shows the longest residence times. This model agrees rather nicely with observed residences times. [7]

The aerosol residence time also influences the variability of concentrations observed. The aerosol variability results from the variability of aerosol sources, the aerosol internal processes and the atmospheric turbulence. [25] Mirme [25] has developed a model connecting the wind velocity and the residence time with a registration time of an aerosol cloud at an observation site. Another possibi-

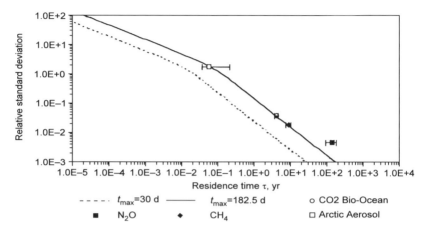

Figure 17. Equation (18) for two different t_{max} and reported variabilities of trace gases [54,55] and aerosols in the Arctic atmosphere, 1994

lity is to use that first-order removal process and connect residence time τ to variability v: [7, 52, 53]

$$v = \sqrt{\frac{1 - e^{-2\frac{t_{max}}{\tau}}}{2\left[1 - e^{-\frac{t_{max}}{\tau}}\right]^2} \frac{t_{max}}{\tau} - 1} \tag{18}$$

where v is the relative standard deviation of the observations and t_{max} a maximum time. For statistical and mathematical reasons, a maximum time t_{max} had to be introduced. [52, 53]

Figure 17 shows a good agreement between observations in the atmosphere for a $t_{max} = 0.5$ years. This time could be understood as a measure of a trace substance identity. The preservation of such an identity depends on the intensity and extent of atmospheric mixing processes. As for the atmospheric aerosol, the Arctic is definitely not a region with intense mixing. However, it is a region far away from sources in more southern latitudes. If under these circumstances such variability is still observed, it indicates the existence of certain identifiable air masses ("clouds") [56] being moved around.

ACKNOWLEDGEMENT

This work has been supported by the German Science Foundation through its Sonderforschungsbereich "Chemie und Dynamik der Hydrometeore".

GLOSSARY OF UNFAMILAR AND SPECIAL TERMS

Aerosol	A suspension of fine solid or liquid (or both mixed) particles in air
Aerosol particle	Solid or liquid (or both) particle mostly consisting of some substance(s) other than water, and without the stable bulk liquid or solid phases of water on it
Atmospheric aerosol	The atmosphere as an aerosol
Chernobyl	Chornobyl different written versions of the location in the Ukraine with the nuclear accident in 1986
Condensation nuclei (CN)	Those particles which will grow to visible size ($\sim 10\,\mu m$) when any liquid condenses on them (usually in a measuring device) at supersaturation just below that necessary to activate small ions
Deliquescence point	The relative humidity at which a water-soluble substance changes phase from crystal to liquid
Hysteresis	In this context, the observation that crystals go into solution at a given relative humidity (the deliquescence point), but for relative humidities becoming lower and lower, the particles remain liquid well below the deliquescence point
Log-normal	Logarithmic normal density distribution
MSA	Methane sulfonate aerosol
nss$-SO_4^{2-}$	Non-sea-salt SO_4^{2-}
Pinatubo	Volcano in the Philippines, erupting in 1991
Residence time	The geochemical concept of first-order removal. It is not only a measure of the time spent in a reservoir, it is also a measure of the time an aerosol particle spent in its size bin
Resuspension	Used mostly only for radioactive material: particles deposited on the ground are resuspended into the air usually by wind stress. That expression, however, could be used in a wider sense for all sorts of particles: mineral and soil dust, aerosol particles incorporated in cloud droplets and evaporated, biological particles resuspended from surfaces, and so on
Small ion	Cluster of atmospheric molecules (often water) carrying at least one electric charge

REFERENCES

1. Pruppacher, H. R., R. Jaenicke (1995). The Procesing of Water Vapor and Aerosols by Atmospheric Clouds, a Global Estimate, *Atmos. Research* **38**, 283
2. Matthias-Maser, S. (1998). Primary Biological Aerosol Particles: Their Significance, Sources, Sampling Methods, and Size Distribution in the Atmosphere, *Environmental Particles:* Vol. IV, *Atmospheric Particles* (eds) R. M. Harrison, R. E. Van Grieken, John Wiley & Sons Ltd, Chichester, Ch. 10
3. Clark, W. E., K. T. Whitby (1967). Concentration and Size Distribution Measurements of Atmospheric Aerosols and a Test of the Theory of Self-Preserving Size Distribution, *J. Atmos. Sci.* **24**, 677
4. Whitby, K. T. (1973). On the Multimodal Nature of Atmospheric Aerosol Size Distribution, Paper presented at the VIIIth International Conference on Nucleation Leningrad
5. Hanusch, T., R. Jaenicke (1993). Simulation of the Optical Particle Counter Forward Scattering Spectrometer Probe 100 (FSSP-100). Consequences for Size Distribution Measurements, *Aerosol Sci. & Techn.* **18**, 309
6. Macher (1993). Biological Aerosols, American Association for Aerosol Research. 12th Annual Meeting, Tutorial Session 10
7. Jaenicke, R. (1988): Aerosol Physics and Chemistry, *Landolt-Börnstein Numerical Data and Functional Relationships in Science and Technology New Series Group V: Geophysics and Space Research* Vol. 4 *Meteorology Subvolume b*, G. Fischer (ed): Physical and Chemical Properties of the Air, Springer-Verlag, Berlin, 391.
8. Kristament, I. S., M. J. Harvey, J. B. Liley (1993). A Seasonal Cycle in the Southwest Pacific Free Tropospheric Aerosol Concentration, *J. Geoph. Res.* **98**, 16829
9. Gras, J. (1991). Southern Hemisphere Tropospheric Aerosol Microphysics, *J. Geophys. Res.* **96**, 5345
10. Junge, C. E. (1952). Gesetzmäßigkeiten in der Größenverteilung atmosphärischer Aerosole über dem Kontinent, *Berichte des DWD in der US-Zone* **35**, 261
11. Deirmendjian, D. (1969): *Electromagnetic Scattering on Polydispersions*, Elsevier, Amsterdam.
12. Marshall, J. S., W. M. Palmer (1948). *J. Meteorol.* **5**, 165
13. Preining, O. (1972). Information Theory Applied to the Acquisition of Size Distributions, *Aerosol Sci.* **3**, 289
14. Davies, C. N. (1974). Size Distribution of Atmospheric Particles, *J. Aerosol Sci.* **5**, 293
15. Junge, C. E. (1963). *Air Chemistry and Radioactivity*, Academic Press, New York.
16. Koutsenogii, P. (1992). Number Concentration and Size Distribution of Atmospheric Aerosol in Siberia, *J. Aerosol Sci.* **25**, 377
17. McCormick, M. P., L. W. Thomason, C. R. Trepte (1995). Atmospheric Effects of the Pinatubo Eruption, *Nature* **373**, 399
18. Borrmann, S., J. E. Dye, D. Baumgardner, J. C. Wilson, H. H. Jonsson, C. A. Brock, M. Loewenstein, J. R. Podolske, G. V. Ferry, K. S. Barr (1993). In-Situ Measurements of Changes in Stratospheric Aerosol and the N_2O–Aerosol Relationship Inside and Outside of the Polar Vortex, *Geophys. Res. Lett.* **20**, 2559
19. Horvath, H., M. Kasahara, P. Pesava (1996). The Size Distribution and Composition of the Atmospheric Aerosol at a Rural and Nearby Urban Location, *J. Aerosol Science* **27**, 417
20. Twomey, S. (1977). *Atmospheric Aerosols*, Elsevier, Amsterdam, 302 pp
21. Injuk, J., L. de Bock, R. Van Grieken (1996). Structural Heterogeneity of Airborne Particles, *Environmental Particles:* Vol. IV, *Atmospheric Particles* (eds) R. M. Harrison, R. E. van Grieken,

22. Heintzenberg, J. (1998). Condensed Water Aerosols, *Environmental Particles:* Vol. IV, *Atmospheric Particles* (eds) R. M. Harrison, R. E. van Grieken, John Wiley & Sons Ltd, Chichester, Ch. 15.

23. Winkler, P., C. E. Junge (1971). Comments on "Anomalous Deliquescence of Sea Spray Aerosols", *J. Appl. Meteor.* **10**, 159

24. Hoyningen-Huene, W. v. (1996). Nonsphericity of Aerosol Particles and Their Contribution to Radiative Forcing, *J. Quant. Spectros. Radiat. Transf.* submitted

25. Mirme, A. (1994). Electric Aerosol Spectrometry, PhD Thesis, University of Tartu

26. Bigg, E. K., J. L. Gras, C. Evans (1984). Origin of Aitken Particles in Remote Regions of the Southern Hemisphere, *J. Atmos. Chem.* **1**, 203

27. Haaf, W., R. Jaenicke (1980). Results of Improved Size Distribution Measurements in the Aitken Range of Atmospheric Aerosols, *J. Aerosol Sci.* **11**, 321

28. Hobbs, P. V., D. A. Bowdle, L. F. Radke (1985). Particles in the Lower Troposphere over the High Plains of the United States. I: Size Distributions, Elemental Compositions and Morphologies, *J. Climate Appl. Meteor.* **24**, 1344

29. Deshler, T., D. J. Hofmann, B. J. Johnson, W. R. Rozier (1992). Balloonborne Measurements of the Pinatubo Aerosol Size Distribution and Volatility at Laramie, Wyoming during the Summer of 1991, *Geophys. Res. Lett.* **19**, 199

30. Leaitch, W. R., G. A. Isaac (1991). Tropospheric Aerosol Size Distributions from 1982 to 1988 over Eastern North America, *Atmos. Envir.* **25A**, 601

31. Dreiling, V., R. Jaenicke (1994). Vertical Aerosol Size Distributions in Remote Regions, Obtained by Integral Parameters, In R. C. Flagan: *4th International Aerosol Conference* Vol. 2, 1050

32. Jaenicke, R. (1993). Tropospheric Aerosols, In Hobbs, P. V. (ed): *Aerosol–Cloud–Climate Interactions*, Academic Press, New York

33. Dreiling, V., B. Friederich (1997). Spatial Distribution of the Arctic Haze Aerosol Size Distribution in Western and Eastern Arctic. *Atmos. Res.* **44**, 133

34. Dreiling, V. (1992). Bestimmung der Aerosolgrößenverteilungen in Raum und Zeit aus integralen Parametern, PhD, University of Mainz

35. Mehlmann, A. (1986). Größenverteilung des Aerosolnitrats und seine Beziehung zur gasförmigen Salpetersäure, PhD University of Mainz

36. Blifford, I. H., L.D. Ringer (1969). The Size and Number Distribution of Aerosols in the Continental Troposphere, *J. Atmos. Sci.* **26**, 716

37. Mehlmann, A., P. Warneck (1995). Atmospheric Gaseous HNO_3, Particulate Nitrate, and Aerosol Size Distributions of Major Ionic Species at a Rural site in Western Germany, *Atmos. Envir.* **29**, 2359

38. Warneck, P. (1988). *Chemistry of the Natural Atmosphere*, Academic Press, New York.

39. Ivey, J. P., G. P. Ayers, T. L. Lewis, R. W. Gillett (1996). High Volume Sampler. In Francey, R. J., A. L. Dick, N. Derek (eds) *Baseline Atmospheric Program Australia 1993*, Department of the Environment, Sport and Technology, Bureau of Meteorology, CSIRO Division of Atmospheric Research, 98

40. Götz, G., E. Mészáros, G. Vali (1991). *Atmospheric Particles and Nuclei*, Akadémiai Kiadó Budapest, 274 pp

41. Staubes, R., B. Schäfer, H. W. Georgii (1993). Dimethylsulphide and Its Reaction Products in the Marine Atmosphere, In Angeletti, G., G. Restelli (eds) *Physico-Chemical Behaviour of Atmospheric Pollutants, Proc. 6th European Symposium*, Varese Vol. 1, 307

42. Eichel, C., M. Krämer, L. Schütz, S. Wurzler (1996). The Water Soluble Fraction of Atmospheric Aerosol Particles and Its Influence on Cloud Microphysic, *J. Geophys. Res.*, submitted

43. Svenningson, I. B., H. C. Hansson, A. Wiedensohler, J. A. Ogren, K. J. Noone, A. Hallberg (1994). Hygroscopic Growth of Aerosol Particles and Its Influence on Nucleation Scavenging in Cloud: Experimental Results from Kleiner Feldberg, *Atmos. Chem.* **19**, 129

44. Becker, K. H., A. Reineking, H. G. Scheibel, J. Porstendörfer (1983). Measurements of Activity Size Distributions of Radon Daughters Indoors and Outdoors, Presented at the International Seminar on Indoor Exposure to Natural Radiation and Related Risk Assessment, Anacapri

45. Siebert, J. J. (1996). Das Beryllium-7 im Mainzer Aerosol, Master Thesis, University of Mainz, Institute for Physics of the Atmosphere

46. Wagenpfeil, F. (1995). Messungen der Resuspension von Riesen-Aerosolpartikeln in der Region von Tschernobyl, PhD, University of Munich

47. Hudson, J. G., A. D. Clarke (1992). Aerosol and Cloud Condensation Nuclei Measurements in the Kuwait Plume, *J. Geoph. Res.* **97**, 14533

48. Kaufman, Y. J., A. Gitelson, A. Karnieli, E. Ganor, R. S. Fraser, T. Nakajima, S. Mattoo, B. N. Holben (1994). Size Distribution and Scattering Phase Function of Aerosol Particles Retrieved from Sky Brightness Measurements, *J. Geophys. Res.* **99**, 10341

49. Vali, G. (1985). Nucleation Terminology, *J. Aerosol Sci.* **16**, 575

50. Jennings, S. G. (1998). Wet Processes Affecting Atmospheric Aerosols, *Environmental Particles:* Vol. IV, *Atmospheric Particles* (eds) R. M. Harrison, R. E. Van Grieken, John Wiley & Sons Ltd, Chichester, Ch. 14

51. Zufall, M. J., C. I. Davidson (1998). Dry Deposition of Particles, *Environmental Particles:* Vol. IV, *Atmospheric Particles* (eds) R. M. Harrison, R. E. Van Grieken, John Wiley & Sons Ltd, Chichester, Ch. 13

52. Slinn, W. G. N. (1988). A Simple Model for Junge's Relationship between Concentration Fluctuations and Residence Time for Tropospheric Trace Gases, *Tellus* **40B**, 229

53. Jaenicke, R. (1989). Comments on an Article of Slinn, *Tellus* **41B**, 560

54. Graedel, T. E., P. J. Crutzen (1993). *Atmospheric Change. An Earth System Perspective*, Freeman, New York, 446 pp

55. Roedel, W. (1992). *Physik in unserer Umwelt: Die Atmosphäre*, Springer, Berlin, 456 pp

56. Nakaya, U., J. Sugaya, M. Shoda (1957). Report of the Mauna Loa Expedition in the Winter of 1956–57, *J. Faculty of Sciences Hokkaido University Japan* **5**, 1

2 Atmospheric Aerosol Sampling

DAVID MARK
University of Birmingham, UK

1 INTRODUCTION

Aerosol particles in the ambient atmosphere arise from a whole range of sources: wind-raised dust from spoil heaps, construction sites and open fields, sea spray, industrial activity, emissions from animals and plants (microbial and fungal spores), traffic, volcanoes, forest fires and combustion processes, residues from the evaporation of sprays, mists and fogs, and photochemical conversion of gas to particles (mainly sulfates and nitrates). Details of the particle size distributions of these aerosols have been given in Chapter 1.

The effects of these particles on humans and the environment are many and varied, as are the measurement techniques that must be employed to monitor them. The very small particles ($\sim0.05\,\mu m$), although not very long-lived due to coalescence, etc., are known to cause the persistent red skies due to Rayleigh scattering of sunlight after large volcanic eruptions. Particles in the range 0.1 to $2-3\,\mu m$ remain in the atmosphere for the longest time (about 8 days in the lower troposphere) and are known to cause visibility problems. For example, particles from the Sahara are known to travel to the Alps where they cause a reduction in

Atmospheric Particles, Edited by R. M. Harrison and R. Van Grieken.
© 1998 John Wiley & Sons Ltd.

sunlight. Particles may deposit on buildings, cars, washing, etc., where they cause a discoloration of the surface, thereby constituting a nuisance. For some of the larger particles, such as acid smuts, physical damage can be caused to the surface. Human health can be affected in a number of ways by atmospheric aerosol particles. The most obvious way is through inhalation, with health effects ranging from bronchitis during the smog episodes in London in the early 1950s, impairment in the development of children from airborne lead from petrol-driven vehicles, and the current concern of increased asthmatic attacks, possibly associated with fine particles. Recent studies using life-size mannequins have shown that particles ranging from submicron to 100 μm (and larger) can enter the body during breathing, and deposit in various regions of the respiratory tract. The other main route is from aerosol particle deposition to crops or to ground where the toxic component (such as a radioactive isotope) may be taken into the plant which is subsequently eaten, either as meat in sheep and cattle or directly as vegetables, etc.

In this chapter we describe briefly the physical processes and properties which influence the behaviour of atmospheric particles, so as to provide the essential background for a discussion on the methods and strategies for the sampling of atmospheric particles. The information given in this chapter relates to methodology chosen to be relevant to the specific problem under consideration. It is gathered from a whole range of sources, mainly from the instrument manufacturers and the scientific literature, because guidance in standards publications (such as ISO Standards and EEC Directives) for the sampling of aerosols in the ambient atmosphere is generally inadequate and out of date. The emphasis of the information is on work carried out in the USA and this reflects the considerable effort that they have expended over the last 15–20 years in improving the method of sampling atmospheric particles for health-related purposes. This chapter does not include the chemical analysis of collected samples, but it does discuss the available sampling methods with an emphasis on the subsequent presentation of the collected sample for the chemical analysis.

2 BASIC PHYSICAL PROPERTIES OF AIRBORNE PARTICLES

The behaviour of atmospheric particles, and their effects on human health and the environment is strongly governed by a series of simple physical processes and particle properties. These differ for each particle according to its size, shape and chemical composition, with particle size being the most important (and dealt with comprehensively in Chapter 1). The main properties of aerosol particles in this context are the aerodynamic properties (involving the motion of particles in air) which determine whether, and for how long, particles remain airborne. This affects the atmospheric concentration and the distance travelled before deposition to ground. These same properties determine whether particles enter the mouth and/or nose during breathing, and how far they penetrate into the respiratory system (see later). A brief outline of these physical properties is given here. A fuller explanation of these properties can be found in a number of well-known texts (e.g. [10, 14, 32]).

2.1 THE MOTION OF AIRBORNE PARTICLES

2.1.1 Drag Force on a Particle

When a particle moves relative to the air, it experiences forces associated with the resistance by the air to its relative motion. For very slow, creeping airflow over the particle, the drag force F_D is given by the well-known Stokes' law

$$F_D = -3\pi d\eta v \tag{1}$$

where d is the geometric diameter of the particle, η the air viscosity and v the mean air velocity. This law is upheld only under certain limited conditions of particle size and aerodynamic conditions where the Reynolds number for the particle, defined as

$$Re_p = \frac{dv\rho_a}{\eta} \tag{2}$$

is very small ($Re_p < 1$), where ρ_a is the density of the air, and where the minus sign indicates that the drag force is acting so as to oppose the motion of the particle. Strictly, equation (1) should be modified by three factors:

1. Cunningham slip correction factor (C_c), which takes account of the fact that the air surrounding the particle is not continuous but is made up of individual molecules. Very small particles (smaller than the mean free path between the gas molecules (~0.06 μm for air at STP)) may slip between successive collisions with the molecules.
2. For large values of Re_p the drag coefficient tends to become constant, a situation known as "non-Stokesian" regime.
3. Generally, particles are not spherical, and particle motion may be affected by their orientation.

Nevertheless, it may—to a first approximation—be considered as a reasonable working assumption for understanding the motion of airborne particles, that Stokes' law holds.

The starting point for all considerations of particle transport is the again well-known Newton's second law (mass × acceleration = net force acting). For this situation the net forces comprise the drag force (described above) and external forces such as gravity, electrical, or some combination of forces. The effect of these forces is to generate and sustain particle motion, and providing that the particle is in motion relative to the air, the drag force will remain finite. For aerosol sampling, however, the nature of the drag force is the predominating influence since the external forces are usually of secondary importance. When the motion of the air, the particle and the forces acting, are considered in the three dimensions, the equations required to predict particle motion can become quite complicated.

2.1.2 Motion under the Influence of Gravity

A simple but very important example of particle motion under the influence of external forces can be found in the motion of particles under the influence of gravity. The forces acting on the particle are shown in Figure 1. The equation of motion for a spherical particle moving in the vertical (y) direction is given by

$$m\frac{\mathrm{d}v_y}{\mathrm{d}t} = 3\pi\eta dv_y - mg \tag{3}$$

where v_y is the particle velocity in the y-direction, m is the particle mass and g is the acceleration due to gravity. This can be reorganized to give

$$\frac{\mathrm{d}v_y}{\mathrm{d}t} + \frac{v_y}{\tau} - g = 0 \tag{4}$$

where τ is known as the particle relaxation time and is given by

$$\tau = \frac{d^2\rho_p}{18\eta} \tag{5}$$

and ρ_p is the particle density. Equation (4) is a simple first-order linear differential equation from which it can be shown that particle velocity under the

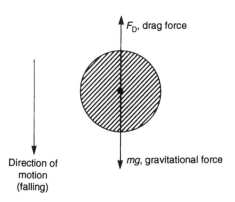

Figure 1 Forces acting on a particle of mass m falling under gravity

influence of gravity tends exponentially towards a terminal value, known as the
sedimentation or *falling speed*, given by

$$v_s = g\tau \tag{6}$$

For example, the falling speeds of spherical particles with the same density as
water $(10^3 \, \text{kg m}^{-3})$ can be obtained as follows: for $d = 1 \, \mu\text{m}$, $v_s =$
$0.003 \, \text{mm s}^{-1}$; $d = 5 \, \mu\text{m}$, $_s = 0.8 \, \text{mm s}^{-1}$; $d = 10 \, \mu\text{m s}^- = 3 \, \text{mm s}^{-1}$; $d =$
$100 \, \mu\text{m}$, $v_s = 240 \, \text{mm s}^{-1}$, etc.

This is the simple case where slip, non-Stokesian conditions and particle nonsphericity are neglected. More generally:

$$\left(v_s = \frac{C_c}{\phi} \right) \left(\frac{24}{C_D Re_p} \right) g\tau \tag{7}$$

Sedimentation is the main mechanism for the deposition of large particles on to
horizontal surfaces such as cars, soil and vegetation. The main complicating
factor is the turbulent fluctuations in wind speed near to the ground which can
include upward motions of velocity as high as $500 \, \text{mm s}^{-1}$. However, these are
rare occurrences and the particles will soon deposit again some time later.

2.1.3 Motion without External Forces

The concept of particle motion without the application of an external force is
also helpful in understanding the behaviour of airborne particles. For example,
consider the simplest case where the air is stationary and a spherical particle is
projected into it with finite initial velocity in the x-direction. Motion is
described by the equation

$$m\frac{dv_x}{dt} = -3\pi\eta dv_x \tag{8}$$

This has the simple solution

$$v_x = v_{x0} \exp\left(-\frac{t}{\tau}\right) \tag{9}$$

where v_{x0} is the initial particle velocity relative to the fluid at time $t = 0$. This
expression has particular relevance to moving air, since it describes how a
particle, initially injected into an airflow with zero velocity, is progressively
pulled along by the drag force exerted by the fluid until it eventually catches up
with it. The particle is then moving at the same velocity as the air and may be
considered to be airborne.

Further integration gives the distance travelled by the particle relative to the
air before it catches up with the airflow, or before it comes to rest if the air is
stationary. This is known as the *particle stop distance* (s) and is given by

$$s = v_{x0}\tau \tag{10}$$

This concept is particularly important when considering how particles behave within moving air that is changing direction. For instance, in the distorted flows entering the human respiratory system and passing through the lung airways or entering aerosol samplers.

For particles moving in these distorted flows, their behaviour may be scaled so that the dimensionless quantity known as the *Stokes number (St)* is constant. *St* is given by

$$St = \frac{d^2 \rho_p U}{18\eta D} \tag{11}$$

where D and U are characteristic dimensional and velocity scales respectively. For particles in the vicinity of aerosol samplers, U is normally the external wind speed and D the dimension of the orifice, the flow into which is responsible for the distortion. Combining equation (5) with equation (10) gives

$$St = \frac{\tau}{(D/U)} = \frac{\tau}{\tau_d} \tag{12}$$

where τ_d is equivalent to a time-scale which is characteristic of the flow distortion. For a very small particle with small particle relaxation time (τ), St will be small, indicating that the particle will tend to respond quickly to changes in the flow and tend to follow the airflow closely. A large particle, with large τ and correspondingly larger St, will tend to respond less effectively to the changing

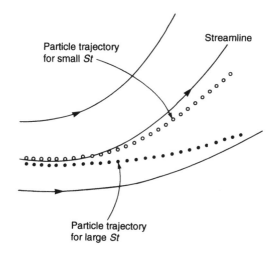

Particle trajectory
for small *St*

Streamline

Particle trajectory
for large *St*

Figure 2 Diagram to illustrate concept of particle inertia

direction. A very large particle will therefore not follow the changes in flow direction and velocity, but will tend to follow the direction of its original motion.

This behaviour can also be followed by combining equations (10) and (12):

$$St = \frac{s}{D} \tag{13}$$

where St is now expressed as the ratio of particle stop distance (s) to the characteristic dimension of the flow distortion (D). Here the particle will tend to follow the airflow when s is small compared to D, and will not follow the airflow when s is large compared to D.

This discussion has led us to understand that St is an important measure of the ability of an airborne particle to respond to the movement of the air around it, and that particle trajectory patterns may differ to an extent dictated largely by the magnitude of St. We therefore have the concept of particle *inertia*, which is a function of both the particle and the airflow in which it is moving. Figure 2 illustrates this concept, which is of great importance in understanding the behaviour of aerosol samplers.

2.1.4 Particle Aerodynamic Diameter

One of the most useful concepts in particle transport is that of the *aerodynamic diameter* (d_{ae}) of the particle. This is defined as the diameter of an equivalent spherical particle of density 10^3 kg m^{-3} (ρ^*) with the same falling speed as the particle in question. It is related to the particle geometrical diameter d by the following:

$$d_{ae} = d \left(\frac{\rho}{\rho^*} \right)^{0.5} \tag{14}$$

For non spherical particles we must introduce an additional quantity known as *dynamic shape factor* (ϕ). Other definitions of particle size are given later in this section (2.5).

2.2 IMPACTION AND INTERCEPTION

Consider what happens when particles are transported in a distorted airflow such as around a bend or about a bluff obstacle (Figure 3). The air itself diverges to pass around the outside of the obstacle. Very small (inertialess) particles would do the same, but larger particles because of the inertial behaviour described above would tend to leave the flow streamlines and continue to travel in the direction of their initial motion. This tendency is greater the more massive the particle, the greater its approach velocity, and the more sharply the

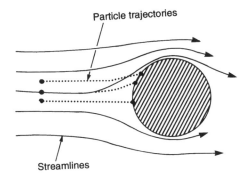

Figure 3 Diagram to illustrate the impaction of particles

flow diverges. This behaviour is consistent with a dependence on Stokes' number (St) described in equation (10), where D is the obstacle dimension and U is the approach velocity. It can be seen in Figure 3 that some particles impact on to the surface of the obstacle, and assuming that they all stick to the surface, the *impaction efficiency* (E) is defined as

$$E = \frac{Number\ of\ particles\ arriving\ by\ impaction}{Number\ of\ particles\ geometrically\ incident\ on\ the\ body} \qquad (15)$$

In addition, the particle trajectories will be dependent upon the Reynolds numbers of both the particle Re_p and the flow around the obstacle (Re), such that

$$E = f(St,\ Re_p,\ Re) \qquad (16)$$

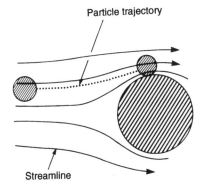

Figure 4 Diagram to illustrate the collection of particles by interception

Impaction is particularly important as a removal mechanism for particles close to the ground as they are carried by the wind around leaves, trees, grass, irregularities in the soil surface, as well as buildings and other man-made obstacles. It is also the main mechanism in the sampling of airborne particles whether by aspiration or by the use of impaction surfaces. In addition it is used as a means of separating particles according to their aerodynamic diameters in cascade impactors (see later).

Particles which are of the same order of size as the obstacle itself may be collected by *interception* if the particle trajectory passes close enough to the surface of the obstacle so that they touch, as shown in Figure 4. This is not an efficient collection process but is enhanced in practice by the presence of fine surface structure (e.g. leaf hairs, surface roughness of soil and building materials). For particle sampling, however, the phenomenon of interception may be largely disregarded.

2.3 DIFFUSION

2.3.1 Molecular Diffusion

So far we have discussed the transport of airborne particles by processes that are assumed to be well ordered and, in theory, deterministic. In practice, however, small particles in still or smooth airflows are seen to move randomly and erratically. This motion results from collisions with the surface of the particles of gas molecules, which are in thermal motion as described by the classical kinetic theory of gases. This movement is independent of any convection associated with the air itself and is known as '*molecular*' or '*Brownian*' *diffusion*. The result of this phenomenon is a net migration of particles from regions of high concentration to regions of low concentration. The resultant local net flux of particles by this process is described by Fick's law of classical diffusion, which in the simplest, one-dimensional case is

$$\text{Local net flux} = -D_B \frac{dc}{dx} \tag{17}$$

where c is the local particle concentration and D_B the coefficient of Brownian diffusion. For small particles in the Stokes' regime, the latter is given by

$$D_B = \frac{k_B T}{3\pi \eta d_v} \tag{18}$$

where T is the air temperature (in K) and k_B the Boltzmann constant $(1.38 \times 10^{-23} \text{J K}^{-1})$. D_B therefore embodies the continual interchange of thermal energy between the gas molecules and the particles and vice versa. The effect of diffusion on particle transport becomes more significant for smaller particles, although even for a particle of diameter 1 μm in air, D_B is only of the order of 10^{-11} m^2 s^{-1}.

From equation (16), the resultant rate of change of local particle concentration is therefore

$$\frac{dc}{dt} = D_B \frac{d^2c}{dx^2} \tag{19}$$

whose solution for the case of N_0 particles released initially at $x = 0$ at time $t = 0$ gives the Gaussian form

$$c(x, t) = \frac{N_0}{2\pi D_B t^{0.5}} \exp\left(\frac{-x^2}{4D_B t}\right) \tag{20}$$

for the concentration distribution along the x-direction at time t. The root mean square displacement of particles from their origin at time t is

$$x' = (2D_B t)^{0.5} \tag{21}$$

The phenomenon of diffusion is not only important in how particles move from one point to another, but also in how they move in relation to one another. It is responsible for collisions between particles which result in the coagulation of small particles to form larger ones.

2.3.2 Turbulent Diffusion

The mixing of particles in a turbulent airflow may be thought of as a form of diffusion over and above the molecular variety described above. In most cases, the flux associated with turbulent diffusion may be described in terms of an expression which is directly analogous to Fick's law as given in equation (17). The only difference is that the turbulent diffusivity of the particles, D_{pt}, replaces D_B. The particle's ability to respond to the eddying, distorted turbulent motions of the surrounding air is dependent upon inertial considerations similar to those discussed earlier. We may describe an inertial parameter (K_{pt}) similar to the Stokes' number already defined:

$$K_{pt} = \frac{\tau u'}{L} \tag{22}$$

where τ is the particle relaxation time, and u' and L are the characteristic turbulence parameters. This leads to

$$\frac{D_{pt}}{D_{ft}} = f(K_{pt}) \tag{23}$$

where D_{ft} is the turbulent diffusivity for the fluid. As K_{pt} increases (i.e. larger particles, greater turbulence intensity, smaller turbulence length scale), the particle responds less well to the fluctuations and so its diffusivity falls. For very large particles or very small length scales the particle does not see the turbulent motions as it travels with the mean flow.

2.4 ASPIRATION

Aspiration concerns the process by which particles are withdrawn from ambient air through an opening in an otherwise enclosed body. It is therefore relevant to aerosol sampling systems, and to the inhalation of aerosols by humans through the nose and/or mouth during breathing. Figure 5 shows a body of arbitrary shape placed in a moving airstream. It has a single orifice positioned at arbitrary orientation to the wind through which air is drawn at a fixed volumetric flow rate by a sampling pump. Particles brought to the vicinity of the sampler by the wind, experience two competing flow regimes, and their behaviour in these regimes is dependent upon the inertial forces as the flow changes direction. Firstly, the external part diverges to pass around the outside of the body, and the particles undergo an impaction process on to the body, similar to that described earlier. Secondly, the particles impact on to the plane of the orifice as they experience the convergent flow into the orifice. For more details see [32].

   Aspiration efficiency (E_A), may be defined for a given particle aerodynamic (d_{ae}) as

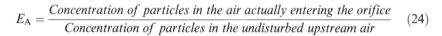

$$E_A = \frac{\textit{Concentration of particles in the air actually entering the orifice}}{\textit{Concentration of particles in the undisturbed upstream air}} \quad (24)$$

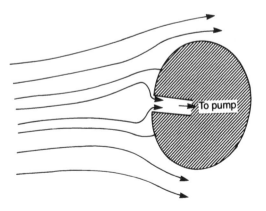

Figure 5 Diagram to illustrate the aspiration of particles into an entry

provided that the airflow and aerosol upsteam of the sampler are uniformly distributed in space. From considerations of particle impaction from one region of the flow to another, a system of equations may be developed which can, in principle, provide estimates for E_A. Generally:

$$E_A = f(St, U/U_s, \delta/D, \theta, B) \qquad (25)$$

where $St(= d_{ae}^2 \rho^* \mu U / 18\eta D)$ is a characteristic Stokes' number for the aspiration system, U the external wind speed, U_s the mean sampling velocity, δ the orifice dimension, D the body dimension, θ the orientation with respect to the wind, and B an aerodynamic bluffness factor.

2.5 PARTICLE SIZE PARAMETERS

Particle size is the most important parameter used to assist in defining the physical characteristics and behaviour of airborne particles. The size distribution of a particular aerosol is generally polydisperse with sometimes up to a 100-fold range between the smallest and largest particles. An appreciation of how aerosol properties can vary with particle size is fundamental to the understanding of their behaviour. Consequently, it is necessary to adopt a microscopic approach and characterize properties on an individual particle basis, with average properties being estimated by integrating over the size distribution.

Aerosol particles are normally sized in terms of a characteristic dimension relating to either the radius or the diameter of the particle, with common units being micrometre (10^{-6} m) and nanometre (10^{-9} m). The simplest case is that of a particle which is perfectly spherical. By definition this has only one dimension—its true *geometric diameter* (d), which is the measurement obtained from evaluation made under a microscope. However, very few particles are spherical (apart from droplets and some combustion particles) and so another index must be found to enable particle "size" to be defined. Use is made of definitions described in terms of one or more "effective" or "equivalent" diameters—not true diameters as such, but dimensions derived from knowledge of some other property (or combination of properties) of the particle.

2.5.1 Geometric Diameters of Non-spherical Particles

Firstly, there are a number of "characteristic" effective geometric diameters derived from two-dimensional pictures of non-spherical particles. The *Feret* (d_F) and *Martin* (d_M) diameters are indices of particle size relating to a dimension of a given particle as seen under the microscope. This measurement is not unique, however, as the choice of orientation of the particle in the field of view is arbitrary. A more usable definition is the *equivalent projected area diameter* (d_P), which is the diameter of an equivalent sphere which (in 2-D) projects the same area as the particle in question. This too, however, is dependent upon the orientation at which the particle is viewed, and so is most appropriate for ensembles of particles.

There are also two other geometrical definitions which are unique for a given individual particle. The *equivalent surface area diameter* (d_A) is the diameter of a sphere that has the same surface area as the particle in question. Both d_P and d_A are relevant to many aspects of the visual and optical appearance of aerosols. Finally, there is the *equivalent volume diameter* (d_V), the diameter of a sphere that has the same volume as that of the non-spherical particle in question, and is relevant to the drag force that the particle experiences as it moves through the air.

2.5.2 Equivalent Diameters based on Behavioural Properties

There are a number of equivalent diameter definitions based upon the behaviour of the particle when subjected to some external force. The most important of these for ambient aerosols is the *aerodynamic diameter* (d_{ae}) which is essential for describing the motion of aerosol particles, and was therefore presented previously. It can also be derived from d_V with a knowledge of the particle density and shape from the following relation:

$$d_{ae} = d_V \left\{ \frac{\rho C_c Re_p^*}{\rho^* C_c^* Re_p \phi} \right\} \tag{26}$$

where ρ_p is the particle density, C_c the Cunningham slip correction factor for the particle, Re_p the particle Reynolds number and ϕ the dynamic shape factor. The terms marked with * refer to the spherical water droplet.

There are many other equivalent diameters for properties such as electrical mobility and diffusion, but an interesting problem arises when considering the complex aggregates formed during combustion (e.g. from car exhausts or chimney stacks). These are made up of large numbers of very small primary particles ($\sim$ tens of nm in size) and the degree of complexity is such that it is difficult to describe the size of the aggregate by any of the geometric definitions. While we can describe aspects of its geometric appearance and use the aerodynamic diameter to describe its aerodynamic behaviour, these do not convey the full nature of the airborne particles. For these particles, Mandelbrot's concept of fractal geometry can convey additional information [21]. This relates to the fact that for certain particles the detailed structures of the chain-like complexes repeat themselves when viewed at progressively larger magnifications. A parameter used to describe these particles is N, the number of primary particles, which is related to the radius of gyration of the particle (R_g), and the fractal dimension (f):

$$N = R_g f \tag{27}$$

3 TYPES OF SAMPLING INSTRUMENT

The assessment of the potential environmental and human health effects of aerosols in the atmosphere requires detailed physical and chemical characterization of the aerosol particles themselves. The size and shape of the particles, their bulk composition, and the distribution of chemical constituents within the particles all have importance. The particle size distribution is very wide and variable and those particles have a wide-ranging effect on both the environment and on humans. The methods used for sampling the aerosols need to be chosen with care. The criteria for the choice are dependent upon answering the simple questions posed above; 'Why do you want to sample aerosols?' and 'How will the results be used?' The answers should provide the sampling rationale relevant to the practical situation in question, and the appropriate sampling methodology can then be prescribed. It is not sufficient to measure what is loosely termed 'total suspended particulate' (TSP), or 'suspended particulate matter' (SPM), as the definitions of these terms are not specific but depend upon the instruments used to measure them. These parameters do not relate to any specific effect that requires the measurement of aerosols. For example, instruments conforming to health-related sampling conventions should be employed to monitor aerosols for health effects; while for dust nuisance, methods related directly to the nuisance are preferred.

There are many reasons for sampling particles in the ambient atmosphere:

- Compliance with air quality standards, set either for the protection of humans from health effects or for the protection of the environment
- Data for epidemiological studies to determine risks to human health from particles
- Study of atmospheric transport of both man-made and natural particles
- Assessment of pollution sources so that strategies can be devised for their abatement and control
- Study of particle deposition for nuisance purposes
- Evaluate phenomenon causing degradation of visibility

For most of the sampling described above, an aspiration sampler involving the extraction of a sample of air for analysis of the particles carried in the air, is the most suitable approach. The other main technique involves the passive collection of particles to represent particle deposition. Before proceeding to describe the sampling methodology in detail, it is instructive to discuss briefly the components of a sampling system relying on aspiration to sample the ambient air.

3.1 THE BASICS OF AN AEROSOL SAMPLING SYSTEM

An aerosol sampling instrument (or sampler) always comprises a number of components which contribute to the overall accuracy with which a sample is

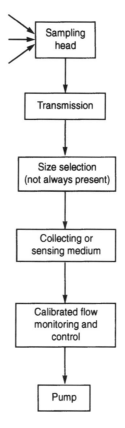

Figure 6 The basics of an aerosol sampling system

taken. These components are: the sampling head, the transmission section, the particle size selector (not always present), the collecting or sensing region, calibrated flow monitoring and control, and the pump. A simple schematic diagram of these essential components is given in Figure 6.

3.1.1 The Sampling Inlet

Many environmentalists have, in the past, overlooked the choice of sampling inlet, instead they have concentrated on the protection of the filter from the rain and snow, the performance of the pump, the choice of filter and the analytical technique to be employed. The use of any cover which in effect defines the sampling inlet (as in the well-known high volume sampler), may lead to large errors in the mass concentration measured, due to variable and inefficient particle sampling. This is especially neglected in the use of cascade impactors to obtain particle size distribution information about the airborne particles. Very often large particles are significantly under-represented in the particle size distributions reported. An inlet should be chosen that meets the requirements of the sampling strategy chosen. Usually a sampling inlet for environmental aerosols must be omnidirectional and have performance independent of wind speed.

3.1.2 Transmission Section

The transmission of sampled particles from the inlet to the particle size selector (if present) and on to the collecting or sensing region is an important part of the sampling system that is again often overlooked. The main problem is to avoid particle losses to the internal walls of the transmission pipe or channel. In samplers where the sampling inlet, size selector and collecting filter are built into a sampling head, then care has normally been taken with the design of the sampler to minimize this problem. However, for instruments where the inlet is remote from the sensing region it is essential to minimize the length of the connecting pipework and to keep sharp bends and horizontal surfaces to a minimum. Wall losses can occur over the whole size spectrum of particles, ranging from inertial and sedimentation losses for the large particles (aerodynamic diameters $> 5 \, \mu m$) to diffusional losses for particles of diameters $< 100 \, nm$. The use of non-conducting plastic piping is not recommended due to possible enhanced deposition from electrostatic forces.

3.1.3 Size Selection

For health-related sampling we wish to select aerosol fractions that relate to particle deposition in various regions of the respiratory system. For this, some form of particle size selector is used to select the relevant portion of the sampled aerosol. Particles are generally selected by aerodynamic means using physical processes similar to those involved in the deposition of particles in the respiratory system. Gravitational sedimentation processes are used to select particles in horizontal and vertical elutriators, centrifugal sedimentation is used in cyclones, inertial forces are used in impactors, while porous foams employ a combination of both sedimentation and inertia forces.

Generally, in samplers for environmental aerosols, impactors or cyclones are used for particle size selection, although porous foams have been used in some prototype samplers.

3.1.4 Filters

A filter is the most common means of collecting the aerosol sample in a form suitable for assessment. That assessment might include gravimetric weighing on an analytical balance before and after sampling to obtain the sampled mass. It might also include visual assessment using an optical or electron microscope and/or a whole range of analytical and chemical techniques. The choice of filter type for a given application depends greatly on how it is proposed to analize the collected sample. Many different filter materials, with markedly different physical and chemical properties, are now available. These include fibrous (e.g. glass, quartz), membrane (e.g. cellulose nitrate, polycarbonate, Teflon) and sintered (e.g. silver) filters. Membrane filters have the advantage that they can retain particles effectively on their surface (good for microscopy), whereas fibrous filters have the advantage of providing in-depth particle collection and hence a high load-carrying capacity (good for gravimetric assessment).

Such filters are available in a range of dimensions (e.g. 25–100 mm diameter) and pore sizes (e.g. from 0.1 to 10 μm). Collection efficiency is usually close to 100% for particles in most size ranges of interest, although sometimes some reduction in efficiency might be traded against the lower pressure drop requirements of a filter with greater pore size. For some types of filter, electrostatic charge can present aerosol collection and handling problems—in which case, the use of a static eliminator may (but not always) provide a solution. For other types, weight variations due to moisture absorption can cause difficulty, especially when being used for the gravimetric assessment of low masses. It is therefore recommended that the stabilization of filters overnight in the laboratory should be carried out before each weighing, together with the use of blank "control" filters to establish the level of variability. It is preferable that temperature and humidity control in the balance room be provided, especially when collected particle weights are low.

The chemical requirements of filters depend on the nature of the analysis which is proposed. As already mentioned, weight stability is important for gravimetric assessment. If particle counting by optical microscopy is required, then the filters used must be capable of being rendered transparent (i.e. cleared). Direct on-filter determination of elemental composition (e.g. by scanning electron microscope and energy-dispersive X-ray analizes, X-ray fluorescence, neutron activation analysis and particle induced X-ray emission) is often required. For these, filters must allow good transmission of the radiation in question, with low background scatter. Collected samples may also be extracted from the filter prior to analysis, using a range of wet chemical methods, ultrasonication, ashing, etc., each of which imposes a range of specific filter requirements.

3.1.5 Pumps

Most samplers require a source of air movement so that particulate-laden air can be aspirated into the instrument. Generally, because aerosol concentrations in the ambient environment are low, flow rates must be high to collect sufficient particles to weigh or to analyze chemically. The actual volumetric flow rate will depend first on sampling considerations (e.g. entry conditions to provide the desired performance), and then the amount of material to be collected for accurate assessment, analytical requirements, etc. Flow rates range from 16.7 to 1200 l min^{-1}, with one sampler (Wide Ranging Aerosol Classifier, WRAC) much higher. Internal flowmeters, usually of the rotameter type or digital counters, are incorporated into most samplers, but these must always be calibrated against a calibrated flow rate standard (e.g. wet gas meter, orifice plate) placed over the sampler inlet. It should also be noted that the flow rate may vary with the resistance imposed by the filter and its collected aerosol mass. For this reason, flow rates should be checked periodically during sampling and adjusted if necessary. Nowadays, however, most pumps employ some form of flow control where a sensor (e.g. pressure or velocity) is built into a feedback loop to eliminate the need for such regular attention during sampling.

3.2 SAMPLERS FOR "TOTAL" AEROSOL

If information is required about the characteristics of all the particles that comprise the environmental aerosol, then a sample must be collected that represents all particle sizes and types in the same relative proportions that they are present in the atmosphere. A sampler achieving this goal would have a sampling efficiency of unity for all particles and provide a true sample of "total aerosol". Only then can chemical and physical analysis of the sampled aerosol particles be considered as an unbiased characterization of atmospheric aerosols. In reality, however, there are no samplers commercially available that are capable of meeting this requirement for all particle sizes and environmental conditions found in the ambient atmosphere.

There are many technical problems involved in designing samplers for "total" aerosol in the ambient atmosphere. They include: (A) the low particle levels found, requiring very high sampling flow rates; (B) the wide range of wind speeds over which they must operate; (C) the requirement to have performance independent of direction to the wind (i.e. must have omnidirectional entry); (D) adverse weather conditions (rain, snow, mist) in which they must operate; and (E) the potentially wide range of particle sizes to be sampled.

European instrument designers in the early 1980s concentrated their efforts on protecting their samplers from the ingress of rain, snow, etc., and they produced similar samplers with flow rates from 4.5 to 45 l min^{-1}, featuring downwards-facing omnidirectional entries, protected by cowls and rain shields. A selection of these samplers is shown in Figure 7. Their sampling efficiencies are given in Figure 8, and it can be seen that performance varies strongly with both particle aerodynamic diameter and wind speed in a manner that is broadly consistent with theory based on vertical elutriation modified to include the effects of cross winds. It is to be expected that any other sampler with similar

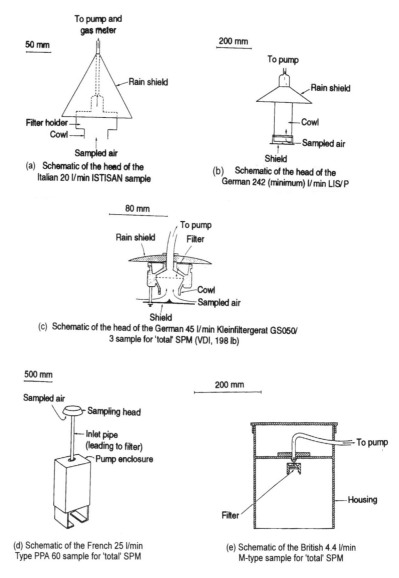

Figure 7 Selecton of early samplers for "total" aerosol concentrations

features would have similar sampling efficiency. Evidently none is satisfactory for sampling true total aerosol.

At the same time in the USA, the well-known Hi-vol TSP sampler was produced. Sampling at flow rates of 1200 l min^{-1}, it features aerosol collection

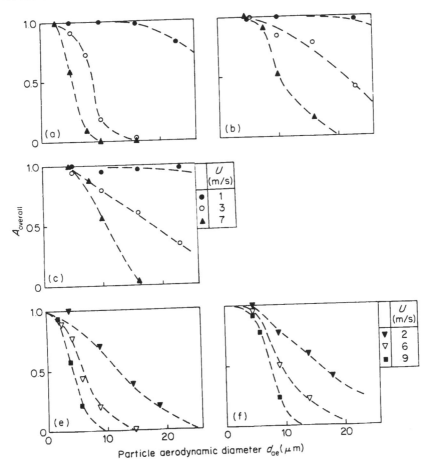

Figure 8 Sampling efficiency for each of the samplers shown in Figure 7

on to a large upwards-facing rectangular filter which is located inside a large, weatherproof housing (see Figure 9). Its performance was broadly similar to those for the samplers featured in Figure 7. In addition due to the rectangular cross-section of the sampler its performance is also dependent upon orientation to the wind (see Figure 9). However, in a recent study of the performance of the Hi-vol PM 10 sampling head (see later) it was found that if the internal impactor size selector is removed, the sampling efficiency is close to that of the Hi-vol TSP sampler, but without the orientation dependence. At this stage, it is important to mention the performance of the entry of widely used Andersen cascade impactor. Both versions of the device (viable and general particles) have a single upwards-facing circular sampling orifice as shown in Figure 10.

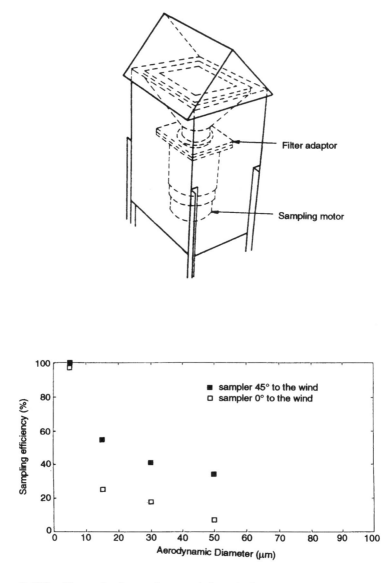

Figure 9 "Hi-vol" sampler for total suspended particulate matter

The performance (also shown in Figure 10) is very similar to those for other samplers mentioned, and is very dependent upon wind speed and particle size. This means, therefore, that the accuracy of the particle size distribution measurements obtained with this instrument is dependent upon the environmental conditions.

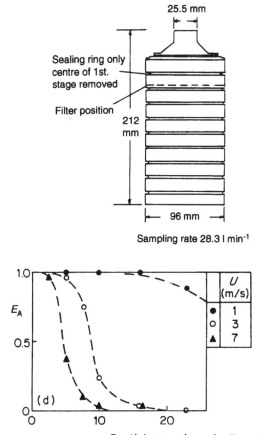

Figure 10 Andersen cascade impactor widely used for particle size distribution measurements of ambient aerosols

However, there are a small number of instruments that have been designed with the specific aim of sampling true total aerosol in the ambient atmosphere. The Wide Ranging Aerosol Classifier (WRAC) [7] makes use of large physical dimensions relative to the stop distance of ambient particles to ensure that particles up to at least 60 μm are sampled with high efficiency independently of wind speed. It has a central inlet of diameter 0.6 m, which is surrounded by a 1.6 m cylindrical shroud designed to act as a wind shield and produce calm-air conditions for the inlet. It is protected with a 1.6 m diameter rain cap as given in Figure 11 which shows the latest version produced by Hollander [16]. The sampling flow rate is $41\,667\,\mathrm{l\,min^{-1}}$, and there are four single-stage impactors

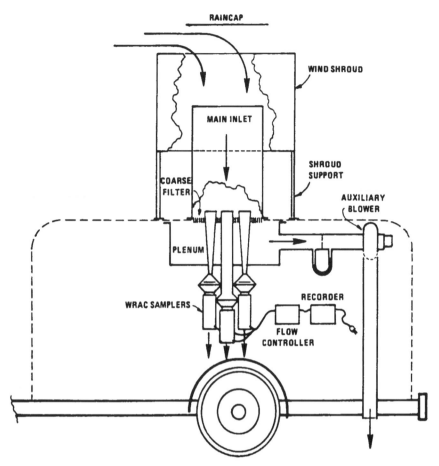

Figure 11 Wide Ranging Aerosol Classifier for ambient aerosols

with cut-points at 10, 20, 40 and 60 μm and total particles filter sampling the air isokinetically inside the central inlet duct. Limited field testing by Hollander [16] has shown that it does sample large particles with high efficiency, but full characterization in the controlled conditions in wind tunnels is required before definitive statements on its performance can be made. It is currently recommended as the reference instrument in a field equivalence testing protocol proposed by the European Union, but because of its large size and power requirements, it is not suitable for general use.

A much smaller and simpler device was first introduced by May. [24] Subsequently referred to as the "aerosol tunnel sampler", the sampler has been developed further by Hofschreuder and Vrins. [15] Shown in Figure 12, it

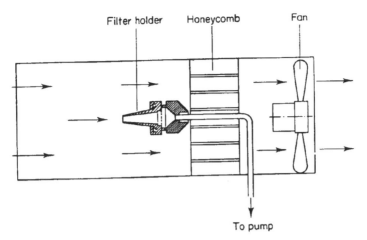

Figure 12 Schematic of the "aerosol tunnel sampler" for the measurement of true total aerosols in the ambient atmosphere

comprises a 150 mm diameter tube through which air is drawn by means of an axial fan to produce a mean air velocity of 9 m s^{-1}. Just in front of a flow-straightening honeycomb, is the 10 mm diameter thin-walled isokinetic sampler, located axially. This takes a representative sample of the aerosol concentration inside the main tube, which is designed to have sampling efficiency close to unity. The net effect is that the aerosol collected in the isokinetic probe should be representative of that in the ambient air outside. Results published by Vrins [15] show this to be the case. The whole arrangement is mounted on a pivot with a wind vane so that, under actual sampling conditions, it is always oriented with the tube mouth facing the wind. It differs from the other sampler mentioned above therefore in that it is not omnidirectional.

3.3 SAMPLERS FOR HUMAN HEALTH EFFECTS

3.3.1 Scientific Framework

For health effects that are suspected to have arisen from particles entering the body through the nose and mouth during breathing, one must use a sampler whose performance mimics the efficiency with which particles enter the nose and mouth and penetrate to the region in the body where the harmful effect occurs. Workers in the occupational hygiene field have realized this for some years and defined the respirable fraction for those particles that penetrate to the alveolar region of the lung and cause diseases such as pneumoconiosis, silicosis, asbestosis, etc.

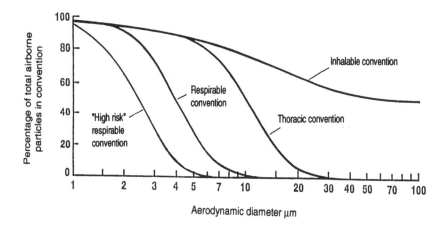

Figure 13 Health-related aerosol sampling conventions (IS 7708, ISO, 1994)

Since the early 1980s an *ad hoc* working group of the International Standards Organization has been formulating health-related sampling conventions for airborne dusts both in the ambient atmosphere and in the workplace. The final agreed conventions have passed through all stages of the approval procedure and should soon become available as International Standard IS 7708. [17] They are defined in Figure 13 and comprise four main fractions:

1. *Inhalable fraction* (E_I) is defined as the mass fraction of total airborne particles which is inhaled through the nose and/or mouth. It was derived from wind tunnel measurements of the sampling efficiency of full-size tailor's mannequins and replaces the very loosely defined "total" aerosol fraction used previously. For ambient atmospheres it is given by

$$E_I = 0.5(1 + \exp[-0.06d_{ae}]) + 10^{-5}U^{2.75}\exp(0.05d_{ae}) \tag{28}$$

   where d_{ae} is the aerodynamic diameter of the particle and U is the wind speed (up to $10\,\mathrm{m\,s^{-1}}$).
2. *Thoracic fraction* (E_T) is defined as the mass fraction of inhaled particles penetrating the respiratory system beyond the larynx. As a function of total airborne particles, it is given by a cumulative lognormal curve, with a median aerodynamic diameter of $10\,\mu\mathrm{m}$ and geometric standard deviation of 1.5.
3. *Respirable fraction* (E_R) is defined as the mass fraction of inhaled particles which penetrates to the unciliated airways of the lung (alveolar region). As a

function of total airborne particles, it is given by a cumulative lognormal curve with a median aerodynamic diameter of 4 μm and a geometric standard deviation of 1.5.

4. *"High risk" respirable fraction* is a definition of the respirable fraction for the sick and infirm, or children. As a function of total airborne particles, it is given by a cumulative lognormal curve with a median aerodynamic diameter of 2.5 μm and a geometric standard deviation of 1.5.

These conventions provide target specifications for the design of health-related sampling instruments, and give a scientific framework for the measurement of airborne dust for correlation with health effects. For example, the inhalable fraction applies to all particles that can enter the body, and is specifically of relevance to those coarser toxic particles that deposit and dissolve in the mouth and nose. The respirable fraction, on the other hand, relates to those diseases of the deep lung, such as the pneumoconioses, while the thoracic fraction may be relevant to incidences of bronchitis, asthma and upper airways diseases. Until recently, this philosophy has not been taken on board by the environmental community for health-related sampling.

In the USA, measurement of an aerosol fraction in the ambient atmosphere specifically related to health effects has been carried out for the last 8–9 years (PM10). This is the USEPA definition of the thoracic aerosol fraction [30], which has been systematically measured with validated instruments, in the USA for the last 8–9 years. It differs from the ISO definition in that it has a 50% penetration at aerodynamic diameter of 10.6 μm and zero penetration at 16 μm as shown in Figure 14. In practice, however, this difference is not significant

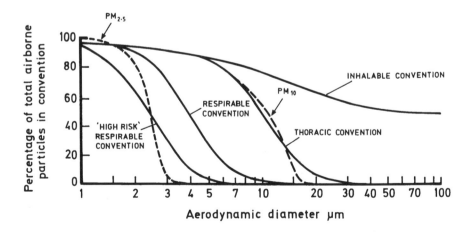

Figure 14 Definitions of PM10 and PM2.5 in relation to the ISO health-related sampling conventions

because most instruments validated for sampling the PM10 fraction will also reliably sample the ISO thoracic fraction.

In the European Union, ambient airborne particle sampling is guided by the Black Smoke and SO_2 Directive (80/779/EEC). [8] This directive makes no reference to any health- related aerosol fraction, and gives inadequate guidance on measurement methods involving either the determination of the "blackness" of collected particles or gravimetric techniques. Following recommendations in a CEC-funded report by Wagner et al. (UK Department of Health), [33] the directive will be soon revised to recommend the measurement of the PM10 particle fraction.

A further aerosol-derived convention is currently being discussed for implementation in the USA in the future for better correlation with some health effects. This is a finer fraction with a median aerodynamic diameter of about 2.5 µm, and separates particles in the accumulation mode from those man-made and wind-blown particles in the coarse mode. The shape of the proposed curve is somewhat sharper than that of the "high risk" respirable fraction of IS 7708, although they both seek to protect those whose health is most at risk from inhalation of airborne particles—children or those with respiratory or cardiovascular disease. Regulatory bodies in Britain are keeping a close eye on these developments with a view to possibly introducing them in the future, especially as a possibly more reliable indicator of levels of vehicle-derived particles which have been shown to be mainly composed of submicron particles.

Currently, for health-related ambient particle sampling, the mass concentration, rather than number concentration, is specified. This is because there is relatively limited epidemiological evidence relating particle exposure in the ambient atmosphere to adverse health effects, and therefore the experience of the occupational field is used. In this environment it is generally mass, rather than the number or surface area of the particles deposited in various regions of the respiratory system that shows reliable correlation with ill health. There are obvious exceptions such as fibrous particles. This means, therefore, that instruments that do not measure mass concentration must be calibrated for the particular aerosol being sampled. The Tapered Element Oscillating Microbalance (TEOM) measures the effect of mass directly, while the response of Beta- gauges is generally independent of the material sampled. However, it has been recently suggested by Seaton [27] that some adverse effects of exposure to ambient particles may be caused by the ultrafine fraction (< 0.1 µm). They suggest that, as these particles may readily penetrate into the interstitium of the alveolar cells, it may be more appropriate to measure the number concentration of particles rather than the mass concentration. The suggestion, however, needs considerable further work before implementation takes place.

3.3.2 Samplers Designed Specifically for Health-related Purposes

With the emphasis over the last 10 years in the USA on measurements of the PM10 fraction, and very little development activity in Europe (including the UK) on health-related sampling, the majority of reliable instruments currently available are for the PM10 fraction. These generally make use of a validated sampling head to select the PM10 fraction of the ambient airborne particles and the collected particles are analized in two main ways:

(A) Gravimetric, cumulative samplers in which the PM10 particles are deposited on a filter over a sampling period of normally 24 hours. The mass of particles collected on the filters is determined by weighing.
(B) Direct-reading monitors in which the selected PM10 particles are either deposited on a filter conducted with continuous assessment of the change of a property of the filter due to their presence, or are conducted to an optical sensing region.

3.3.2.1 Gravimetric, cumulative samplers   These samplers basically comprise an omnidirectional rain-protected entry followed by a size-selective stage (normally an impactor) to select the PM10 particles which are collected on the filter. There are a number of different samplers available, ranging in flow rate from 16.7 to $11301\,\mathrm{min}^{-1}$ and cost from about £7500 to £15 000. A summary of the important features of the samplers, together with the cost (in the UK) is given in Table 1 and a brief description of the design features and relative performance of the sampling heads is given below.

High flow rate samplers such as the Graseby Andersen PM10 Hi-Vol Sampler (see Figure 15) have been the mainstay of routine PM10 measurements in the USA. The high flow rate of $11301\,\mathrm{min}^{-1}$ has the advantage of providing sufficient sample both for gravimetric and chemical analysis over the specified 24-hour sampling period. They are especially useful in studies to determine the low levels of dioxins, PCBs and PAHs found in the ambient atmosphere. The important features of the sampling head are an omnidirectional narrow slot entry, which samples particles independently of wind speed up to $10\,\mathrm{m\,s}^{-1}$, and a multi-orifice single-stage impactor, which allows the PM10 fraction to penetrate to a $25 \times 20\,\mathrm{cm}$ filter. The sampling efficiency of the sampler, demonstrating full agreement with the USEPA PM10 Convention, is also shown in Figure 14. The sampler is mains powered with either volume or mass flow rate automatically controlled. Similar systems are available from other suppliers such as ECOTEK (and a high volume virtual impactor is available from MSP Corporation, USA) to insert between the impactor and the filter to provide the PM 2.5 fraction of the PM10 particles.

A number of low flow rate PM10 samplers are available. They all make use of the SA246b PM10 inlet developed and validated by Graseby Andersen. The

Table 1. Examples of gravimetric, cumulative PM10 particle samplers for the ambient atmosphere currently available in the UK

Name	Flow rate ($1\,min^{-1}$)	Filter diameter (mm)	Comments	Approximate cost (£ 000)
PQ167 Portable PM10 Sampling Unit	16.7	47	Uses validated PM10 inlet (US EPA Protocol) connected to microprocessor flow-controlled pump. Battery powered—lasts for over 24 hours using quartz or glass fibre filters	9
Partisol Model 2000 Air Sampler	16.7	47	Uses validated PM10 inlet connected to microprocessor-controlled pump. Supplied as stand-alone unit or with three additional satellites controlled by hub unit dependent upon wind speed and/or direction conditions	9 for hub unit 3 for each satellite
PM10 Dichotomous Sampler	16.7	2 off 37	Uses validated SA246b PM10 inlet followed by virtual impactor to give two fractions collected on filters; 10–2.5 μm and < 2.5 μm	15
PM10 Medium Flow Sampler	113	102	Medium flow rate sampler uses Teflon or quartz filters, specially used for X-ray fluorescence and other compositional analyses	13
SA 1200 PM10 High volume Ambient Air Sampler	1200	200 × 250	Standard sampler used in USA for PM10 aerosol in ambient atmosphere—high flow rate means short sampling times and large masses for gravimetric and chemical analysis	8

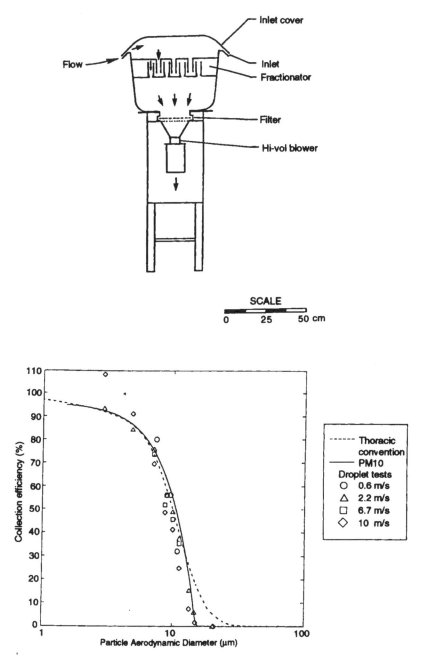

Figure 15 Graseby–Andersen high flowrate PM10 sampler

entry (see Figure 16) consists of a flanged downwards-pointing circular entry with a disc rain cap held some distance above. This forms the omnidirectional entry through which particles enter, followed by a single-stage impactor which allows the PM10 fraction to penetrate to a filter. The sampling efficiency of the entry, demonstrating full agreement with the USEPA PM10 Convention, is also shown in Figure 16. Three devices are included in Table 1, each having separate features that makes them to a certain extent complementary.

1. The Casella PQ 167 Portable PM10 Sampling Unit uses a specially designed battery-powered pump which lasts for 24 hours using 47 mm diameter quartz or glass fibre filters. This makes the sampler very useful at sites where mains power is not easily available.

2. The Rupprecht and Patashnick Partisol Model 2000 Air Sampler can either be supplied as a stand-alone unit or with three additional satellite PM10 sampling head and filter units. The satellites are connected to the main hub unit (at a maximum distance of about 3 m) through simple airflow lines which are switched via solenoid valves controlled by the sampling programme defined by the user. This enables four daily samples to be taken without attention. In addition, separate samples can be taken for different wind speeds, wind directions and at specified times, thereby providing useful evidence for source apportionment studies and for correlation with meteorological conditions. For all units the PM10 head can be replaced with a small cyclone inlet made by URG which has a 50% penetration at 2.5 μm. The sampler runs off mains power only. New versions of this sampler will soon be available that use a filter carousel system to enable up to 16 samples to be taken sequentially from one sampling head.

3. Finally, the Graseby Andersen PM10 Dichotomous Sampler separates particles into two distinct fractions—2.5 to 10 μm and < 2.5 μm. This is achieved by following the PM10 inlet with a virtual impactor with a 50% penetration at 2.5 μm, and collecting both fractions on 37 mm diameter Teflon membrane filters. This gives some size distribution information of particles within the PM10 fraction, and provides separate samples for chemical species studies.

It has been reported by van der Meulen (1992) that the SA 246b, 16.7 l min⁻¹ PM10 sampling head systematically under-samples the PM10 aerosol fraction by 20–30% when compared to either the WRAC (see above) or the Hi-vol PM10 sampling head in side-by-side field tests. This has then been used to question the applicability of wind tunnel-based performance testing. However, this difference may not be due to problems with wind tunnel testing *per se*, rather to the range of test conditions specified. More information on the relative performances of samplers for the PM10 aerosol fraction will soon become available from the results of a pilot study carried out for the European Union under the auspices of CEN/TC264/WG6 to evaluate a protocol for the field equivalence testing of PM10 samplers.

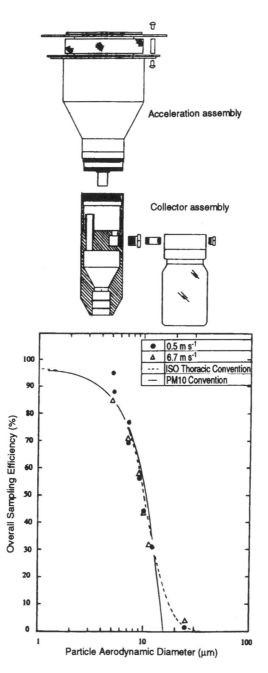

Figure 16 Graseby–Andersen low flowrate PM10 sampler as fitted to direct-reading instruments

Table 2. Examples of direct-reading monitors for PM10 particles in the ambient atmosphere

Name	Measurement technique	Flow rate ($l\,min^{-1}$)	Particle fraction	Concentration range ($\mu g\,m^{-3}$*)	Precision ($\mu g\,m^{-3}$*)		Comments	Approx. cost ($'000)
					1 hour	24 hours		
TEOM Series 1400a Ambient Particulate Monitor	Tapered element oscillating micro balance	16.7 through inlet with 3 through filter/detector	PM10 (validated head) with possible 2.5 or 1.0 μm	$0.06\text{-}1.5 \times 10^3$	1.5	0.5	Only direct-reading monitor in which output directly related to mass. Empl oyed as particle monitor at EUN sites	30
W & A Beta Gauge Automated Particle Sampler	Attenuation of beta rays by particles collected on a filter	18.9	PM10 with EPA validated head	$4\text{-}10^4$	4	0.1	One of a number of filter tape-based beta gauges—measurement cycle 1 hour	?
Airborne Particle Monitor APM1	Attenuation of beta rays by particles collected on a filter	15–30	PM10 (non-validated)	$2\text{-}10^7$	56	2	Cassette system with 30 filters in sequential loader. Integrity of each sample maintained for compositional analysis	?
GRIMM Model 1.104 Dust Monitor	Light-scattering photometer	16.7 through inlet with 1.26 through detector	PM10 with EPA validated head	$1\text{-}5 \times 10^4$ indicated	Not given	Not given	Optical particle counter with in-built filter for on-site calibration, as response may be dependent upon refractive index and size of particles	21
Data RAM Portable Real-Time Aerosol Monitor	Light-scattering photometer	2	PM10 or PM 2.5 (non-validated)	$0.1\text{-}10^3$ indicated	1.0	Not given	Optical device calibrated with AC fine test dust. May need on-site calibration to give reliable mass measurements as response dependent upon refractive index and size of particles. Entry dependent upon wind speed	18

* Manufacturer's figures.

3.3.2.2 Direct-reading monitors For these instruments, sampling and analysis are carried out within the instrument and the concentration can be obtained almost immediately. Like the low flow rate cumulative samplers described above, these instruments generally use the validated SA 246b PM10 inlet to select the PM10 particles which either deposit on a special filter stage or penetrate into a particle sensing region. Instead of direct weighing the presence of the particles either on the filter or in the sensing region gives rise to a change in some property of the zone, which can be related by calibration to the mass of particles present. A number of different instruments are available and these may be classified into three main categories: optical, resonance oscillation, and beta particle attenuation. A summary of the main features of some examples of the direct-reading instruments for health-related purposes is given in Table 2, and the principles of operation will be described below.

Optical These instruments employ the interaction between airborne particles and visible light in a sensing region, and generally their response is dependent upon the size distribution and refractive index of the particles. They therefore require calibration to give results in terms of mass or number concentrations. This calibration only holds provided that the nature of the particles does not change. Two instruments are currently available:

1. The Data RAM Portable Real-Time Aerosol Monitor uses a near infrared source near forward angle scattering to detect the concentration of particles in the range $0.1–1000\ \mu g\,m^{-3}$. It uses a scaled-down version of the SA 246b PM10 inlet to select the PM10 fraction at a flow rate of $2\,l\,min^{-1}$, although this has not been experimentally validated. The impactor stage can be replaced to give a nominal 50% penetration at $2.5\ \mu m$. The instrument is powered by a 6 V lead acid battery which lasts for 20 hours, making it very useful at sites where mains power is not easily available. It is calibrated with AC fine test dust (mass median aerodynamic diameter $2–3\ \mu m$, standard deviation 2.5), and may need on-site calibration to ensure valid results in terms of mass concentration.

2. The Grimm Stationary Environmental Dust Analyser 1.200 comprises basically two samplers in one unit. The so-called reference unit uses a standard $16.7\,l\,min^{-1}$ PM10 inlet connected to a 25 mm diameter glass fibre filter. This is used to obtain the cumulative mass concentration by gravimetric analysis of the filter. An airflow splitter is then used to extract air at $1.26\ l\ min^{-1}$ containing the PM10 particles which is fed into a light scattering optical particle counter. The details in Table 2 are just those of the optical counter. The optical counter uses peak height analysis to separate the particles into eight channels in the range $0.3–15\ \mu m$, from which the mass concentration of particles within the PM10 fraction can be determined. An in-built back-up filter also enables the response to be calibrated on-site for the specific particles. The instrument has the

potential to be a very versatile tool providing both mass and number size distributions of the PM10 fraction from which other health-related fractions can be calculated. It is currently undergoing further field evaluation. It is small enough to be carried by hand and has the option of either battery or mains power.

Oscillating microbalance   The frequency of mechanical oscillation of an element such as a tapered glass tube is directly proportional to the mass of the tube. Change in effective mass of the tube, such as that due to deposition of particles on the surface of a filter at the free end of the tube, is reflected in a change in its resonant frequency.

This is the principle of operation behind the Rupprecht and Patashnick Tapered Element Oscillating Microbalance (TEOM). A standard $16.7\,l\,min^{-1}$ PM10 inlet selects the PM10 particles which pass through a flow splitter in which $3\,min^{-1}$ pass through a 16 mm diameter filter connected to the top of the narrow end of a hollow tapered glass tube, and $13.7\,l\,min^{-1}$ are led away for other purposes. As the particles increasingly collect on the filter, the tube's natural frequency of oscillation decreases. The change in this frequency is directly proportional to the added mass. The inlet including the sensing system is kept at a steady $50\,^{\circ}C$ to drive off any sampled water droplets. The instrument is microprocessor controlled and the mass concentration values are updated every 13 seconds with average concentrations provided every 30 minutes or every hour. Once teething problems associated with the effects of external vibration were resolved, it has proved to be a very reliable monitor, and has been used by many organizations throughout the world. The filter is removed after a number of weeks sampling (dependent upon the ambient particle levels) and the chemical composition of the deposited material analysed. This will provide compositional information integrated over the sampling period. Unfortunately, at approximately $30\,000 for a complete unit very few small companies or local governments have the finance to buy one—a cheaper version would be most welcome. In addition, some concern has been expressed about the potential loss of volatile material in the stable temperature of $50\,^{\circ}C$ (Weiner 1995, personal communication).

Beta particle attenuation   This involves the measurement of the reduction in intensity of beta particles passing through a dust-laden filter or collection substrate. In such instruments, the change in attenuation reflects the rate at which particles are collected on the filter and hence on the concentration of the sampled particles. For most substances encountered the attenuation of the beta particles is directly dependent on the mass of particles deposited. Two main types of instrument have been developed: one using filter tape to collect the particles, and the other using a stack of conventional filters in a sequential loader.

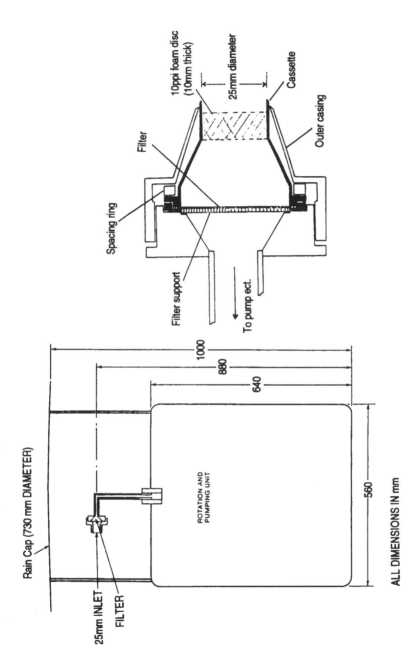

Figure 17 Prototype sampler for the measurement of inhalable aerosols in the ambient atmosphere

There are many tape-based beta particle attenuation devices available, and details of just one typical example are given here. The Wedding & Associates Beta Gauge Automated Particle Sampler uses their version of the low volume, validated PM10 inlet at 18.9 $lmin^{-1}$ to select the PM10 particles which are deposited on to either a glass fibre or PTFE filter tape. A beta particle attenuation system comprising a 3.7 MBq C-14 source and a fast response silicon semiconductor detector is used to detect the presence of dust on the filter with a one hour measurement cycle.

The Elecos Airborne Particle Monitor APM-1 is an interesting alternative to the tape-based beta particle attenuation monitors. Instead of the filter tape, it deposits PM10 particles on to one of thirty 47 mm diameter filters held in a sequential loader system. This prevents violation of the integrity of the deposited particles enabling subsequent unambiguous gravimetric and chemical analizes to be made.

With the introduction of the TEOM instrument, the future availability of beta particle attenuation monitors may be somewhat limited, as they have no significant advantages over the TEOM. In addition, the relationship with mass is not linear for some materials, and is also affected by the uniformity of particle deposition on the filter.

3.3.2.3 Comparability of methods There is a potential, as yet unresolved, problem of the comparability of the data from the two methods. The filter of the TEOM is held at stable temperature of 50 °C, while the filter stages used in the optical and beta particle monitors and the gravimetric samplers have no specific temperature control. This means that while the TEOM can be considered to underestimate aerosol concentrations due to evaporation of the volatile components on the particles, beta particle monitors may over-sample due to condensation of water vapour on to the filters in conditions of high humidity, and similarly optical monitors may register fine water droplets. The comparison of direct-reading methods with the gravimetric samplers is further complicated by the filter conditioning procedures necessary to obtain accurate weighing. It is not yet known to what extent this procedure gives similar filter conditions to those for the two types of direct-reading instruments.

3.3.3 Health-related Samplers for Coarse Particles

It is only recently that the environmental protection community has realized that particles as large as 100 μm can enter the human nose and mouth during breathing. While the occurrence of these large particles may be rare, they can occur close to industrial processes and during episodes of high wind speeds. Once inhaled they will deposit in the nasopharyngeal region and if toxic (such as lead, radioactive particles, etc.) may enter the blood system there or in the gut. The relevant ISO health-related fraction is the *inhalable fraction*, for which there is currently no commercially available instrument. The old "total suspended particulate" (TSP) was effectively defined by the "high-volume" sampler used to measure this parameter. The performance of this well-known device was found to be both orientation and wind speed dependent, with its efficiency falling well below the inhalable aerosol convention, especially for large particles. [34]

As a potential European reference method, the European Community and the UK Department of the Environment provided limited funds for the development of a sampler designed specifically to match the requirements of the ISO inhalable aerosol fraction. [22] It was designed to mimic the essential features of humans by achieving omnidirectionality using a single orifice that rotates through 360° (see Figure 17). It comprises a detachable sampling head which sits on top of a rectangular cabinet which houses the head rotation mechanism (drive motor, gearbox and rotating seal), the pump, an automatic flow control system, and associated switch gear. A protective canopy is positioned over the sampling head to prevent the entry of unwanted rain and snow. The sampling efficiency of the sampler, as tested in a large wind tunnel for a range of particle sizes and types in a range of wind speeds, is also given in Figure 18. [22] It is now available for commercial exploitation, although its future use is somewhat in doubt as more emphasis for correlation with human health effects is focused on the smaller particles in the ambient atmosphere.

3.4 PARTICLE SIZE MEASUREMENT

Knowledge of the size distribution of particles suspended in air is essential for the understanding of many aspects of the behaviour and effects of aerosols. For example, the aerodynamic diameter of a particle governs whether it will be inhaled and where in the respiratory tract it is deposited. The transport of particles in the atmosphere, their deposition to surfaces and resuspension of deposited particles are all size-dependent processes, and cannot be fully understood without information on the particle size distribution.

However, as discussed earlier, atmospheric aerosols comprise particles from a wide range of sources both natural and man-made, giving nearly a four-decade range of particle sizes. The well-known trimodal size distribution presented initially by Whitby [35] and reproduced in Figure 19, suggests particles ranging from several nanometres (nm) to hundreds of micrometres (μm). To produce this complete size distribution, Whitby used a combination of electrical aerosol analyzers for the fine particles, and two optical particle counters for the coarser ones. Despite significant improvements in measurement technologies since those early days, to cover the complete size range still requires at least two complementary instruments.

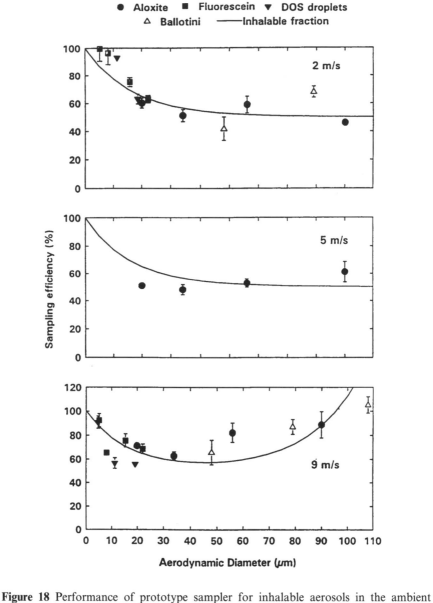

Figure 18 Performance of prototype sampler for inhalable aerosols in the ambient atmosphere

The information given in Figure 19 is nominally in the form of the *volume* differential distribution, describing the volume of particulate material as a function of particle size. Other important measures of the amount of particu-

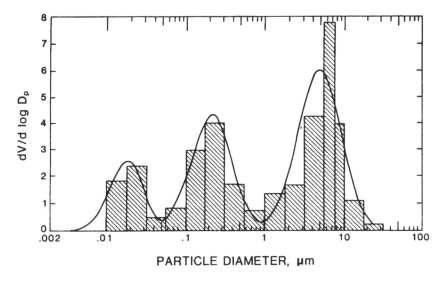

PARTICLE DIAMETER, µm

Figure 19 Size distribution of ambient aerosols reported by Whitby [35]

late material are *mass, number, surface area, chemical composition, biological activity*, etc., dependent upon what is required, and the instrument used. There is a wide number of instruments available for particle size measurements, a brief review will be given below.

Instruments used for aerosol size distribution measurements can be classified according to the physical principle by which they classify the particles. Three types of instrument will be discussed here, operating according to the following principles: (i) inertial classification, (ii) light scattering, (iii) electrical mobility. Other principles such as diffusional classification may also be used but the resulting instrumentation is not considered to be practical for this application.

3.4.1 Inertial Classification

This family of instruments includes those devices that use inertial and gravitational forces to separate airborne particles according to their aerodynamic diameters. The main devices of this type used for ambient aerosols are *cascade impactors*. These are multi-stage impaction devices in which airborne particles are separated into aerodynamic size classes. A schematic diagram of a cascade impactor is given in Figure 20. In use, an aerosol is drawn through a series of progressively narrower jets, each followed by an impaction surface. When the aerosol stream curves to flow around the impaction surface, those particles with high inertia cannot follow the flow and impact on the obstructing surface.

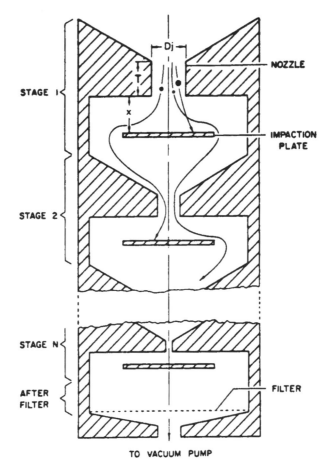

Figure 20 Schematic of a cascade impactor for inertial classification of aerosols

Assuming that any particle that impacts sticks to the surface, then this process divides the aerosol into two groups: the larger particles that have deposited, and the smaller ones which remain airborne. This process is repeated as the aerosol flows through stages with jets of decreasing diameters at increasing velocities. After passing through the last stage, the stream is usually drawn through a final filter as it exits the device.

A large variety of cascade impactors have been designed and built for use with ambient aerosols. A summary of the main features of some of the most widely used devices is given in Table 3. Flow rates vary from 5 to 1130 l min^{-1}, and they segregate particles in up to 14 size classifications with stage cut-points from 32 to 0.035 μm.

Table 3. Examples of cascade impactors for ambient aerosols

Name of cascade impactor	Nominal flow rate (l min^{-1})	Number of stages	Stage d_{50}-cut points (μm)	Comments
Andersen Ambient Cascade Impactor	28	8 + filter	9–0.4	Original multi-orifice impactor—version for bioaerosols uses agar-filled dishes as collection surfaces
May/RE "Ideal" Cascade Impactor	5	7 + filter	32–0.5	Slot jet design, with low wall losses. Collection on to glass microscope slides for microscopic analysis. May be operated at 20 l min^{-1}
MSP MOUDI	30	10	10–0.056	Micro-Orifice Uniform Deposit Impactor. Last two stages have 2000 orifices. Alternate stages rotate to produce uniform deposit
Berner Low-Pressure Impactor	30	9	16–0.06	Three low-pressure stages. Specially built to provide sufficient sample form ambient environment for gravimetric and chemical analysis
QCM C-1000 Cascade Impactor	0.25	10	25–0.05	Dual crystal, direct-reading monitor. Each collection stage is frequency controlling element of a quartz crystal microbalance
Dekati Electrical Low-Pressure Impactor	10.5	12	10–0.047	Direct-reading device. Electrically charged particles collect on to metal foil. Current carried by charged particles on each stage measured with electrometer

One of the first, and still widely used, cascade impactors is the multi-jet Andersen sampler [1]. This is available in two models, one for collecting airborne particles on glass or stainless steel collection plates; the other for bacterial particles, using a nutrient agar plate as the impaction surface. The non-bacterial particle sampler has eight stages with cut-points from 9 to 0.4 μm at the specified flow rate of 28 l min^{-1}. Later impactors have incorporated improved designs to give sharp cut-off characteristics and to minimize inter-stage particle wall losses.

Several cascade impactors incorporate a number of low-pressure stages to separate particles as small as 0.05 μm. One example is the Berner low-pressure impactor [3], which has nine stages with cut-points from 16 to 0.06 μm at a flow rate of 30 l min^{-1}. It has been used for a variety of atmospheric measurements, including ambient size distributions for sulfate, nitrate, ammonium, acid, and traffic-derived aerosols. An alternative way of achieving stage cut-points below 0.5 μm is to use stages with a large number of very small jets. The Micro-Orifice Uniform Deposit Impactor (MOUDI) [23] has two lower stages with 2000 jets per stage and jet diameters down to 0.048 mm. At the operational flow rate of 30 l min^{-1} the instrument's 10 stages give cut-points from 10 to 0.056 μm. This very useful device has the additional feature of rotating collection substrates which provides uniform deposits, ideal for direct analysis by X-ray fluorescence, and allowing the collection of larger samples before becoming overloaded.

All the cascade impactors described so far are used to collect long-term size-separated samples for subsequent gravimetric, microscopic and/or chemical analysis. Two instruments have been developed which provide real-time information on the amount of particulate material on each impactor stage. The impactors in the Quartz Crystal Microbalance QCM C-1000 use piezo-electric quartz crystals as collection substrates, which resonate as a function of the mass loading to provide real-time determinations of mass particle size distributions in the range 0.05–30 μm. The instrument is capable of measuring very low concentration aerosols ($< 1 \mu g \, m^{-3}$), but this high sensitivity leads to problems with stage overloading (max. load $\sim 1 \mu g$ per stage) limiting their use for long-term monitoring. A more recent development is the Electrical Low Pressure Impactor (ELPI) [18], which uses a Berner-type low-pressure impactor to size separate electrically charged aerosols. The electrical current drawn from each impactor stage is proportional to the number of particles collected on that stage. The instrument operates at a flow rate of 10.5 l min^{-1}, with 12 impaction stages providing size distribution information in the range 0.034–10 μm. An additional benefit with the ELPI is that the metal foil collection substrates are easily removed for subsequent gravimetric and chemical analysis.

3.4.1.1   Limitations with cascade impactors   The main problem with the use of cascade impactors for particle size measurements is degradation of size separation due to particle bounce. This occurs when hard, solid particles have sufficient kinetic energy to bounce off the collection surfaces to be deposited on lower stages or on the inter-stage walls of the impactor. The effects of particle bounce are minimized by the application of a sticky coating to the collection surfaces to absorb the kinetic energy of the impacting particles. Examples include Apiezon L and Vaseline greases, and oil, although care must be taken in the choice of coating so that interference with subsequent chemical analysis is avoided. For further useful information on the use and misuse of cascade impactors the reader is referred to the AIHA Monograph entitled *Cascade Impactor: Sampling and Data Analysis.* [19]

3.4.2 Light Scattering Instruments

A large group of aerosol analysers operate on the basis of light interaction with airborne particles. The most useful for particle size distribution measurements of ambient aerosols are *optical particle counters*. These measure the quantity of light scattered by individual particles as they pass through a focused beam of light. Each particle produces a pulse of scattered light, which is received by a photosensitive detector. The magnitude of the electrical signal that is fed to a multi-channel analyser is proportional to the size of the particle. Several types of optical particle counter are commercially available from companies such as Royco, Climet TSI, Polytec, PMS, etc. The minimum particle size limit of this family of devices is about 0.1 µm for open-cavity laser-based instruments of the Knollenberg type and the maximum is over 50 µm for the Polytec instrument. Maximum particle concentrations are typically of the order of 10 000 particles cm^{-3}.

3.4.2.1   Limitations with optical particle counters   The ability of an optical particle counter to give particle size information is affected by both the refractive index and the shape of the particles, and the size parameter they give is known as the light scattering diameter. The information given therefore is not unambiguous as a transparent particle will be sized differently from a black absorbing particle of nominally the same geometric size. In addition, optical particle counters provide information on the *number* size distribution of particles. While this parameter is definitely of interest for ambient aerosols, optical particle counters with a minimum detectable particle size of 0.1 µm will not detect the smaller ultra-fine particles. Although these are insignificant in mass terms, they are an appreciable proportion of the total number of particles found in both urban and rural atmospheres (see Figure 21).

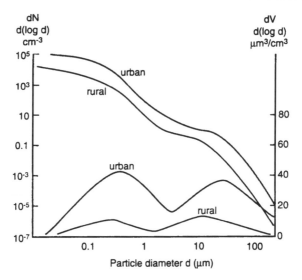

Figure 21 Mass and number size distributions for aerosols in urban and rural atmospheres

3.4.3 Electrical Mobility

The electrical mobility of an aerosol particle is a measure of the velocity of motion resulting from the force it experiences when placed in an electric field. For particle motion in the Stokes region, the electrical mobility, Z of a particle of charge, ne, in a electric field of unit strength is given by

$$Z = \frac{neC_c}{3\pi\eta d} \tag{29}$$

where C_c is the Cunningham slip correction factor. This relationship is similar to that used to determine the mechanical mobility, with the electrical force replacing the gravitational force.

A family of instruments has been produced by TSI using the principle of electrical mobility. The latest, and most useful for the measurement of ambient aerosols, is the scanning mobility particle sizer (SMPS). This instrument comprises: (A) a single-stage impactor, (B) a bipolar aerosol particle charger, (C) a differential electrical mobility analizer, and (D) a condensation particle counter. A schematic is given in Figure 22. Aerosols are sampled at a flow rate of $21\,\text{min}^{-1}$ into the single-stage impactor which removes particles larger than $1\,\mu\text{m}$ aerodynamic diameter. The penetrating particles are transported through the bipolar charger (Kr-85 source) where they acquire a well-defined and predictable electrostatic charge. The charged particles are then introduced at the outside of a electrical mobility classifier, consisting of two concentric metal

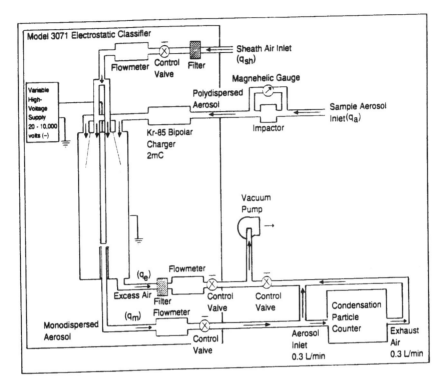

Figure 22 Schematic of the Scanning Mobility Particle Sizer (SMPS)

cylinders. Clean sheath air is passed through the central cylinder which surrounds a central metal electrode. When a negative voltage is applied to this electrode, an electric field is set up and positively charged particles are attracted. Those with high electrical mobility are deposited on the electrode's surface, while those with lower mobility pass though the cylinders and exit through a narrow slit at the bottom of the rod. From here they are passed into a condensation particle counter where they grow in an atmosphere of butyl alcohol vapour to a size (2–3 μm) large enough for detection by an optical particle counter. Those with much lower mobility pass through the instrument with the sheath air. By changing the voltage applied to the electrode, different monodisperse particles are counted.

The instrument is fully automated and can measure the size distribution of particles in the range from 1 μm to 5 nm. It therefore has the capability of measuring the number concentration and size distribution of the *ultra-fine* particle fraction, currently under suspicion of having a major contribution to health effects associated with the inhalation of ambient aerosols. [27] It is currently the only instrument capable of automatic measurement of size distributions down to 5 nm.

3.4.4 Discussion of Particle Size Measurements

Measurements of aerosol size distributions, while seemingly simple in principle, are notoriously difficult to achieve reliably in practice. The following two main points should be addressed carefully when using any of the above devices to obtain size distribution measurements of ambient aerosols:

1. *Sampler entry and transmission*: very few of the instruments described above are supplied with carefully designed inlets and transmission sections to ensure that the aerosol size distribution measurements obtained are representative of the ambient aerosols. Generally, the instruments have just an inlet port and the user is left to connect his/her own entry and pipework. The problems associated with the choice of entry and transmission pipework are described in section 2.3.1, and essentially a weather-protected omnidirectional entry is required with a conducting metal transmission pipe of minimum length.

2. *Maximum number concentrations*: for optical particle counters, including condensation nucleus counters, care must be taken to ensure that the maximum particle concentrations are not exceeded. When this happens, two or more particles are present in the sensitive volume at the same time, resulting in a false signal that leads to an underestimation of the total particle count and an overestimation of the average particle size. This is known as coincidence and may be overcome by the use of a properly designed diluter system.

3.5 SAMPLING DEPOSITED PARTICLES

3.5.1 Background

The measurement of deposited particles was among the first studies of air pollution, starting around the beginning of the century. It is mainly used to provide information in response to complaints about the perceived nuisance from particles deposited on cars, buildings, windows and washing, etc. It is also important to provide information that helps to estimate the likely risk to health from toxic material (radioactive particles, dioxins, etc.) depositing on plants or soil that are either directly eaten or are found in the food chain by secondary pathways.

Deposition of particles to ground is found in one of two ways, depending upon whether the particles are considered to be "large" or "small". The boundary between large and small is considered to be at a particle aerodynamic diameter of about 20 μm. Large particles are presumed to fall to the ground like other forms of precipitation (rain, snow) and the mass of particles deposited in time t, (M_d) is given by

$$M_d = A_d \int M v_s \mathrm{d}t \tag{30}$$

where A is the surface area over which the deposition occurs, M is the ambient mass concentration of the particles and v_s is the sedimentation velocity of the particles. The deposition of large particles is usually measured with passive deposition gauges, which are essentially upwards-facing bowls. These will be described in more detail later.

For small particles there are additional processes by which they are deposited to the ground—by impaction or trapping on to surface roughness (e.g. grass), and by diffusion to the surface. The deposition rate is defined in the same way as for large particles, but the sedimentation velocity is replaced by an effective deposition velocity which takes account of these other processes. The effective deposition velocity of small particles is always higher than their gravitational sedimentation velocities, with typical values in many practical cases of around $1 \mathrm{cm} \mathrm{s}^{-1}$. The deposition of small particles is estimated by measuring the airborne concentration and assuming an effective mean deposition velocity derived from experimental measurements. [28]

The flux of particles (F), blown by the wind past a point, is sometimes measured and quoted as particle deposition. However, it is not directly related to deposition, being defined as the product of the ambient mass concentration of particles (M) and the mean wind speed (U). The flux of particles in time t is given by

$$F = A_F \int M U \mathrm{d}t \tag{31}$$

Flux can be related to deposition only if the relationship between particle size (and therefore sedimentation velocity), wind speed and ambient mass concentration is known. This is never known reliably. Particle flux is measured using passive flux gauges, which are essentially containers with vertical openings intended to collect the horizontal flux of particles past a measuring station. They may be used to assess the source strength of wind-blown particles and its depletion rate with distance from the source. A brief description of a number of devices is given later.

3.5.2 Ambient Mass Concentration Measurements for Small Particle Deposition

There are no samplers designed specially for this purpose, but if a certain size range of particles is of interest, then a sampler can be selected from those described above for total and health- related sampling. For example, a sampler selecting the thoracic fraction would provide useful measurements of mass concentration of the particle size range (below 20 μm) for which the small particle deposition criteria apply. If a sample is required of the whole of the ambient particle size distribution for deposition estimates, then one of the instruments mentioned in the section on samplers for "total" particles may be used. An alternative could be the sampler for the inhalable fraction which has a defined sampling characteristic for particles of aerodynamic diameters up to 100 μm.

3.5.3 Particle Deposit Gauges

There are many designs of particle deposit gauges, as they were derived from rain gauges for which each country has its own favourite design. They all basically consist of an upwards-facing bowl mounted on a stand at sufficient height (typically between 1 and 2 m) to avoid collection of saltated material from the surface. Some examples of deposit gauges used in various countries are given in Figure 23. Using these devices and designs very similar, some countries have introduced standard criteria for deposition rates as annual or monthly deposition averages. For insoluble dusts, deposition rates exceeding 200 mg m^{-2} day^{-1} on a monthly average are generally considered to be high enough to cause complaints in residential areas.

A common feature is the relatively high aspect ratio of depth to diameter of the bowl used. This feature has been widely criticized by Hall and co-workers (at the former Warren Spring Laboratory) in numerous papers on deposit gauge performance. He showed that the aerodynamic blockage of the bowl caused the airflow to rise and accelerate over the gauge, and particles descending under gravity in the wind tend to be displaced away from the opening, thus reducing the catch. In addition, a circulating flow is driven inside the bowl by the wind, resulting in the removal of particles already collected. The performance of the British Standard Deposit Gauge [47] given in Figure 24 is typical of that for all the deep bowl gauges shown in Figure 23. The collection efficiency is very dependent upon wind speed and particle size, with very low efficiencies for particles below 200 μm and wind speeds above 2–3 m s^{-1}.

After a programme of testing and development, Hall and colleagues [11, 13] produced a new deposit gauge with low aerodynamic blockage and good retention characteristics. Shown in Figure 25, it is based around the shape of an inverted "frisbee", and comprises a shallow aluminium dish with sufficient curvature to allow efficient drainage of rainwater. The collection performance of this first prototype gauge is also given in Figure 25, and considerable improvement over the deep-bowl-type gauge can be seen. Further recent improvements have included the insertion of a porous polyester foam disc to improve retention characteristics and eliminate the splashing of raindrops. The top of the dish has a turned-in lip to minimize particle blow-out, and a flow-deflecting ring is fitted

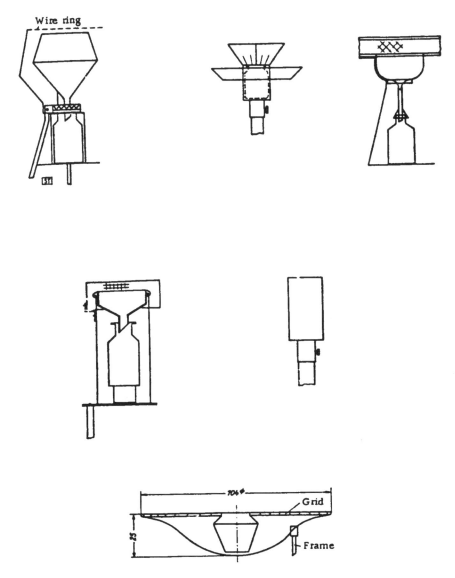

Figure 23 Some example of particle deposit gauges

around the gauge to provide better control of the airflow. This design (shown in Figure 26) is now commercially available and it is intended that it should be the basis of a new standard gauge design for use in Britain, and internationally in the long term.

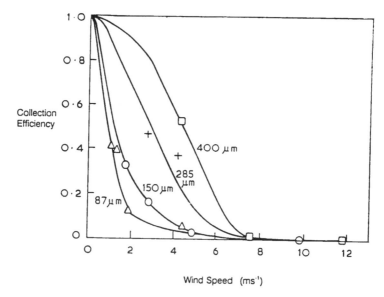

Figure 24 Typical collection efficiency for deep bowl deposit gauges shown in Figure 23

3.5.3.1 Operational aspects Most of the operational guidance given in BS 1747 Part 1 remains valid for the new design of deposit gauge, and reference should be made to that document and subsequent successor for detailed operational instructions. Some of the more important aspects will be briefly outlined here.

Siting of deposit gauge The deposit gauge is a passive device, and therefore is not constrained to sites where electrical power is available. Nevertheless, in order to obtain unambiguous results, careful consideration should be given to the siting of the gauges. While trying to ensure that the gauge is as close as possible to the area of the complaint, the distance between the gauge and any large obstruction should be at least twice the height of the obstruction. This ensures that errors due to localized separations and swirls in the airflow are avoided. The smoothest possible surface should be chosen so that the plane of the gauge opening is maintained horizontal. In addition, the height of the gauge opening above the ground should be chosen so as to avoid the entry of saltating material into the gauge, while limiting the height to ensure that the particles collected are representative of those that deposit to ground. A height of 1.5–2 m has been found to be a good compromise.

Operation When set up, an algicide may need to be added to the collecting bottle in order to prevent the growth of algae in any rainfall collected when the

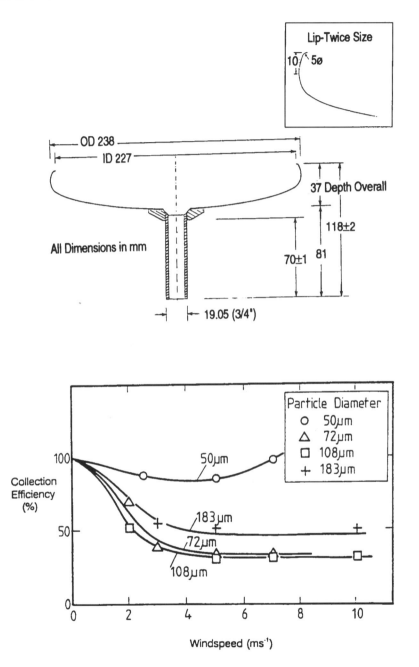

Figure 25. Shallow bowl dust deposit gauge (inverted Frisbee) together with its performance

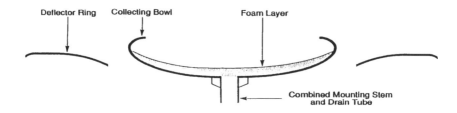

Figure 26 Improved design of shallow bowl gauge incorporating aerodynamic deflectors

sampling period is long (1 month or more). At the end of the sampling period the contents of the gauge must be washed down into the collecting bottle. This is done using a measured volume of distilled water, with stubborn deposits being removed from the surface with a plastic scraper. The volume of distilled water can be subsequently subtracted from the total volume in the collecting bottle to give a measure of the rainfall.

The collected material is separated from the water in the bottle by gentle vacuum filtration. A two-stage filtration process is employed: (A) leaves and insects are removed with a coarse wire mesh (tea strainer), and (B) particles are collected on a 47 mm diameter filter. Glass fibre (Whatman GF/A) filters are used for gravimetric determination only, while for particle size analysis membrane filters (Millipore) are used. A careful drying and conditioning procedure is required to enable the weight of collected particles to be determined. This includes the use of a control procedure in which a number of blank filters are treated in the same way as the sample filters, so that any changes in weight introduced by the filtration process can be taken into account.

The resultant weight of collected particles is expressed as deposition rate by dividing by the collection area of the gauge and the number of days over which the sample has been collected (units are $mg\,m^{-2}\,d^{-1}$).

3.5.3.2 Relationship to dust nuisance   The measurements obtained from the deposit gauge relate to the rate at which particles deposit on to a surface (the ground). In order to relate that measurement to the likely nuisance caused by the dust, use must be made of the past-deposited dust measurements and the frequency of complaints. A number of countries have introduced standard criteria for deposition rates as annual or monthly deposition averages. For insoluble dusts they have found that deposition rates exceeding $200\,mg\,m^{-2}\,d^{-1}$ on a monthly average will be high enough to create a nuisance to residential property. For the nuisance from deposited dust to be perceived, there must be a significant increase in the dust deposition rate over that which is normal for the area. Bate and Coppin [2] report average background deposition rates for different locations and suggest that the measured level must be at least twice the average background deposition rate before complaints are likely to ensue.

For particles that are likely to damage the surface upon which they deposit, compositional analysis of the particles and acidity analysis of the water would assist in determining the likely nuisance.

3.5.4 Flux Gauges

Flux gauges are used to provide a measurement of the horizontal flux of particles passing a given point, and when arranged in four orthogonal directions give an indication of the direction of the source of the wind-blown particles. An example of the latter use is the BS Directional Gauge, [6] which uses four flux gauges arranged orthogonally on a vertical stand (see Figure 27). Each flux gauge is a cylindrical tube with a vertical opening for the collection of particles passing horizontally by the device. However, the collection efficiency (also given in Figure 27) of the flux gauge is low and dependent upon both wind speed and particle size.

A new flux gauge has been developed [12] in which air is passed through the device and the particles caught on a particle trap (see Figure 28). It uses a wedge shape to provide airflow through the gauge, and a plug of porous polyester foam to trap the particles. Its collection efficiency in relation to the BS Directional Gauge is also given in Figure 28.

It must be remembered that this device does not give information on deposited particles, but is a very useful tool in source apportionment and fugitive emission studies.

4 GENERAL SAMPLING PROCEDURES

4.1 SAMPLING STRATEGY

4.1.1 Programme Types

When formulating a strategy for the health-related sampling of aerosols in the ambient atmosphere it is necessary, first of all, to decide what the information is to be used for. There are two main types of programme:

1. Long-term studies at a number of fixed sites throughout the country to monitor pollution trends. At these sites a whole range of pollutants (including aerosols) are measured by continuous monitors with the results being sent by telemetry to a central data acquisition centre. Besides providing information to meet guide and limit values in government and European Directives, these programmes also provide invaluable data for public information and epidemiological studies.

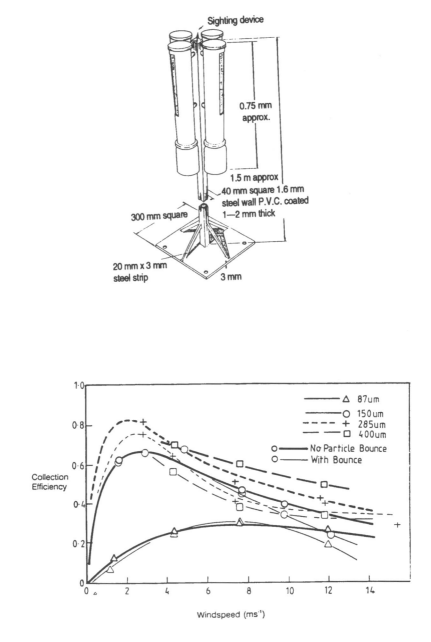

Figure 27 British Standard Directional Gauge together with the collection efficiency for one flux gauge tuble

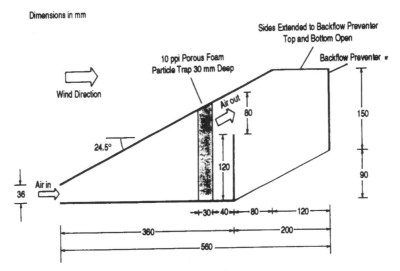

Cross Section through the Improved Design of Flux Gauge.

☐ New Design

▨ British Standard Directional Gauge

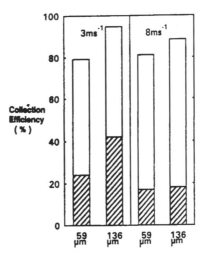

Figure 28 Wedge-shaped flux gauge together with its performance

2. Shorter-term studies designed to address a specific pollution problem. In these studies a large number of samplers are concentrated in a small area around a particular aerosol source, and measurements are normally taken before and after a specific operation has taken place. For instance, the construction of a new road or a major industrial plant could result in significant dust exposures to the general public during construction but not before and after.

4.1.2 Sampler Location

Recommended criteria for the siting of monitoring stations have been developed to cover all situations where human exposures to aerosols could arise. [30] A useful, brief guide to the types of location to be included is given by Ott. [25] A number of considerations need to be addressed which will affect the aerosol measurement when selecting the location of outdoor monitoring sites. These include:

1. The proximity of point sources, which could result in highly variable concentration gradients.
2. Obstructions or changes to airflow caused by tall buildings, trees, etc., and abrupt changes in terrain, which could introduce localized separations and swirls into the airflow, again causing highly variable concentration gradients.
3. The height of the sampler entry above the ground, which is a compromise between siting the entry at head height and ensuring that it does not get damaged or receive any extraneous, non-sampled material. For some samplers the sampling head must be aligned vertically.

Additionally, for research studies where attempts are being made to correlate the symptoms of disease with exposure to aerosols, it may not be sufficient just to monitor outdoor concentrations. In a recent study carried out in the USA [9] respirable aerosol levels indoors were found to be nearly twice as high as those measured outdoors.

4.1.3 Sampling Frequency and Duration

Ideally, to ensure accurate estimates of the likely exposures of humans to aerosols in the ambient atmosphere, measurements should be taken continuously for the whole period of the study. However, this would be prohibitively expensive and so statistically based sampling programmes have to be developed.

For long-term network studies, it has been common practice to use the high-volume gravimetric samplers that provide the mean concentration over an integrated period of say, 24 hours. In the USA, samples for the PM10 fraction are carried out every third day, and in areas where the levels exceed the specified limits, it is anticipated that the sampling frequency will be increased to a daily rate. The use of gravimetric samplers in this programme requires the filters to be changed and weighed for each sample, which is very costly. This has been the driving force behind the development of the continuous samplers mentioned above, to enable more comprehensive coverage of aerosol concentrations to be made, provided that these systems are fully tested.

The shorter-term, more specific studies to determine, for example, the likely exposures from particular operations should be undertaken for long enough to enable the measurements to reflect changes in emissions due to changes in weather conditions. These effects are usually seasonal in nature and so the programme should cover at least a year, with an intensive period for gathering baseline information, and a second phase activated when the weather conditions have changed.

4.2 OPERATIONAL ASPECTS

Most manufacturers supply comprehensive operational instructions with the instruments that they sell, and these should be read carefully before planning the details of the sampling programme. There are some general aspects, however, that merit mention here.

4.2.1 Choice of Filters

The choice of suitable filters for the gravimetric samplers mentioned above depends upon the sampling and analytical requirements. The most commonly used filters for sampling aerosols in the ambient atmosphere are glass fibre filters. These robust filters have low moisture retention, and high collection efficiencies at relatively low pressure drops. As particles are collected in the depth of the fibre bed, glass fibre filters also have the ability to collect and retain large sample masses with a low pressure build-up rate (and small change in flow rate). However, if compositional analysis or microscopic investigation of the collected particles is required, membrane filters are more suitable. In addition, glass fibre filters suffer from positive artefact mass addition due to the *in situ* conversion of sulfur dioxide to sulfate, and negative artefact mass due to volatilization of ammonium nitrate and ammonium chloride particles. Glass fibre filters also have very high chemical blank values and are not suitable for some trace element analysis such as zinc and barium. Quartz fibre filters have superior resistance to artefact effects and moisture absorption, but they are very fragile and so require extra care when handling.

Membrane filters are made from a number of different materials with pore sizes ranging from 0.03 to 8 μm. They provide samples better suited for trace elemental analysis studies such as instrumental neutron activation analysis (INAA), energy-dispersive X-ray fluorescence (ED-XRF) or microscopic analysis. The membrane filters commonly used for particle sampling include polycarbonate (e.g. Nuclepore) and PTFE (commonly known as Teflon). Polyester filters are also available, but are becoming less widely used. However, both the polycarbonate and the PTFE filters are difficult to handle as they are very thin and susceptible to electrostatic charge build-up. This can result in the loss of coarse particles during transport. A similar problem occurs with fluorocarbon filters, which are the most suitable filter for inorganic analysis. Pure silver membrane filters are ideal for situations where impurities in the filter or high weight losses render normal membranes unsuitable. An example of this is in the sampling of coal tar pitch volatiles.

The main problem with all membrane filters is that, unlike glass fibre filters, particle collection takes place at the surface of the membrane filter. This severely limits the amount of sample that can be collected because when more than a single layer of particles is deposited on the surface, the resistance to airflow increases rapidly and there is a tendency for the deposit to be dislodged from the filter, especially during transport back to the laboratory.

Filter choice for continuous monitors is very limited. The β-gauges can be used with either glass fibre or PTFE filters, with the 30 separate filters contained in the cassette of the Elecos APM-1 device being especially suitable for analysis.

In summary, quartz fibre filters are recommended for routine particle sampling when mass concentration is determined by weighing. In general, other types of filter (Nuclepore, PTFE, fluorocarbon) are recommended if specific chemical analizes are required. However, glass filters can be used for atomic emission spectroscopy, and for other chemical analizes provided care is taken to check for impurities.

4.2.2 Filter Handling, Conditioning and Weighing

Important aspects that are often overlooked in the methodology of sampling are the handling and conditioning of the filters. When taking new filters from their boxes it is important to inspect the state of the surfaces. For instance, glass fibre filters often have loose fibres from the cutting process that could be easily lost at any stage of the sampling process. It is essential, therefore, that these are removed prior to the first weighing. In addition, some membrane filters have been seen with a very fine powder coating when new. Such filters should be rejected and a new batch used. When laden with dust, it is essential that the filters are supported and kept upright to prevent the loss of material that could arise from either sharp impact or touching the internal walls of the containers.

There are a number of schools of thought about the conditioning of filters prior to weighing, and it is currently the subject of much debate by a committee of the International Standards Organization. It is recommended that all weighing is carried out under temperature and humidity control, and that the filters are allowed to condition in that atmosphere for at least 24 hours both before and after sampling. The use of desiccators is generally not recommended because when the filter is removed from the desiccator the weight will be unstable due to the absorption of moisture from the balance room atmosphere. However, dessicators may be used provided that the filters are removed at least 24 hours prior to weighing. A set of control filters to take account of changing conditions is highly recommended; one control for every five sample filters is a reasonable number to choose.

With mass concentrations of PM10 in the ambient atmosphere generally about 20–30 $\mu g\,m^{-3}$, care must be taken to ensure that the balance used has sufficient accuracy. This normally requires a five-or six-place balance, available from a number of balance manufacturers. This should be placed on a solid, vibration-free bench in a room specially designated for this purpose, with temperature, and preferably humidity, control.

4.2.3 Flow rate Calibration

Besides the measurement of the mass of the collected sample, the most important parameter to determine accurately is the flow rate. Fortunately, most aerosol samplers are fitted with automatic flow rate control or compensating systems, in which the air velocity or pressure drop is continuously monitored and signals are sent via a feedback loop to either open a valve or to increase the speed of the motor when they have changed by 5%. In the instruments made by Wedding and Associates, flow control is provided by means of a "choked" flow venturi system, which limits the flow rate to a maximum value provided that sufficient vacuum potential is maintained.

While flow control devices provide major improvements in maintaining the set flow rate throughout the sampling period, it is still necessary to set the flow rate before sampling can begin. Ideally this should be carried out by measuring the flow entering the sampling inlet using a calibrated wet gasmeter, but this is only possible for samplers with single unidirectional entries. For those omni-directional samplers with horizontal slit entries, setting the flow rate is achieved by removing the entry and using a special fitting with an orifice plate over the installed filter. Once set to the correct value, the size selective entry is replaced and the flow control systems maintain the flow rate to within the specified limits (normally 5%).

5 CONCLUDING REMARKS AND FUTURE REQUIREMENTS

This chapter has given a brief outline of the physical processes involved, and practical instruments available, for the sampling of particles in the ambient atmosphere. By necessity it has been impossible to mention all samplers currently used for this purpose, and as a means of reducing numbers, only those samplers are mentioned whose sampling efficiencies are known, and come close to the required specifications. This strategy is proposed with the aim of improving the accuracy (bias + precision) of the sampling of airborne particles, so that results obtained by different teams in different countries can be reliably combined and compared. With many more accurate, validated samplers available, or soon to be available, there is no excuse to use an uncharacterized or unsuitable particle sampler.

When faced with decisions as to the choice of sampler and sampler methodology for a particular application, it is worth while reiterating the two questions posed at the start of the chapter—"Why do you want to sample airborne particles?" and "How will the results be used?" The answers should provide the sampling rationale relevant to the practical situation in question, and the appropriate sampling methodology can then be prescribed.

5.1 WHAT OF THE FUTURE?

The main motivation for the sampling of ambient airborne particles remains for the prediction and prevention of adverse health effects to humans. The future requirements for health- related sampling of particles in the ambient atmosphere should be set by the perceived or predicted risk to the health of people from inhalation of the particles. Currently, it is believed that the PM10 fraction (representing those particles that penetrate below the larynx) gives the most relevant indicator of risk to health. However, with the increase in the incidence of asthma thought by some to be associated with an increase in particulate emissions from increased use of diesel vehicles, a finer fraction is being suggested. For this purpose, cyclone-based sampling heads developed in the USA for fractions with 50% penetration at 2.5 μm (PM2.5) and 1 μm (PM1) are available as direct substitutes for the PM10 sampling heads. The PM10 Dichotomous Sampler, which provides two samples, 10 to 2.5 μm and < 2.5μμ, should be a useful sampler for this purpose.

For epidemiological studies, it may be necessary to use personal samplers to provide a reliable estimate of individual exposures to ambient particles. This has been shown to be essential in the occupational field, where measurements with personal samplers can be up to 1000 times those from suitable static samplers sited in the workplace. There are a number of samplers, developed initially for workplace sampling, that could be used to monitor individual exposures to all three health- related fractions. One such pilot study is currently under way in Birmingham, UK. Prototype personal samplers for the thoracic fraction are used. Airflow is provided by a small sampling pump carried in a small camera case worn around the shoulders. This has been designed to be as light and quiet as possible in order to minimize any discomfort to the wearer.

One of the most difficult problems in relating exposure to a specific particulate pollutant to adverse health effects is the unambiguous identification of that pollutant in the soup of particles found in the ambient atmosphere. This is especially relevant for particulate emissions from diesel vehicles. One interesting new development is the Rupprecht and Patashnick Ambient Carbon Particulate Monitor. [26] This device automatically determines hourly averages of the organic and elemental carbon particulate concentration.

Finally, Seaton et al. [27] have proposed that the many ultra-fine particles ($< 0.1 \mu m$) that are present in the ambient atmosphere may explain the observed association between particulate air pollution and exacerbations of illness in people with respiratory disease and rises in the numbers of deaths from cardiovascular and respiratory disease among older people. They suggest that ultra-fine particles are able to readily penetrate to the interstitial tissue of the lung where they provoke a marked inflammatory response not seen for larger particles of the same chemical composition. They further suggest that the number and composition of the particles should more readily relate to disease rather than the mass of the particles deposited. In order to provide this information about the ambient particles use must be made of devices such as condensation nucleus counters which give the number concentration of particles down to $0.03 \mu m$.

ACKNOWLEDGEMENTS

The author would like to thank Professor James H. Vincent (University of Minnesota, USA) and Dr David Hall (Building Research Establishment, UK) for the many hours of fruitful discussion on aerosol sampling.

REFERENCES

1. Andersen A. A. (1958) *J. Bacteriol.*, **76**: 471–484.
2. Bate K. J. and Coppin N. J. (1991) *Mine and Quarry*, March 1991....
3. Berner A., Lürzer C., Pohl F., Preining O., Wagner P. (1979) *Sci. Tot. Envir.* **13**: 245–261.
4. BS 1747 (1969) Part 1, BSI, London.
5. BS 1747 (1969) Part 2, BSI, London.
6. BS 1747 (1972) Part 5, BSI, London.
7. Burton R. M. and Lundgren D. A. (1987), *Aerosol Sci. Technol.* **6**: 289–301.
8. Council of the European Communities (1980) *Directive 80/779/EEC*. OJ L229. 30.8.80, pp. 30–39.
9. Ferris B. G. and Spengler J. D. (1985) *Tokai J. Exp. Clin. Med.* **10**: 263.
10. Fuchs N. A. (1964) *The Mechanics of Aerosols*. Pergamon Press, Oxford.
11. Hall D. J. and Upton S. L. (1988) *Atmos. Envir.* **22**: 1383–1394.
12. Hall D. J., Upton S. L. and Marsland G. W. (1994) *Atmos. Environ.*, **28**: 2963–2979.

13. Hall D. J. and Waters R. A. (1986) *Atmos. Envir.* **20**: 219–222.
14. Hinds W. C. (1982) *Aerosol Technology.* Wiley & Sons, New York.
15. Hofschreuder P. and Vrins E. (1986) In *Aerosols: Formation and Reactivity*, Pergamon, Oxford, pp. 491–494.
16. Hollander W., Blomesath W. and Beyer A. (1990) *J. Aerosol Sci.* **21**: 41–46.
17. ISO (International Standards Organization) (1994) IS 7708, ISO, Geneva.
18. Keskinen J., Pietarinen M., Lehtimaki M. (1992) *J. Aerosol Sci.* **23**: 353–360.
19. Lodge J. P. and Chan T. L. (eds) (1986) *Cascade Impactor: Sampling and Data Analysis.* AIHA, Akron, Ohio, USA.
20. McFarland A. R. and Ortiz C. A. (1984) Dept. of Civil Engineering, Texas A&M University. *Air Quality Lab Report No 4716/01/02/84/ARM.*
21. Mandelbrot B.B. (1983) *The Fractal Geometry of Nature*, Freeman, New York.
22. Mark D., Vincent J. H., Aitken R. J., Botham R. A., Lynch G., van Elzakker B. G., van der Meulen A. and Zierock K-H. (1990) *Institute of Occupational Medicine Report No. TM/90/14....*
23. Marple V. A., Liu B. Y. H. and Kuhlmey G. A. (1981) *J. Aerosol Sci.* **11**: 333–337.
24. May K. R., Pomeroy N. P. and Hibbs S. (1976) *J. Aerosol Sci.* 7: 53–62.
25. Ott W. (1977) *J. Air Pollution Control Assoc.* **27**: 543.
26. Rupprecht, E., Meyer, M. and Patashnick, H. (1992) *J. Aerosol Sci.*, **23**, Suppl. 1, S635–S638.
27. Seaton, A., MacNee, W., Donaldson, K. and Godden, D. (1995) *The Lancet*, **345**, Jan. 21, 176–178.
28. Slinn W. G. N. (1982) *Atmos. Envir.* **16**: 1785–1794.
29. Upton S. L. and Barrett C. F. (1985) *Warren Spring Laboratory Report No 526(AP)M.*
30. USEPA (United States Environmental Protection Agency) (1987) *Federal Register 40 CFR Part 53....*
31. Van der Meulen A. (1993) *Measurement of Airborne Pollutants*, Butterworth-Heinemann, Oxford, pp. 17–22.
32. Vincent, J. H. (1989) *Aerosol Sampling: Science and Practice.* Wiley & Sons, Chichester.
33. Wagner M., von Nieding G. and Waller R. E. (1988) *Final Report on CEC Contract No BU (84) 146 (491)....*
34. Wedding, J. B., McFarland, A. R. and Cermak, J. E. (1977) *Env. Sci. and Technol.* **11**(4).
35. Whitby K. R. (1978) *Atmos. Environ.*, **12**: 135–159.

FURTHER READING

Cohen B. S. and Hering S. V. (eds) (1994) *American Conference of Governmental Hygienists: Air Sampling Instruments* (8th edition). ACGIH, Cincinnati, Ohio, USA.

Garland J. A. and Nicholson K. W. (1991) A Review of Methods for Sampling Large Airborne Particles and Associated Radioactivity. *J. Aerosol Sci.*, **22**(4):479–499.

Hinds W. C. (1982) *Aerosol Technology: Properties, Behaviour and Measurement of Airborne Particles.* Wiley-Interscience, New York.

UNEP/WHO (1994) GEMS/AIR Methodology Review Handbook Series, Volume 3, *Measurement of Particulate Matter in Ambient Air.* UNEP, Nairobi.

Vincent J.H. (1989) *Aerosol Sampling: Science and Practice.* Wiley & Sons, Chichester.

NOMENCLATURE

A_d	Surface area over which particles deposit to ground
A_F	Cross-sectional area through which particles pass in flux measurements
B	Aerodynamic bluffness
C_C	Cunningham slip correction factor
C_D	Drag coefficient
d	Particle geometric diameter
d_{ae}	Particle aerodynamic diameter
d_A	Particle equivalent surface area diameter
d_{em}	Particle electrical mobility diameter
d_F	Particle Feret diameter
d_M	Particle Martin diameter
d_p	Particle equivalent projected area diameter
d_v	Particle equivalent volume diameter
D	Characteristic dimension in a flow system
D_B	Coefficient of Brownian diffusion
D_{ft}	Coefficient of turbulent diffusion for fluid
D_{pt}	Coefficient of turbulent diffusion for particle
e	Fundamental electronic charge
E	Efficiency of particle impaction on to a bluff body
E_A	Aspiration efficiency of sampler
E_I	Inhalable aerosol fraction
E_R	Respirable aerosol fraction
E_T	Thoracic aerosol fraction
F	Flux of particles passing through a vertical area A_F
F_D	Drag force on a particle
g	Acceleration due to gravity
k_B	Boltzmann's constant
K_{pt}	Inertial parameter related to turbulence
m	Particle mass
M_d	Mass of particles deposited to ground
n	Number of electronic charges on a particle
N	Number of particles
Re	Reynolds number for flow system
Re_p	Reynolds number for flow about a particle
s	Particle stop distance
St	Stokes number
U	Free-stream air velocity
v_x, v_y	Mean velocity in the x- or y- direction
Z	Particle electrical mobility
δ	Dimension of sampler orifice
ϕ	Particle dynamic shape factor

η Air viscosity
θ Orientation of sampler to wind
ρ_A, ρ_p Densities of air and particle
τ Particle relaxation time

3 Inorganic Composition of Atmospheric Aerosols

MARTINE CLAES, KRISTIN GYSELS and RENÉ VAN GRIEKEN
University of Antwerp (U.I.A.), Belgium

and

ROY M. HARRISON
The University of Birmingham, UK

Atmospheric Particles, Edited by R. M. Harrison and R. Van Grieken.
© 1998 John Wiley & Sons Ltd.

1 INTRODUCTION

In the past few decades, there has been a growing concern about environmental contamination by heavy metals. Industrial production, energy generation and vehicular traffic have increased significantly the levels of these pollutants in the atmosphere. The multi-elemental composition of the aerosol referring to sources, source types and/or source regions of the various aerosol constituents are briefly described in this chapter; for more detailed information, Chapter 12 is highly recommended. Furthermore, this chapter contains information on atmospheric behaviour and transport of the various aerosol constituents and information on dry and wet deposition velocities and fluxes from air to sea or earth surface; more details are well described in Chapter 1.

In order to assess the impact of heavy metals on our health, high quality and quantity data on trace metals in aerosols are to be provided by adequate analytical techniques. In the first part of chapter, we will focus our attention on those analytical techniques that can be used for measuring the bulk composition of atmospheric aerosol samples, including nuclear techniques, X-ray emission techniques, optical atomic spectrometric and mass spectrometric techniques and voltammetry. The characteristics of the various techniques are discussed and their advantages and drawbacks for analysing various types of aerosol samples are compared. The analysis of samples obtained from total and size-fractionating filter collectors as well as from cascade impactors or other collection devices is addressed. Recent publications of applications of each of the presented techniques to atmospheric pollution studies are also included.

The second part of this chapter will focus on secondary inorganic aerosols comprising particulate matter formed by chemical conversion of gases to less volatile compounds. For the major species, including sulfate, nitrate and chloride, the formation, concentration, size distribution and analysis techniques will be discussed.

2 METALS AND THEIR COMPOUNDS

2.1 SOURCES

The processes which emit trace metals into the atmosphere are very diverse, like volcanic activity, vegetation emissions, soil erosion, man-made pollution and aerosol formation from sea-spray. Measurements of natural sources at remote locations have not been very abundant. The accuracy of calculations differs depending on the type of source and the elements in question. Anthropogenic sources have been studied with measurements and calculations done either for a certain country, for a given element or for an individual source. To determine the source of a given element in aerosols, three main approaches exist, namely direct measurements, calculation of enrichment factors and source receptor modelling. Those techniques are well described by Fergusson. [1] In the following sections, natural and anthropogenic sources of metal releases into the atmosphere, will briefly be discussed.

2.1.1 Natural Sources

An accurate assessment of natural source strengths is quite difficult but also important, as for many elements, natural emissions exceed those from anthropogenic sources. In many cases, the estimates of emission rates or emission fluxes are based on educated guesses and they are not confirmed by measurements. Even in cases of available data for a certain natural source (e.g. volcanic eruption), disagreements of measurements of different authors make this information hard to utilize. Among the natural sources of trace elements, the wind-blown dust and volcanic eruptions are considered the most important. An overview of the world-wide emissions of trace elements from natural sources is given in Table 2 in Chapter 12. The emission factors for wind-blown dust are calculated from the concentration of each metal in different soils and the amount of continental dust annually brought into the atmosphere. Wind-blown soil dust is known to bring 20×10^6 kg yr^{-1} of Pb into the atmosphere globally. [2] Natural Zn emissions (43.5×10^6 kg yr^{-1}), with wind-blown dust as major contributor, are estimated to be almost 10 times lower than the combined industrial emissions (314×10^6 kg yr^{-1}). Some Saharan dusts are known to contain high levels of Cr, up to 3000 ppm. Long-range transport of such soils was expected to contribute significantly to Cr release, even in Europe, [3] together with Cr condensation on airborne soil dust.

The most extensively studied source appears to be volcanic eruptions. During eruptions, volatile elements (such as As, Cd and Se) evaporate and then condense on the very small particles. Among the important metals released by volcanoes are Zn, Pb and Ni. Data on trace element emissions during volcanic eruptions are being collected by researchers around the world. A comparison of volcanic emission estimates is given by Pacyna. [2] Generally, the emission rates of individual volcanoes can vary to a great degree, which makes detailed calculations rather difficult.

For forest wildfires, the emission factors are estimated from the average acreage that is burned and the concentrations of trace elements in forest stock. In certain parts of the world forest fires are the major emission sources.

Another possible natural process which could be important for some metals is marine emissions in the gas phase (e.g. for As) and sea-salt particles (e.g. for Cd, Cu, Ni, Pb and Zn) due to the gas bubbling breaking process at the sea surface. Due to breaking waves, air is captured and released as bubbles at the sea surface; in this way jet and film drops containing metals are ejected into the atmosphere. Reinjection by sea-spray of previously deposited particles has been proposed by Duce *et al.* [4] as a possible source for marine-derived particles containing trace metal. The reality and quantification of this possible recycled marine component are still controversial. [5] The calculations for bubble bursting have been performed using the trace element concentrations in surface ocean waters and the enrichment of trace elements in the atmospheric sea-salt particles.

2.1.2 Anthropogenic Sources

Anthropogenic source processes are mainly related to the volatility of elements at high temperatures of fossil fuel combustion, and many other high-temperature industrial processes, particularly the extraction of non-ferrous metals from

Table 1. World-wide anthropogenic emissions of trace elements during 1975 (10^9g yr^{-1}), after Pacyna[2]

Source	As	Cd	Cu	Ni	Pb	Se	Zn
Mining, non-ferrous metals	0.013	0.002	0.8	—	8.2	0.005	1.6
Primary non-ferrous metal production	15.2	4.71	20.8	9.4	76.5	0.28	106.7
Secondary non-ferrous metal production	—	0.60	0.33	0.2	0.8	—	9.5
Iron and steel production	4.2	0.07	5.9	1.2	50	0.01	35
Industrial applications	0.02	0.05	4.9	1.9	7.4	0.06	26
Coal combusion	0.55	0.06	4.7	0.7	14	0.68	15
Oil combustion (including gasoline)	0.004	0.003	0.74	27	273	0.06	0.1
Wood combustion	0.60	0.20	12	3.0	4.5	—	75
Waste incineration	0.43	1.40	5.3	3.4	8.9	—	37
Manufacture, phosphate fertilizer	2.66	0.21	0.6	0.6	0.05	—	1.8
Miscellaneous	—	—	—	—	5.9	—	6.7
Total	23.6	7.3	56	47	449	1.1	314

sulfides. To compare these emissions with trace element releases from natural sources, one must take into account the scale of pollutant perturbations. For Se, Hg and Mn global natural emissions exceed total releases from anthropogenic sources. One can distinguish three main emission types: global, regional and local. Statistical information on the world consumption of ores, rocks and fuels and on the world production of various types of industrial goods are available [2] and global emissions of trace elements from anthropogenic sources have been calculated. [6] Table 1 represents one of the first world-wide anthropogenic emissions of trace elements after Pacyna. [2] More detailed and recent data are given in Chapter 12 of this volume.

The global emission inventories show that most anthropogenic Be, Co, Mo, Sb and Se are emitted from coal combustion. Large amounts of As, Cd and Cu are emitted by smelters and secondary non-ferrous metal plants. Mn and Cr are released mainly from factories producing iron, steel and ferro-alloys. Mn is the most widely used element in ferro-alloys, followed by Si, Cr and P. Mn particles with Zn and Fe are derived from the ferrous manganese furnace process. Yearly, 10×10^9 kg of chromite ($FeOCr_2O_3$) is used globally, mainly for metallurgical and chemical applications. [7] Cr vapours from steel production can condense as oxides on the surface of different sorts of airborne particles. [8] Cr was found to be significantly concentrated on the surface of combustion particles, [9] a process that forms a wide variety of Cr-containing particles and complicates source identification. The cement industry is another potential source of atmospheric Cr as they use in their high-temperature process more than 30 raw materials containing Cr concentrations in the ppm range. [5] Metallic Cr is primarily used in non-ferrous alloys, where less expensive ferrochromium alloys can give undesirable amounts of Fe. [5] The emission of Fe-rich particles mainly results from metallurgical processes. Fe–S particles can be formed by reaction between iron oxide and sulfuric acid, during their release in ferrous metallurgy related combustion processes. Most Zn production originates from ZnS minerals, which are converted into metallic Zn in a series of processes. The main use of Zn is in galvanizing iron and steel products, coated with Zn to increase the corrosion resistance. Zn is also used in Ni–Zn batteries, in plumbing materials, in paints and in cosmetic and pharmaceutical products. Main anthropogenic sources are iron and steel industries, industrial applications, coal and wood combustion and waste incineration. [10] ZnS is related to wind-blown Zn ores, which are mostly stored uncovered and therefore easily taken up by the wind. Emission of Pb particles in the atmosphere occur from a wide range of applications. The average emission for all sources was estimated by Pacyna [2] to be around 123×10^6 kg yr^{-1} in Europe. The most important source is the combustion of gasoline contributing for more than 50% of the total Pb emission. This number should have been drastically reduced by the introduction of unleaded gasoline. Other major emitters are coal combustion, steel industry and cement factories. [10] According to Fergusson, [1] metallic Pb comes mainly from emissions of coal-fired power stations, cement and fertilizer production, base metal smelting and even automotive exhaust. Particle types of Pb together with Cl and Br originate from car exhaust emissions and have been studied extensively, [11] while particles of Pb associated with Ca or P are linked to emissions of fertilizer production or cement manufacturing. Elements like Ni and V are released by oil combustion processes in power plants due to oil-fired furnaces. The contribution of natural sources to the total emission of Ni and V is very low, respectively 14 and 16%, [12] and therefore these elements are used as indicator elements of oil-fired power plants and oil combustion sources in numerous industries. The main source for Ti-release into the atmosphere is pigment spray, but also minor pollution processes and sources such as soil dispersion, asphalt production and coal-fired boilers and power plants are known. [13]

Inventories of trace element emissions on a regional basis refer to a scale of 10–1000 km. For this study, emissions of the trace elements As, Be, Cd, Co, Cr, Cu, Mn, Mo, Ni, Pb, Se, Sb, V, Zn and Zr to the atmosphere were estimated for all European countries by Pacyna. [12] For certain sources, information on emission factors was difficult to obtain. Cadmium is emitted from stabilizers, alloys and batteries, but the trace element emissions from these sources are relatively small and can be neglected. Mn emission from man-made sources in Europe is several times larger than from natural sources in this part of the world. [2] Pacyna [2] also pointed out that the steel industry is the most important emitter of Cr in Europe.

Local scale of emission surveys covers the effluents of pollutants from single point sources, as power plants and individual factories. Such inventories are used to study the consequences of pollutant emissions in the environment around the emission source and are often applied to calculate human exposure to trace elements [2] and then to analyze dose–response relationships.

Although changes in emission patterns are related to changes in energy generation and industrial activities, emissions can be reduced, for example, by replacement or introduction of new and more efficient control devices, by new coal technologies. A recent overview of global particle emissions for natural and anthropogenic sources is given by Hidy and Wolf. [14] These calculations note the global importance of wind-borne dust and sea-salt, as well as the secondary aerosol contributions.

2.2 SIZE DISTRIBUTIONS

The knowledge of aerosol size distributions is essential because the particle size significantly affects ambient transport and deposition processes as well as uptake in the respiratory system, which is especially important for heavy metals. Moreover, elemental size distributions can give an indication of the source of the element. For theoretical considerations and more details concerning size distribution, reference is made to Chapter 1.

Experimentally, size distributions can be traced using cascade impactors for aerosol sampling. Unfortunately size distributions measured this way show some inaccuracy due to errors inherent to cascade impactors. To obtain representative aerosol samples it is necessary to sample isokinetically, [15] but even then, the collection of large particles is limited due to deposition in the impactor inlet. [16] Moreover, the size distribution can be distorted because of bounce-off effects which cause particles to be deposited on subsequent impaction stages. This effect can be minimized by coating the impaction surface with a layer of grease. [17] Besides the large uncertainty for large particles, sampling without a back-up filter after the last impactor stage can cause significant losses of small particles. Detailed information on aerosol sampling, including size segregated sampling, is given in Chapter 2 of this volume.

The typical characteristics of size distributions vary from element to element and from sampling location to sampling location, but generally they show a bimodal shape with one maximum in the submicron and a second maximum in the supermicron range. In many cases one of these maxima is predominant, which can be explained by the emission sources. Aerosols generated by high-temperature processes (e.g. combustion) usually have a peak below 1 µm aerodynamic diameter and aerosols originating from mechanical sources (bubble bursting, soil suspension) above 1 µm, [18] so generally submicrometre particles can be associated with anthropogenic sources while supermicrometre particles can be considered to be mainly natural in origin. Davidson and Osborne [19] have reviewed the literature on metal size distributions and were able to conclude that airborne Pb is primarily of submicrometre size while Mn and Fe are generally associated with particles larger than 1 µm. Cd, Cu and Zn are detected in both size classes.

The size distributions of several elements in an urban and a suburban region (Vienna, Austria) have been measured by Horvath et al. [20] and their results were consistent with the results obtained by Davidson and Osborne, except for Mn, whose size distribution showed a clear peak at 0.7 µm. These small Mn particles could result from emission by coal-fired power plants. Fe was found to be predominantly of supermicrometre size, although submicrometre Fe was also present. This can be attributed to industrial sources, while the coarse fraction is attributed to soil and road dust. The size distribution of Pb showed a maximum at 0.4 µm, being anthropogenic in origin, while Cu and Ni have a very broad size distribution without a clear maximum, which can be attributed to a wider variety of sources for these elements. The size distributions of Cd, Cu, Pb and Zn above the North Sea were measured by Injuk et al. [21] They also found Pb to be associated with submicrometre sizes, while the size distribution of Zn showed a maximum above 1 µm. Cd and Cu were detected in both size classes.

The calculation of enrichment factors allows us to decide whether a certain element has additional sources in comparison with soil erosion, for example. Lannefors et al. [22] calculated enrichment factors for several elements relative to Ti, which can be considered to be mainly soil-derived. An enrichment factor > 1 indicates the presence of anthropogenic or natural sources other than soil suspension. This was found to be the case for submicrometre Mn, while super-micrometre Mn as well as K, Ca and V were found to be mainly soil-derived. The enrichment factors of Cu, Fe, Zn, Ni and Pb in the supermicrometric range were very high, indicating the possible occurrence of condensation or coagulation reactions or fractionating effects.

As already mentioned above, atmospheric size distributions are important because of the influence of particle size on transport and deposition processes. Two types of deposition processes exist: dry and wet deposition. The dry deposition flux is a result of gravitational settling, turbulent diffusion, Brownian diffusion or impaction and can be calculated as the product of the atmospheric concentration and the dry deposition velocity, which is a function of particle diameter. This is also an important factor in rain-out and wash-out procedures (wet deposition). Wash-out can be described as the interception of atmospheric particles by falling raindrops while rain-out is the condensation of water vapour on to aerosols (condensation nuclei). The ratio of dry to wet deposition depends on the amount of precipitation, but both mechanisms can be considered equally important for trace elements. [1] Dry and wet aerosol removal processes are discussed in detail in Chapters 13 and 14, respectively, of this volume.

A model of deposition fluxes is described by Slinn. [23] This can be used to calculate deposition velocities and fluxes after measuring atmospheric size distributions. Dry and wet deposition fluxes of various elements above the North Sea have been calculated this way, for example by Injuk et al. [24] Total deposition values of 690 and 3400 t yr^{-1} have been obtained for Cu and Zn respectively, which is more than half of the estimated riverine input. [25] For Pb, the atmospheric input (1970 t yr^{-1}) value is even higher than the riverine input. Atmospheric input can thus be considered to be an important source of pollutants in the North Sea.

In addition to theoretical determination of deposition fluxes, total deposition can also be measured directly when analysing the material deposited on to collection plates or into bottles. [26]

2.3 SPATIAL DISTRIBUTIONS

On a global scale the highest concentrations of anthropogenic aerosols occur above the continents of North America and Europe and locally in the former Soviet Union and the coasts of China and Japan. [27] The heavy metal concentrations can be very high in industrial regions, and depend highly on the type of industrial activity. In rural and remote areas, the concentrations of heavy elements depend largely on the closeness of industrial or natural emission sources but are much lower than in urban areas. [1] Spatial distributions of certain elements can be estimated based on information concerning industrial activities in the region of interest. This method was used by Pacyna and Münch [28] in order to estimate the European distributions of As, Cd, Hg, Pb and Zn. A more detailed description can also be found in this book in Chapter 12.

Next to peak concentrations in highly polluted areas, background marine and continental aerosol concentrations occur. The dominant aerosol type in the marine environment is sea-salt, formed by the bubble-bursting mechanism, but in some areas a clear anthropogenic or continental influence is also noticed. This occurs mostly in the tropical North Atlantic, the Carribean Sea, the Mediterranean Sea, the Black Sea, the Indian Ocean and the Bay of Bengal, areas influenced by dry land masses. [29]

In remote dry areas (Canada, Siberia) large soil dust particles form an important constituent of the background aerosol. Aerosols in the tropical regions arise from biogenic production and biomass burning. [29]

The highest continental background aerosol concentrations exist over the North Pole. This phenomenon is known as Arctic haze. The main source of this pollution is Europe and the former Soviet Union. [30] In the winter meteorological conditions favour transportation from these regions to the Arctic, so the Arctic haze occurs most in this season. [31]

2 4 CHEMICAL ANALYSIS

2.4.1 X-ray Fluorescence Analysis

2.4.1.1 Principle X-ray fluorescence (XRF) employs electromagnetic radiation for generating inner shell vacancies in the atoms of the analyte elements in the sample and thus producing the characteristic X-rays, resulting from the deexcitation of the induced vacancies. Numerous variants of the basic process have been proposed. They differ both in the type and source of ionizing radiation and in the method employed to measure the characteristic X-ray emission. [32–34] For routine XRF analysis, either a wavelength-dispersive (WD) or an energy-dispersive (ED) detection system may be used. The common EDXRF mode provides simultaneous determination of numerous elements, but its use in trace analysis is hampered by insufficient sensitivity for low-Z elements and by inter-element spectral effects. The WDXRF version has a much better spectral resolution and allows us to assess more elements, because it usually has lower detection limits for low-Z elements and also permits us to measure very light and heavy elements because of a superior spectral resolution. Yet the WDXRF instrument is more expensive. Both established techniques are multi-elemental, provide non-destructive analysis of a variety of sample types, including environmental samples, and detection limits are often in the 1–10 µg g^{-1} range. In XRF, matrix effect correction has to be performed because both the exciting and characteristic radiation interact significantly with matter, while passing through the sample. However, in contrast to some other atomic spectrometric techniques, XRF has the advantage that the physics of the matrix effects are well understood. Many books [35–37] and review papers, [38–42] discussing fundamentals, new developments and trends have been published. Although XRF is a mature technique, advances are still made; fundamental parameter approaches for quantification, microfocus X-ray tubes, etc. [43] However, the most significant advances in XRF for aerosol analysis occur in the areas of total reflection XRF (TXRF) and synchrotron radiation XRF (SR-XRF). TXRF is based on the principle that, when an exciting X-ray beam impinges on a thin sample, positioned on a very flat support, below a critical angle, this radiation does not penetrate in the support but is totally reflected. As a consequence, the support material does not produce any significant interfering background radiation, leading to extremely low detection limits of a few pg. A precondition is that the sample must be present as a very thin layer, otherwise the principle of total reflection will be lost. [44] SR-XRF makes use of the intense tunable X-rays produced by a relativistic electron beam of some GeV generated in a synchrotron accelerator to perform X-ray analysis and microanalysis. Major characteristics of SR are its wide and continuous spectral range of very high intensity, a high degree of polarization in the plane of the electron orbit and the natural collimation. Because of the high intensity of the photon source, very small and thin samples can be analyzed which cannot be done by conventional XRF. Detection limits in the range of $0.1–1\ \mu g\,g^{-1}$ for bulk analyses have been reported. [45]

2.4.1.2 Aerosol analysis   Aerosols are preferably collected by total filter samplers, stacked filter units or dichotomous samplers which all provide uniform aerosol layers on filters, an ideal target for XRF analysis. The analized sample area is typically of the order of a few cm [2]. Nuclepore membranes are very suitable for aerosol collection prior to XRF analysis because of their properties; they are very thin ($1\ mg\,cm^{-2}$), pure and essentially collect the aerosol at the surface only. Cellulose fibre filters, like Whatman 41, are thicker ($8.5\ mg\,cm^{-2}$) and hence produce a larger X-ray scatter background, resulting in higher detection limits. The main disadvantage of glass fibre filters in addition to their thickness, is their high inorganic impurities, although their reduced hygroscopicity, ability to withstand higher temperatures and high collection efficiencies make them ideal for organic analysis for example. Consequently, for techniques such as XRF, where absorption of the incident radiation and of the generated characteristic X-rays has to be considered, membrane filters have a definite advantage over cellulose filters. Limits of detection obtainable are a few ng cm^{-2}. [46] For certain applications, Teflon membrane filters may be preferable to Nuclepore filters. Stacked filter units with two subsequent membranes with different size cut-offs are suitable for size-fractionated collection. Also, dichotomous samplers (virtual impactors) can be used for such collection. Detailed size distribution measurements of the atmospheric trace elements can be obtained by using cascade impactors. For analysis with XRF, single-jet cascade impactors (Batelle type [47]) are to be preferred. To avoid bounce-off effects, the surfaces can be coated with paraffin or vaseline, although this is not always possible because the process may interfere with later chemical analysis. Cascade impactors suffer from inlet and wall losses due to the deposition of larger particles on surfaces other than the impaction plate; they can be minimized by maintaining a small flow into the coarse particle receptor.

As mentioned above, some serious problems stem primarily from particle size effects and X-ray absorption effects due to the filter media (filter penetration effect) in the case of fibre filters like Whatman 41. [37] Once the particle size distribution is known for a particular element, appropriate corrections can be made; however, this is not easy. The particle depth distribution influences the absorption of X-rays. Assuming an exponential function for the depth distribution, the average penetration depth of the element can be calculated. [48] For Nuclepore, Teflon and Millipore filters, X-ray absorption effects are negligible for the energy range above 5 keV and one can assume that particles stay on the surface. Corrections have to be made for small-particle elements and for the lower energy range.

Because of its characteristics, TXRF is most suitable for analysing aqueous samples with low concentrations of dissolved material. However, the analysis of atmospheric aerosols has been gaining strongly in importance over the last few years; it was firstly described by Leland et al. [49] In TXRF, aerosol sampling can be done by using cascade impactors with polished and totally reflecting supports as impaction surfaces, as was first demonstrated by Schneider. [50] The consequences, however, are the impossibility to apply an adhesive coating to minimize particle bounce-off effects and an internal standard solution still has to be pipetted on to each reflector after sampling. The limits of detection are in the order of 0.2 ng for most heavy metals. In contrast to the typical aerosol investigations using widespread filters and their digestion, the sample preparation step involving direct aerosol collection on the carriers is reduced considerably, from a few hours to a few minutes. Also, the low costs of the proposed method should be recognized. [50–52] Dixkens et al. [53] developed an electrostatic precipitator (ESP) in order to determine directly the trace elements of deposited particles on quartz glass supports by TXRF. The performed experiments demonstrated that the designed ESP is a useful tool for direct analysis of submicron particles. The collecting efficiency of the particles with diameters less than 1 μm is about 100%. When aerosol collection is done on filters, destruction of the filter is necessary. For aerosol samples collected on Nuclepore filters, microwave acid digestion can be applied for TXRF analysis. The recovery for all elements is quantitative within the usual standard deviation of 10–15%. [52]

A comparative study of the two different sampling methods (filters and impactors) was carried out by Salva *et al.* [51] in 1993. Recent work on results of trace element characterization (bulk analysis) of air particulates by XRF techniques has been reviewed by Valkovic *et al.* [54]

2.4.1.3 Applications of XRF   Since the early 1970s, the time when XRF lent itself to the analysis of aerosol samples, the number of publications in this field has increased and in the last decade reached between 35 and 50 per year. [55]

Most often, tube-excited XRF is used, in either the WD or ED mode. The use of portable instruments for on-site air pollution measurements is gaining enormously in importance, after the application in industrial hygiene problems described by Rhodes *et al.* [56, 57] and the *in situ* analysis potential for sulfur and other elements in aerosols described by Gilfrich and Birks. [58] Scanning of millimetre-sized depositions from a multi-orifice low-pressure impactor plate [59] and the optimization of a process for thin-layer and thin-tablet methods [60] were both used for the analysis of atmospheric pollutants in aerosol samples with energy dispersive XRF spectrometry. The most important recent advances are in the application of TXRF, [49, 61] polarized beam XRF [62] and of SR-XRF [63] for aerosol analysis. These techniques were used for the analysis of urban and rural aerosols from different cities. [64] Airborne fly-ash particles and other industrial emissions have also been analyzed. [64] Klockenkämper *et al.* [61] pointed out the need for a suitable impactor material for collection of air dust to be analyzed with TXRF, in order to prevent high blank values, collection losses and memory effects.

As for other multi-elemental techniques, remarkable advances have been made in the handling of data generated by XRF. It is now common use to apply receptor models such as principal factor analysis, [65] chemical mass balance or multiple regression analysis for processing the large numbers of aerosol elemental concentrations measured by XRF. The variability of the elemental concentrations can then be used to identify the aerosol sources. [66] During recent years, several papers covering particle characterization have been published. The elemental concentrations in fine particles over two periods of several months were determined by the Fine Particle Network of the USA by means of WDXRF, [65] leading to a database of fine particle composition to be developed and seasonal averages to be calculated. XRF was used to determine trace metals in aerosol samples in China. [67] Several investigators used XRF techniques for Pb in aerosols, [50] for S [68] and for multi-element investigations. [69] Otten *et al.* [70] determined elemental concentrations in atmospheric particulate matter sampled on the North Sea and the English Channel.

Direct analysis of aerosol-loaded quartz fibre filters by XRF requires suitable calibration standards. Results of a comparison between two different preparation techniques for quartz fibre filter standards for EDXRF analysis were published by Haupt *et al.* [71] Filter types were evaluated for the preparation of ambient aerosol reference materials for elemental analysis. [72] Kunugi [73] reviewed the use of artificial collecting membranes for the determination of carbon in suspended particles in air by XRF spectroscopy. Procedures for preparing customized particulate-loaden glass fibre filter standards for XRF have also been described. [74]

For recent applications and developments in analysis of airborne particulate matter, the proceedings of the international TXRF and European EDXRS conferences, which are held every two years, as well as those of the annual Denver X-ray analysis conference, are a valuable source. A recent overview of aerosol analysis with XRF (WD and ED) and TXRF has been given by Török *et al.* [75]

2.4.2 Particle-induced X-ray Emission Analysis

2.4.2.1 Principle Particle-induced X-ray emission (PIXE) shows most resemblance with XRF, and it is also the strongest competitor of XRF. In PIXE, heavy charged particles (usually protons of 1–4 MeV) are used to produce the characteristic X-rays of the analyte elements in the sample, and the generated X-rays are measured with an ED Si(Li) detector. The projectiles are provided by small particle accelerators, such as a van de Graaff accelerator or a compact cyclotron. This hampered the growth of PIXE due to the need to have access to such an accelerator so that the technique is less widespread than XRF. Ultra-high vacuum is not mandatory in PIXE and the common vacuum shared by beam-line and specimen chamber is typically about 10^{-6} Torr. Motivation and technology for non-vacuum analysis as well as principles and fundamentals of PIXE have recently been well described by Johansson *et al.* [76] Compared to conventional ED-XRF, PIXE allows us to analyze smaller sample masses and offers detection limits that are one order of magnitude better. PIXE is a fast, non-destructive multi-element technique with detection limits of 0.1–1 ppm or of 10^{-11} g in absolute terms. The sub-ppm detection limits in PIXE are only obtained when the sample matrix consists of light elements. For heavier element matrices, the detection limits are worsened because of the increased continuum background and the presence of intense X-ray lines from the matrix elements in the PIXE spectrum. It appears that PIXE is the method of choice for multi-element analysis only when very many samples and/or very small sample amounts are to be analyzed. Another favourable feature is that PIXE can be complemented with other ion beam analysis (IBA) techniques, such as elastic (and non-elastic) scattering, and proton-induced gamma-ray emission (PIGE). In this way a simultaneous measurement of light elements becomes feasible. Limitations of the technique, which are also shared by conventional XRF, are that it suffers from spectral interferences and that matrix effects have to be accounted for. Nowadays, the microbeam variant of PIXE provides a spatial resolution of the order of 1 μm and detection limits down to the $pg\,g^{-1}$ level, which makes elemental determinations down to $10^{-17}\,g\,\mu m^{-2}$ possible. [77] A major trend is the combination of micro-PIXE with nuclear microprobe techniques such as charged particle reaction analysis, Rutherford backscattering analysis and scanning transmission microscopy. [64]

2.4.2.2 Aerosol analysis The characteristics of PIXE make it a very suitable technique for the direct, non-destructive analysis of small atmospheric aerosol samples. Samplers that collect the particulate material on the surface of a filter or substrate film, used in XRF, are also suitable in PIXE. [78] Although it is highly preferable for EDXRF that the particles are present as a uniform layer, PIXE can also easily handle samples that are non-uniform within the beam area, if the proton beam intensity is homogeneous in this area. The same three types of matrix effects have to be considered in PIXE as in XRF, i.e. particle size effects, matrix effects in accumulated layers of particles and matrix effects in the collection substrate. Corrections are generally carried out by relying on the fundamental parameter approach or on semi-empirical approaches. In order to keep the background minimal and to avoid absorption effects in the collection substrate, thin organic filters, such as Nuclepore or Millipore filters, or films are preferred. Teflon membrane filters can also be used but their high fluorine content gives rise to high pronounced prompt γ-ray background in the PIXE spectrum. Also total filter samplers and stacked filter units can be used for collecting samples to be analyzed by PIXE. A detailed overview of the various samplers has been made. [37] The optimum mass loading of a filter is a monolayer of particulates of about $100\,\mu g\,cm^{-2}$, depending on the size of the particles. With such values, the corrections for the X-ray attenuation and ion energy loss are so small that they do not introduce any significant errors. In many cases, no corrections are needed. Beam diameters of 1–10 mm are used by most PIXE researchers, so that the irradiated area of the sample is typically less than 1 cm², versus a few cm² in XRF. Therefore, in the case of aerosol filter samples, which are often much larger than the beam size, lateral uniformity of the particulate deposit is required. The accuracy of analysis of air pollution samples with PIXE is better than 5–10%. Limits of detection are in the region of a few ng cm⁻². [79] A serious limitation of PIXE is that only one-third of the mass of a typical ambient aerosol can be determined, since the major part is composed of light elements. The analysis of deposition samples by PIXE is less straightforward. Dry deposition samplers rarely collect the fall-out dust on a surface which is suitable for direct analysis. Particle size effects would be more important and difficult to estimate, so that PIXE has virtually never been applied to such samples. On the other hand, analysis of wet deposition is more favourable with PIXE; examples of analysis of wet deposition are given in refs [80] and [81].

In PIXE, the possibility of losing volatile elements or unstable compounds from the target is greater than in XRF. This is partly due to the vacuum used and the highly focused intense charged particle beam, leading to negative errors of e.g. 50% when sulfur is determined by PIXE. Also chemical reactions lead to significant losses when 2 MeV protons, 18 MeV α-particles or other types of high-energy excitation radiation are used. [82]

2.4.2.3 Applications of PIXE Johansson *et al.*, [83] pioneers of PIXE, already knew the potential of PIXE for sensitive multi-element analysis of air particulate matter. PIXE has been used very frequently for aerosol deposits, for:

1. Very remote areas, where the airborne trace metal concentrations are very low. [64]
2. Impactor deposits with size resolution or samplers with time resolution, in which the particulate is spread out over several subsamples. [84, 85]
3. Indoor and work environment aerosols.
4. Large aerosol sampling networks, where thousands of samples have to be studied. [85, 86–92]

In atmospheric aerosol research, the need to collect enough material for analysis, putting severe constraints on sampling devices and pumps, has led to the development of small, inexpensive and lightweight samplers with high time and/or size resolution. A detailed overview of these innovative sampler designs has been given by Annegarn *et al.* [93]

Tabacniks *et al.* [94] developed a three-step extraction procedure to allow for PIXE analysis of particulate matter collected on glass fibre filters. Analyses of airborne samples collected with a micro-orifice eight-stage impactor are reported by Maenhaut *et al.* [95] To understand source–receptor relationships, a new interpretation technique for PIXE results has been given by Annegarn *et al.* [96] For example, Romo-Kroger [97] described the advantages and disadvantages of using deuteron excitation as an alternative in PIXE analysis.

The potential of PIXE combined with other ion beam analytical techniques for an almost complete characterization of the atmospheric aerosol, has been pointed out by Swietlicki *et al.* [98]

Malmqvist [99] reviewed all PIXE research on work environment aerosols up to 1983, which was later updated by Annegarn *et al.* [100] Since then, more recent and detailed reviews [46, 78, 101–103] have been published on the status of PIXE in aerosol research. The performance of PIXE for aerosol samples was evaluated thoroughly in a comparison with other techniques and has been published elsewhere. [104]

2.4.3 Nuclear Activation Analysis

2.4.3.1 Principle   In nuclear activation analysis, the production of a radio-nuclide from an analyte element is used for the identification and quantitative determination of the element. Neutron activation analysis (NAA) is the most common variant, since normally, thermal neutrons in a nuclear reactor are used to irradiate the sample; charged particles or energetic photons are applied exceptionally. As a result of a nuclear reaction between projectile and isotope of the element of interest, radionuclides are produced. Some time after the end of the irradiation, the radiation emitted by the decaying radionuclides is measured with a suitable detector. Nearly all multi-element instrumental neutron activation analysis (INAA) is nowadays based on the measurement of the emitted γ-rays using high-resolution γ-ray spectro-meters with semiconductor detectors. The elements present are identified by the energies and relative intensities of the γ-ray lines in the spectrum. The amount of the element is related to the area of the appropriate photo-peak. For further detailed information about the fundamentals of charged particle activation analysis, the book by Vandecasteele [105] is highly recom-mended.

To favour the detectability of as many elements as possible, it is common in NAA to subject the samples to consecutive irradiations and to perform con-secutive measurements at different intervals after each irradiation. By doing this, about 40 elements can be determined in many environmental samples. For ultra-trace element determinations down to $pg\,g^{-1}$ level, it may be necessary to resort to post-irradiation radiochemical separations (radiochemical NAA or RNAA). [106] However, several elements (e.g. Li, Be, B, C, N, O, Si, P, S, Sn, Tl and Pb) are difficult to measure with NAA due to their poor sensitivity and detection limits. Despite the high resolution of semiconductor detectors, the detectability of a γ-line in the spectrum of an activated sample depends strongly on the presence of other radionuclides. INAA is only possible when the activity induced in the matrix is not very high and when no major activity is produced that overshadows the other radionuclides. A major drawback of INAA is the long counting times used, up to several hours. Moreover, each sample is measured four to five times over a period which may extend up to several weeks. Another disadvantage is that the technique may be labour-intensive and that a nuclear reactor is required.

Recent developments in NAA are the increased automatization [107] in all stages of the analysis and the use of cold neutron beams for neutron-capture prompt γ-ray activation analysis (PGNAA). [108] PGNAA has been used successfully in the past, but few such facilities exist and they cannot be con-sidered suitable for routine environmental analysis since only one sample at a time is to be irradiated and usually for several hours. The methods of Compton suppression and gamma-gamma coincidence counting, applicable for airborne particulate samples, are promising. [109]

2.4.3.2 Aerosol analysis

For techniques such as NAA, it is necessary to collect larger amounts of particulate matter per stage than may be obtained with single jet cascade impactors. The most often used device is then a six-stage, multi-slotted Sierra Hi Vol cascade impactor. The standard slotted glass fibre filters are often replaced by similar cellulose filter substrates.

In INAA, aerosol samples can be measured non-destructively after very little sample preparation. Atmospheric aerosols collected on an organic filter are very well suited, as the major elements of the sample and of the filter, C, O and H, do not give significant γ-activities. The entire sample or a representative fraction of a filter is brought back to a small volume or pelletized, and placed in a precleaned polyethylene bag or vial for irradiation. The irradiated sample is afterwards transferred to a new vial for counting. From the viewpoint of aerosol collection substrates, INAA offers greater flexibility than XRF or PIXE; the latter two techniques need preferably thin sample surfaces and substrates. The sample size required for INAA is typically in the order of several mg of particulate matter, depending on the neutron flux available. Reviews of sampling for airborne particulate matter with NAA have appeared. [109–112]

Quantification can be performed either absolutely, leading to some problems in determining the source of the experimental parameters, or with the more popular k_0 method. [46] In this method, nuclear constants (k_0 values) are used for the elements to be quantified, and flux and flux-ratio monitors are co-irradiated with the sample. The most reliable method, however, is still co-irradiation of multi-element standards, containing all elements to be measured, together with the sample. Very accurate results for aerosol samples can be provided by INAA; since neither neutrons nor γ-rays are strongly absorbed by matter, corrections for absorption effects are small and usually negligible. Still, in some cases interfering nuclear reactions may occur which are rarely very important. Other sources of error, like γ- line overlap and dead time losses, can be assessed, overcome or corrected for. Precisions of 1–2% are feasible. [113]

The detection limits for INAA of aerosols, particularly in remote areas, depend on the variability of the elemental concentrations in the blank filters or collection surfaces. Therefore, valid values for detection limits cannot be given.

2.4.3.3 Applications of INAA

In marine aerosols, the high Na concentration leads to ^{24}Na($t_{1/2} = 14.96$ h), while in continental aerosols the main radio-nuclides are ^{28}Al($t_{1/2} = 2.24$ min) and ^{82}Br($t_{1/2} = 35.3$ h). [114] Due to these activities, INAA of a number of elements is not possible when very low detection limits are required. However, in environmental studies, extreme sensitivity is often not required. More important are multi-element possibilities, ease of analysis of large numbers of samples and the possibility of automation. Typically, 30–40 elements can be detected in airborne particulate matter; a suitable irradiation and counting scheme has already been described by Dams *et al.* [113] in 1970.

One of the techniques used by Larsen *et al.* [115] to assess the elemental composition of airborne dust, was INAA. Izumikawa *et al.* [116] discussed the accuracy and precision of this technique for the determination of metals in airborne particles. Kitto *et al.* [117] used INAA to determine several metals collected by a five-stage sampling device. The determination of trace elements presented in dry atmospheric depositions by INAA was reported by Morselli *et al.* [118] Specific methodologies for an analytical protocol for the use of NAA of airborne particulate matter is discussed in detail by Landsberger. [109] Topics covered include available samplers, choice and preparation of filters, laboratory design and specific conditions for irradiation, decay and counting procedures, choice of gamma-rays and spectral and nuclear interferences.

NAA for Ce, Yb, Se and Cr in the atmosphere of some major urban and industrial areas of Punjab (Pakistan) was performed by Qadir *et al.* [119] Also analyses of elemental composition of aerosol samples in Ontario, [120] over the Pacific Ocean [121] and in India [122, 123] were carried out by means of NAA. Ondov and Divita [124] were able to establish size spectra of trace elements in urban aerosol particles by analyzing samples from micro-orifice impactors with instrumental NAA.

Future analytical developments are expected in the use of epithermal NAA to categorize the distinct advantages for certain elements in air particulate matter. Previous work has been done in this area; [109] however, much more detailed investigation is needed.

2.4.4 Anodic Stripping Voltammetry Analysis

2.4.4.1 Principle Voltammetry involves the measurement of current–potential curves at a solid or stationary liquid electrode. In anodic stripping voltammetry (ASV), the analyte ion is electrodeposited within the electrode by reduction of metal cations to metal, which then forms an amalgam with the electrode matrix, often mercury. Besides ASV, other stripping analysis approaches are available such as cathodic stripping voltammetry (CSV) and adsorptive or cathode film stripping voltammetry. They are all based on the same three steps. First, preconcentration of the analyte within the electrode is achieved while stirring. After stirring or rotation, the solution is allowed to come to rest. Finally, an applied potential is ramped and the resultant current is measured. The read-out obtained is thus stripping current as a function of the electrode potential. The stripping current due to oxidation of each analyte is proportional to the concentration of that analyte on or in the electrode and thus in the analytical solution. The principles, instrumentation and applications of stripping voltammetry are described in more detail by Wang. [125]

In pollutant and environmental analysis, voltammetric techniques are used in order to take advantage of the extreme sensitivity offered by stripping analysis. The sensitivity is optimized by preconcentrating the analyte within or on to the working electrode. Detection limits as low as 10^{-10} mol l^{-1} can be achieved with little or no sample pretreatment. Electroanalytical techniques are readily amenable to speciation analysis, which can be acquired with regard to the oxidation states of an element in solution. ASV has been used as part of comprehensive speciation schemes, which are used to characterize the distribution of the analyte within the sample.

The main limitation of stripping analysis is its restriction to about 30 metals at maximum. Interference problems include the formation of intermetallic compounds, adsorption of organic substances on the electrode, overlapping peaks, contamination of samples and losses of analyte, which can be minimized by proper selection of the experimental conditions. In practice, very often only Cu, Zn, Cd and Pb are determined simultaneously by ASV.

2.4.4.2 Aerosol analysis and applications of ASV   In the early years, destruction of the organic collection filters (Millipore) and sample, achieved by low-temperature ashing process, was applied for analysis of airborne particulates with ASV. [126] Nowadays, airborne particulates are mostly collected on membrane filters using high-volume samplers and these are treated with a mixture of hydrochloric, nitric and perchloric acids, prior to analysis with ASV. The preferred media for stripping voltammetry is ammonium tartrate buffer (pH 4.6).

The use of ASV for measuring metals in aerosols has briefly been discussed in a review paper by Harrison. [122]

In 1973, an ASV procedure for the determination of Cd, Zn and Pb in airborne particulate matter was developed by Wilson *et al.* [126] The method involved the destruction, by a low-temperature ashing process, of the particulate matter collected by filtration. The high sensitivity of the method permitted the use of a relatively small sample, an advantage for size distribution studies. In 1991, Casassas *et al.* [128] improved this procedure by using a mixed technique for potential scanning in the stripping step and a small volume electrolysis cell requiring small amounts of aerosol sample.

Further studies on metals in airborne particulates collected at various locations were performed by Khandekar *et al.* [129] ASV was preferred over NAA and atomic absorption spectrometry due to its ability to measure simultaneously various metals at ultratrace levels without prior chemical separation. MacLeod and Lee [130] reported on Pb, Cd and Cu measurements in ambient air samples adjacent to highways, while Harrison and Winchester [131] determined these elements in aerosol particles in Chicago. Pb was measured in particles and aerosols by differential pulse ASV(DPASV) by Wonders et al. [132] Nam et al. [133] determined Pb and Cd collected on glass fibre filters by DPASV; the same technique was used by Scholz et al. [134] Results on atmospheric concentrations and size distributions of Cd, Cu, Pb and Zn above the North Sea, measured with DPASV, were published recently. [135, 136]

Special attention has been given to the quantitation of heavy metals because of their emission with coal fly ash and subsequent deposition into natural water. Kryger et al. [137] applied voltammetry for the determination of Pb, Cd and Tl in fly ash.

In general, ASV suffers from problems in dissolving the aerosol material so completely that no organic interference or masking occurs.

2.4.5 Atomic Absorption Spectrometry Analysis

2.4.5.1 Principle Atomic absorption spectrometry (AAS) is based on the measurement of the absorption of radiation by free atoms in the ground state. Several types of AAS exist, according to the different atomization techniques which can be used. The flame (FAAS) and electrothermal (ETAAS) or graphite furnace (GFAAS) techniques are the most common. Next to these, hydride generation (HGAAS) and cold vapour (CVAAS) techniques, useful for the analysis of gaseous covalent hydrides and for the determination of Hg, respectively, also deserve mentioning.

The light source, usually a hollow cathode lamp or an electrodeless discharge lamp, emits radiation of a frequency characteristic of the elements present in the sample. A part of this radiation will be absorbed by the atomic vapour according to the Lambert–Beer law. Usually, for analytical purposes, the absorbance A is used:

$$A = \log I_0/I = abc$$

where I_0 is the intensity of incident radiation, I the intensity after absorption, a the absorption coefficient, b the path length of the radiation through the sample and c the concentration. The concentration of the analyte can be determined by constructing calibration curves. Unfortunately the linear region is limited to about 0.5–1.0 absorbances. [114] The accuracy depends on the ability to cope with interferences and is limited to 0.5–5%.

Interferences are the result of the presence of other compounds besides the analyte in the sample and can be divided into spectral and non-spectral interferences. [138] Spectral interferences produce a signal independent of the analyte concentration and can arise from the overlap of the analyte line with resonant lines of other elements present, which can easily be avoided by choosing another resonance line of the analyte. Absorption by molecular species and scattering are especially problematic with electrothermal atomization and can be corrected by instrumental techniques like deuterium background correction, Zeeman correction or Smith–Hieftje background correction. [46, 114]

Non-spectral or matrix interferences may be proportional to the magnitude of the analyte signal. Chemical interferences occur when compounds of low volatility are formed with the analyte element and can generally be compensated for by means of the standard addition technique or by adding protective agents. Ionization interferences, the altering of the degree of ionization of the analyte, can be reduced by using cooler cell temperatures or by the addition of ion suppressors. When analysing halides with GFAAS, the use of hydrochloric acid for sample dissolution should be avoided, in order to prevent the formation of volatile salts which can be lost during the ashing or atomization stage. [114]

Detection limits are about two orders of magnitude lower with GFAAS than with FAAS. [46, 114]

2.4.5.2 Aerosol analysis Sampling of aerosols can be done by means of filters or cascade impactors. Direct analysis of solids and slurries can be performed by GFAAS, and the preparation of slurries from aerosol samples has been studied recently. [139]

Normally, however, samples are analyzed in the form of liquids in AAS. Therefore, the sampled particulate matter has to be brought into solution, by dissolving either both the collection substrate and the aerosol or only the aerosol. The procedure used depends on the composition of the collection substrate, the nature of the aerosol and the requirement of the analysis. [140] Reviews by Sneddon [141] and by Cresser *et al.* [65, 142] give an overview of sample preparation techniques used for different elements. Acid digestion and dry ashing are traditional decomposition methods, but they can be quite complex and time-consuming and are considered to form the rate-limiting step in the analysis of environmental samples. [65] Recently, microwave-assisted digestion has been found to reduce sample preparation time in comparison with more traditional techniques. [143]

2.4.5.3 Applications of AAS In spite of the rather complex sample preparation techniques, FAAS as well as GFAAS are commonly used techniques for the analysis of aerosols. Numerous examples can be found in the literature. Different sampling techniques and sample preparation methods are described.

Yamashige *et al.* [144] compared several sample decomposition techniques and proposed an optimum method for the determination of various elements. Ba, Cr, Si and Ti required the alkali fusion method which consists of ashing in an electrical furnace prior to fusion with Na_2CO_3 and boric acid on a Bunsen burner. The sample is then dissolved in HCl. This method was also found to give good results for the determination of Al, Fe, Mn, Mg, Ca, Co, Ni and V, as well as three acid decomposition methods, using mixtures of $HF-HNO_3-HCl$, $HF-HNO_3-HClO_4$ and $HF-HNO_3-H_2O_2$, respectively. These acid decomposition methods also gave good results for the determination of Cd, Cu, K, Li, Na, Pb and Zn, while the best results for As and Sb were obtained with the $HF-HNO_3-H_2SO_4-KMnO_4$ method. Analysis was performed by FAAS and HGAAS.

For the determination of Be, the traditional sample preparation technique requires the use of the explosive $HClO_4$. Miller [145] proposed an alternative technique, using a mixture of HNO_3 and $KHSO_4$ allowing for a reduction of corrosion and explosion hazards.

Dissolution with HNO_3 has been used by Pakkanen and Hillamo [146] for the GFAAS-analysis of aerosols collected by means of a Berner impactor in which Nuclepore filters were used as collection substrates, and this method was found to be excellent for the determination of Cd and Pb, and good for Mn and Cu. This sample preparation technique may provide a routine method for the analysis of metals in Berner impactor aerosol samples. Mineral soil dust particles, however, were found to be incompletely solubilized. [147]

Ottley and Harrison [148] have sampled marine aerosols with a May cascade impactor, using Whatman 541 filter paper as impaction substrates, as well as with an open-faced filter pack filled with Whatman Teflon membrane filters. The samples were digested with 70% HNO_3 in a steel digestion bomb. Analysis was done by FAAS as well as by GFAAS. To minimalize matrix interferences, the standard addition method has been used for calibration.

When analyzing aerosols, the small amount of sample collected might constitute a problem. Therefore, Frenzel [149] developed a multi-stage microanalytical procedure for the determination of 16 elements (Al, As, Cd, Ca, Cu, Fe, K, Mg, Mn, Na, Ni, Pb, Se, Si, V and Zn) in minute samples of atmospheric particulate matter collected on Teflon membrane filters, using FAAS as well as GFAAS. The sample preparation technique consisted of extraction by ultrasonic agitation prior to digestion with 65% HNO_3, 40% HF and 30% H_2O_2 in a domestic microwave oven. This method can be useful for the analysis of minute samples of aerosols.

A less time-consuming method than the ones previously described is the slurry sample introduction technique as used by Fernández et al. [150] The slurry, obtained by ultrasonic agitation of the filters in nitric acid, can be introduced directly into the atomizers. The reproducibility varied from 1 to 3%, which is similar to acid dissolution methods. Carneiro et al. [139] obtained detection limits of 15, 36 and 38 ng cm^{-2} filter material for Sb, Ni and V, respectively.

In addition to aerosol analysis using traditional sample preparation techniques, aerosols can also be analyzed directly by AAS. Lüdke et al. [151] used a special porous graphite tube as the filter. This acts not only as an efficient collector, which also allows for size-fractionated collection of air particulates, but it can also be directly employed as an electrothermal atomizer. The procedure is quick and easy, and can be calibrated with reference solutions. The agreement with comparative measurements by established methods (filtering with quartz fibre filter tissues and analysis by ETAAS and ICP-AES) was satisfactory.

Direct analysis of filters is possible as demonstrated by Low and Hsu [152] who applied this technique for Pb in atmospheric samples. The cellulose nitrate membrane filters used for the collection were transferred to a pyro-graphite platform boat which can be introduced into a graphite furnace for drying, ashing and atomization of the sample. These graphite platform boats could also be used as direct samplers in dry deposition studies. This method is fast and simple and can be used for the analysis of aerosols collected on membrane filters, for example in cascade impactors, and might be extended to other metals such as Cd and to other collection surfaces.

A similar technique was used by Isozaki et al. [153] They have determined Pb in aerosols by direct analysis of a small disc cut from the collection substrate, cellulose nitrate filters, using tungsten furnace AAS. Palladium chloride solution was used as a matrix modifier.

Alternatively, direct analysis of trace metals (Pb, Mn and Hg) in laboratory air has been performed by Sneddon. [154] The particulate matter is precipitated electrostatically on to a tungsten rod which is injected into an electrothermal atomizer prior to AAS analysis. The detection limits were found to be 0.04 $ng\,m^{-3}$ for Pb, 0.05 $ng\,m^{-3}$ for Mn and 0.1 $ng\,m^{-3}$ for Hg. The accuracy, however, was poor.

Disadvantages of these direct analysis methods are high matrix effects and interferences.

2.4.6 Atomic Emission Spectrometry Analysis

2.4.6.1 Principle Several atomic emission spectrometry (AES) techniques exist, according to the source which is used for the element excitation. Classical sources are the flame and the arc and spark discharges. The most widely used AES technique, however, is now inductively coupled plasma atomic emission spectrometry (ICP-AES), in which the sample is atomized, ionized and excited in a plasma, an electrically neutral, highly ionized gas composed of ions, electrons and neutral particles.

Due to the high temperature in the plasma, the atomization yield will be higher than in a classical flame. The longer residence times of the sample in the source and the inert environment (noble gas), together with the higher temperature, also cause matrix effects, chemical interferences and solute vaporization interferences to be lower. On the other hand, the high temperature is responsible for the very complex nature of the emission line spectra, which leads to line overlap and spectral interferences. This can be avoided by choosing a specific analyte line free of interferences. [155] The source consists of a torch made up of three concentric quartz tubes, through which argon is forced to flow. The central gas stream carries and nebulizes the sample. The outer stream is a source of ions and isolates the plasma from the outer quartz tube. The third stream is used to control the position of the plasma inside the torch. [156] After the argon gas is made electrically conductive by a Tesla spark, the formed plasma is sustained by an electromagnetic field induced by an RF current. [46] The characteristic radiation emitted by the excited atoms or ions, formed in the plasma source, is resolved by a diffraction grating and detected by a photo-multiplier.

ICP-AES exhibits a linear dynamic range of three to six orders of magnitude. Detection limits are lower than those of FAAS, but poorer than those of GFAAS. [114, 157]. Thomas and Jehl [158] have found ICP-AES to be more sensitive than AAS for the determination of metals. Most metals could be determined to one-tenth of their threshold limit value (TLV). Moreover much less sample solution was required than for conventional AAS.

Two types of instruments exist: a simultaneous and a sequential spectrometer. The first one is a multi-element technique, ideal for routine analysis but not very flexible. The sequential system uses only one channel at a time but is more flexible and allows for background correction.

2.4.6.2   Aerosol analysis and applications of AES   For the analysis of aerosols with ICP-AES, the same sample preparation techniques can be used as those used for AAS-analysis. [65] Examples of evaluation of different sample preparation techniques can be found in the literature. Infante and Acosta [159] have compared various acid digestion procedures. Extraction methods using mixtures of HCl and HNO_3, HNO_3 and $HClO_4$; and aqua regia and HF were applied to NBS standard materials. The latter offered the best extraction efficiencies and was less complicated than the other procedures. A similar investigation was conducted by Wang *et al.* [160] They used different acid procedures for the dissolution of aerosols collected on glass fibre filters. Digestion with a mixture of HNO_3 and $HClO_4$, followed by addition of HF, gave the best results. HNO_3 is necessary for the removal of organic material, $HClO_4$ for the oxidation and dissolution of metals and HF destroys unresolved matrix like silicate.

Instead of the conventional nebulization sample introduction technique, an alternative and more sensitive technique has been used by Valdés-Hevia *et al.* [161] They determined Pb by a hydride generation technique similar to that used in AAS. They obtained detection limits of 2–13 $\mu g\,l^{-1}$, compared to 20 $\mu g\,l^{-1}$ using conventional nebulization.

Recently ICP-MS has been used more widely for aerosol analysis than ICP-AES.

2.4.7 Inductively Coupled Plasma Mass Spectrometry Analysis

2.4.7.1 Principle The inductively coupled plasma is a suitable source of singly charged, monatomic positive ions. When the ICP is used in a horizontal configuration, the ions can be directed into a mass spectrometer and detected according to their *m/z* ratio. The production of singly charged ions is very efficient: 90% or more for over 50 elements. On the other hand, relatively few doubly charged positive ions are formed.

An extraction interface is used to transport the ions from the hot plasma, which operates at atmospheric pressure, to the high vacuum of the mass spectrometer. The ions are focused and transmitted to a quadrupole type mass analyzer by an electrostatic lens. An entire mass spectrum can be obtained by operating the mass spectrometer in the scanning mode. [46]

For the calibration of ICP-MS, matrix matched reference solutions are used. The technique possesses a linear dynamic range of five to six orders of magnitude. Detection limits are of the same order as those of GFAAS. When spectral interferences or matrix effects can be avoided, the accuracy of ICP-MS is high.

Spectral interferences arise from overlap between a peak from a monopositive analyte ion and that from a polyatomic and/or doubly charged ion with the same *m/z* ratio. Major sources of interfering ionic species could be the argon, the water and the acid from the solution. Nitric acid is the preferred solvent in ICP-MS because nitrogen yields only small peaks from positive ions. Spectral overlap may also occur between isobaric peaks of neighbouring elements in the periodic table. By choosing another analyte mass peak, spectral interferences may be overcome. [46]

Matrix effects, which form an important limitation to ICP- MS, occur in the presence of high salt concentrations. Therefore solutions containing more than $10 \, \text{g} \, \text{l}^{-1}$ of dissolved solids are rarely analyzed by ICP-MS. Both signal suppression and signal enhancement can occur. [162] Matrix effects may be overcome by dilution, matrix matching, use of internal standards, standard addition, chemical separation or isotope dilution. [114]

2.4.7.2 Aerosol analysis and applications of ICP- MS The samples have to be brought into solution, to which end similar techniques can be used as for AAS and ICP-AES. [65, 142] An alternative technique is laser sampling ICP-MS, as used by Denoyer. [163] The samples were mixed with a cellulose-based binding agent and pressed into pellets for direct solid sampling with the laser and semi-quantitative analysis.

In other recent studies more traditional sample preparation techniques were used. Katoh *et al.* [164] used complexation combined with solvent extraction on aerosol samples, collected on cellulose nitrate membrane filters. Ni, Cu, Zn, Cd, Tl and Pb were determined by isotope dilution ICP-MS. Detection limits ranged from 0.15 pg m^{-3} for Cd to 30 pg m^{-3} for Zn based on three times the blank standard deviations.

For the determination of Pt in airborne particulate matter, Mukai et al. [165] used flow injection ICP-MS which uses a cation exchange resin column as introduction system. In this way Pt, dissolved as an anion, can be separated from various cations which can cause interferences. The samples were collected on quartz fibre filters which were digested with aqua regia, HF, HClO$_4$, and HCl. The combination of this digestion method with the previously described sample introduction method is very useful for the monitoring of very low levels of Pt in the atmosphere.

Acid dissolution, using a 3 : 1 mixture of HNO$_3$ and HF, was also used by Jalkanen and Häsänen [166] on coal fly ash and urban aerosol samples. The method is very simple, and the recoveries vary between 80 and 98% for most elements. Comparison with NAA results showed good agreement.

3 SECONDARY INORGANIC AEROSOLS

3.1 INTRODUCTION

Secondary aerosol comprises particulate matter which has formed within the atmosphere by chemical conversion of gases to less volatile species which condense into a particulate form, and may comprise either liquid droplets or solid. As will be discussed later, the condensation process may lead to new particle formation (termed homogeneous nucleation) or involve solely condensation on to existing particles. While the actual pathway may have relatively little influence on the bulk particle chemistry, it will however profoundly influence the aerosol size distribution, and hence the distinction is an important one.

There are relatively few inorganic components of the atmospheric aerosol which are wholly or even predominantly secondary. Frequently the most abun-

Table 2. Major secondary inorganic aerosol components

Aerosol species	Chemical formula	Relative abundance*
Sulfuric acid	H$_2$SO$_4$	Major
Ammonium hydrogen sulfate	NH$_4$HSO$_4$	Major
Letovicite	(NH$_4$)$_2$SO$_4$ · NH$_4$HSO$_4$	Major
Ammonium sulfate	(NH$_4$)$_2$SO$_4$	Major
Metal ammonium sulfates	(NH$_4$)$_2$SO$_4$ · MSO$_4$ (M = metal, e.g.Pb)	Minor
Ammonium nitrate	NH$_4$NO$_3$	Major
Sodium nitrate	NaNO$_3$	Minor
Ammonium chloride	NH$_4$Cl	Minor

* Relative abundance in a typical polluted atmosphere.

dant is sulfate, which arises from the oxidation of sulfur dioxide to sulfuric acid, with varying degrees of subsequent neutralization by ammonia. The major stoichiometric species observable in the atmosphere appear in Table 2. It should, however, be borne in mind that ammonia neutralization is progressive and hence the atmosphere may contain a mixture of more than one of the species in Table 2, and indeed, individual particles may contain non-stoichiometric ammonium species intermediate between those in the table. X-ray powder diffraction studies of atmospheric particles have also revealed the presence of metal ammonium sulfate double salts which are thought to have arisen as a result of the coagulation of ammonium sulfate with metal sulfate particles. The metal sulfates most probably originate from reactions of sulfur dioxide and its oxidation products with metal oxide and halide particles, as explained further in the next section, and hence these sulfates should also be considered as secondary.

Typically, the second most abundant secondary aerosol component in polluted atmospheres is nitrate. This arises from the oxidation of nitrogen dioxide to nitric acid which may either react reversibly with ammonia forming ammonium nitrate, or irreversibly with sodium chloride to form sodium nitrate. Ammonium nitrate, like ammonium chloride, is potentially volatile and may redissociate into the gas phase precursors if the ambient concentrations of nitric acid and ammonia fall below those required to sustain the existence of particulate ammonium nitrate. In parts of the western United States, ammonium nitrate is a dominant constituent of the atmospheric aerosol and it is abundant over much of western Europe. Projections of the impacts of control policies upon concentrations of sulfate and nitrate aerosol in the United Kingdom indicate that by the year 2010 the aerosol will be dominated by nitrate, whereas at present the concentration of sulfate generally exceeds that of nitrate. [167]

The third commonly observed inorganic secondary aerosol component is chloride in the form of ammonium chloride. In marine-influenced locations the dominant form of chloride is usually primary sodium chloride. However, when appreciable concentrations of hydrogen chloride gas arise from emissions (e.g. coal combustion, refuse incineration) or displacement from marine sodium chloride by other strong acids, in the presence of sufficient ammonia, then ammonium chloride particles can form, and may be a significant component of the aerosol. This, together with other typical secondary aerosol components, is listed in Table 2.

3.2 FORMATION MECHANISMS AND AEROSOL CHEMISTRY

3.2.1 Sulfates

Primary emissions of sulfate arise from industries such as sulfuric acid manufacture and from production and use of sulfate minerals such as gypsum. In general, however, these represent only a small proportion of atmospheric sulfate, by far the major proportion arising from the oxidation of sulfur dioxide.

There are many reaction mechanisms by which sulfur dioxide can be oxidized to sulfate in the atmosphere. In practice, only a limited number of processes involving a small number of oxidizing species take place at a rate sufficient to influence the atmospheric burden of sulfate. [168] The reaction mechanisms fall into two groupings, either homogeneous reactions taking place in the gas phase, or heterogeneous processes in cloud, fog water or aerosol droplets involving transfer of sulfur dioxide and oxidant species to the droplet phase with subsequent liquid phase oxidation. While many oxidants have been suggested as playing an important role in homogeneous gas phase oxidation of sulfur dioxide, in practice the rate of oxidation by the hydroxyl radical usually exceeds the rate of other pathways by an order of magnitude or more. The mechanism [169] is as follows:

$$SO_2 + OH \rightarrow HSO_3$$
$$HSO_3 + O_2 \rightarrow SO_3 + HO_2$$
$$SO_3 + H_2O \rightarrow H_2SO_4$$

The rate of oxidation of sulfur dioxide is dependent critically on the concentration of hydroxyl radical. This species is predominantly formed by photochemistry and hence night-time concentrations are low, while daytime concentrations rise typically to a peak in the middle of the day with a seasonal and latitudinal dependence relating to solar intensities. Typical concentrations of hydroxyl radical are in the range $(0.5$–$5)$ $\times 10^6 \, cm^{-3}$ daytime mean and $(0.3$–$3)$ $\times 10^6 \, cm^{-3}$ 24-hour mean, with seasonal variation of about threefold suggested by model studies. At a daytime concentration of $5 \times 10^6 \, cm^{-3}$ of hydroxyl radical, sulfur dioxide is oxidized at a rate of 1.6% per hour to sulfuric acid.

Sulfur dioxide is a highly water-soluble gas and hence in the presence of liquid water will tend to partition appreciably into the solution phase. On doing so, the following chemical equilibria are established:

$$SO_2(g) + H_2O \Leftrightarrow SO_2.H_2O$$
$$SO_2.H_2O \Leftrightarrow HSO_3^- + H^+$$
$$HSO_3^- \Leftrightarrow SO_3^{2-} + H^+$$

These equilibria and the solubility of SO_2 itself are highly sensitive to pH, but over the range pH 2–7, HSO_3^- is the predominant species which is subject to oxidation. Under near neutral conditions, pH 6, oxidation of hydrogen sulfite ion by dissolved ozone, hydrogen peroxide, or dioxygen with metal catalysis can be rather rapid. However, oxidation causes conversion of the weak sulfurous acid to strong sulfuric acid and the pH very rapidly drops. At pH 3 the only oxidation process which is still rapid is that due to hydrogen peroxide. [170]

Indeed, oxidation by this compound is extremely fast and the amount of sulfate produced is typically limited by the availability of hydrogen peroxide which is rarely present in air at concentrations above about 1 ppb. Thus, in clouds, continuous entrainment of cleaner air containing hydrogen peroxide can lead to appreciable sulfate production, but in more polluted ground-level air such as occurs in fogs, very little hydrogen peroxide is available, and it is the slower oxidation processes due to ozone and dioxygen with metal catalysis which are important. Some workers have also reported relatively rapid oxidation of sulfur dioxide on the surface of carbonaceous particles or in liquid droplets containing particulate carbon, [171] but the oxidation rates are highly dependent on the form of carbon present [172] and extrapolation to the atmosphere is very difficult. Hygroscopic aerosols such as sodium chloride in the marine boundary layer can also provide an aqueous medium for sulfur dioxide oxidation. [173] These may be of some importance where liquid phase aerosol abundance is appreciable, and/or other sulfur dioxide oxidation pathways are relatively slow.

Sulfate formed from SO_2 oxidation is initially in the form of sulfuric acid. However, ammonia is a ubiquitous component of tropospheric air and in the lower troposphere some ammonia neutralization inevitably progresses. The rate of this process has been studied through atmospheric measurements by Harrison and Kitto [174] who determined the change in strong acid aerosol content of air advected off the ammonia-depleted atmosphere of the North Sea on to land in eastern England where appreciable ammonia emissions occur. A rate constant for the ammonia/acid sulfate reaction was defined in terms of the following expression:

$$R = kC_i$$

where R is the reaction rate $(\text{neq m}^{-3}\,\text{s}^{-1})$, C_i the mean concentration of ammonia (neq m^{-3}) and k the rate constant (s^{-1}). The research revealed the following expression for the value of k:

$$k = 2.3 \times 10^{-5}(\text{H}^+/\text{NH}_4^+) + 4 \times 10^{-5}\ \text{s}^{-1}$$

Thus, the more acidic the aerosol as expressed by the H^+/NH_4^+ mole ratio, the more rapid the ammonia neutralization process. This value was found to be broadly consistent with rate constants estimated in other studies. [175, 176] In the presence of an excess of ammonia, full neutralization to $(\text{NH}_4)_2\text{SO}_4$ is likely to proceed and this compound has been identified in the atmosphere in a number of studies using X-ray powder diffraction to determine crystalline salts. [177]

While the above discussion has considered sulfur dioxide as the only precursor of sulfate, there are clear indications that dimethyl sulfide (DMS) produced by photoplankton in the marine environment can be oxidized to sulfate by two pathways; one involving the intermediacy of sulfur dioxide, the other entailing sulfate production without intermediate formation of SO_2. [178] The rate of sulfate production from marine DMS is dependent on the rate of DMS oxidation, itself a function of OH radical concentration in daytime and NO_3 concentration at night, together with the concentration of DMS. [179]

3.2.2 Nitrates

There is little primary particulate nitrate in the atmosphere other than close to industrial sources such as ammonium nitrate fertilizer works which can provide appreciable point sources. [180] Generally, atmospheric nitrate arises from the oxidation of nitrogen dioxide to nitric acid, which forms particles as a result of reaction with ammonia or with sodium chloride. [168]

The main pathway for nitric acid formation in daylight is from the reaction of NO_2 with the hydroxyl radical. [170]

$$NO_2 + OH \rightarrow HNO_3$$

At a high daytime hydroxyl radical concentration of $5 \times 10^6 \, cm^{-3}$, the rate of oxidation of NO_2 is almost 20% per hour [168] and thus appreciably faster than the rate of sulfate formation from SO_2. Not only does nitrate have a different rate of atmospheric formation than sulfate, it has sink mechanisms which operate at different rates and thus while sulfur dioxide and nitrogen dioxide have common source areas, the atmospheric distribution of sulfate and nitrate can differ substantially. At night-time, due to much reduced concentrations of hydroxyl, other mechanisms take over to convert nitrogen dioxide to nitric acid. [170] Probably the most important involves reaction of NO_2 with ozone to form the NO_3 radical which is unstable in daytime due to photolysis:

$$NO_2 + O_3 \rightarrow NO_3 + O_2$$

Subsequent conversion of the nitrate radical to nitric acid can occur by a number of chemical mechanisms. Probably the most important involves further reaction with NO_2 to form dinitrogen pentoxide which is converted to nitric acid by reaction with water, generally in a process enhanced by the presence of liquid water:

$$NO_3 + NO_2 \Leftrightarrow N_2O_5$$
$$N_2O_5 + H_2O \Leftrightarrow 2HNO_3$$

Other reactions of NO_3 also contribute to nitric acid formation, particularly hydrogen atom abstraction from volatile organic compounds. Reactions with aldehydes are particularly favourable: [170]

$$NO_3 + RH \rightarrow HNO_3 + R$$

A study of NO_2 oxidation in the North Sea atmosphere showed that daytime oxidation rates were consistent with hydroxyl radical concentrations expected for the season, and that night-time rates of NO_2 oxidation were consistent with the rate of the NO_2 reaction with ozone without subsequent amplification by the N_2O_5 pathway. [181]

In the marine atmosphere, and also in marine-influenced terrestrial atmospheres with low ammonia concentrations, conversion of HNO_3 to particulate nitrate is predominantly through reaction with sodium chloride particles: [182]

$$HNO_3 + NaCl \rightarrow NaNO_3 + HCl$$

The rate of this reaction is dependent on the available particle surface area as well as the size distribution of particles and the accommodation coefficient for HNO_3 on marine aerosol particles. Prediction of the reaction rate from theoretical considerations is feasible if the above information is available, but determination of the rates from field experiments is extremely difficult. However, the fact that observed size distributions for nitrate in the marine atmosphere closely parallel the surface area distribution of sodium [182] is a clear indication of the importance of the pathway.

Nitric acid vapour also reacts reversibly with ammonia. If the necessary concentration product $[HNO_3][NH_3]$ for NH_4NO_3 particle formation is exceeded, theoretically NH_4NO_3 aerosol should exist [183] (see also Chapter 6). A number of experimental studies have sought to establish whether the atmospheric concentration product of ammonia and nitric acid is consistent with that predicted by chemical thermodynamics for equilibrium with NH_4NO_3 particles. The results of one such study appear in Figure 1 in which the lines represent the $[HNO_3][NH_3]$ concentration product as a function of temperature for specific relative humidities. The points represent individual experimental measurements of the concentration product, which in general conform reasonably well to the theoretical lines. While it is possible that some discrepancies arise from uncertainties in the thermodynamic calculations, particularly for concentrated solution droplets where ion activity coefficients may be inaccurately estimated, the major cause of discrepancies appears to be associated with kinetic limitations to the achievement of equilibrium. Harrison and MacKenzie [184] have sought to investigate the impact of kinetic limitations, showing that the outcomes are broadly consistent with experimental observations. Wexler and Seinfeld [185] have examined from a theoretical viewpoint the time-scales involved in the approach to equilibrium. In the case of high number densities of fine particles, characteristic times are of the order of a few seconds and equilibrium is reached extremely rapidly. However, at low number densities and large particle sizes, characteristic times for the chemical reaction can be of the order of hours leading to long equilibration times. This appears to be consistent with field observations. [183, 184, 186]

The volatility of ammonium nitrate has a number of consequences. Firstly, if the atmospheric concentrations of ammonia and HNO_3 are insufficient to exceed the necessary concentration product, then ammonium nitrate should not be present. Thus, for example, in marine atmospheres where both HNO_3

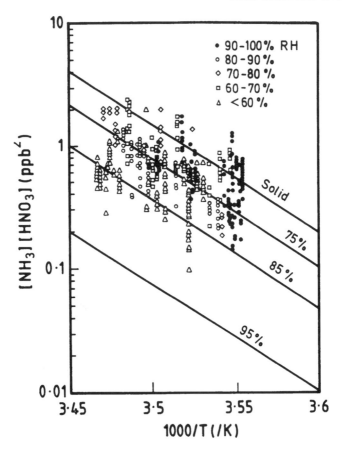

Figure 1. Theoretical relationship between $[NH_3][HNO_3]$ concentration product and inverse temperature (lines) for different relative humidities, with points derived from field measurements superimposed[186]

and NH_3 concentrations are low, it is not normal to observe ammonium nitrate. [140] Secondly, solid ammonium nitrate collected on filters is potentially unstable, especially if the filter is at a temperature above that of the ambient airstream. Thus, reliable sampling of nitrate aerosol is potentially difficult and this topic is addressed below.

3.2.3 Chloride

As discussed above, in atmospheres with abundant ammonia and hydrogen chloride, secondary ammonium chloride aerosol can form. Its thermodynamic properties are remarkably similar to those of ammonium nitrate and the required concentration products are of a similar order. [187, 188] There has been much less experimental research into the properties of secondary chlorides in the atmosphere than for nitrates, but Figure 2 shows the results of one study [183] giving both theoretically predicted concentration products and observational data for [HCl][NH₃].

3.2.4 Mixed Salts

As a result of atmospheric coagulation processes between particles of different composition, and of the uptake of gases into pre-existing particles, it is quite

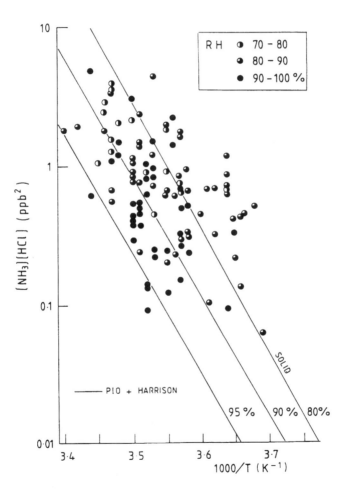

Figure 2. Theoretical relationship between [NH₃][HCl] concentration product and inverse temperature (lines) for different relative humidities, with points derived from field measurements superimposed[183]

Table 3. Sulfate double salts detected in the atmosphere

$(NH_4)_2SO_4 . 2NH_4NO_3$
$(NH_4)_2SO_4 . 3NH_4NO_3$
$CaSO_4 . (NH_4)_2SO_4 . H_2O$
$PbSO_4 . (NH_4)_2SO_4$
$3(NH_4)_2SO_4 . NH_4HSO_4$
$Na_2SO_4 . NaNO_3 . H_2O$
$Na_2SO_4 . (NH_4)_2SO_4 . 4H_2O$
$ZnSO_4 . (NH_4)_2SO_4 . 6H_2O$
$Fe_2(SO_4)_3 . 3(NH_4)_2SO_4$

possible for salts of mixed composition to be formed within the atmospheric aerosol. While it is possible in some studies that such mixed salts are an artefact of the sampling procedures, other studies have taken great care to avoid the possibility of collection of liquid droplets which subsequently coalesce leading to reaction, and it seems very probable that such processes are occurring in the atmosphere rather than within sample collection devices. [177] A commonly observed compound at the time that lead was heavily used in gasoline additives and emitted from vehicles as PbBrCl is $PbSO_4(NH_4)_2SO_4$. This compound has been widely observed in city air [177] and it was proposed to have been formed by the following reaction: [189]

$$2PbBrCl + 2(NH_4)_2SO_4 \rightarrow PbSO_4(NH_4)_2SO_4 + PbBrCl(NH_4)_2BrCl$$

Table 3 includes some double salts observed in the atmosphere by X-ray powder diffraction. These include two double salts of ammonium sulfate and ammonium nitrate, both of which are reported to be thermodynamically favoured under certain conditions [190] (see also Chapter 6).

3.3 GAS-TO-PARTICLE CONVERSION

As indicated above, formation of secondary aerosol generally involves a chemical conversion process to form an involatile species followed by condensation of that species. If condensation forms wholly new particles it is termed homogeneous nucleation. There is a large energy barrier to new particle formation due to the large excess energy associated with the very strongly curved surfaces of nanometre-sized particles (the Kelvin effect). Thus, a substantial energy barrier has to be overcome, and only if a substantial supersaturation with respect to the involatile component is achieved, will homogeneous nucleation take place. [191]

There are rather few species formed in the atmosphere of sufficiently low volatility to undergo homogeneous nucleation. The most commonly cited example is that of sulfuric acid which after formation from SO_2 oxidation can undergo binary nucleation with water. However, this process is not inevitable and its occurrence depends upon favourable conditions. Those conditions which favour binary nucleation of H_2SO_4/H_2O are as follows:

- high rate of gas phase oxidation of SO_2 and thus high H_2SO_4 formation rate
- high humidity
- low pre-existing particle surface area available for condensation of newly formed H_2SO_4
- low temperature

Several attempts at theoretical prediction of the binary nucleation process [192, 193] have been reported, but there has currently been rather little experimental validation of the theoretical predictions. Field observations suggest that new particle formation may occur in the air outflowing from clouds where conditions of low temperature, high humidity and low pre-existing particle surface are all favourable. [194]

Recent theoretical work suggests that the presence of ammonia may act to enhance considerably the rate of homogeneous nucleation of sulfate particles. [195] As yet, there has been no experimental testing of this observation. Field measurements with tandem particle counters capable of identifying new particles in the 3–8 nm size range indicate that bursts of new particle formation are readily observable in the coastal environment and tend to correlate with the rate of ozone photolysis to excited state oxygen atoms, the process which initiates formation of hydroxyl radical. [196] Such observations do, however, demonstrate new particle formation even in the urban atmosphere where pre-existing particle concentrations are appreciable.

The alternative pathway for sulfuric acid formed from gas-to-particle conversion is to condense on to pre-existing aerosol surfaces, generally within the accumulation range (approx. 100 nm–2.5 μm) where most of the aerosol surface area is present in polluted atmospheres. This process leads to particle growth, but no change in the total number of particles. In the case of ammonium nitrate and ammonium chloride, it seems most unlikely that sufficient supersaturation is achieved to cause homogeneous nucleation, and most probably these compounds condense onto existing particles.

As noted above, liquid droplets, which may be cloud water, fog or hygroscopic aerosol, provide an aqueous medium into which soluble trace gases may partition. Thus, sulfur dioxide may be taken up and oxidized to sulfuric acid, and ammonia, nitric acid and hydrogen chloride can be absorbed leading to production of ammonium nitrate and ammonium chloride within the solution. Thus, atmospheric solution droplets containing secondary components arise and may convert to dry particles if the humidity regime changes. Detailed numerical models of gas–aerosol interactions have been proposed which account for the chemistry and microphysics of these processes. [197, 198] These are valuable in predicting both the rate of production of secondary aerosol components, but also the particle size distribution of secondary material (see also Chapter 1).

3.4 SIZE DISTRIBUTIONS (SEE ALSO CHAPTER 1)

Examples of size distributions of sulfate, nitrate, chloride and ammonium measured in California [199] appear in Figure 3 in which some constituents exhibit a trimodal distribution. Nitrate was typically bimodal with a fine mode associated with ammonium and a coarse mode associated with sodium. The greatest abundance of ammonium, nitrate and sulfate is in the accumulation mode around 0.6 µm aerodynamic diameter. This corresponds to particles which have grown through condensation and coagulation processes and the major part of the sulfate particle mass is in this mode. A coarse particle mode is centred around 3 µm and contains appreciable nitrate, ammonium and sodium, with some chloride, corresponding in part to nitrate coatings on marine sodium chloride particles. Sulfate was primarily in two submicrometre modes; strong acid (H^+) was associated with the smaller sulfate mode.

Almost all published work on continental aerosols shows nitrate to be associated with larger particles than sulfate. The reason for this relates to the

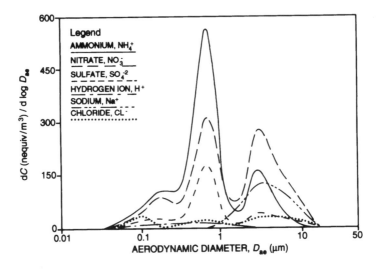

Figure 3.   Aerosol size distribution measured in southern California for a period of strong marine influence[199]

volatility of nitric acid which has two consequences. Firstly, nitric acid vapour can react with particle surfaces, such as those of sodium chloride as described above. Secondly, submicron ammonium nitrate particles are subject to a significant Kelvin effect which destabilizes them relative to larger particles. This leads to evaporation and recondensation on to particles of larger sizes. Sulfate can show multimodal behaviour for several reasons. As noted above, secondary sulfate is normally associated with submicrometre particles, but primary sulfate associated with minerals such as gypsum will generally be in the coarse mode. Another primary form of coarse sulfate is that in marine aerosol. Thus, marine particles typically exhibit a fine particle sulfate mode due to homogeneous oxidation of SO_2 and DMS, and a coarse mode due to the sulfate in sea-water. When the sea-water sulfate is subtracted using the sulfate/sodium ratio of sea-water as a measure of sea-water sulphate, a coarse particle non-sea-salt sulfate ($NSS-SO_4^{2-}$) peak may remain in the size distribution. [182] This is likely to be secondary sulfate which has arisen from the ozone oxidation of sulfur dioxide in sea-salt droplets which are partly buffered against acid formation as a result of their carbonate content. [173]

3.5 CONCENTRATIONS OF SECONDARY SULFATE, NITRATE AND CHLORIDE IN THE ATMOSPHERE

Ion balance studies such as that due to Harrison and Pio [200] have shown that the major part of both sulfate and nitrate is normally secondary in origin and associated in the aerosol with either NH_4^+, or in the case of sulfate, H^+. This, of course, may not be the case in heavily marine-influenced particles where sodium nitrate may be present. Table 4 illustrates a range of measured concentrations of sulfate and nitrate. In continental air the major part of sulfate is secondary ammonium sulfate, sulfuric acid and its partially neutralized forms,

Table 4. Reported concentrations of secondary aerosols constituents from around the world ($\mu g\ m^{-3}$)

Location	$NSS-SO_4^{2-}$	NO_3^-	NH_4^+	Study
Azores, N. Atlantic				
clean air	0.40 ± 0.27	0.24 ± 0.12	0.08 ± 0.04	Harrison et al.[201]
clean and polluted air	0.85 ± 1.06	0.47 ± 0.39	0.19 ± 0.22	Harrison et al.[201]
Central Pacific	1.72 ± 1.15			Quinn et al.[202]
Equatorial Pacific		$0.08 - 0.31$		Huebert[203]
Miami, USA	2.34	1.87	0.61	Prospero and Jennings[204]
Essex, UK	6.46	6.06	4.15	QUARG[167]
Santa Barbara, USA	2.9	3.0	1.0	Chow et al.[205]

while in marine air sea-salt sulfate is important. Fitzgerald [206] suggests that the non-sea-salt sulfate contributed by oxidation of DMS in marine air amounts to about 0.2–1.5 µg m^{-3}, which is seasonally variable due to the temporal variation in DMS production from algal blooms. As may be seen from Table 4, marine air contains generally very low concentrations of aerosol nitrate consistent with the lack of major sources of gaseous nitrogen in the marine environment and the highly efficient dry deposition of nitric acid vapour and coarse sodium nitrate particles. Continental air, however, can contain highly elevated concentrations of nitrate.

A time series of sulfate and nitrate in air data has been collected in southern England [167] since 1954 and appears in Figure 4. This shows clearly a rising trend in sulfate to 1976 which has since fallen, and a trend in nitrate which continued to the last of the measurements in 1988.

In contrast to sulfate and nitrate, the contribution of secondary chloride to chloride concentrations in the atmosphere is frequently rather small. Size-discriminated measurements of chloride may allow an estimation of the secondary chloride which will be predominantly in the fine particle fraction, whereas primary chloride from sea-salt and road de-icing salt is mainly coarse. Using ion balance studies, Harrison and Pio [200] regressed chloride not associated with marine sources on ammonium not required for neutralization of sulfate and nitrate, finding a good relationship corresponding to the atmospheric ammonium chloride. Typically, this accounted for only about 0–3 µg m^{-3} of secondary chloride, which is probably in excess of current concentrations in the UK atmosphere due to enhanced pollution control measures and a reduction in coal burning since the time of those measurements.

The spatial distribution of sulfate aerosol across western Europe is rather well known because of the monitoring activities of EMEP (the European

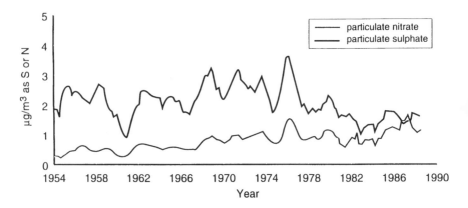

Figure 4.   Three-month running mean concentration of sulfate and nitrate at Chilton, UK[167] from 1954 to 1988

Monitoring and Evaluation Programme). A map of the sulfate distribution [207] measured in 1993 is shown in Figure 5. Additionally, because of the global climate impacts of fine particulate matter in the atmosphere, which for simplicity has often been regarded as wholly sulfate, climate modellers have taken a great interest in the spatial distribution of atmospheric fine particle sulfate aerosol. While model simulations of the global distribution of atmospheric sulfate have been created, validation against field measurements is, as yet, inadequate.

3.6 MEASUREMENT OF SULFATE, NITRATE AND CHLORIDE IN THE ATMOSPHERE

These are among the more commonly measured atmospheric aerosol constituents, but one of the most problematic measurements. Sampling of all of the secondary species can be subject to major artefact problems. These are outlined below:

3.6.1 Filter Artefacts

It was discovered in the 1970s that measurements of atmospheric sulfate, which to that point in time had generally used glass fibre high-volume filters for sampling, were subject to substantial positive artefacts due to the reaction of acidic sulfur dioxide with the basic glass fibre material. [208] This could account for several $\mu g\ m^{-3}$ of positive artefact. Artefact nitrate could also arise on these filters from reactions of nitric acid, and to a lesser extent nitrogen dioxide. [209] As a result of these findings there was a move either to acidify the surface of glass fibre filters or to use intrinsically less acid filter media. The ideal solution is to use non-reactive PTFE filters which are thought to be free of this kind of artefact, but use is limited due to relatively high cost.

3.6.2 Volatilization Artefacts

Both ammonium nitrate and ammonium chloride are potentially volatile, and may be lost if, as often happens, the sampling filter is within a housing which is warmer than the ambient atmosphere. Additionally, if the salts are collected from the air at a relatively low temperature and sampling is continued to periods of higher temperature, volatilization losses may once again take place. Thus, some published studies have demonstrated substantial inaccuracies in measurement of the partition of nitrate between particulate and gaseous species when using conventional filter-pack sampling methodology. [210] Other studies have, however, shown that with care and under ideal circumstances good quality data can be obtained. [211]

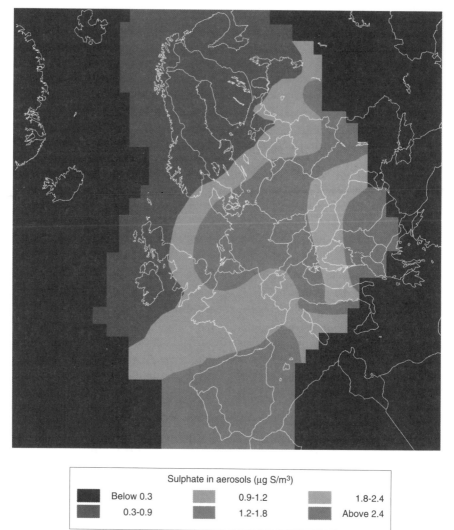

Sulphate in aerosols (µg S/m³)

■	Below 0.3	■	0.9-1.2	■	1.8-2.4
■	0.3-0.9	■	1.2-1.8	■	Above 2.4

Figure 5.   Annual mean sulfate in aerosols across Europe, 1993[207]

3.6.3 Chemical Reaction Artefacts

Particles collected on an air filter form a potential substrate for reaction with atmospheric gases drawn though the filter. Thus, for example, sulfuric acid aerosol can react with ammonia leading to enhanced ammonium concentrations on the filter. Marine sodium chloride on a filter can react with nitric acid vapour leading to a gain in apparent particle nitrate and a reduction in apparent particle chloride. Sulfuric acid droplets contacting nitrate or chloride particles can lead to displacement of nitrate and chloride as their acid gases. There is thus tremendous potential for acid–base reactions to alter the composition of airborne particulate matter collected on a filter. While many workers have chosen to ignore this problem, others have addressed it by pre-removal of acidic and basic trace gases prior to the particle filter using denuders. Thus, ammonia may be collected on a citric acid or phosphorous acid-coated denuder, HNO_3 and HCl on a sodium carbonate-coated denuder, which is followed by a particle filter. [212] The consequence of pre-removal of ammonia and acid gases is to destabilize ammonium nitrate and ammonium chloride which are then liable to evaporate from the particle filter. This is accounted for by backing up the particle filter with impregnated or nylon filters capable of collecting the evaporating ammonia, HCl and HNO_3. This method [213] may even identify the result of reaction of sulfuric acid with chloride and nitrate particles as this will create an imbalance in the chemical equivalents of NH_4^+ on the one hand and $(Cl^- + NO_3^-)$ on the other, which would be in balance if they arose solely from the evaporation of ammonium nitrate and ammonium chloride.

Thus, it is possible to collect artefact-free samples, although only by adding considerable complexity to the sampling apparatus which inevitably leads to reduced precision because of the greater number of analyses which must be conducted on smaller quantities of analyte. Once a sample is collected it is usual to extract into an aqueous medium, if necessary using a small quantity of a short chain alcohol to wet the hydrophobic PTFE filter. The now almost standard technique for analysis of sulfate, nitrate and chloride is ion chromatography, usually with suppressed conductivity detection. Ammonium is determined in the same solution. Commonly the indophenol blue colorimetric procedure [214] or a liquid phase fluorescence method of which there are several variants [215,216] is used.

A range of techniques have been reported in the literature for speciation of individual compounds of sulfate, nitrate and chloride. It is not appropriate to deal with these in detail here, and the reader is referred to Chapter 4. Crystalline salts such as ammonium sulfate, ammonium chloride and the ammonium sulfate–ammonium nitrate double salts may be determined semi-quantitatively using X-ray powder... diffraction techniques. Identification of pure ammonium nitrate has generally proved difficult, however. Sulfuric acid has been widely measured through determination of strong acid associated with the particulate fraction, generally using samplers which pre-remove ammonia by a denuder. The collected strong acid may be determined by micro-Gran titration or by pH measurement under carefully defined circumstances. [217] Such measurements have generally been associated with studies of the health effects of airborne strong acid and are more concerned with the measurement of H^+ in aerosol than of sulfuric acid *per se*.

REFERENCES

1. Fergusson, J. E. (1990). *The Heavy Elements: Chemistry, Environmental Impact and Health Effects*, Pergamon Press, Oxford.
2. Pacyna, J. M. (1986). Atmospheric trace elements from natural and anthropogenic sources. In *Toxic Metals in the Atmosphere*, Wiley Series in Advances in Environmental Science and Technology Vol. 17, ed. J. O. Nriagu and C. I. Davidson, Wiley, New York, 33.
3. Pacyna, J. M. and Nriagu, J. O. (1988). Atmospheric emissions of chromium from natural and anthropogenic sources. In *Chromium in the Natural and Human Environments*, Advances in Environmental Science and Technology Vol. 20, ed. J. O. Nriagu and E. Nieboer, Wiley, New York, 105.
4. Duce, R. A., Liss, P. S., Merrill, J. T., Atlas, E. L., Buat-Menard, P., Hicks, B. B., Miller, J. M., Prospero, J. M., Arimoto, R., Church, T. M., Ellis, W., Galloway, J. N., Hansen, L., Jickels, T. D., Knap, A. H., Reinhardt, K. H., Schneider, B., Soudine, A., Tokas, J. J., Tsunogai, S., Wollast, R. and Zhou, M. (1991). The atmospheric input of trace species to the world ocean, *Global Biogeochem. Cycles*, **5**, 193.
5. Ottley, C. J. and Harrison, R. M. (1993). Atmospheric dry deposition flux of metallic species to the North Sea, *Atmos. Environ.*, **27A**, 685.
6. Pacyna J. M. (1983). Emission factors of atmospheric elements. In *Toxic Metals in the Air*, in Environmental Science and Technology Series, ed. J. O. Nriagu, Wiley, New York, 1.
7. Nriagu, J. O. (1988). Production and uses of chromium. In *Chromium in the Natural and Human Environments*, Advances in Environmental Science and Technology Vol. 20, ed. J. O. Nriagu and E. Nieboer, Wiley, New York, 81.
8. Moore, J. W. and Ramamoorthy, S. (1984). *Heavy Metals in Natural Waters*, Springer-Verlag, New York.
9. Natusch, D. F. and Wallace, J. R. (1974). Urban aerosol toxicity: The influence of particle size, *Science*, **86**, 695.
10. Nriagu, J. O. and Pacyna, J. M. (1988). Quantitative assessment of world wide contamination of air, water and soils by trace metals, *Nature*, **333**, 134.
11. Harrison, R. M. and Sturges, W. T. (1983). The measurement and interpretation of Br/Pb ratios in airborne particles, *Atmos. Environ.*, **17**, 311.
12. Pacyna, J. M. (1983). *Trace Element Emission from Anthropogenic Sources in Europe*, NILU Report 10/82, The Norwegian Institute for Air Research, Lillestrom, Norway.
13. Hopke, P. K. (1985). *Receptor Modelling in Environmental Chemistry*, Wiley, New York.
14. Hidy, G. M. and Wolf, M. (1995). In *Aerosol Forcing of Climate*, Environmental Science Research Report ES 17, ed. R. J. Charlson and J. Heintzenberg, Wiley, New York, 171.
15. Durham, M. D. and Lundgren, D. A. (1980). Evaluation of aerosol aspiration efficiency as a function of Stokes number, velocity ratio and nozzle angle, *J. Aerosol Sci.*, **11**, 179–188.
16. Okazaki, K., Wiener, R. W. and Willeke, K. (1987). Isoaxial aerosol sampling: Nondimensional representation of overall sampling efficiency, *Atmos. Envir.*, **21**, 1181–1185.
17. Dzubay, T. G., Hines, L. E. and Stevens, R. K. (1976). Particle bounce errors in cascade impactors, *Atmos. Envir.*, **10**, 229–234.

18. Martinsson, B. G., Hansson, H.-C. and Lannefors, H. O. (1984). Southern Scandinavian aerosol composition and elemental size distribution characteristics. Dependence on air-mass history, *Atmos. Envir.*, **18**, 2167–2182.

19. Davidson, C. I. and Osborne, J. F. (1986). The size of airborne trace metal containing particles, in *Toxic Metals in the Atmosphere*, Wiley Series in Advances in Environmental Science and Technology, Vol. 17, Nriagu, J. O. and Davidson, C. I., Eds., Wiley, New York, 355–390.

20. Horvath, H., Kasahara, M. and Pesava, P. (1996). The size distribution and composition of the atmospheric aerosol at a rural and nearby urban location, *J. Aerosol Sci.*, **27**, 417–435.

21. Injuk, J., Otten, Ph., Laane, R., Maenhaut, W. and Van Grieken, R. (1992). Atmospheric concentrations and size distributions of aircraft-sampled Cd, Cu, Pb and Zn over the southern bight of the North Sea, *Atmos. Envir.*, **26A**, 2499–2508.

22. Lannefors, H., Hansson, H.-C. and Granat, L. (1983). Background aerosol composition in Southern Sweden—Fourteen micro and macro constituents measured in seven particle size intervals at one site during one year, *Atmos. Envir.*, **17**, 87–101.

23. Slinn, W. G. N. (1983), Air-to-sea transfer of particles, in *Air–Sea Exchange of Gases and Particles*, Liss, P. S. and Slinn, W. G. N., Eds., D. Reidel Publishing Company, Dordrecht, 299–405.

24. Injuk, J., de Leeuw, G. and Van Grieken, R. Assessment of atmospheric pollutant fluxes to the North Sea by TXRF, *Atmos. Environ.*, in press

25. OSPARCOM (1993), *North Sea Assessment Report 2b*, North Sea Task Force. Oslo and Paris Commissions, London. Ollsen and Olsen, Fredensburg, Denmark.

26. Baeyens, W., Dehairs, F. and Dedeurwaerder, H. (1990). Wet and dry deposition fluxes above the North Sea, *Atmos. Envir.*, **24A**, 1693–1703.

27. Kellogg, W. W. (1978). Global influence of mankind on the climate, in *Climatic Change*, Gribben, J., Ed., Cambridge University Press, London, 205–227.

28. Pacyna, J. M. and Münch, J. (1989). European inventory of trace metal emissions to the atmosphere, in *Heavy Metals in the Environment*, Vol. 1, Vernet, J.-P. Ed., CEP Consultants, Edinburgh, 144–147.

29. Bridgeman, H. (1990). *Global Air Pollution*, Wiley, Chichester.

30. Shaw, G. E. (1982). Evidence for a central Eurasian source of Arctic haze in Alaska, *Nature*, **299**, 815–818.

31. Barrie, L. A. (1986). Arctic air pollution: an overview of current knowledge, *Atmos. Envir.*, **20**, 643–663.

32. Jenkins, R. (1974). *An Introduction to X-Ray Spectrometry*, Heyden and Sons, New York.

33. Leibhatsky, H. A., Pfeiffer, H. G., Winslow, E. H. and Zemany, P. D. (1972). *X-Rays, Electrons, and Analytical Chemistry*, Wiley-Interscience, New York.

34. Bertin, E. P. (1975). *Principles and Practice of X-Ray Spectrometry Analysis*, 2nd Ed., Plenum Press, New York.

35. Williams, K. L. (1987). *An Introduction to X-Ray Spectrometry*, Allen and Unwin, London.

36. Jenkins, R. (1988). *X-Ray Fluorescence Spectrometry*, Wiley, New York.

37. Van Grieken, R. E. and Markowicz, A. A. (1993). *Handbook of X-Ray Spectrometry: Methods and Techniques*, Marcel Dekker Inc., New York.

38. Klockenkämper, R. (1987). X-ray spectral analysis—present status and trends, *Spectrochim. Acta*, **42B**, 423.

39. Markowicz, A. A. and Van Grieken, R. E. (1988). X-ray spectrometry, *Anal. Chem.*, **60**, 28R.

40. Ure, A. M., Ellis, A. T. and Williams, J. G. (1988). Atomic spectrometry update. Inorganic mass spectrometry and X-ray fluorescence spectrometry, *J. Anal. At. Spectrosc.*, **3**, 175R.
41. Wobrauschek, P., Aiginger, H., Owesny, G. and Streli, C. (1988). Progress in X-ray fluorescence analysis, *J. Trace Microprobe Techn.*, **6**, 295.
42. Janssens, K. H. and Adams, F. C. (1989). New trends in elemental analysis using X-ray fluorescence spectrometry, *J. Anal. At. Spectrosc.*, **4**, 123.
43. Nichols, M. C., Boehme, D. R., Ryon, R. W., Wherry, D. C., Cross, B. J. and Aden, G. D. (1987). Parameters affecting X- ray microfluorescence (XRMF) analysis, *Adv. X-Ray Anal.*, **30**, 45.
44. Klockenkämper, R. and von Bohlen, A. (1989). Determination of critical thickness and the sensitivity for thin film analysis by total reflection X-ray fluorescence spectrometry, *Spectrochim. Acta*, **44B**, 461.
45. Maenhaut, W. (1990). Recent advances in nuclear and atomic spectrometric techniques for trace element analysis. A new look at the position of PIXE, *Nucl. Instrum. Meth. Phys. Res.*, **49B**, 518.
46. Maenhaut, W. (1989). Analytical techniques for atmospheric trace elements. In *Control and Fate of Atmospheric Trace Metals*, ed. J. M. Pacyna, and B. Ottar, Kluwer Academic Publishers, Dordrecht, The Netherlands, 259.
47. Mitchell, R. I. and Pilcher, K. M. (1959). Improved cascade impactor for measuring aerosol particle sizes, *Ind. Eng. Chem.*, **51**, 1039.
48. Adams, F. and Van Grieken, R. (1975). Absorption correction for X-ray fluorescence analysis of aerosol loaded filters, *Anal. Chem.*, **47**, 1767.
49. Leland, D. J., Bilbrey, D. B., Leyden, D. E., Wobrauschek, P., Aiginger, H. and Puxbaum, H. (1987). Analysis of aerosols using total reflection X-ray spectrometry, *Anal. Chem.*, **59**, 1911.
50. Schneider, B. (1989). The determination of atmospheric trace metal concentrations by collection of aerosol particles on sample holders for total reflection X-ray fluorescence, *Spectrochim. Acta*, **44B**, 519.
51. Salva, A., von Bohlen, A., Klockenkämper, R. and Klockow, D. (1993). Multi-element analysis of airborne particulate matter by total reflection X-ray fluorescence, *Quimica Analitica*, **12**, 57.
52. Injuk, J. and Van Grieken, R. (1995). Optimisation of total reflection X-ray fluorescence for aerosol analysis, *Spectrochim. Acta*, **50B**, 1787.
53. Dixkens, J., Fissan, H. and Dose, T. (1993). A new particle sampling technique for direct analysis using total reflection X-ray fluorescence spectrometry, *Spectrochim. Acta*, **48B**, 231.
54. Valkovic, V., Dargie, M., Jaksic, M., Markowicz, A., Tajani, A. and Valkovic, O. (1996). X-ray emission spectroscopy applied for bulk and individual analysis of airborne particulates, *Nucl. Instr. Meth. Phys. Res.*, **113B**, 363.
55. Van Grieken, R., Araujo, F., Rojas, C. and Veny, P. (1990). In *XRF and PIXE Applications in Life Science*, ed. R. Moro, and R. Cesareo, World Scientific, Singapore, New York, London, Hong Kong, 79.
56. Rhodes, J. R., Stout, J. A., Schindler, J. S. and Piorek, S. (1983). *Portable X-ray Survey Meters for In Situ Trace Element Monitoring of Air Particulates*, ASTM Special Technical Publication 786, Philadelphia, PA.
57. Von Alfthan, C., Rautala, P. and Rhodes, J. R. (1980). Applications of a new multi-element portable X-ray spectrometer to materials analysis, *Adv. X-Ray Anal.*, **23**, 27.
58. Gilfrich, J. V. and Birks, L. S. (1978). *Portable Vacuum X-Ray spectrometer: Instrument for On-site analysis of Airborne Particulate Sulfur and Other Elements*, EPA Report 600/7-78-103, Research Triangle Park, NC.

59. Standzenieks, P., Teeyasoontranont, V. and Oeblad, M. (1992). An improved X-ray technique for scanning the millimeter sized deposits from a multi orifice low pressure impactor, *J. Aerosol Sci.*, **23**, S719.
60. Matherny, M. and Balgava, V. (1993). Optimization of thin layer methods in the energy dispersive fluorescence X-ray spectroscopy of inorganic atmospheric pollutants, *Fresenius' J. Anal. Chem.*, **346**, 162.
61. Klockenkämper, R., Bayer, H. and von Bohlen A. (1995). Total reflection X-ray fluorescence analysis of airborne particulate matter, *Adv. X-ray Chem. Anal.*, **26s**, 41.
62. Wobrauschek, P. and Aiginger, H. (1986). Analytical application of total reflection and polarized X-rays, *Fresenius' Z. Anal. Chem.*, **324**, 865.
63. Gilfrich, J. V., Skelton, E. F., Qadri, S. B., Kirkland, J. P. and Nagel, D. J. (1983). Synchrotron radiation X-ray fluorescence analysis, *Anal. Chem.*, **55**, 187.
64. Hewitt, C. N. (1991). *Instrumental Analysis of Pollutants*, Elsevier Applied Science, London and New York.
65. Cresser, M. S., Garden, L. M., Armstrong, J., Dean, J. R., Watkins, P. and Cave, M. (1996). *J. Anal. At. Spectrom.*, **11**, 19R.
66. Rojas, C. M., Artaxo, P. and Van Grieken, R. (1990). Aerosols in Santiago de Chile: a study using receptor modeling with X-ray fluorescence and single particle analysis, *Atmos. Environ.*, **24B**, 227.
67. Cheng, Y. (1991). *Huanjing Huaxue*, **10**, 56.
68. Spitzer, Z., Lisy, J., Kobr, M. and Santroch, J. (1989). *Ochr. Ovdusi*, **3**, 17.
69. Sadasivan, S. and Negi, B. S. (1990). Elemental characterization of atmospheric aerosols, *Sci. Total Environ.*, **96**, 269.
70. Otten, P., Injuk, J. and Van Grieken, R. (1994). Elemental concentrations in atmospheric particulate matter sampled on the North Sea and the English Channel, *Sci. Total Environ.*, **155**, 131.
71. Haupt, O., Klaue, B., Schaefer C. and Dannecker, W. (1995). Preparation of quartz fibre filter standards for X-ray fluorescence analysis of aerosol samples, *X-ray Spectrom.*, **24**, 267.
72. Kriews, M., Dannecker, W., Naumann, K. and Waetjen, U. (1988). On the preparation of an ambient aerosol reference material for elemental analysis, *J. Aerosol Sci.*, **19**, 1295.
73. Kunungi, M. (1989). *Bunseki*, **5**, 380.
74. Ehman, D. L. and Anselmo, V. C. (1989). Preparation of customized particulate-laden glass fiber filter standards for X-ray fluorescence spectrometric analyses. *Spectroscopy (Eugene, Oreg.)*, **4**, 46.
75. Török, S. B., Labar, J., Injuk, J. and Van Grieken, R. (1996). X-ray spectrometry, *Anal. Chem.*, **68**, 467R.
76. Johansson, S. A. E., Campbell, J. L. and Malmqvist, K. G. (1995). *Particle Induced X-ray Emission Spectrometry (PIXE)*, Wiley, New York.
77. Watt, F., Grime, G., Blower, G. D., Takacs, J. and Vaux, D. J. T. (1982). The Oxford 1 μm proton microprobe. *Nucl. Instrum. Methods Phys. Res.*, **197**, 65.
78. Johansson, S. A. E. and Campbell, J. L. (1988). *PIXE: A Novel Technique for Elemental Analysis*, Wiley, Chichester, UK.
79. S. Bamford, M. Dargie, E. Greaves, B. Holynska, A. Markowicz, R. Van Grieken, R.R. Vis and R.L. Walsh (1997) *Sampling, Storage and Sample Preparation Procedures for X-ray Fluorescence Analysis of Environmental Samples*, Report IAEA, TECDOC-950, IAEA, Vienna.
80. Jervis, E., Landsberger, S., Lecomte, R., Paradis, P. and Monaro, S. (1982). Determination of trace pollutants in urban snow using PIXE techniques, *Nucl. Instrum. Methods*, **193**, 323.

81. Hansson, C., Ekholm, A. K. C. and Ross, H. B. (1988). Rainwater analysis: a comparison between proton-induced X-ray emission and graphite furnace atomic absorption spectroscopy, *Environ. Sci. Technol.*, **22**, 527.

82. Hansen, D., Ryder, J. F., Mangelson, N. F., Hill, M. W., Faucette, K. J. and Eatough, D. J. (1980). Inaccuracies encountered in sulfur determination by particle induced X-ray emission, *Anal. Chem.*, **52**, 821.

83. Johansson, B., Akselsson, R. and Johansson, S. A. E. (1970). Elemental trace analysis at the 10^{-12} g level, *Nucl. Instrum. Methods*, **84**, 141.

84. Aldape, F., Flores, J., Garcia, R. and Nelson, J. W. (1996). PIXE analysis of atmospheric aerosols from a simultaneous three site sampling during the autumn of 1993 in Mexico City, *Nucl. Instr. Meth. Phys. Res.*, **109B**, 502.

85. Salma, I., Maenhaut, W., Cafmeyer, J., Annegarn H. J. and Andreae, M. O. (1994). PIXE analysis of cascade impactor samples collected at the Kruger National Park, South Africa, *Nucl. Instr. Meth. Phys. Res.*, **85B**, 849.

86. Flocchini, R. G., Cahill, T. A., Shadoan, D. J., Lange, S. J., Eldred, R. A., Feeney, P. J., Wolfe, G. W., Simmeroth, D. C. and Suder, J. K. (1976). Monitoring California's aerosols by size and elemental composition, *Environ. Sci Technol.*, **10**, 76.

87. Flocchini, R. G., Cahill, T. A., Pitchford, M. L., Eldred, R. A., Feeney, P. J. and Asbaugh, L. L. (1981). Characterization of particles in the arid West, *Atmos. Environ.*, **15**, 2017.

88. Eldred, R. A., Cahill, T. A. and Feeney, P. J. (1987). Particulate monitoring at U.S. National Parks using PIXE, *Nucl. Instrum. Meth.*, **22B**, 289.

89. Rojas, C. M., Van Grieken, R. and Maenhaut, W. (1993). Elemental composition of aircraft-sampled aerosols above the Southern Bight of the North Sea, *Water Air Soil Pollut.*, **71**, 319.

90. Miranda, J. (1996). Studies of atmospheric aerosols in large urban areas using PIXE: An overview, *Nucl. Instrum. Meth. Phys. Res.*, **109B**, 439.

91. Koltay, E. (1994). The application of PIXE and PIGE techniques in the analytics of atmospheric aerosols, *Nucl. Instr. Meth. Phys. Res.*, **85B**, 75.

92. Artaxo, P., Gerab, F. and Rabello, M. L. C. (1993). Elemental composition of aerosol particles from two atmospheric monitoring stations in the Amazon Basin, *Nucl. Instrum. Methods Phys. Res.*, **75B**, 277.

93. Annegarn, H. J., Cahill, T. A., Sellschop, J. P. F. and Zucchiatti, A. (1986). Time sequence particulate sampling and nuclear analysis, *Physica Scripta*, **37**, 282.

94. Tabacniks, M. H., Orsini, C. Q. and Maenhaut, W. (1993). PIXE analysis of atmospheric particulate matter in glass fiber filters, *Nucl. Instrum. Methods Phys. Res.*, **75B**, 262.

95. Maenhaut, W., Ducastel, G., Hillamo, R. E. and Pakkanen, T. (1993). Evaluation of the applicability of the MOUDI impactor for aerosol collections with subsequent multielement analysis by PIXE, *Nucl. Instrum. Methods Phys. Res.*, **75B**, 24.

96. Annegarn, H. J., Zucchiatti, A., Cereda, E. and Barga- Marczazzan, G. M. (1990). Source profiles by unique ratios (spur) analysis: interpretation of time-sequence PIXE aerosol data, *Nucl. Instrum. Methods Phys. Res.*, **49B**, 372.

97. Romo-Kroger, C. M. (1990). Deuteron excitation as an alternative in PIXE nalysis, *Nucl. Instrum. Methods Phys. Res.*, **49B**, 29.

98. Swietlicki, E., Martinsson, B. G. and Kristiansson, P. (1996). The use of PIXE and complementary ion beam analytical techniques for studies of atmospheric aerosols, *Nucl. Instr. Meth. Phys. Res.*, **109B**, 385.

99. Malmqvist, K. G. (1984). PIXE—A useful tool in studies of work environment aerosols, *Nucl. Instr. Meth.*, **3B**, 259.

100. Annegarn, H. J., Zucchiatti, A., Sellschop, J. P. F. and Booth-Jones, P. (1987). PIXE characterization of airborne dust in the mining environment, *Nucl. Instr. Meth.*, **22B**, 235.
101. Cahill, T. A. (1990). Analysis of air pollutants by PIXE: the second decade, *Nucl. Instr. Meth.*, **49B**, 345.
102. Cahill, T. A., Miranda, J. and Morales, R. (1991). Survey of PIXE programs, *Int. J. PIXE*, **1**, 297.
103. Isabelle, D. B. (1994). The PIXE analytical technique: principle and applications, *Radiat. Phys. Chem.*, **44**, 25.
104. Maenhaut, W., Raemdonck, H. and Andreae, M. O. (1987). PIXE analysis of marine aerosol samples: accuracy and artifacts, *Nucl. Instrum. Meth.*, **22B**, 248.
105. Vandecasteele, C. (1988). *Activation Analysis with Charged Particles*, Ellis Horwood, Chichester.
106. Erdtmann, C. and Petri, H. (1986). In *Treatise on Analytical Chemistry*, second ed., Part I, Vol. 14, ed. I. M. Kolthoff, P. J. Elving and V. Krivan, Wiley, New York, 419.
107. Bode, P., Korthoven, P. J. M. and De Bruin, M. (1987). Microprocessor-controlled facility for INAA using short half- life nuclides, *J. Radioanal. Nucl. Chem.*, **113**, 371.
108. Lindstrom, R. M., Zeisler, R. and Rossbach, M. (1987). *J. Radioanal. Nucl. Chem.*, **110**, 33.
109. Landsberger, S. (1992). Analytical methodologies for instrumental neutron activation analysis of airborne particulate matter, *J. Trace Microprobe Tech.*, **10**, 1.
110. Mizohata, A. and Otoshi, T. (1994). Elemental composition of atmospheric aerosol particles, *Radioisotopes*, **43**, 236.
111. Steinnes, E. (1992). *Appl. Int. Radiat. Conserv. Environ. Pro Symp.*, 387.
112. Dams, R. (1992). Nuclear activation techniques for the determination of trace elements in atmospheric aerosols, particulates and sludge samples, *Pure Appl. Chem.*, **64**, 991.
113. Dams, R., Robbins, J. A., Rahn, K. A. and Winchester, J.W. (1970). Nondestructive neutron activation analysis of air pollution particulates, *Anal. Chem.*, **42**, 861.
114. Vandecasteele, C. and Block, C. B. (1993). *Modern Methods for Trace Element Determination*, Wiley, Chichester.
115. Larsen, E., Hansen, A. B., Kristensen, L. V., Solgorazol, P., Damsgaard, E. and Heydorn, K. (1992). ICP-MS, INAA and PIXE analysis of airborne dust samples, *J. Trace Microprobe Tech.*, **10**, 43.
116. Izumikawa, Y., Nezu, T. and Otoshi, T. (1992). *Nippon Kankyo Esei Senta Shoho*, **19**, 60.
117. Kitto, M. E., Anderson D. L. and Zoller, W. H. (1988). Simultaneous collection of particles and gases followed by multielement analysis using nuclear techniques, *J. Atmos. Chem.*, **7**, 241.
118. Morselli, L., Zappoli, S., Gallorini, M. and Rizzio, E. (1988). Characterisation of trace elements in dry depositions by instrumental neutron activation analysis, *Analyst (London)*, **113**, 1575.
119. Qadir, M. A., Iqbal, M. Z. and Zaidi, J. H. (1995). Neutron activation analysis of Ce, Yb, Se and Cr in the atmosphere of some major urban and industrial areas of Punjab, Pakistan, *J. Radioanal. Nucl. Chem.*, **201**, 347.
120. Landsberger, S. and Wu, D. (1993). Improvement of analytical sensitivities for the determination of antimony, arsenic, cadmium, indium, iodine, molybdenum, silicon and uranium in airborne particulate matter by epithermal neutron activation analysis, *J. Radioanal. Nucl. Chem.*, **167**, 219.

121. Yang, S. and Zhang, Y. (1992). *Nucl. Sci. Tech.*, **3**, 205.
122. Weginwar, R. G. and Garg, A. N. (1992). Multielement neutron activation analysis of fugitive dust particulates from cement factories in central India, *J. Radioanal. Nucl. Chem.*, **162**, 381.
123. Chutke, N. L., Ambulker, M. N., Aggarwal, A. L. and Garg, A. N. (1994). Instrumental neutron activation analysis of ambient air dust particulates from metropolitan cities in India, *Environ. Pollut.*, **85**, 67.
124. Ondov, J. M. and Divita, F. Jr (1993). Size spectra for trace elements in urban aerosol particles by instrumental neutron activation analysis, *J. Radioanal. Nucl. Chem.*, **167**, 247.
125. Wang, J. (1985). *Stripping Analysis: Principles, Instrumentation and Applications*, VCH Publishers, Deerfield Beach.
126. Wilson, G. S., Colovos, G. and Moyers, J. (1973). The determination of trace amounts of zinc, cadmium, lead and copper in airborne particulate matter by anodic stripping voltammetry, *Anal. Chim. Acta*, **64**, 457.
127. Harrison, R. M. (1984). Recent advances in air pollution analysis, *CRC Crit. Rev. Anal. Chem.*, **15**, 1.
128. Cassassas, E., Perez-Vendrell, A. M. and Puignou, L. (1991). An improved voltammetric procedure for the determination of Zn, Pb, Cd and Cu in the atmospheric aerosols, *Int. J. Environ. Anal. Chem.*, **45**, 55.
129. Khandekar, R. N., Dhaneshwar, R. G., Palrecha, M. M. and Zarapkar, L. R. Z. (1981). *Anal. Chem.*, **307**, 365.
130. MacLeod, K. E. and Lee, R. E. (1973). Selected trace metal determination of spot tape samples by anodic stripping voltammetry, *Anal. Chem.*, **45**, 2380.
131. Harrison, R. P. and Winchester, J. W. (1971). Area-wide distribution of lead, copper, and cadmium in air particulates from Chicago and northwest Indiana, *Atmos. Environ.*, **5**, 863.
132. Wonders, J. H. A. M., Houweling, S., De Bont, F. A. J., Van Leeuwen, H. P., Eeckhaoudt, S. M. and Van Grieken, R. (1994). *Int. J. Environ. Anal. Chem.*, **56**, 193.
133. Nam, D. Q., Skacel, F. and Buryan, P. (1994). Determination of airborne lead and cadmium collected on glass fibre filters by differential-pulse anodic stripping voltammetry, *Sci. Total Environ.*, **144**, 87.
134. Scholz, F., Nitschke, L. and Henrion, G. (1987). Determination of mercury traces by differential pulse stripping voltammetry after sorption of mercury vapour on a gold-plated electrode, *Anal. Chim. Acta*, **199**, 167.
135. Injuk, J., Otten, P., Laane, R., Maenhaut, W. and Van Grieken, R. (1992). Atmospheric concentrations and size distributions of aircraft-sampled Cd, Cu, Pb and Zn over the Southern Bight of the North Sea, *Atmos. Environ.*, **26**, 2499.
136. Otten, P., Injuk, J. and Van Grieken, R. (1994). Vertical sulfur dioxide, ozone, and heavy metal concentration profiles above the Southern Bight of the North Sea, *Isr. J. Chem.*, **34**, 411.
137. Christensen, J. K, Kryger, L. and Pind, N. (1982). The determination of traces of cadmium, lead and thallium in fly ash by potentiometric stripping analysis, *Anal. Chim. Acta.*, **141**, 131.
138. Welz, B. (1985). *Atomic Absorption Spectrometry*, VCH Verlagsgesellschaft mBH, Weinheim.
139. Carneiro, M. C., Campos, R. C. and Curtius, A. J. (1993). Determination of Sb, Ni and V in slurry from airborne particulate material collected on filter by graphite furnace atomic absorption spectrometry, *Talanta*, **40**, 1815–1822.
140. Van Loon, J. C. (1980). *Analytical Atomic Absorption Spectroscopy, Selected Methods*, Academic Press, New York.

141. Sneddon, J. (1983). Collection and atomic spectrometric measurement of metal compounds in the atmosphere, *Talanta*, **30**, 641–648.
142. Cresser, M. S., Armstrong, J., Cook, J. M., Dean, J. R., Watkins, P. and Cave, M. (1995). Atomic spectrometry update—environmental analysis, *J. Anal. At. Spectrom.*, **10**, 9R–48R.
143. Feng, Y. and Barratt, R. S. (1994). Digestion of dust samples in a microwave oven, *Sci. Tot. Envir.*, **143**, 157–161.
144. Yamashige, T., Yamamoto, M. and Sunahara, H. (1989). Comparison of decomposition methods for the analysis of atmospheric particulates by atomic absorption spectrometry, *Analyst*, **114**, 1071–1077.
145. Miller, T. J. (1991). Development of a rapid, economical and safe method, for the dissolution of particulate beryllium, for atomic spectrophotometry, *Anal. Lett.*, **24**, 2075–2081.
146. Pakkanen, T. A. and Hillamo, R. E. (1988). Analysis of low metal concentrations from atmospheric aerosol samples collected with a multi-stage impactor, *J. Aerosol Sci.*, **19**, 1303–1306.
147. Pakkanen, T. A., Hillamo, R. E. and Maenhaut, W. (1993). Simple nitric acid dissolution method for electrothermal atomic absorption spectrometric analysis of atmospheric aerosol samples collected by a Berner-type low-pressure impactor, *J. Anal. At. Spectrom.*, **8**, 79–84.
148. Ottley, C. J. and Harrison, R. M. (1993). Atmospheric dry deposition flux of metallic species to the North Sea, *Atmos. Environ.*, **27A**, 685–695.
149. Frenzel, W. (1991). Microanalytical concept for multicomponent analysis of airborne particulate matter, *Fresenius J. Anal. Chem.*, **340**, 525–533.
150. Fernández C. A., Fernández, R., Carrión, N., Loreto, D., Benzo, Z. and Fraile, R. (1991). Metals determination in atmospheric particulates by atomic absorption spectrometry with slurry sample introduction, *At. Spectrscopy*, **12**, 111–118.
151. Lüdke, C., Hoffmann, E. and Skole, J. (1994). Studies on the determination of the metal content on airborne particulates by furnace atomization non-thermal excitation spectrometry, *J. Anal. At. Spectrom.*, **9**, 685–689.
152. Low, P. S. and Hsu, G. J. (1990). Direct determination of atmospheric lead by Zeeman solid sampling graphite furnace atomic absorption spectrometry (GFAAS), *Fresenius J. Anal. Chem.*, **337**, 299–305.
153. Isozaki, A., Morita, Y., Okutani, T. and Matsumura, T. (1996). Direct determination of lead in suspended particulate matter in air by tungsten furnace-atomic absorption spectrometry, *Analytical Sciences*, **12**, 755–759.
154. Sneddon, J. (1991). Direct and near-real-time determination of lead, manganese and mercury in laboratory air by electrostatic precipitation-atomic absorption spectrometry, *Anal. Chim. Acta*, **245**, 203–206.
155. Schramel, P. (1988). ICP and DCP emission spectrometry for trace element analysis in biomedical and environmental samples. A review, *Spectrochimica Acta*, **43B**, 881–896.
156. Robinson, J. W. (1990). *Atomic Spectroscopy*, Marcel Dekker, New York.
157. Cresser, M. S. and Marr, I. L. (1991). Optical spectrometry in the analysis of pollutants, in *Instrumental Analysis of Pollutants*, Hewitt, C. N., Ed., Elsevier Applied Science, London, 99.
158. Thomas, T. C. and Jehl, L. J. (1988). Metal exposure evaluation: a rapid multi-element analysis technique using ICP- AES, *At. Spectroscopy*, **9**, 154–156.
159. Infante, R. N. and Acosta, I. L. (1988). Comparison of extraction procedures for the determination of heavy metals in airborne particulate matter by

inductively coupled plasma-atomic emission spectroscopy, *At. Spectroscopy*, **9**, 191–194.

160. Wang, C.-F., Miau, T. T., Perng, J. Y., Chiang, P. C., Tsai, H. T. and Yang, M. H. (1989). Multi-element analysis of airborne particulate matter by inductively coupled plasma atomic emission spectrometry, *Analyst*, **114**, 1067–1070.

161. Valdés-Hevia y Temprano, M. C., Fernández de la Campa, M. R. and Sanz-Medel, A. (1993). Sensitive method for the determination of lead by potassium dichromate–lactic acid hydride generation inductively coupled plasma atomic emission spectrometry, *J. Anal. At. Spectrom.*, **8**, 821–825.

162. Vanhaecke, F., Vanhoe, H., Dams, R. and Vandecasteele, C. (1992). The use of internal standards in ICP-MS, *Talanta*, **39**, 737–742.

163. Denoyer, E. R. (1992). Semiquantitative analysis of environmental materials by laser sampling inductively coupled plasma mass spectrometry, *J. Anal. At. Spectrom.*, **7**, 1187–1193.

164. Katoh, T., Akiyama, M., Ohtsuka, H., Nakamura, S., Haraguchi, K. and Akatsuka, K. (1996). Determination of atmospheric trace metal concentrations by isotope dilution inductively coupled plasma mass spectrometry after separation from interfering elements by solvent extraction, *J. Anal. At. Spectrom.*, **11**, 69–71.

165. Mukai, H., Ambe, Y. and Morita, M. (1990). Flow injection inductively coupled plasma mass spectrometry for the determination of platinum in airborne particulate matter, *J. Anal. At. Spectrom.*, **5**, 75–80.

166. Jalkanen, L. and Häsänen, E. (1996). Simple method for the dissolution of atmospheric aerosol samples for analysis by inductively coupled plasma mass spectrometry, *J. Anal. At. Spectrom.*, **11**, 365–369.

167. Quality of Urban Air Review Group, (1996). *Airborne Particulate Matter in the United Kingdom*, QUARG, 176 pp.

168. Harrison, R. M. (editor) (1996). *Pollution, Causes, Effects and Control*, 3rd Edition, Royal Society of Chemistry, London.

169. Stockwell, W. R. and Calvert, J. G. (1983). *Atmos. Environ.*, **17**, 2231.

170. Finlayson-Pitts, B. J. and Pits, J. N. Jr (1986). *Atmospheric Chemistry*, Wiley, New York.

171. Chiang, S. G., Toosi, R. and Novakov, T. (1981). *Atmos. Environ.*, **15**, 1287.

172. Harrison, R. M. and Pio, C. A. (1983). *Atmos. Environ.*, **17**, 1261–1275.

173. Chamiedes, W. L. and Stelson, A. W. (1992). *J. Geophys. Res.*, **97**, 20,565–20,580.

174. Harrison, R. M. and Kitto, A.-M. N. (1992). *J. Atmos. Chemistry*, **15**, 133–143.

175. Vermetten, A. W. M., Asman, W. A. H., Buijsman, E., Mulder, W., Slanina, J. and Waijers-Ijpelaan, A. (1985). *VDI Berichte*, **560**, 241–251.

176. Asman, W. A. G. and van Jaarsveld, H. A. (1990). *Report No. 228471007*, National Institute of Public Health and Environmental Protection, Bilthoven, The Netherlands.

177. Barrie, L. A., Harrison, R. M., and Sturges, W. T. (1989). *Atmos. Environ.*, **23**, 1083–1098.

178. Bandy, A., Scott, D. L., Blonquist, B. W., Chen, S. M. and Thornton, D. C. (1992). *Geophys. Res. Lett.*, **19**, 1125–1127.

179. Winer, A. M., Atkinson, R. and Pitts, J. N. Jr (1984). *Science*, **224**, 156.

180. Harrison, R. M. and McCartney, H. A. (1979). *Atmos. Environ.*, **13**, 1105–1120.

181. Harrison, R. M., Msibi, M. I., Kitto, A.-M.N. and Yamulki, S. (1994). *Atmos. Environ.*, **28**, 1593–1599.

182. Ottley, C. J. and Harrison, R. M. (1992). *Atmos. Environ.*, **26A**, 1689–1699.

183. Allen, A. G., Harrison, R. M. and Erisman, J. W. (1989). *Atmos. Environ.*, **23**, 1591–1599.
184. Harrison, R. M. and Mackenzie, A. R. (1990). *Atmos. Environ.*, **24A**, 91–102.
185. Wexler, A. S. and Seinfeld, J. H. *Atmos. Environ.*, (1990). **24A**, 1231–1246.
186. Harrison, R. M. and Msibi, I. M. *Atmos. Environ.*, (1994). **28**, 247–255.
187. Pio, C. A. and Harrison, R. M. (1987). *Atmos. Environ.*, **21**, 1243–1246.
188. Pio, C. A. and Harrison, R. M. (1987). *Atmos. Environ.*, **21**, 2711–2715.
189. Harrison, R. M. and Biggins, P. D. E. (1979). *Environ. Sci. Technol.*, **13**, 558–565.
190. Bassett, M. and Seinfeld, J. H. (1983). *Atmos. Environ.*, **17**, 2237–2252.
191. Friedlander, S. K. (1977). *Smoke, Dust and Haze*, Wiley, New York.
192. Middleton, P. and Kiang, C. S. (1978). *J. Aerosol Sci.*, **9**, 359–385.
193. Jaecker-Voirol, A. and Mirabel, P. (1989). *Atmos. Environ.*, **23**, 2053–2057.
194. Hoppel, W. A., Frick, G. M., Fitzgerald, J. W. and Larson, R. E. (1994). *J. Geophys. Res.*, **99**, 14 443–14 459.
195. Coffmann, D. J. and Hegg, D. A. (1995). *J. Geophys. Res.*, **100**, 7147–7160.
196. Harrison, R. M. *7th Symposium on Physico-Chemical Behaviour of Atmospheric Pollutants*, Venice, submitted (1996).
197. Pilinis, C. and Seinfeld, J. H. (1987). *Atmos. Environ.*, **21**, 2453–2466.
198. Pilinis, C. and Seinfeld, J. H. (1988). *Atmos. Environ.*, **22**, 1985–2001.
199. Wall, S. M., John, W. and Ondo, J. L. (1988). *Atmos. Environ.*, **22**, 1649–1656.
200. Harrison, R. M. and Pio, C. A. (1983). *Environ. Sci. Technology.*, **17**, 169–174.
201. Harrison, R. M., Peak, J. D. and Msibi, M. I. (1996). *Atmos. Environ.*, **30**, 133–143.
202. Quinn, P. K., Bates, T. S., Johnson, J. E., Covert, D. S. and Charlson, R. J. (1990). *J. Geophys. Res.*, **95**, 16 405–16 416.
203. Huebert, B. J. (1988). *Geophys. Res. Lett.*, **7**, 325–328.
204. Prospero, J. M., and Jennings, S. G. (1996). *IGAC Newsletter* No. 7, December.
205. Chow, J. C., Watson, J. G., Lowenthal, D. H. and Countess, R. J. (1996). *Atmos. Environ.*, **30**, 1489–1499.
206. Fitzgerald, J. W. (1991). *Atmos. Environ.*, **25**, 533–545.
207. Hjellbrekke, A. G., Lovblad, G., Sjoberg, K., Schaug, J. and Skjelmoeu, J. E. (1995). *EMEP Data Report 1993*, EMEP, Norway.
208. Pierson, W. R., Hammerle, R. H. and Brachaczek, W. W. (1976). *Anal. Chem.*, **48**, 1808.
209. Pierson, W. R., Brachaczek, W. W., Kornishi, T. J., Truex, T. J. and Butler, J. W. (1980). *J. Air Pollut. Contr. Assoc.*, **30**, 30–34.
210. Mulawa, P. and Cadle, S. H. (1985). *Atmos. Environ.*, **19**, 1317–1324.
211. Harrison, R. M. and Kitto, A.M.-N. (1990). *Atmos. Environ.*, **24A**, 2633–2640.
212. Allegrini, I., de Santis, F., di Palo, V., Febo, A., Perrino, C. and Possanzini, M. (1987). *Sci. Tot. Environ.*, **67**, 1–16.
213. Koutrakis, P., Wolfson, J. M., Slater, J. L., Braver, M., Spengler, J. D., Stevens, R. K. and Stone, C. L. (1988). *Environ. Sci. Technol.*, **22**, 1463–1468.
214. Weatherburn, M. W. (1967). *Anal. Chem.*, **39**, 971.
215. Rapsomanikis, S., Wake, M., Kitto, A.M.-N. and Harrison, R. M. (1988). *Environ. Sci. Technol.*, **22**, 948–951.
216. Genfa, Z. and Dasgupta, P. K. (1989). *Anal. Chem.*, **61**, 408–412.
217. Kitto, A.-M. N. and Harrison, R. M. (1992). *Atmos. Environ.*, **26A**, 2389–2399.

4 Speciation Techniques for Fine Atmospheric Aerosols

ROGER L. TANNER
Tennessee Valley Authority, Muscle Shoals, USA

Atmospheric Particles, Edited by R. M. Harrison and R. Van Grieken.
© 1998 John Wiley & Sons Ltd.

1 INTRODUCTION

Particulate mass (in the size range $< 10\,\mu m$ in diameter) and the lead content of aerosols are the only regulated properties or constituents of ambient atmospheric particles in the USA. This is true despite the fact that the toxicity of other species which may be present in atmospheric particles is recognized, and the ability to measure some of those toxic species at atmospheric levels is established. The basis for regulation, though, is the establishment of a causative link between the presence of a species in, for example, airborne particles, and the development of a deleterious condition in humans or other organisms. Health effects of chemical constituents of atmospheric aerosols have received much attention, e.g. strong acid content, sulfate, trace metals such as Cr(VI), Hg, and organic species such as polynuclear aromatic hydrocarbons (PAHs) and dioxins. But the practical and extraordinarily difficult task of establishing the etiology of an effect and attributing it to a particular species or group of species in aerosol particles for which population exposure can be documented, has not been accomplished in most cases. Thus, given the historical limitations in analytical chemical and epidemiological techniques, only particle mass in a defined particle size range and particulate lead have been regulated in the USA. The justification that has been given is that (lead excepted) health effects show no greater correlation with specific compounds or species in particles than they do with particle mass, and data for the latter are much more abundant and much easier to obtain.

Recently the National Ambient Air Quality Standard (NAAQS) for particulate matter has been reviewed in the USA. Information is available suggesting that deleterious effects of fine particles ($<\sim 2.5\,\mu m$) on organisms are more serious than for coarse particles and, as a result, the addition of a fine particulate mass standard to the present PM10 standard has been made, although the scientific basis for this hypothesis remains controversial. Further, the question as to whether it is particulate mass or the mass of one or more its chemical constituents which are apparently related to increased respiratory disease and other health effects is not resolved. This question of quantitating and controlling the health effects of airborne particles is also of major concern in other industrialized areas of North and South America, Europe and Asia. Thus it appears timely to review the currently available techniques for determining individual chemical species present in atmospheric particles, to indicate the limitations of those methods, to identify those species for which routine data can be acquired with reasonably modest resources, and to suggest areas of technique development which may be needed in the future. Those readers interested in the broader question of monitoring for airborne particles for compliance with ambient air quality standards should consult the recent comprehensive review by Chow. [1]

Information on the nature and amounts of chemical constituents in atmospheric aerosols is also important in studies of aerosol growth and scavenging processes, since the rate of growth of atmospheric fine particles with increasing humidity depends on their chemical nature (McMurry and Stolzenburg[2]). Fine particles of two or more chemical compositions in ambient air may grow by accretion of water at high humidities at dramatically different rates [3,4]. Hence, since particle growth, incorporation into clouds and dry deposition are the major loss mechanisms for atmospheric particles, knowledge of their chemical composition is important in determining both their atmospheric lifetimes as well as their effect on atmospheric visibility and the global energy balance.

In this chapter, I shall use the term "speciation" to mean the determination of particular chemical species which may be present in airborne particles. In addition, the term "chemical species" will mean distinct, identifiable chemical compounds, such as ammonium nitrate, sulfuric acid and benzo(a)pyrene, or compound groups (e.g. $C_{20} - C_{40}$ n-alkanes, methoxyphenols, aluminosilicates). This will serve to distinguish chemical species potentially identifiable in airborne particles from bulk chemical quantities such as mass of soluble sulfate, elemental carbon or droplet pH, which may be present in airborne particulate matter from a variety of sources. Because of considerations of length, this chapter will focus on chemical speciation techniques of atmospheric fine particles (as defined above), and will be illustrative with respect to methodologies, rather than strictly comprehensive.

In the sections below, I will first discuss the complications introduced in chemical speciation of atmospheric fine particles by the existence of distributions of species between gaseous and particulate or droplet phases in the atmosphere, before continuing with discussion of specific chemical speciation techniques. Identification methods based on the morphology or crystal structure of individual particles and their surface reactions will be briefly reviewed. Thermal evolution or decomposition methods, widely used to detect chemical species based on their relative volatility and known evolution products, will be discussed in the next section, followed by discussion of spectroscopic techniques for particles, focusing almost exclusively on Fourier transform infrared (FTIR) methods. Gas chromatography-mass spectrometry (GC-MS) and related methods for aerosol constituents (principally thermally stable organic species in complex mixtures) will then be covered. A brief discussion of identification methods based on deliquescence properties and particle growth as a function of relative humidity will then follow. I will then discuss the problem of internal vs external mixtures in the speciation of particulate constituents, and follow this with some examples from ambient studies comparing chemical compositions obtained from direct methods with that inferred from bulk analizes of ionic or elemental species. The chapter will conclude with a brief summary and prospects for future improvements in speciation techniques. Methods for analysis of the chemical composition of single particles—electron microprobe analysis, particle-induced X-ray emission, laser microprobe spectroscopy, secondary ion mass spectrometry, micro-Raman spectroscopy and photoelectron spectroscopy—have been treated in a previous series volume [5], and will not be considered here.

2 CONSIDERATIONS FOR CHEMICAL SPECIES DISTRIBUTED BETWEEN ATMOSPHERIC PHASES

A phenomenon of particular relevance—so much so that it will be reviewed prior to discussion of individual speciation methods—is the observed distribution of certain groups of atmospheric chemical species between gaseous and condensed phases. There are several types of phase-partitioning phenomena which may be relevant to atmospheric analytical chemistry: physical sorption or (irreversible) chemisorption of gases on to solid particles; take-up of gases into liquid droplets consistent with Henry's law; and distribution between phases according to multi-phase chemical equilibria. Examples of each of these phenomena, along with their effects on the measurement of specific chemical species in aerosol particles and droplets, will be given in the following paragraphs.

2.1 PHYSICAL SORPTION OR (IRREVERSIBLE) CHEMISORPTION OF GASES ON TO SOLID PARTICLES OR INTO DROPLETS

Weak, physical sorption of gases on to atmospheric particles usually does not markedly affect particle composition under ambient atmospheric conditions, since rarely is the composition of either the particle phase (of interest in this chapter) or of the gas phase significantly changed thereby. Irreversible physical sorption, often with subsequent neutralization or reaction of the sorbed gas on the surface or within the bulk droplet phase, does play a significant role in heterogeneous atmospheric chemistry, since it may lead to formation of significant amounts of, for example, sulfites, nitrites and organic acid salts, which would not otherwise be present in particles. It is also a possible route for the conversion of SO_2 to sulfate by catalytic oxidation on the particle surface or in the bulk aerosol droplet following sorption from the gas phase.

2.2 TAKE-UP OF GASES INTO LIQUID DROPLETS ACCORDING TO HENRY'S LAW

This phenomenon may influence chemical speciation in atmospheric particles, since particles containing hygroscopic or deliquescent species or salts may take up water at higher humidities, and the resulting droplets may convert soluble ionic salts into electrolyte-containing droplets (e.g. ammonium, sulfate, bisulfate, nitrate). Uptake of water may also promote the acid–base dissociation of soluble strong and weak acids, forming, for example, acetate ions in solution from acetic acid vapour *originally* in the gas phase. It may also, as noted above, facilitate reactions of atmospheric significance that take place in the liquid phase, for example, oxidation of dissolved SO_2 by peroxides.

2.3 DISTRIBUTION BETWEEN PHASES ACCORDING TO MULTI-PHASE CHEMICAL EQUILIBRIA

This phenomenon creates significant problems in the sampling of atmospheric aerosol particles for the purpose of identifying and quantifying chemical species therein, but also can affect the integrity of sample composition during storage and in some analysis techniques. The most intensely studied example of this phenomenon is the ammonia–nitric acid–ammonium nitrate system. The equilibrium constant describing the equilibrium between gaseous ammonia and nitric acid and both solid-phase and aqueous solutions of ammonium nitrate (reaction (1)) is known as a function of temperature:

$$NH_4NO_3 \text{ (s or l)} \Leftrightarrow NH_3(g) + HNO_3(g) \tag{1}$$

Some studies of its behaviour in mixed salts (with ammonium sulfate and/or ammonium chloride) have also been conducted. [6] In addition, a substantial number of studies have compared calculated values of the gaseous species in the presence of ammonium nitrate with ambient observations, with moderately good agreement being observed in most of those studies. [7] Models of aerosol composition now also consider the effects of kinetic limitations on attaining equilibrium in reaction (1), as well as the temperature and particle-size dependencies of those rates. [8] Developments in this area have been reviewed in a previous series volume. [9]

In determining particular species in aerosol particles, one must thus be aware that even inorganic ionic species can be labile under atmospheric conditions, and employ sampling and analysis procedures which do not change the chemical composition of the particles. Organic aerosol species are frequently labile, and in some cases, semi-volatile, with the fraction present in the particle phase highly dependent on temperature, pressure and the presence of other chemical species in the particle, thereby affecting methods for their chemical analysis (cf. Eatough *et al.* [10]).

3 IDENTIFICATION BY THE MORPHOLOGY OR CRYSTAL STRUCTURE OF INDIVIDUAL PARTICLES

The morphology of particles and agglomerated clusters has been observed by optical microscopy and various forms of electron microscopy, for example in combustion aerosols containing elemental carbon with traces of sorbed compounds, in fly-ash particles of characteristic morphology, in sea-salt particles with nitrate on the surface, and in solid graphitic or other core particles

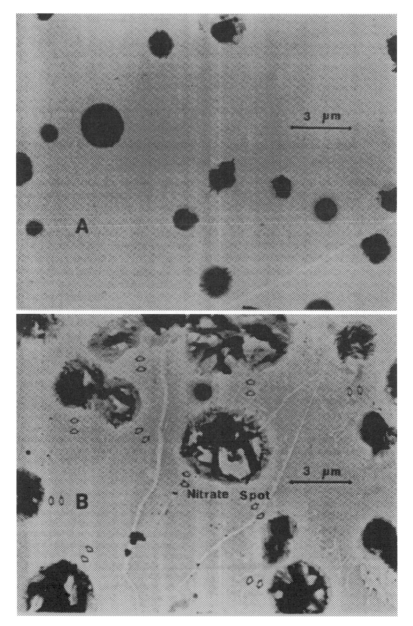

Figure 1.   Typical photomicrographs of particles collected on stage 3 of the Casella impactor in Haifa, on a summer's day. (A) Before any treatment. (B) After the nitron treatment. Almost all particles in (B) reacted with nitron. The 'fibre' appearance is typical of that of sodium nitrates

surrounded by sulfate-containing droplets or crystalline layers. Direct observation of individual particles, their physical shape, and the heterogeneous nature of the distribution of species has provided both for identification of chemical species and some indication of the mechanisms by which they were formed.

Individual particle identification and characterization have been developed mainly by the application of electron microscopy for the study of changes in morphology of individual particles upon exposure to the beam of a transmission electron microscope (TEM). Surface (so-called "spot") techniques may identify the constituents on the outside of a particle (e.g. sulfate and nitrate) containing a core of a different composition, for example, sulfate or nitrate surrounding an elemental carbon or fly-ash particle core, [11] Differences in chemical composition of fine sulfate particles, and the presence of external and internal mixtures with sulfate with other primary particles, were elucidated in airborne data reported by Ferek et al. [12] Direct observation of nitrate coatings on the surface of sea-salt particles have been made, [13] based on a combination of spot tests and TEM as shown in Figure 1. The mechanism of formation presumably involves the reaction of nitric acid vapour with sea-salt aerosol, as indicated by laboratory exposures followed by bulk and individual particle analizes reported by Mamane and Gottlieb. [14]

Additional studies include the characterization of individual fly-ash particles from power plant emissions, [15] with both morphological and elemental abundance differences observed in particles emitted from coal-fired vis-à-vis oil-fired power plants. Recently differences in morphology and elemental composition have been observed in continental particles separated by size and hygroscopicity, [16] and marine aerosol particles have been characterized by a combination of individual particle (electron microprobe) and bulk analyses. [17]

X-ray crystallography has been used by several investigators to identify specific compounds found in atmospheric aerosols. For example, Tani et al. [18] used the Debye–Scherrer method to identify the chemical identity of samples collected with a modified Lundgren impactor, after transferring the particles to an X-ray capillary. Numerous compounds were identified including, in the 0.3–1.0 μm size range, silica, $CaCO_3$, $CaSO_4$, mixed ammonium–lead and ammonium–calcium sulfates, ammonium sulfate and bisulfate, mixed ammonium nitrate–sulfate salts, NaCl and letovicite [$(NH_4)_3H(SO_4)_2$], the 1 : 1 mixed salt of ammonium bisulfate and ammonium sulfate. This latter compound was found to be responsible for most of the acidity observed in the samples by infrared spectroscopy. Amounts of each identified phase were semi-quantitatively determined from the diffracted X-ray intensities, as "major", "medium", "minor" or "trace" constituents of the aerosol sample.

Other earlier studies in which atmospheric particulate constituents were identified by X-ray diffraction (XRD) include Brosset et al., [19] who described the use of a Guinier camera to analyze samples scraped from filters. This same technique was used by Biggins and Harrison, [20, 21] only with ultrasonic removal of particles to the mounting substrate. A diffractometer was used by Davis and co-workers [22–24] to identify samples on filters; likewise O'Connor and Jaklevic [25, 26] used a diffractometer to identify samples collected with a dichotomous sampler. It is the crystalline phases present at the time of analysis which are identified by XRD, of course, and there may have been chemical reactions which changed the composition. Also, it is difficult and time-consuming to obtain more than semi-quantitative information concerning the amounts of each phase present in the sample. However, it is a useful approach to confirming that identifiable crystalline phases are present in a sample, and not have to depend on inferences from bulk analyses of constituent ions.

4 THERMAL EVOLUTION/DECOMPOSITION METHODS

The thermal properties of common inorganic and organic constituents of atmospheric particles are known and vary widely. For example, the chemical form of sulfates in dry atmospheric particles may be as sulfuric acid, ammonium bisulfate, ammonium sulfate, the crystallographically distinct 1 : 1 mixture of sulfate and bisulfate, letovicite ($[NH_4]_3H[SO_4]_2$), or less soluble alkaline and alkaline earth sulfates sometimes called "refractory" sulfates. Organic and elemental carbon in airborne particles is commonly quantified by thermal methods [1, and references therein], but discussion thereof is beyond the scope of this chapter.

Sulfuric acid can be volatilized from aerosol particles or droplets by heating them to about 120–130 °C, [27] and the ammonium salts can likewise be volatilized by increasing the temperature to about 220 °C, [28] Other sulfate salts such as sodium or calcium sulfate are still non-volatile at the latter temperature. Early work in this area using both thermal and specific extraction techniques is discussed by Leahy et al. [29] and later work is reviewed comprehensively by Forrest and Newman [30] and for atmospheric strong acids by Tanner. [31] By heating the air transporting these sulfate-containing particles for sufficient duration, volatile species may be converted to the gas phase, rendering them not collectable by traditional particulate filters, and thereby distinguishing between sulfuric acid, ammonium sulfates and "refractory" sulfates. [32] A similar approach was reported by Huntzicker et al. [33] and has more recently been updated by Clarke [34], operating as a thermo-optic aerosol discriminator (TOAD) for in situ analysis of size-resolved aerosols. Application of this approach to aerosol studies in the remote marine troposphere has been reported. [35, 36]

In all thermal-based techniques, care must be taken to avoid converting the strong acid species, H_2SO_4, to a non-volatile salt by reaction with metal or glass surfaces. The volatilization products, ammonia and hydrated sulfuric acid vapour may be collected using a denuder or selective filter surface, or the quantity of sulfuric acid and volatile ammonium salts may be determined by difference. One limitation of this general approach is that ammonium sulfate and bisulfate have nearly identical volatilization temperatures, hence they cannot be distinguished by this approach, and the degree of neutralization of sulfuric acid by ambient ammonia cannot be assessed by thermal-based techniques when ammonium-to-sulfate molar ratios exceed one. A continuous sulfuric acid/sulfate monitor with enhanced sensitivity [37] suitable for ambient applications has been reported by Allen et al. [38]

Another approach to ammonium acid sulfate speciation was reported by Slanina et al., [39] in which ambient air, denuded of SO_2 at room temperature, is passed through two sequential CuO-coated at 120 and 220 °C, respectively, causing sulfuric acid and ammonium acid sulfates $[(NH_4)_x(H)_y(SO_4)_2]$ to be volatilized and deposited on the sequential denuder tubes. The deposited sulfates are released as SO_2 by heating the tubes to 800 °C, and their composition quantified by flame chemiluminescence produced in a H_2-rich flame (flame photometric detection). This apparatus has been automated and has a detection limit of about 0.1 $\mu g\,m^{-3}$ for sampling times of a few minutes.

A simpler, but procedurally similar approach was taken by Yoshizumi and Hoshi [40] to distinguish sodium and ammonium nitrates collected on filter media by heating the deposit in an oven at temperatures at which ammonium nitrate will volatilize but sodium nitrate will not. This approach has been criticized by Sturges and Harrison [41] based on the likelihood of chemical interactions and interferences when analysing real-world aerosols.

5 SPECTROSCOPIC METHODS FOR AEROSOL PARTICLES

Spectroscopic methods of analysis (principally FTIR methods) have been used widely in atmospheric chemistry, and especially in smog chamber studies of the formation of excess ozone and other photochemical oxidants in the presence of hydrocarbons, nitrogen oxides and sunlight. The advent of FTIR absorption technology and the availability of multiple-reflection, long-path-length cells have led to frequent use of this technology to identify products of photochemical oxidation of ozone precursors. In only a few cases has the FTIR technique been extended to the identification and quantitation of the constituents of atmospheric particulate matter using attenuated total reflectance (ATR) techniques and other technologies.

Cunningham, Johnson and co-workers developed a technique for airborne particles, analogous to the multiple-reflection cell in gaseous measurements, using an ATR cell to collect airborne particles [42, 43] with multiple reflectances to increase sensitivity. [44, 45] This approach was used mainly to detect inorganic constituents—sulfate, bisulfate and nitrate. Frequency shifts in the

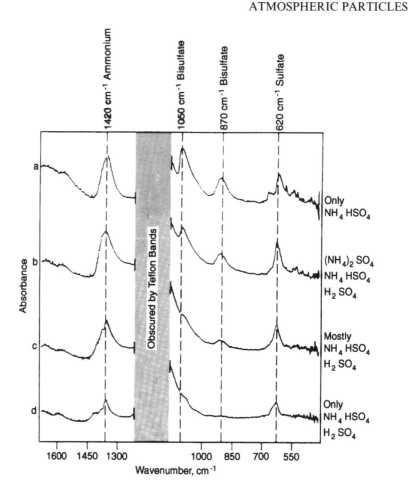

Figure 2. Absorbance spectra of particulate samples corresponding to different sulfate speciation: (a) only ammonium bisulfate; (b) coexisting ammonium sulfate, sulfuric acid and ammonium bisulfate; (c) mostly ammonium sulfate and sulfuric acid; (d) only ammonium sulfate and sulfuric acid

sulfate absorbance band were used to evaluate, at least semi-quantitatively, the presence of sulfate more acidic than ammonium bisulfate (i.e. sulfuric acid droplets). Various refinements have been made in recent years; see, for example, Figure 2 for absorbance spectra from. [46] These refinements include a method for the direct, quantitative measurement by FTIR absorption spectroscopy of ammonium bisulfate collected from ambient air containing other sulfates on Teflon filters. [46–49] This approach has been used alongside more conventional fine particle collection on Teflon filters, with extraction and single-ion analyses of sulfate, nitrate, ammonium and H^+, and the results are in reasonably good agreement. [46]

Another approach has also used infrared absorption to identify constituents of size-segregated samples collected with a low-pressure impactor. [50–54] Both organic and inorganic constituents have been identified and in some cases quantified. This technique is limited in application due to the slow sampling rate of the impactor used, but the approach has more recently been extended to samples size segregated at higher flow rates. Other related approaches include the use of FTIR for chromatographic extracts of atmospheric aerosols [55], the application of FTIR spectroscopy to individual particles and clusters [56], and the use of diffuse reflectance FTIR for speciation of atmospheric particles by functionality. [57]

6 GAS CHROMATOGRAPHY-MASS SPECTROMETRY (GC-MS) AND RELATED METHODS

A wide variety of analyses have been done on atmospheric particulate matter using gas chromatography for separation and quantitation of individual constituents (principally organic species) and usually mass spectrometry for species identification. This approach received a major boost when a wide variety of capillary columns became available in the mid to late 1980s, since atmospheric particulate samples may contain hundreds of organic species and usually require the resolution of capillary columns. Often it has been possible to quantify separated compounds using a mass-selective detector. More reliable identification is possible if the GC effluent is analysed by infrared spectroscopy (see section 5), but the limit of detection continues to be of the order of fivefold less by infrared detection than with a mass-selective detector.

The most fruitful application of this approach has come from the efforts of Simoneit, Mazurek and co-workers to identify and quantitate organic aerosol species which are tracers of particulate types of combustion of biomass and anthropogenic hydrocarbon emissions. In Mazurek and Simoneit [58] a method is described for the comprehensive analysis of solvent- extractable lipids in airborne particulate samples from a variety of locations in the western USA, Africa and Australia. A means of distinguishing organic particulate material (e.g. fatty acids) of biogenic origin from that of man-made origin is described, based on the relative amounts of odd-and even-carbon compounds in homologous series. This approach was expanded in Mazurek et al. [59] to include high-resolution GC quantification of concentrated, dual-solvent-extracted atmospheric organic mixtures, with identification using HRGC-MS. Innovative use of a perdeuterated mixture of compounds not found in ambient air to assess percentage recovery of different compound types has been added to allow improved determination of low-microgram amounts of natural and man-made organic species in particulate samples taken in the Los Angeles basin.

Refinement of this approach led to the detailed determination of source contributions of contemporary biological material to visibility-reducing particles in the remote, arid south-west USA [60]. Elaborate blank procedures allowed for the contributions from the sampling apparatus and from the analysis procedures to be accurately accounted for at the microgram-per-filter level. Then biogenic aerosol markers—n-alkanoic acids, n-alkanes and conifer resin-derived species—could be accurately quantified and their seasonal variation determined in these samples taken from the *S*ubregional *C*ooperative *E*lectric Utility, Department of Defense, *N*ational Park Service and *E*nvironmental Protection Agency *S*tudy (SCENES).

The HRGC approach was used by Hildemann *et al.* [61] to characterize organic aerosol emissions from individual combustion sources in the Los Angeles basin, a prerequisite to accurate allocation of the dozen or more source types contributing to primary organic emissions in the basin. By comparing several distinguishing features of the HRGC data from acidic and neutral fractions, total vs eluatable organics, unresolved (petroleum-based) to resolved peak ratios, and organic mass distributions (n-alkane equivalents), several distinct emission sources could be identified: oil-fired boilers, motor vehicles, roofing tar, tire dust, fireplace wood combustion, cigarette smoking, meat-cooking, natural gas home appliances, urban leaf detritus and paved road dust. Further work using HRGC of ambient samples from four sites in the LA basin [62] has shown that the observed seasonal variation in higher molecular weight neutral organics derived from motor vehicle exhaust, tire dust, road dust and vegetative detritus is entirely driven by changes in meteorology and atmospheric dilution. By combining HRGC and GC-MS analysis of samples from specific sources of primary organic aerosol, Rogge and colleagues [63–65] developed sets of molecular marker compounds which are useful in quantifying the contributions of three major sources: meat-cooking operations; non- catalyst and catalyst-equipped automobiles and heavy-duty diesel trucks; and road dust, tire debris and (automobile) brake lining dust.

Although most methods development to specifically identify and quantify aerosol organic species has used solvent extraction, there has been some effort to use thermodesorption of particle-bound organics in combination with GC-MS for this purpose. For example, Helmig *et al.* [66] analyzed particle-phase organics collected on glass fibre filters from a forest atmosphere by thermodesorption–GC-MS, and identified a series of long-chain alkanes, fatty acids and esters, (man-made) phthalate esters, and a few specific polynuclear aromatic hydrocarbons (PAHs). This technique is based on previous work by Schuetzle *et al.* [67] and Cautreels and van Cauwenberghe, [68] addressing specific sources of particle-phase organics, and work of a more survey nature reported by, for example, Schuetzle *et al.* [69] and Cronn *et al.* [70] In this approach, it is simpler to identify

constituents of a wide range of polarity; however, there may be significant
analytical problems for particulate organic species of limited thermal stability.

7 IDENTIFICATION METHODS BASED ON WATER UPTAKE PROPERTIES

Information on the nature and amounts of chemical constituents in atmospheric
aerosols is also important in studies of aerosol growth and scavenging processes,
since the rate of growth of atmospheric fine particles with increasing humidity is
highly dependent on their chemical nature. [2] Early on, Charlson *et al.* [71]
reported a method in which the scattering coefficient of ambient fine particles is
measured as a function of relative humidity using a nephelometer with measured
scattering normalized to that of the "dry" particle (30% RH). The chemical
composition of sulfate-dominated particles was inferred by addition of ammonia
(intermittently) before the light-scattering measurement. Cases with significant
sulfuric acid content changed from a hygroscopic dependence of scattering to the
deliquescent curve with a sharp increase near 80% that is typical of ammonium
sulfate. Aerosol particles of ammonium bisulfate exhibited a less striking change
upon ammonia addition. Future work refined the "humidographic" technique,
and some short-time resolution changes in composition could be identified by
the apparatus shown in Figure 3. [72] However, for multi-component internal

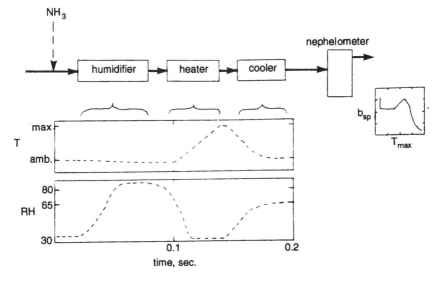

Figure 3. Block diagram of thermidograph. Data are scattering vs peak temperature.
Note that the RH is controlled to be above 80% at the inlet to the heater and near 65% at
the inlet of the nephelometer

mixtures, or for fine particles in which the contribution of sulfate to fine particle mass was lower, this technique was less useful.

Indeed, in some recent studies, fine particles of two or more chemical compositions have been found in ambient air, one of which type grows rapidly by accretion of water at high humidities, and a second type which grows much less rapidly. [3, 4] Hence, since particle growth, incorporation into clouds and dry deposition are the major loss mechanisms for atmospheric particles, the knowledge of their chemical composition derived from light-scattering techniques may be important in determining both their atmospheric lifetimes as well as their effect on atmospheric visibility and the global energy balance, even though the method is of limited utility for chemical speciation.

8 THE PROBLEM OF INTERNAL VS EXTERNAL MIXTURES

"External mixture", in the context of airborne particles in the atmosphere, refers to non-homogeneity in the chemical composition of individual particles. That is, particles of, for example, soot coated with trace elements and PAHs, may be mixed with ammonium sulfate particles in the gas phase of the aerosol such that each cubic metre of air contains the same mass of carbon, sulfate, etc., but each particle in the aerosol does not have the same chemical composition (or necessarily the same size distribution). In contrast, an internally mixed aerosol contains equal masses of chemical species per cubic metre of air *and* each particle has approximately the same chemical composition. Obviously, whether an aerosol is internally or just externally mixed has major implications for chemically speciated analysis; for example, sulfuric acid- and ammonium sulfate-containing particles may coexist for a time in an external mixture, but will react rapidly to form principally ammonium bisulfate in an internally mixed aerosol at relative humidities above about 40%. The existence of internal *vis-à-vis* external mixtures also strongly affects the optical properties of atmospheric particles, their behaviour in ambient environments in which temperature and relative humidity are changing, and even their physical properties (e.g. their nucleation, coagulation and deposition rates).

Conventional wisdom has been that most aerosols in the fine particle size range are internal mixtures, since the constituent particles are formed by nucleation and growth of secondary particles rather than by direct emission from primary sources. In contrast, most of the particles in the coarse particle size range are formed by mechanical processes, and are believed to be chemically distinct although often externally mixed. This view is a simplified one but widely used in models of aerosol behaviour. Chemical (elemental) analyses of aerodynamically sized particles have for at least two decades shown major differences between the size distributions of different elements (cf. Leaderer *et al.* [73]). Nevertheless, fine particles were often considered to be of about the same chemical composition (internally mixed) for modelling purposes, until the discovery that nitrate, formed in the fine particle regime by the reverse of reaction (1), was frequently found in the same size regime (mostly in coarse particles) as sea-salt aerosol. [74–76] This is discussed further below.

Recent efforts to chart the growth of individual particles or narrow size ranges of ambient particles as a function of relative humidity have shown that even in the fine particle range, particles frequently can have widely varying chemical compositions and grow at different rates as the relative humidity increases within the normal ambient range. [2, 3, 77] The work of John et al. [78] suggests that there are two fine particle modes, a long-normal condensation mode centred at 0.2 μm and a droplet mode at 0.7 μm, and the latter mode, containing internally mixed ammonium sulfate and nitrate, could have grown from the former by addition of water (and sulfate). In contrast, McMurry and Stolzenburg [2] found clear evidence of external mixtures in fine particles during the same study. This controversy has particular relevance when comparing the bulk composition of particles based on ionic analyses with that based on chemically speciated analyses. Fine aerosol particles are known to absorb water due to the presence of soluble inorganic salts such as ammonium sulfate, and recent work has shown that the water content of fine aerosols can be influenced (increased or decreased) by the presence of organic species in the particles. [79] Clearly, it is an oversimplification to assume that all fine particles have the same chemical composition, and inferences of chemical composition based on bulk analyses must be made with great caution.

9 COMPARISON OF CHEMICAL COMPOSITION FROM DIRECT METHODS WITH THAT INFERRED FROM BULK ANALYSES

There is a paucity of fine particle studies from which comparisons of data from bulk analyses of ions or classes of compounds can be made with quantitative information on individual chemical species. There are several good reasons why this is so: particles usually consist of both inorganic and organic species mixed with a dynamically varying amount of water; particles from both internal and external mixtures may be present; a wide variety of organic species (man-made and natural) may be mixed with elemental (sooty) carbon; and particle composition may vary with particle size. Nevertheless, there are a few comparisons from the literature which can be used to demonstrate the value of this approach.

9.1 SPECIATION OF H_2SO_4 AND OTHER ACID SULFATES BY VARIOUS TECHNIQUES

As noted above, the measurement of the strong acid content of atmospheric aerosols and the determination of which species are responsible for the measured

H^+ (sulfuric acid, bisulfates and/or other species) has remained an active research area. Thermal methods of quantifying H_2SO_4 in airborne particles were reviewed above and the problems with distinguishing ammonium bisulfate from $NH_4)_2SO_4$ noted. Methods for selective, sequential extraction of sulfuric acid and ammonium bisulfate have been reported, but with limited success. [10, 29, 80] Whereas extraction of pure sulfuric acid on inert filters with benzaldehyde was demonstrated, the application of this approach to ambient samples is problematical because of problems with solvent impurities and various inter-ferences. [29, 81] In general, these and later studies support the view that acidic sulfates are most frequently found as internal mixtures since significant amounts of sulfuric acid are present only when the ammonium-to-sulfate ratio exceeds one. Cases evaluated were those in which the sum of ammonium and H^+ equivalent concentrations were approximately equal to the sum of sulfate and nitrate equivalent concentrations; nitrate levels are much less than sulfate under conditions in which significant strong acidity is present. Therefore, the principal sources of acidity in aerosols are deduced to be ammonium bisulfate and/or the 1 : 1 mixed salt of ammonium sulfate and ammonium bisulfate, letovicite. Direct measurements of sulfuric acid and ammonium sulfate coexisting in fine particles are difficult and relatively rare; however, Sheridan *et al.* [82] use a combination of electron microscopy and XRF based elemental analysis to distinguish indivi-dual fine particles which are less than, or more acidic than NH_4HSO_4, and to demonstrate that the sulfate particles were distinct (externally mixed) from carbonaceous particles. For application of FTIR in this regard, see next para-graph and reference. [46]

Cunningham *et al.* [83] report data from a network of five sites in which sulfate, ammonium, nitrate and the molar ratio, H^+/SO_4^{2-}, from the 0.3–1.0 µm stage of a Lundgren impactor were measured by FTIR spectroscopy. Compar-ison of the molar deficiency of cations, considering the aerosol to be a mixture of ammonium acid sulfates and ammonium nitrate, with the measured acid-to-sulfate ratios gave reasonable agreement. Of course, the coexistence of ammonium nitrate with acidic ammonium sulfates in the same particles is not likely, but the comparison was made with diurnally and seasonally averaged data.

9.2 COMPOSITION AND SIZE DISTRIBUTION OF ATMOSPHERIC NITRATE

Much effort has been expended on the development of equilibrium models predicting the formation of ammonium nitrate from gaseous precursors, NH_3 and HNO_3. Indirect evidence is compelling that fine particle nitrate is predominantly in the chemical form of ammonium nitrate or ammonium nitrate–sulfate mixtures, but the direct measurement of ammonium nitrate in atmospheric fine particles remains elusive. Direct observations appear to be limited to the X-ray diffraction data reported by Harrison and Sturges [84] and the thermal speciation data of Ruprecht and Sigg. [85]

Observations of the amount and chemical nature of atmospheric nitrates have proved difficult not only because of the ammonium nitrate–nitric acid, ammonia equilibrium noted above, but also because nitric acid may react with atmospheric particulate matter to change both the composition and the size distribution. There is evidence that nitrate present in fine particles as ammonium nitrate can be transferred to larger particles because it is in quasi-equilibrium with nitric acid vapour and the latter can react with, for example, sea-salt aerosol particles to form "non-volatile" sodium nitrate, [78] thereby releasing HCl(g). [86] This process has been the subject of laboratory studies as well, and HNO_3-particle interactions with coarse soil dust may also explain attributes of continental aerosols. [87] Therefore, direct evidence of the presence of ammonium nitrate in fine particles whose inorganic ion content (sulfate, nitrate, ammonium, H^+, at minimum) is known is not generally available but must be inferred from bulk ion data and their agreement with equilibrium models. [88]

9.3 ORGANIC CARBON VS SUM OF SPECIATED ORGANICS IN FINE AEROSOLS

As noted above, the determination of "bulk" organic and elemental (black) content of fine aerosols is now widespread, and the details of its application by thermal speciation methods have been reviewed by Chow et al. [89] Attempts to characterize the species which make up the observed organic carbon content have proved more challenging. Generally, no more than half the mass of organic carbon has been characterized by species or chemical functional group in atmospheric particle samples. As discussed above, in a few cases the chemical characterization has identified the presence of molecular markers of sources which can be used to allocate the particle mass and its organic fraction to a definable series of emission sources. A particularly informative breakdown of the organic fraction and its seasonal variability is given by Rogge et al. [90] for urban Los Angeles aerosols. For example, the annual mean of total fine particle mass for a secondary aerosol-dominated site was about $42 \, \mu g \, m^{-3}$, of which about 15% was "organic carbon", and about 7% elemental carbon. Of the organic carbon ($6.2 \, \mu g \, m^{-3}$), 60% consisted of elutable organics (on the HRGC column used), while 40% consisted of non-extractable, non-elutable organics. Only about 28% of the elutable organic compounds ($1.07 \, \mu g \, m^{-3}$) were resolved by HRGC, and of these, specific n-alkanes, n-alkanoic and n-alkenoic acids, aromatic polycarboxylic acids, diterpenoid acids and PAHs totalling about $760 \, ng \, m^{-3} (\approx 70\%)$ were actually identified by GC-MS (see Table 1). So we see that the challenge of arriving at a full characterization of organic species in atmospheric fine particle samples remains, both because of the complexity of the mixtures involved, but also

Table 1. Fine aerosol emission rates for single organic compounds from non-catalyst and catalyst automobiles and from heavy-duty diesel trucks

| | Emission rates (μg km^{-1}) | | | |
| | Autos | | Heavy-duty diesel trucks | Compd ID* |
	Non-catalyst	Catalyst		
Polycyclic aromatic hydrocarbons				
Phenanthrene	17.3	0.88	12.2	A
Anthracene	5.1	0.11	1.6	A
Methylphenanthrenes, -anthracenes	42.2	1.9	31.3	B
BDimethylphenanthrenes, -anthracenes	75.5	3.0	56.9	B
Fluoranthene	48.3	2.0	13.0	A
Pyrene	31.0	2.5	22.6	A
Benzacenaphthylene	16.5	0.77	2.9	B
2-Phenylnaphthalene	11.9	0.57	3.5	B
2-Benzylnaphthalene	4.3	0.72	nd	
BMethylfluoranthenes, -pyrenes	106.9	4.2	12.8	B
Benzo[a]fluorene/ benzo[b]fluorene	31.0	0.72	1.9	A
Benzo[ghi]fluoranthene	24.7	1.3	6.9	B
Cyclopenta[cd]pyrene	49.3	1.7	1.4	A
Benzo[a]anthracene	73.8	1.9	3.6	A

* A = positive; B=probable.

because of the presence of mixtures of multifunctional, very polar compounds at trace levels.

10 CONCLUSIONS AND FUTURE PROSPECTS

Much of the effort to monitor aerosols in the atmosphere for regulatory purposes will continue to focus on bulk analyses of major ionic species which are implicated in negative health or ecosystem effects or in visibility impairment. However, as information continues to accumulate that external mixtures of aerosol particles are common, even in the fine particle regime, the need for individual compound identification and quantitation will increase. Improvements in methods for identification of species in individual particles, in some cases with *in situ* methods, [91] will prove to be important. Some of the methods for speciation developed 10–25 years ago have not been widely used, but may have some potential in specialized applications. Of particular note are methods for automated speciation of SO_2 and various sulfate compounds with short time resolution (cf. Slanina *et al.* [39]) which could be refined and simplified using advances in computer-based data logging and analysis methods developed over the last 10 years. A direct approach to quantifying acidic sulfate species (sulfuric acid and ammonium bisulfate) in particles has been proposed, combining the FTIR spectroscopic measurement of individual acidic sulfate species with the immediate analysis of strong acidity of extracts from the same filter sample, thus avoiding many of the ambiguities of parallel filter measurement protocols.

Additional analytical needs which require further research activities include the analysis of multifunctional hazardous and/or toxic air pollutants, many in the particle phase or distributed between particle and gas phase. For example, metallic or trace element compounds emitted from combustion sources may have widely differing toxicity depending on the oxidation state of the element and the chemical properties of the actual species present in ambient particles. Means of efficiently evaluating the distribution of pollutant species (e.g., metal oxides or halides from emitted species) in different oxidation states and between particulate and gaseous phases will need to be improved and refined. The health effects of toxic organic species may be misrepresented based on epidemiological studies in which the exposure to the particle-phase portion of phase-distributed species was not accurately measured. The goal for measurement science in all these studies is to define and quantify the actual species and its concentration present in ambient air at the time and place of exposure.

GLOSSARY

External mixture   A chemically non-homogeneous mixture of particles in air, such that each cubic metre of air contains the same mass of chemically distinct species (carbon, sulfate, etc.), but individual particles in the aerosol do not have the same chemical composition or necessarily the same size distribution

Fine particle mass   Mass of atmospheric particles in the size range less than a few μm in diameter, as distinguished from the mass of larger particles (coarse particle mass). The division between coarse and fine particles, based on both their settling velocities and their abundance in atmospheric particles, is controversial but generally set in the range of 1–3.5 μm

Internal mixture   A chemically homogeneous mixture of particles in air containing equal masses of chemical species per cubic metre of air *and* with each particle having about the same chemical composition

Speciation   Differentiation and quantitation of the components of a mixture by chemical species.

LIST OF SYMBOLS

FTIR   Fourier transform infrared (spectroscopy)
GC-MS   Gas chromatography/mass spectrometry
HRGC   High resolution (capillary) gas chromatography
PM2.5   Particulate mass in particles less than 2.5 μm in diameter
TEM   Transmission electron microscopy
XRD   X-ray diffraction spectrometry
XRF   X-ray Flourescence Spectrometry

REFERENCES

1. J. C. Chow, Measurement methods to determine compliance with ambient air quality standards for suspended particles, *J. Air & Waste Manage. Assoc.* **45**, 320 (1995).
2. P. H. McMurry and M. R. Stolzenburg, On the sensitivity of particle size to relative humidity for Los Angeles aerosols, *Atmos. Environ.* **23**, 497 (1989).
3. X. Q. Zhang, P. H. McMurry, S. V. Hering and G. S. Casuccio, Mixing characteristics and water content of submicron aerosols measured in Los Angeles and at the Grand Canyon, *Atmos. Environ.* **27A**, 1593 (1993).
4. M. L. Pitchford and P. H. McMurry, Relationship between measured water vapor growth and chemistry of atmospheric aerosol for Grand Canyon, Arizona, in winter 1990, *Atmos. Environ.* **28**, 837 (1994).
5. C. Xhoffer, L. Wouters, P. Artaxo, A. van Put and R. van Grieken, Characterization of individual environmental particles by beam techniques, in *Environmental Particles*, Vol. 1, ed. by J. Buffle and H. P. van Leeuwen, Lewis Publishers, Inc., Chelsea, MI (1992), p. 107.
6. A. G. Allen, R. M. Harrison and J. W. Erisman, Field measurements of the dissociation of ammonium nitrate and ammonium chloride aerosols, *Atmos. Environ.* **23**, 1591 (1989).
7. R. M. Harrison and I. M. Msibi, Validation of techniques for fast response measurement of HNO_3 and NH_3 and determination of the $[NH_3][HNO_3]$ product, *Atmos. Environ.* **28**, 247 (1994).
8. A. S. Wexler and J. H. Seinfeld, Analysis of aerosol ammonium nitrate: departures from equilibrium during SCAQS, *Atmos. Environ.* **26A**, 579 (1992).
9. R. L. Tanner and R. M. Harrison, Acid–base equilibria of aerosols and gases in the atmosphere, in *Environmental Particles*, Vol. 1, ed. by J. Buffle and H. P. Van Leeuwen, Lewis Publishers, Inc., Chelsea, MI (1992), p. 75.
10. D. J. Eatough, A. Wadsworth, D. A. Eatough, J. W. Crawford, L. D. Hansen and E. A. Lewis, A multiple-system, multi- channel diffusion denuder sampler for the determination of fine particulate organic material in the atmosphere, *Atmos. Environ.* **27A**, 1213 (1993).
11. Y. Mamane, in *Proc. Symp. Heterogeneous Processes in Source-Dominated Atmospheres*, ed. by T. Novakov, Report LBL-20261, Lawrence Berkeley Laboratory, Berkeley, CA (1985), p. 44.
12. R. J. Ferek, A. L. Lazrus and J. W. Winchester, Electron microscopy of acidic aerosols collected over the northeastern United States, *Atmos. Environ.* **17**, 1545 (1983).

13. Y. Mamane and M. Mehler, On the nature of nitrate particles in a coastal urban area, *Atmos. Environ.* **21**, 1989 (1987).
14. Y. Mamane and J. Gottlieb, Nitrate formation on sea-salt and mineral particles—a single particle approach, *Atmos. Environ.* **26A**, 1763 (1992).
15. Y. Mamane, J. L. Miller and T. G. Dzubay, Characterization of individual fly ash particles emitted from coal- and oil-fired power plants, *Atmos. Environ.* **20**, 2125 (1986).
16. P. H. McMurry, M. Litchy, P.-F. Huang, Z. Cai, B. J. Turpin, W. D. Dick and A. Hanson, Elemental composition and morphology of individual particles separated by size and hygroscopicity with the TDMA, *Atmos. Environ.* **30**, 101 (1996).
17. J. R. Anderson, P. R. Buseck, T. L. Patterson and R. Arimoto, Characterization of the Bermuda tropospheric aerosol by combined individual-particle and bulk-aerosol analysis, *Atmos. Environ.* **30**, 319 (1996).
18. B. Tani, S. Siegel, S. A. Johnson and R. Kumar, X-Ray diffraction investigation of atmospheric aerosols in the 0.3–1.0 μm aerodynamic size range, *Atmos. Environ.* **17**, 2277 (1983).
19. C. Brosset, K. Andreasson and M. Ferm, The nature and possible origin of acid particles observed at the Swedish west coast, *Atmos. Environ.* **9**, 631 (1975).
20. P. D. E. Biggins and R. M. Harrison, The identification of specific chemical compounds in size-fractionated atmospheric particulates collected at roadside sites, *Atmos. Environ.* **13**, 1213 (1979).
21. P. D. E. Biggins and R. M. Harrison, Characterization and classification of atmospheric sulfates, *J. Air Pollut. Contr. Assoc.* **29**, 838 (1979).
22. B. L. Davis and N. K. Cho, Theory and application of X- ray diffraction compound analysis to high-volume filter samples, *Atmos. Environ.* **11**, 73 (1977).
23. B. L. Davis, Additional suggestions for X-ray quantitative analysis of high-volume filter samples, *Atmos. Environ.* **12**, 2403 (1978).
24. B. L. Davis and L. R. Johnson, On the use of various filter substrates for quantitative particulate analysis by X-ray diffraction, *Atmos. Environ.* **16**, 273 (1982).
25. B. H. O'Connor and J. M. Jaklevic, X-ray diffractometry of airborne particulates deposited on membrane filters, *X-Ray Spectrometry* **9**, 60 (1980).
26. B. H. O'Connor and J. M. Jaklevic, Characterization of ambient aerosol particulate samples from the St. Louis area by X-ray powder diffractometry, *Atmos. Environ.* **15**, 1681 (1981).
27. L. W. Richards, K. R. Johnson and L. S. Shepard, *Sulfate Aerosol Study*, Report AMC-8000.13 FR, Rockwell International Air Monitoring Center, Newbury Park, CA (1978).
28. W. G. Cobourn, R. B. Husar and J. D. Husar, Continuous *in situ* monitoring of ambient particulate sulfur using flame photometry and thermal analysis, *Atmos. Environ.* **12**, 89 (1978).
29. D. Leahy, R. Siegel, P. Klotz and L. Newman, The separation and characterization of sulfate aerosol, *Atmos. Environ.* **9**, 219 (1975).
30. J. Forrest and L. Newman, Determination and speciation of ambient particulate sulfur compounds, in *Methods of Air Sampling and Analysis*, 3rd Ed., ed. by J. P. Lodge, Jr., Lewis Publishers, Inc., Chelsea, MI (1989), p. 715.
31. R. L. Tanner, The measurement of strong acids in atmospheric samples, in *Methods of Air Sampling and Analysis*, 3rd Ed., ed. by J. P. Lodge, Jr., Lewis Publishers, Inc., Chelsea, MI (1989), p. 703.
32. R. L. Tanner, T. D'Ottavio, R. W. Garber and L. Newman, Determination of ambient aerosol sulfur using a continuous flame photometric detection system. I. Sampling system for aerosol sulfate and sulfuric acid, *Atmos. Environ.* **14**, 121 (1980).

33. J. J. Huntzicker, R. S. Hoffman and C. S. Ling, Continuous measurement and speciation of sulfur-containing aerosols by flame photometry, *Atmos. Environ.* **12**, 83 (1978).
34. A. D. Clarke, A thermo-optic technique for in situ analysis of size-resolved aerosol physico-chemistry, *Atmos. Environ.* **25A**, 635 (1991).
35. A. D. Clarke, N. C. Ahlquist and D. S. Covert, The Pacific marine aerosol: evidence for natural acid sulfates, *J. Geophys. Res.* **92**, 4179 (1987).
36. A. D. Clarke, J. N. Porter, F. Valero and P. Pilewski, Aerosol measurements and optical extinction in the troposphere, in *Abstracts of the 4th Intern. Aerosol Conf.*, Los Angeles, CA (1994), pp. 1046–1047.
37. T. D'Ottavio, R. Garber, R. L. Tanner and L. Newman, Determination of ambient aerosol sulfur using a continuous flame photometric detection system. II. The measurement of low-level sulfur concentrations under varying atmospheric conditions, *Atmos. Environ.* **15**, 197 (1981).
38. G. A. Allen, W. A. Turner, J. M. Wolfson and J. D. Spengler, Description of a continuous sulfuric acid/sulfate monitor, in *Proc. 4th Annual National Symp. on Recent Advances in Pollutant Monitoring of Ambient Air and Stationary Sources*, Raleigh, NC, US Environmental Protection Agency Rpt. EPA-600/9-84-019 (1984), p. 140.
39. J. Slanina, C. I. F. Schoonebeek, D. Klockow and R. Niessner, Determination of ambient aerosol sulfuric acid and ammonium sulfates by means of a computer-controlled thermodenuder system, *Anal. Chem.* **57**, 1955 (1985).
40. K. Yoshizumi and A. Hoshi, Size distributions of ammonium nitrate and sodium nitrate in atmospheric aerosols, *Environ. Sci. Technol.* **19**, 258 (1985).
41. W. T. Sturges and R. M. Harrison, Thermal speciation of atmospheric nitrate and chloride: a critical evaluation, *Environ. Sci. Technol.* **22**, 1305 (1988).
42. P. T. Cunningham, S. A. Johnson and R. T. Yang, Variations in chemistry of airborne particulate material with particle size and time, *Environ. Sci. Technol.* **8**, 131 (1974).
43. P. T. Cunningham and S. A. Johnson, Spectroscopic observation of acid sulfate in atmospheric particulate samples, *Science*, **191**, 77 (1976).
44. S. A. Johnson, R. Kumar and P. T. Cunningham, Airborne detection of acidic sulfate aerosol using an ATR-impactor, *Aerosol Sci. Technol.* **2**, 401 (1983).
45. S. A. Johnson, G. T. Reedy and R. Kumar, Presented at the 8th Annual AAAR Meeting, Reno, NV, 9–13 October 1989, *Abstracts*, p. 329.
46. W. A. McClenny, K. J. Krost, E. H. Daughtrey, Jr., D. D. Williams and G. A. Allen, Speciation of ambient sulfate particulate using FT-IR-based absorption to complement wet chemical and thermal speciation measurements, *Appl. Spectrosc.* **48**, 706 (1994).
47. W. A. McClenny, J. W. Childers, R. Rühl and R. A. Palmer, FTIR transmission spectrometry for the nondestructive determination of ammonium and sulfate in ambient aerosols collected on Teflon filters, *Atmos. Environ.* **19**, 1891 (1985).
48. K. J. Krost and W. A. McClenny, Fourier transform infrared spectrometric analysis for particle-associated ammonium sulfate, *Appl. Spectrosc.* **46**, 1737 (1992).
49. K. J. Krost and W. A. McClenny, FT-IR transmission spectroscopy for quantitation of ammonium bisulfate in fine-particulate matter collected on Teflon filters, *Appl. Spectrosc.* **48**, 702 (1994).
50. M. Dangler, S. Burke, S. Hering, and D. Allen, A direct method of identifying functional groups in size segregated atmospheric aerosols, *Atmos. Environ.* **21**, 1001 (1987).
51. S. Brown, M. C. Dangler, S. R. Burke, S. V. Hering and D. T. Allen, Direct Fourier transform infrared analysis of size-segregated aerosols: results from the Carbonaceous Species Methods Intercomparison Study, *Aerosol Sci. Technol.* **12**, 172 (1990).

52. T. Pickle, D. T. Allen and S. E. Pratsinis, The sources and size distributions of aliphatic and carbonyl carbon in Los Angeles aerosol, *Atmos. Environ.* **24A**, 2221 (1990).

53. E. J. Palen, D. T. Allen, S. N. Pandis, S. E. Paulson, J. H. Seinfeld and R. C. Flagan, Fourier transform infrared analysis of aerosol found in the photo-oxidation of isoprene and β-pinene, *Atmos. Environ.* **26A**, 1239 (1992).

54. E. J. Palen, D. T. Allen, S. N. Pandis, S. Paulson, J. H. Seinfeld and R. C. Flagan, Fourier transform infrared analysis of aerosol found in the photo-oxidation of 1-octene, *Atmos. Environ.* **27A**, 1471 (1993).

55. R. J. Gordon, N. J. Trivedi, B. P. Singh and E. C. Ellis, Characterization of aerosol organics by diffuse reflectance Fourier transform infrared spectroscopy, *Environ. Sci. Technol.* **22**, 672 (1988).

56. R. Kellner and H. Malissa, Fourier transform infrared microscopy—a tool for speciation of impactor-sampled single particles or particle clusters, *Aerosol Sci. Technol.* **10**, 397 (1989).

57. I. R. Kaplan and R. J. Gordon, Non-fossil-fuel fine-particle organic carbon aerosols in southern California determined during the Los Angeles aerosol characterization and source apportionment study, *Aerosol Sci. Technol.* **21**, 343 (1994).

58. M. A. Mazurek and B. R. T. Simoneit, in *Identification and Analysis of Organic Pollutants in Air*, ed. by L. H. Keith, Ann Arbor Sci., Boston (1984), pp. 353–370.

59. M. A. Mazurek, B. R. T. Simoneit, G. R. Cass and H. A. Gray, Quantitative high-resolution gas chromatography/mass spectrometry analizes of carbonaceous fine aerosol particles, *Intern. J. Environ. Anal. Chem.* **29**, 119 (1987).

60. M. A. Mazurek, G. R. Cass and B. R. T. Simoneit, Biologic input to visibility-reducing aerosol particles in the remote arid southwestern United States, *Environ. Sci. Technol.* **25**, 684 (1991).

61. L. M. Hildemann, M. A. Mazurek, G. R. Cass and B. R. T. Simoneit, Quantitative characterization of urban sources of organic aerosol by high-resolution gas chromatography, *Environ. Sci. Technol.* **25**, 1311 (1991).

62. L. M. Hildemann, M. A. Mazurek, G. R. Cass and B. R. T. Simoneit, Seasonal trends in Los Angeles ambient organic aerosol observed by high-resolution gas chromatography, *Aerosol Sci. Technol.* **20**, 303 (1994).

63. W. F. Rogge, L. M. Hildemann, M. A. Mazurek, G. R. Cass and B. R. T. Simoneit, Sources of fine organic aerosol. I. Charbroilers and meat cooking operations, *Environ. Sci. Technol.* **25**, 1112 (1991).

64. W. F. Rogge, L. M. Hildemann, M. A. Mazurek, G. R. Cass and B. R. T. Simoneit, Sources of fine organic aerosol II. Noncatalyst and catalyst-equipped automobiles and heavy-duty diesel trucks, *Environ. Sci. Technol.* **27**, 636 (1993).

65. W. F. Rogge, L. M. Hildemann, M. A. Mazurek, G. R. Cass and B. R. T. Simoneit, Sources of fine organic aerosol. III. Road dust, tire debris, and organometallic brake lining dust: roads as sources and sinks, *Environ. Sci. Technol.* **27**, 1892 (1993).

66. D. Helmig, A. Bauer, J. Muller and W. Klein, Analysis of particulate organics in a forest atmosphere by thermodesorption GC/MS, *Atmos. Environ.* **24A**, 179 (1990).

67. D. Schuetzle, A. L. Crittenden and R. J. Charlson, Application of computer controlled high resolution mass spectrometry to the analysis of air pollutants, *J. Air Pollut. Contr. Assoc.* **23**, 704 (1973).

68. W. Cautreels and K. van Cauwenberghe, Determination of organic compounds in airborne particulate matter by gas chromatography-mass spectrometry, *Atmos. Environ.* **10**, 447 (1976).

69. D. Schuetzle, D. Cronn, A. L. Crittenden and R. J. Charlson, Molecular composition of secondary aerosol and its possible origin, *Environ. Sci. Technol.* **9**, 838 (1975).

70. D. Cronn, R. J. Charlson, R. L. Knights, A. L. Crittenden and B. R. Appel, A survey of the molecular nature of primary and secondary components of particles in urban air by high-resolution mass spectrometry, *Atmos. Environ.* **11**, 929 (1977).
71. R. J. Charlson, A. H. Vanderpol, D. S. Covert, A. P. Waggoner and N. C. Ahlquist, Sulfuric acid–ammonium sulfate aerosol: optical detection in the St. Louis region, *Science* **184**, 156 (1974).
72. T. V. Larson, N. C. Ahlquist, R. E. Weiss, D. S. Covert and A. P. Waggoner, Chemical speciation of H_2SO_4–$(NH_4)_2SO_4$ particles using temperature and humidity controlled nephelometry, *Atmos. Environ.* **16**, 1587 (1982).
73. B. P. Leaderer, with D. M. Bernstein, J. M. Daisey, M. T. Kleinman, T. J. Kneip, E. O. Knutson, M. Lippmann, P. J. Lioy, K. A. Rahn, D. Sinclair, R. L. Tanner and G. T. Wolff, Summary of the New York Summer Aerosol Study (NYSAS), *J. Air Pollut. Contr. Assoc.* **28**, 321 (1978).
74. R. M. Harrison and C. A. Pio, Size differentiated composition of inorganic atmospheric aerosols of both marine and polluted continental origin, *Atmos. Environ.* **17**, 1733 (1983).
75. F. Bruynseels and R. van Grieken, Direct detection of sulphate and nitrate layers on sampled marine aerosols by laser microprobe mass analysis, *Atmos. Environ.* **19**, 1969 (1985).
76. S. M. Wall, W. John and J. L. Ondo, Measurements of aerosol size distributions for nitrate and major ionic species, *Atmos. Environ.* **22**, 1649 (1988).
77. X. Q. Zhang, B. J. Turpin, P. H. McMurry, S. V. Hering and M. R. Stolzenburg, Mie theory evaluation of species contributions to 1990 wintertime visibility reduction in the Grand Canyon, *J. Air Waste Manage. Assoc.* **44**, 153 (1994).
78. W. John, S. M. Wall, J. L. Ondo and W. Winklmayr, Modes in the size distributions of atmospheric inorganic aerosol, *Atmos. Environ.* **24A**, 2349 (1990).
79. P. Saxena, L. M. Hildemann, P. H. McMurry and J. H. Seinfeld, Organics alter hygroscopic behavior of atmospheric particles, *J. Geophys. Res.* **100**, 18 755 (1995).
80. R. L. Tanner, R. Cederwall, R. Garber, D. Leahy, W. Marlow, R. Meyers, M. Phillips and L. Newman, Separation and analysis of aerosol sulfate species at ambient concentrations, *Atmos. Environ.* **11**, 955 (1977).
81. B. R. Appel, S. M. Wall, M. Haik, E. L. Kothny and Y. Tokiwa, Evaluation of techniques for sulfuric acid and particulate strong acidity measurements in ambient air, *Atmos. Environ.* **14**, 559 (1980).
82. P. J. Sheridan, R. C. Schnell, J. D. Kahl, J. F. Boatman and D. M. Garvey, Microanalysis of the aerosol collected over south-central New Mexico during the ALIVE field experiment, *Atmos. Environ.* **27A**, 1169 (1993).
83. P. T. Cunningham, B. D. Holt, S. A. Johnson, D. L. Drapcho and R. Kumar, in *Chemistry of Particles, Fogs, and Rain*, ed. by J. L. Durham, Butterworth Pub., Boston (1984), pp. 53–130.
84. R. M. Harrison and W. T. Sturges, Physico-chemical speciation and transformation reactions of atmospheric nitrogen and sulfur compounds, *Atmos. Environ.* **18**, 1829 (1984).
85. H. Ruprecht and L. Sigg, Interactions of aerosols (ammonium sulfate, ammonium nitrate and ammonium chloride) and of gases (HCl, HNO_3) with fogwater, *Atmos. Environ.* **24A**, 573 (1990).
86. C. A. Pio, T. V. Nunes and R. M. Leal, Kinetic and thermodynamic behavior of volatile ammonium compounds in industrial and marine atmospheres, *Atmos. Environ.* **26A**, 505 (1992).
87. G. T. Wolff, On the nature of nitrate in coarse continental aerosols, *Atmos. Environ.* **18**, 977 (1984).

88. J. G. Watson, J. C. Chow, F. W. Lurmann and S. P. Musarra, Ammonium nitrate, nitric acid, and ammonia equilibrium in wintertime Phoenix, Arizona, *J. Air & Waste Manage. Assoc.*, **44**, 405 (1994).

89. J. C. Chow, J. G. Watson, L. C. Pritchett, W. R. Pierson, C. A. Frazier and R. G. Purcell, The DRI thermal/optical reflectance carbon analysis system: description, evaluation and applications to U.S. air quality studies, *Atmos. Environ.* **27A**, 1185 (1993).

90. W. F. Rogge, M. A. Mazurek, L. M. Hildemann, G. R. Cass and B. R. T. Simoneit, Quantification of urban organic aerosols at a molecular level: identification, abundance and seasonal variation, *Atmos. Environ.* **27A**, 1309 (1993).

91. B. A. Mansoori, M. V. Johnston and A. S. Wexler, Quantitation of ionic species in single microdroplets by on-line laser desorption/ionization, *Anal. Chem.* **66**, 3681 (1994).

5 Structural Heterogeneity within Airborne Particles

JASNA INJUK, LIEVE DE BOCK AND RENÉ VAN GRIEKEN
University of Antwerp (UIA), Belgium

1 INTRODUCTION

Particle characterization is of a great importance for a wide variety of fields related to technological progress and environmental pollution, and at the same time one of the most demanding issues of microbeam analysis. Through the availability of sophisticated image analysis systems and associated software,

Atmospheric Particles, Edited by R. M. Harrison and R. Van Grieken.

direct particulate measurements, at a microscopic scale, are nowadays becoming more and more important.

The number of studies of environmental particles has expanded rapidly during the last 20 years, as a result of technological advances in instrumentation and in improved knowledge of electron, proton and ion optics, and electronics. [1] Various microbeam techniques like electron probe X-ray microanalysis (EPXMA), scanning electron microscopy (SEM-EDX), scanning proton microprobe (SPM), laser microprobe mass analysis (LAMMS), electron energy-loss spectroscopy (EELS), secondary ion mass spectrometry (SIMS) and Auger electron spectroscopy (AES) have proven their success in the field of aerosol research. Such microanalyses can reveal whether a specific element or compound is uniformly distributed over all the particles of a population or whether it is a component of only a specific group of particles. Even more, the elemental lateral and depth distribution within a particle can sometimes be inferred. However, each of the above techniques is limited by either poor resolution, or high detection limits, or poor low Z-elemental detection or lack of quantization.

The present chapter discusses exclusively the measurement of the compositional heterogeneities within a single aerosol particle. Some examples of heterogeneity investigations from the literature and from recent research by the authors are given. Also, based on a literature study, the importance of particle surfaces with respect to toxicity and atmospheric reactions is outlined. The analysis of whole individual aerosol particles will not be treated since a chapter was addressed to it in a previous book of this series. [2] Rather, a brief overview will be given of the techniques that have proven, so far, their success in studying the chemical heterogeneties within individual particles, in terms of quantification, elemental mapping, lateral and depth resolution and detection limits.

2 MICROANALYSIS TECHNIQUES

2.1 ELECTRON PROBE X-RAY MICROANALYSIS AND SCANNING ELECTRON MICROSCOPY

In EPXMA as well as SEM, a nanometre-sized electron probe generates various signals upon sample interaction. The detection of the emitted characteristic X-rays by wavelength- or energy-dispersive spectrometers (WDX and EDX, respectively) as well as the backscattered electron image provide compositional information on the sample. Morphological studies are based on the detection of backscattered and secondary electrons. The differences between EPXMA and SEM-EDX have been reduced over the years to only a slightly different instrumental set-up, so that both techniques are nowadays to some extent used for chemical as well as morphological studies.

In spite of their unfavourable detection limits (0.1%), EPXMA and SEM-EDX are considered to be the most popular non-destructive microanalytical techniques. Part of this popularity can be attributed to the availability of statistically relevant information, since the automation of these techniques allows the analysis of several hundreds of particles in a few hours with a relative accuracy of about 5% and a lateral resolution of 0.1–5 μm. The in-depth resolution of EPXMA/SEM-EDX is about 0.5–5 μm. Manual analysis, on the other hand, offers the possibility of detailed morphological studies and of element-mapping within individual (giant) aerosol particles.

A variety of programs have been developed over the years to solve the problem of quantization with electron beam instruments. In the field of single particle analysis, the small dimensions of the particles, leading to uncertainties in the determination of the interaction volume, together with a lack of suitable standards still make quantitative analysis rather difficult. However, the recent CITZAF correction procedure package, proposed by Armstrong, [3] claims to provide reasonable quantitative results in the analysis of unconventional samples, such as individual unpolished particles.

Other specific problems are being solved by the new generation of scanning electron microscopes which permit probing down to low-Z elements, like C, N and O, analyze any sample wet or dry, insulating or conductive, [4] and offer a higher lateral resolution while working with a high-magnification secondary electron image. [5] Moreover they provide much enhanced image analysis capabilities.

2.2 ELECTRON ENERGY LOSS SPECTROSCOPY

By combining electron spectroscopy and transmission electron microscopy, the analytical power of EELS is coupled with the ability to select, image and obtain diffraction patterns from small areas (< 0.01 μm lateral resolution, 0.01–0.1 μm in-depth resolution). Apparatus for EELS has recently become commercially available and as yet very few publications have dealt with individual particle analysis. [6–8]

The principle of EELS is based on the theory of electron–solid interactions. As the electrons of the beam (initial energy of E_0) interact with electrons of the atoms present in the sample, they can lose an amount of energy, ΔE, characteristic for these atoms. The initial electrons will leave the sample with reduced energy, $E_0 - \Delta E$. Thus the electron energy-loss spectrum represents the graphical display of the energy, ΔE, lost by the scattered electrons and its corresponding intensity. Accordingly, EELS detects directly the electron beam interaction with the atoms, which produces inner-shell ionizations or the excitation of oscillations of the electrons in the valence band of a solid (plasmon oscillation). The transmission signal consists of electrons in an energy range from E_0 down to a loss of several thousand eV. After dispersion by an energy analyzer, electrons with the same energy loss are focused on the same point. By scanning this dispersion plane over a detector, the intensity $I(E)$ of the transmitted signal can be plotted as a function of energy loss. Various interaction processes can be related to the properties of the specimen by examining the features of a typical energy-loss spectrum.

Just as for X-ray spectroscopy, the characteristic absorption edges are of major importance for the microanalytical approach of EELS. The rapid increase in the spectrum intensity at the edge makes it fairly easy to distinguish among the elements. EELS can observe edges of all elements from Li to U (they have at least one edge in the energy-loss range of 0–2000 eV that can be used for analysis). One complication in energy-loss spectra is the variation in edge shapes that can be observed.

Quantitative analysis is simply conversion of the measured number of inner-shell excitation events to an atomic concentration. Quantification without the use of reference materials can be performed only for very thin samples of uniform thickness and homogeneous element distributions.

The techniques has some advantages compared to X-ray analysis like: low-Z elements can be detected with a high sensitivity, the detectable mass is much lower and information about the electronic state and chemical bonding of the sample can be provided. EELS has been used to identify quantities of less than 10^{-20} g and concentrations of less than 100 ppm of elements such as P and Ca in an organic matrix. However, the accuracy of quantitative analysis is often no better than 20%. The main sources of error are in the background subtraction and inner-shell cross-section determination, in lens aberration and radiation damage problems.

2.3 AUGER ELECTRON SPECTROSCOPY

AES is a surface-sensitive technique in which the electrons, emitted by a radiationless transition from an excited state, produced upon impaction of an electron beam, are investigated. Since the energy of the Auger electrons is purely a function of the atomic energy levels and there exist no two elements with the same set of atomic binding energies, AES leads to direct element identification. The acquired Auger spectra reveal, besides the small Auger peaks, other peaks due to elastic scattering, backscattering of primary electrons, etc., making the identification and location of the peaks of interest sometimes very difficult. To overcome this problem, the Auger spectra are usually smoothed, to reduce the noise in the spectrum, as well as differentiated to remove the continuum. The AES technique is particularly useful for the detection of low Z-elements ($Z < 13$) because, for heavier elements, the emission of an X-ray photon or so-called fluorescent decay, becomes the more significant relaxation process. The penetration depth is in the range of 2–10 atomic layers for the typical energies (50–2000 eV) used in AES. Elemental detection limits lie in the range of 0.1–1% within the analytical volume, and the variation in sensitivity between the low-Z elements is less than a factor of 10. Sputter-cleaning of the specimen surface and in-depth profiling is achieved by an Ar^+ ion beam. Rastering of the electron beam over the sample surface becomes available in scanning Auger microscopy (SAM).

Although this technique appears to be very useful for investigating surface coatings on environmental particles, its low sensitivity to high Z-elements, its low detection power and the charging effects caused by non-conductive samples, are still major drawbacks for its frequent application in this field.

2.4 LASER MICROPROBE MASS SPECTROMETRY

LAMMS was originally designed for the analysis of biomedical samples, especially thin sections, with high lateral resolution and extreme detection sensitivity (10^{-18}–10^{-20} g). Since then, various instrumental set-ups were developed for different analytical purposes.

The principle of LAMMS is based on the excitation of a microvolume of the sample to an ionized state by a focused laser beam. The basic components of the instrument are a laser system, an optical instrument and a mass spectrometer. The ionized part of the evaporated material consists of positive and negative elemental and molecular ions. Usually a time-of-flight mass spectrometer separates the ions with different mass-to-charge ratios according to their flight times. The analytical information is derived from mass spectrometry of these ions.

Quantification by LAMMS is not yet straightforward. The way things stand now, since no theoretical model can predict the ion yield for a specific specimen as a function of target and laser beam parameters, quantification is based on the use of empirical procedures mostly involving standards. As it is very troublesome to determine the sampled mass, quantification must always rely on the use of either internal reference elements or standard samples which closely resemble the unknown one. Particle standards can be prepared by nebulization of aqueous solutions or by crushing material to micrometre size. Once suitable standards are available, quantification can be carried out with calibration curves or with relative sensitivity factors. The precision for quantitative particle analysis is at best 20–40% for elements of low ionization potential, with an accuracy of a factor of two. For elements of high ionization potential, it worsens to 80%. The best precision is obtained for thin samples: between 15 and 30% for all elements.

LAMMS is in principle capable of detecting all elements and organic compounds and of generating stoichiometric information. The spatial resolution of the analysis is 0.5–3 μm, the in-depth resolution is > 1 μm and the detection limits are of a few ppm.

Since physicochemical properties of airborne particles have been stated to control different atmospheric processes, the necessity of analysing single airborne particles in their native state, i.e. by on-line analysis, evolved. Off-line analysis methods are often limited to the non-volatile part of the particle and, due to sample–surface interactions, chemical alterations of the sample can occur leading to a possible loss of valuable information. *In situ* characterization of the chemical composition and size of individual airborne particles becomes available by performing on-line laser microprobe mass spectrometry (on-line LAMMS) or rapid single particle mass spectrometry (RSMS). Different experimental set-ups have been developed over the years, [9–21] but all of them are based on the same principle of laser desorption/ionization and analysis using mass spectrometry. When applying this technique aerosol particles are collected immediately into the mass spectrometer. The scattered radiation obtained from the interaction of each individual aerosol particle with the He–Ne laser beam reveals information on the particle size and triggers the laser which vaporizes the particle and ionizes the fragments. For each single particle a complete mass spectrum is recorded with a time-of-flight mass spectrometer. Further research in this field will improve and optimize the different aspects of the on-line particle analysis. However, recent developments by Prather *et al.* [15] revealed the capability of characterizing simultaneously the size and chemical composition of individual aerosol particles down to 50–100 nm in real time. Reents *et al.* [20] reported on the ability to detect particles as small as 20 nm using a different instrumental configuration. Moreover, the simultaneous detection of positive and negative ions from the same airborne particle by real-time LAMMS was accomplished for the first time by Hinz *et al.* [21]

2.5 SECONDARY ION MASS SPECTROMETRY

In SIMS, an ion beam (O_2^+, O^-, Ar^+, Cs^+, Ga^+, ...) or molecules generated in a duoplasmatron, with an energy between 1 and 20 keV, are used to bombard the surface of a solid sample. The primary ions set up a collision cascade and are implanted into the solid. The penetration depth (about 10 nm) is determined by the energy and mass of the primary ion and the composition of the material. Due to the ion bombardment secondary particles such as ions, neutrals and clusters are emitted from the surface. This process is referred to as sputtering. The information depth (the depth where most of the secondary ions are emitted from) is of the order of one to several nanometres. The positive or negative secondary ions are extracted into a mass spectrometer (quadrupole, magnetic sector, time-of-flight) where they are separated according to their m/z ratio. After mass separation the ion current is recorded with a suitable detector. Different types of information can be obtained such as the surface composition, depth distribution and lateral distributions. A mass spectrometer provides elemental coverage from H to U, isotopic characterization, and detection limits of 10^{-15}–10^{-19} g. Because of the continuous erosion of the sample surface by the ion bombardment, depth profiling of elements becomes possible. A depth resolution in the order of a few nanometres can be achieved. Most SIMS instruments are also capable of visualizing the lateral distribution ($< 1 \mu$m) of the elements in the form of ion images. By the combination of ion images and depth profiling, a three-dimensional analysis of the sample surface becomes possible.

SIMS offers special capabilities for particle analysis. Compositional heterogeneities within single particles are frequently observed. SIMS, as a microanalysis technique, is capable of obtaining signals from a minimum sample volume of 0.001μm^3. Depth profiling can be successfully applied to individual particles, although irregular topography of particles can degrade the depth resolution. Quantification is possible by the use of relative elemental sensitivity factors estimated from analysing standard samples under conditions identical to the unknown sample. The detection limit is about 1 ppm, with an accuracy of 10%.

2.6 SCANNING PROTON MICROPROBE

In SPM an intense radiation of MeV heavy ions is scanned across the sample to reveal microstructures in a wide range of materials. Heavy ions are attractive for single particle analysis since they penetrate deep into the material without being deflected much (the microbeam is almost free of scattered particles). Development of a focusing device for MeV ions did not create much interest in the 1970s, but during the 1980s and 1990s nuclear microprobes have come into widespread use. One reason might be that commercial systems have become available. [22] It is only recently that SPM was applied in environmental science on a single particle level. A reason for this was the requirement of a micron or submicron stable beam for several hours of radiation, a bright and stable ion source and a fast and flexible data acquisition and analysis system. Today, there are only a few nuclear microprobes in the world which meet these requirements where the successful analyses of individual particles can be carried out.

Several analytical techniques could be implemented in the SPM, but μ-PIXE (micro-proton induced X-ray emission) is the most important one. Due to its large reaction cross-sections, SPM provides concentrations of 20–30 elements at ppm levels in small samples of a few μm in diameter and weighing only 1 pg or less. As a consequence, the absolute mass detectable by SPM is of the order of 10^{-17}–10^{-18} g. This puts SPM in a unique position of being able to provide trace element concentrations on a single particle level and at the same time retaining a high spatial resolution (from 0.5 to 10 μm). With its unique ability to simultaneously use several complementary nuclear analytical techniques like RBS (Rutherford backscattering spectrometry) and PIGE (particle induced gamma-ray emission), it can provide high quality analytical data for almost all elements of the periodic table. As the proton beam goes through the sample and is collected in a Faraday cup, quantification is relatively easy to perform, with an accuracy of 10–20%.

3 ELEMENTAL HETEROGENEITY INVESTIGATIONS WITHIN SINGLE AEROSOL PARTICLES

The recent recognition of aerosols and, for example, their capacity to affect the global climate, immediately set off a great demand for a high-quality data on fine aerosol concentrations from the entire world. Many micro- and trace analytical methods have already produced a very good database on background aerosol concentration world-wide, but in contrast, not much attention was focused on internal heterogeneity studies of aerosols. By focusing attention exclusively on one individual particle, detailed information can be achieved for aerosols with respect to: (A) their "life cycle" and evolution, (B) their behaviour during physicochemical interactions with the transporting medium (e.g. gas-to-particle conversion or coagulation processes), (C) the enrichments at the particle surface, (D) distribution of constituents with depth in the particle and (E) the particle size distribution. In the process of source identification, the analysis of a single aerosol particle has an indispensable value and this is where single particle techniques can make a significant contribution.

3.1 ANTHROPOGENIC AEROSOLS

Elements of anthropogenic origin are mainly released to the atmosphere in gaseous form as a result of fossil fuel combustion. Such elements (e.g. S, Cr, Pb, Br) are present in the submicrometre aerosol fractions and they usually form a thin film on larger particles of natural origin (e.g. Pb or Ti layer on a gypsum particle or S shell around a quartz particle). The bioavailability of such elements is largely dependent on their concentration profile.

Anthropogenic aerosols, especially sulfates and soot particles, were found to be significant in the context of climate changes. The global increase of cloud reflectivity due to anthropogenic aerosols was estimated to be 2%. [23]

3.1.1 Fly-ash Particles

Fly-ash particles have been one of the most widely studied types of airborne pollutant matter. It is well known that these particles are of a complex structure and morphology and a broad size range; their surface layer is of special interest, since it contains certain toxic trace elements, often in a soluble form. Thus, fly-ash particles interacting with biological systems may produce, as will be further discussed in more detail, enhanced concentrations of toxic species at the point of direct contact.

Over the last 20 years the structure of fly-ash particles has been the subject of different studies involving various analytical techniques like EPXMA, SEM-EDX, LAMMS, SIMS, SPM and AES.

About 20 years ago Linton et al. [24,25] demonstrated the elemental surface predominance in coal fly-ash particles of elements like Li, Na, S, K, V, Cr, Mn, Fe, Tl and Pb. This was attributed to the condensation of species previously volatilized in the high-temperature combustion zone of a particulate emission source. Potentially toxic elements like Pb and Tl were present at bulk concentrations of only 620 and 30 ppm respectively, while their surface concentrations were 4% and 4500 ppm, respectively. The predominant surface region for such elements was substantially soluble, thus being very likely accessible to the environment either by wash-out processes in the atmosphere or the groundwater or by solubilization with biological systems. This problem was further investigated by Cox et al. [26] who also applied SIMS to characterize single fly-ash particles in terms of intra-particle concentration variations. A downward trend with depth for trace elements as Tl, Pb and U, as opposed to the flat profiles for Si and Ti, indicated that the observed profile shapes were not an artefact generated by loss of analytical volume or change in total yields.

In 1983, Vis et al. [27] scanned systematically, by a proton microbeam, fly-ash particles from the magnetic fraction, to find elemental distributions across their cross-sections. In general, magnetic particles were characterized by high surface concentrations of K, Ti, V, Cr and S, while the Fe distribution showed a maximum at the centre of the particle. Experimental evidence on surface enrichments of toxic elements in the 10 μm and less size fraction of fly-ash was reported by Jakšić et al. [28] in 1991. Based on elemental analysis, two main groups were recognized: aluminosilicate particles and Ca-rich or Fe-rich particles with a high level of Ti. Line scan across the particle diameter showed that for Ca- and Fe-rich particles, most of the elements appeared to have a homogeneous distribution; no significant surface or core enrichment was found. In contrast, the aluminosilicate fly-ash fraction was characterized by a compositional heterogeneity. Most of the analyzed particles formed an aggregate with usually one or a few larger spherical aluminosilicate particles surrounded by other submicron (usually Ca-rich) particles. An areal scan across one aluminosilicate particle, showing a complex surface structure, is given in Figure 1. How an elemental composition affects the electrical properties of fly-ash particles was demonstrated by Cereda et al. [29] About 200 particles, in the 1–5 μm size range, collected with high collection efficiency (99.8%) at the inlet and outlet of the electrostatic precipitator of a coal-fired power plant, were analyzed and compared. An increasing concentration trend of trace elements with decreasing particle size was clearly observed. Particularly in the case of S, enhanced concentrations, between 1600 and 2300 ppm, were found at the electrofilter outlet compared to the S "inlet values" of 500–1300 ppm. This result was accounted for by the condensation of sulfates on the particle surfaces when particles are passing through the electrofilter. The electrostatic precipitator acts selectively with the particle composition: particles with higher concentrations of minor and trace elements are more likely to escape the electrofilter. The analysis of particles from the inlet of the precipitators showed a non-uniform distribution of minor (P, Ca, Ti, Fe) and trace elements (S, Mn, Ni, Cu, Zn, Ga). These results reflected a poor degree of coalescence upon combustion, in the case of minor elements, and surface segregation and evaporation–condensation mechanisms, in the case of the trace elements.

According to Van Malderen *et al.*, [30] fly-ash particles are one of the most abundant particle types in Siberian winter air masses. Despite the presence of a major hydroelectrical plant at Novosibirsk, most of the electrical production in

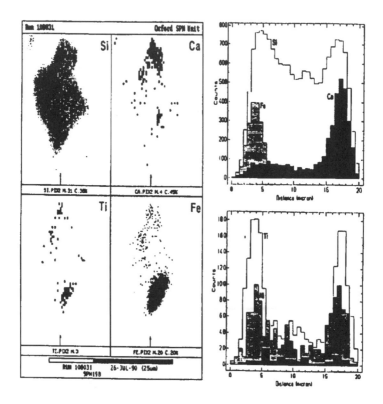

Figure 1. Results of line scans on an aluminosilicate fly-ash particle. (Reprinted from Jakšić *et al.*, Nucl. Instr. Meth. B56/57:699–703 (1991), with permission of Elsevier)

the region is based on coal-burned power plants. The combined result of the burning of coal and the presence of a snow cover, which prevents soil from being blown into the air during the winter months, explains the enormous relative occurrence (up to 30% of the total abundance) of the fly ash. Aerosol particles were analyzed by proton microbeam, positioned first in the middle and then at the edges of a particle in order to check its homogeneity. In Figure 2. μ-PIXE and RBS spectra for three different points in a fly-ash particle are given. The RBS spectrum provides information on the lighter and some major elements of the investigated particle (so-called matrix composition), while its trace elemental composition is apparent from the μ-PIXE spectrum. Deconvolution of the RBS spectrum involves correction for the composition and thickness of the support layer (in this case: the Nuclepore filter, $C_{16}O_3H_{16}$). The corresponding composition of a "typical Siberian" fly-ash particle is given in Table 1. Complete thick target corrections for beam stopping, X-ray attenuation and X-ray fluorescence were implemented. Some compositional variabilities were shown, although not so drastic when compared to the above-mentioned studies.

The majority of coal fly-ash particles seem to consist, as a result of the coal combustion process, of a mineral core containing a surface layer of volatile elements like S. [31] Evidence of the presence of S as a coating on individual coal fly-ash particles was recently demonstrated by plotting the S/Si X-ray intensity ratio, acquired by SEM-EDX analysis, as a function of the geometrical particle diameter. [32] These results showed the smallest coal fly-ash particles contain the highest S-enrichments. Aerosol models developed by Biermann and Ondov [33] can be applied to determine the concentration of the thin surface layer deposited on the solid core of the particle. On the other hand, in the fly-ash particles emitted by oil-fired power plants, three different particle modes could be distinguished. Chemical and physical surface reactions taking place during and after combustion could be responsible for the S content in the coarse and large particle mode. These modes were composed of organic material and mineral elements, respectively. Based on the $BaCl_2$ microspot technique the

Table 1. The main composition (wt%) of a giant "Siberian" fly-ash particle as determined by RBS and μ-PIXE at different position within the particle

Fly-ash particle	C	O	Al	Si	Fe
Centre (RBS)	35	26	27	12	0.92
Centre (PIXE)	—	—	15	10	0.94
Side 1 (RBS)	65	19	3.6	12	0.75
Side 1 (PIXE)	—	—	11	25	0.92
Side 2 (RBS)	41	22	33	3.5	0.69
Side 2 (PIXE)	—	—	—	—	0.99

— = not detected.

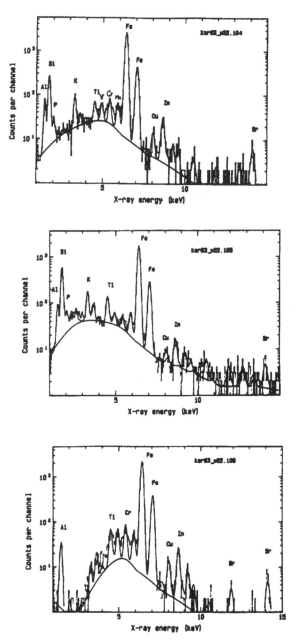

Figure 2. The PIXE (left) and proton backscattered spectra (right) of the Siberian fly-ash particle, taken at the three different positions: (a) centre, (b) upper and (c) bottom side of the particle

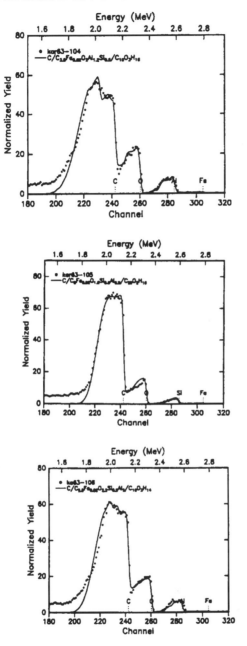

Figure 2 (*Contd.*)

small particle mode was identified as water-soluble sulfates which occur by nucleation and coagulation of sulfuric acid produced by burning S-containing fuel.

A detailed AES study on the surface elemental composition of coal fly-ash particles was also performed by Linton *et al.* [25] Qualitative elemental depth profiles of Al, Ca, Fe, K, Na, S and Si revealed that only K, Na and S exhibited major surface enhancement although potentially toxic elements like Pb, Ti, Mn and Cr were also present. The results obtained were qualitatively in agreement with the SPM data, suggesting that both techniques can be applied to investigate the actual elemental composition of the surface region. The overall results of this study suggest the elemental surface predominance is a general phenomenon in particles produced by high-temperature processes for which volatilization and subsequent condensation of elements probably occur. Similar research was performed by Hock and Lichtman [34] on nanometre surface coatings of individual coal fly-ash particles collected in-plume and in-stack.

Fly-ash particles formed by shells with different composition and structure are often observed. [35] In Figure 3 the Auger depth profile of the fly-ash particle is shown. The shell structure of this particle is clearly demonstrated. Three different shells can be observed. The topmost shell is mainly composed of aluminosilicates. Underneath this shell, an aluminosilicate and iron oxide shell are distinguished, while the inner shell is mainly composed of iron oxides.

In a study to investigate the potential effect of the Kuwait oil fire plumes, in May 1991, on the local and global environment, the $BaCl_2$ chemical spot test in

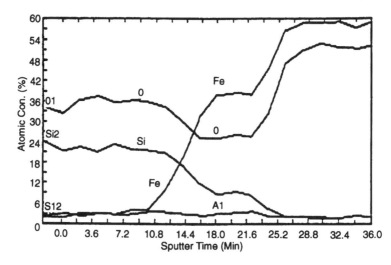

Figure 3. Auger depth profile of the fly-ash particle; sputtering rate 25 Å min^{-1}. (Reprinted from Boni *et al.*, in: *Combustion Efficiency and Air Quality*, pp. 213–240, I. Hargittai and T. Vidoczy, Eds., with permission of Plenum Press)

combination with EM (transmission and scanning) was also performed to distinguish sulfate-containing particles. [36] The aerosol samples, collected on board an aircraft at different altitudes in the plumes, were coated with $BaCl_2$ which formed particles <50 nm on the background. Introducing the coated samples into a humid chamber for 2 hours provided the correct environment for the reaction of $BaCl_2$ with sulfate. By applying this method, which was first introduced by Bigg et al., [37] sulfate-containing particles were surrounded by a hollow ring and pure sulfate particles could be distinguished from sulfate-coated particles based on the shape of the particles' centre core. The number of sulfate-coated soot, salt and dust particles appeared to increase with the distance from the fire plumes, and the thickness of the S coating increased with altitude. The presence of these particle types indicate a gas-to-particles conversion, by means of catalytic oxidation in combination with heterogeneous nucleation, during plume dispersion. The increase of small sulfate particle concentrations with height and distance from the plume can be attributed to the transformation of SO_2 gas into sulfate particles via oxidation followed by homogeneous nucleation. The majority of particles in the plume behaved as cloud condensation nuclei since their surface characteristics changed from hydrophobic to hydrophilic by the coating with sulfate. In this way these particles are able to initiate haze, smog, fog or cloud droplets. Air trajectory analysis suggested the existence of periods in which aerosols generated by the fires could have reached China and influenced the severe rainfall at the end of May and the beginning of June.

The characterization of individual oil-soot particles on a Ta-substrate has been accomplished using SIMS by McHugh and Stevens. [38] On the average, high levels of O, V, C, Na, Ca and K could be identified; the total composition can vary strongly between different oil-soot particles since a wide variety exists in the inorganic trace elemental composition of crude oils and in the combustion conditions. Moreover, the SIMS images of major elements present in the oil-soot particles showed that local differences in the elemental distribution within individual oil-soot aerosol particles were present for Na, K, Ca, Mg, Al, Si, Ti and B. A homogeneous distribution was usually found for V and C, the two major compounds.

Besides different on-line LAMMS studies on artificially generated aerosols, [11–16] a report was recently published on the possibilities of the LAMPAS system (laser mass analysis of particles in the airborne state), developed at the Institute of Laser Medicine, University of Düsseldorf, Germany, for the analysis of single atmospheric particles by on-line laser time-of-flight MS. [17] Using this set-up aerosol particles from ambient air could be detected with a high repetition rate, a superior mass resolution which remains constant over the entire mass range and a very high signal to noise ratio. The comparison between the on-line spectrum of an aerosol particle collected from the laboratory room air, showing the typical carbon cluster ion distributions of soot particles, and a candle soot particle analyzed by LAMMS, revealed the quality of the on-line spectrum to be as good as the LAMMS off- line spectrum. Both spectra appeared to be similar although additional information indicating the presence of an aqueous phase surrounding the soot particle could only be identified in the on-line spectrum. The capability of the instrument to distinguish dry and wet particles was previously demonstrated by analysis of aqueous salt droplets. [17,19]

3.1.2 Urban Aerosols

Outdoor aerosol particles in the urban areas stem to a great extent from anthropogenic emission. In many cases these particles contain higher concentrations of several potentially hazardous trace elements which are normally found in natural crustal dust, soil and sediment. These elements, like S, V, Cr, Mn, Co, Ni, Zn, Se, Cd, Sb, Tl and Pb, are preferentially associated with particle surfaces as a result of condensation from the vapour phase or adsorption from solution.

Fumes arising from iron and steel industries are considered to be the major source of airborne particles containing metal oxides which may be carcinogenic. The toxicity of these metals seemed to be related to their chemical oxidation state. LAMMS analysis on particles emitted as dust by the steel industry revealed a correlation between the Cr oxidation state and the particle size. [39] Particles smaller than 1.1 μm were composed of P- and Na-matrices, whereas large particles (>6 μm) mainly contained Ca. Cr was found to be present in the hexavalent state for both size fractions. Trivalent Cr could only be detected in particles of intermediate size. For the verification of the Cr oxidation state and nature of the Cr-compounds present in particles the micro-Raman technique is also desirable.

Airborne microparticles, collected from outdoor air near the city of Karlsruhe, in the Rhine river valley, have a clear shell structure. [40] The majority of the particles were of respirable size with diameter <2 μm. The core of the coarse particles (2–5 μm) contained geogenic material, which was covered by a double layer structure. The geogenic origin was confirmed by the presence of typical soil elements such as Na, K, Mg and Fe. The topmost layer of some 60 nm was of nitrate and ammonium sulfate, while a deeper layer of 400 nm primarily consisted of carbonate and organic carbon compounds. Moreover, Na and Mg (as carbonate and nitrate) were found in this surface region. Submicron particles (0.2–0.8 μm) contained a core dominated by graphitic or organic carbon, almost completely covered by a 15 nm thick layer of ammonium sulfate. These particles basically originate from vehicle traffic. Ammonium sulfate is possibly directly formed from gaseous precursors at the surface of the soot (this reaction can even take place in the exhaust pipe of catalyst-equipped cars) or agglomerates as fine particles on to the soot particles after emission. The observed layer structures are of importance for the removal of the soot from the atmosphere, since the conversion of the soot surface from originally hydrophobic into hydrophilic behaviour should ease wet precipitation. Another study [41] of large car exhaust particles indicated surface enrichment (down to a few μm) by S, Cl, Br and Pb (with absence of Fe, although Fe was one of the major constituents). Also, the trace elements Cr, Ni and Tl were found to be highly surface enriched. This finding was attributed to the deposition of volatile Pb and S species (like PbBrCl and SO_2) on to the surfaces of refractory Fe-containing particles as the temperature decreases in the automobile exhaust system. The Fe-rich particles were derived from ablation of Fe from the exhaust system. The observed trace amounts of Cr, Ni and Tl have been introduced by impurities of Pb-gasoline additive or, in the case of Cr and Ni, by engine wear. The small particles (<1 μm) were found to be composed, almost exclusively, of Cl, Br and Pb (no Fe).

The overall composition and depth distributions of the components of combustion particles generated in household fires were investigated by Bentz et al. [42] The results showed that the combustion particles have a shell structure with organic halogen compounds and PAH-content of about 1 ppm in a surface layer. Cl and hydrocarbons were enriched in a surface region of some 100 nm thickness. This layer was of mainly aliphatic character. The concentration of hydrocarbons decreased towards the core. Obviously, a condensation of low-volatile hydrocarbons with Cl-compounds or a surface reaction with gaseous Cl or HCl is responsible for the organic surface layer enriched with Cl.

In the huge area of Siberia, there are many remote areas but also there are some regions where massive industrial complexes are located with pollution levels several times higher than in any Western European or North American country. Our recent sampling and research program, for some areas of Siberia, have provided a detailed characterization on the airborne particulate matter representative of this part of the world. Typical aerosol particles, resulting from emissions by the heavy metal industry, were found to be K- and Cl-rich particles containing significant amounts of heavy metals such as Cr, Ni, Zn, As and Pb in a concentration range of 0.2% (Ti) to 1.8% (Fe). Some elements like Cl, K, Ca and Fe were found to be homogeneously distributed over the particles, while Ti, Ni, Cu, As and Pb were dominantly concentrated at one edge of the particles. Frequently, particle agglomerates were also observed. [30]

Recently, the Singapore group has reported on the analysis of more than 300 urban aerosol particles by SPM. [43] Each particle was characterized by its size, thickness, total mass and concentration of up to 29 elements. Particles were characterized into several groups, according to their major components: marine and soil particles were found to be the most abundant, representing more than 86% of the total mass, followed by Fe (3.6%), Pb (3.1%) and Ti (1.8%). "Exotic" particles with high concentrations of Sn, Sb and Ba were also found (2%), but their source was not identified. S, as one of major anthropogenic pollutants in urban areas, was found to be quite evenly distributed in all particles. This was expected, because S is released to the atmosphere in gaseous form which "condenses" on solid particles through reactions with water vapour and sunlight forming sulfuric acids. Also, it reacts with virtually "harmless" marine particles forming hydrochloric acids. Most of the S on Cl-rich particles was concentrated on the surface of larger particles. This was proven by "washing" one sample in a warm (40 °C) solution of 5% HNO_3 for about 2 minutes, followed by rinsing in distilled water and proton irradiation once again. The result was quite surprising: most of the aerosol particles remained intact on pioloform substrate but PIXE spectra showed almost complete absence of marine components (Na, Mg, Cl) as well as a high depletion of S, Pb and Br. Such a shell around particles limits the PIXE quantitative analysis; if, for example, S forms a shell around a Si-rich particle and matrix corrections are performed by a "classical approach", assuming a homogeneous distribution of all elements and a flat target, the obtained mass will be overestimated by a factor of two or even more, depending on the particle size and composition. Therefore, a precise concentration determination requires a more sophisticated approach.

Carbonaceous particles in the atmosphere contribute significantly to the total suspended particulate matter and can sometimes comprise more than 50% of the total submicron mass fraction. They exist mainly in the particle size range below 3 μm with the major amounts in the size range below 1 μm. Atmospheric carbonaceous material consists of different types of particulate carbon fractions, namely: carbonates, elemental carbon and organic carbon. Although observed in atmospheric particles, carbonates form a minor component in the fine fraction of aerosols, at least in polluted urban atmospheres. Mueller et al. [44] reported that the carbonate carbon present in aerosols from Pasadena, California, represented systematically less than 5% of the total carbon. In contrast, elemental carbon is a prominent constituent of atmospheric aerosols and is present in urban areas but also in remote environments, such as above tropical oceans and in the Arctic region. In urban areas, the elemental carbon fraction accounts for less than 40% of the total carbonaceous content in the atmosphere. [45] Elemental carbon is especially of interest since it is an extremely efficient absorber of visible light. Organic carbon aerosol present in the Los Angeles area was found to account for roughly 60–70% of the fine particulate carbonaceous aerosol. [46] Up to now, analysis of carbonaceous particles has mostly been limited to bulk analysis and only recently have carbonaceous particles been subjected to a single particle approach. [47] Carbonaceous particles in the Phoenix, Arizona, urban atmosphere have widely diverse origins and as a consequence, a considerable range in structures and morphologies. A significant coagulation between particles of anthropogenic and non-anthropogenic origin was found in ambient air samples. The graphitic structures were interpreted as a part of the primary soot spherules and the amorphous areas as hydrocarbons condensed during aggregation.

Secondary transformation reactions of carbonaceous particles in the presence of SO_2/NO_2 mixtures were also observed. The coated aggregates have carbon structural variations similar to those in uncoated aggregates, suggesting both types of particles form from similar pre-emission sources. The use of EELS has provided visual evidence of S and C coatings present on the surfaces of carbonaceous and quartz particles identified in the Phoenix environment.

A combination of a bulk analytical technique, differential pulse anodic stripping voltametry (DPASV), and a microanalytical technique, LAMMS, was applied to identify Pb in standard particles and natural aerosols. [48] The aerosol particles were collected both in the city of Wageningen and at the downwind side of a highway at Ede, in The Netherlands. Discrimination between Pb present as a coating or homogeneously distributed over the particle appeared to be possible for both techniques. LAMMS results revealed, moreover, that most of the particles could be identified as mixed salts of ammonium nitrate and ammonium sulfate with a variable content of metals, especially V and Pb.

3.2 COAL-MINE DUST PARTICLES

There are many reports of the inhibition of quartz toxicity in various quartz-containing dusts. [49] For many years, coal mine workers' pneumoconiosis has been related to the presence of quartz in the respirable coal dust fraction. However, the experimental toxicity of coal mine dusts was usually found not to be consistently correlated with the gross quartz content as determined by bulk analysis. Such a lack of distinct correlation can be widely ascribed to the heterogeneity of such materials. Mixed dusts, such as authentic coal mine dusts, represent rather heterogeneous mixtures of elements down to the individual particle level. [50] Comparative studies of the toxic potency of authentic coal mine dusts and artificial quartz–coal mixtures have suggested that quartz toxicity may be inhibited by protective impurities on the quartz surface. [51] This became more apparent after LAMMS studies of natural dusts and authentic coal mine dusts were performed. [52] These analyses revealed that only a minute fraction of "pure" quartz (about 1% of the total abundance) may be detected in coal mine dusts. Most of the quartz surface in such highly heterogeneous particulate matter appeared to be covered by aluminosilicates. LAMMS investigation has shown that fibrogenicity of coal dust samples was not correlated with the incidence of the pure quartz particles but with the incidence of Fe, Mg(Ca)-containing particles. These particles seem to have a major role in harmfulness of coal mine dusts.

3.3 MARINE AEROSOLS

Andreae et al. [53] reported in 1986 on the heterogeneous nature of remote marine aerosols collected during a cruise on the equatorial Pacific Ocean.

Several thousand particles were subjected to automated EPXMA and imaging of silicon-containing particles was performed by SEM. Beam focusing at high magnification revealed information on the individual components of composite particles and mineral standards were utilized to calibrate the acquired X-ray spectra. They concluded that the internal mixing with sea-salt, detected in a large fraction of the silicate mineral component, as well as the enrichment of excess sulfate on sea-salt particles could probably be attributed to processes occurring within clouds.

Individual North Sea aerosol particles collected from an aircraft at six different heights were analyzed by LAMMS. [54] The different particle types, earlier distinguished by EPXMA, were also found with LAMMS but many particles seemed to appear as internal mixtures. The interpretation of the particle type abundances as a function of wind direction and particle size made it possible to apportion the different types to their source. Besides a decrease in relative sea-salt concentration above the inversion layer, no clear differences were found in the particle abundances as a function of height.

SEM and TEM together with thin-film chemical tests were applied to study the nature, composition and morphology of methanesulfonic acid (MSA) particles, sampled over Sakushima Island in Japan, near the Pacific Ocean. [55] MSA and SO_2 can be identified as the major reaction products obtained from the oxidation of marine dimethyl sulfide. At high relative humidity the coarse particle mode appeared to be dominated by mixed particles produced through heterogeneous nucleation reactions of gaseous MSA with sea-salt or soil particles. The fine particle mode was characterized by sulfate particles. Nitrate was present in both modes.

Concerning quantification, Pardess et al. [56] developed a new method, with a reasonable accuracy, for the SEM-EDX determination of the sulfur mass content in single aerosol particles larger than 0.4 μm with a minimum of 10^{-13} g sulfur. Calibration was achieved by analysing particles of known composition and size. Moreover quantitative estimations can be acquired for the sulfur concentration inside heterogeneous aerosol particles.

Recent laboratory investigations by Cheng [57] revealed the existence of a shell structure composed of chlorides ($MgCl_2$ and KCl) on the surface of marine aerosol particles. Moreover the majority of sea-water drops which evaporate and become salt-saturated during free fall in the air (RH < 60%) seemed to change phase to produce hollow sea-salt particles. [58] These results were obtained by close examination of the different stages in the evaporation process of an individual sea-water droplet, using a polarizing microscope and an SEM. The shell structure is very valuable for the understanding of nucleation processes in the atmosphere, since the presence of a chloride film provides a highly hygroscopic surface, which already initiates the condensation of water vapour at an RH of 40%.

Direct evidence of the formation of nitrate on marine particles, which was already frequently stated by different scientists, [59–61] was provided by Mamane and Gottlieb. [62] They studied the heterogeneous reactions of NO_2 and HNO_3 on individual sea-salt and mineral particles, under controlled conditions inside a Teflon reaction chamber, by bulk and individual particle analysis. Different particles did react with nitrogen oxides and HNO_3 to form nitrates. The formation of these nitrates seemed to depend on the presence of UV radiation. Under dark conditions, 0.1–8 mg NO_3^- g^{-1} aerosol was produced compared to 1.4–28 mg NO_3^- g^{-1} aerosol under UV radiation. Microscopic investigations associated with a specific microspot technique, [63] by which the particles after exposure to NO_2 or NO_3 are coated with Nitron, proved that in 50% of the soil dust particles studied, mixed nitrate particles were formed upon reaction with NO_2 and HNO_3 and that both gases reacted with 95% of all sea-salt particles to produce a surface coating of nitrates. Based on these results, the heterogeneous nitrate formation could be classified as an important removal process for NO_x under solar radiation in the ambient atmosphere.

On-line RSMS investigations performed by Carson et al. [18] on the behaviour of NaCl particles upon exposure to NH_3 and HNO_3 vapour prior to sampling into the MS also provide evidence for the formation of an NH_4NO_3 coating on the particles' surface. Since no change in particle size distribution could be observed by an aerodynamic particle sizer, the coating thickness seemed to be less than 200 nm. RSMS was also performed for the speciation of semi-volatile and non-volatile sulfur compounds as well as for the measurement of isotopic ratios. [18]

The composition of giant North Sea aerosol particles was investigated by Injuk et al., [64,65] using the SPM analytical facilities of the Institute of Reference Materials and Methods (IRMM) in Belgium and of the Lund Institute of Technology in Sweden. Together with other microanalytical research centres they offer the possibility of a total quantitative scanning analysis (TQSA), a new way of data acquisition and processing. Under these conditions, all the incoming data are recorded in the original time sequence of events or stored as sorted spectra in three-dimensional data sets, without any limitation on the amount of data taken up due to predefined energy regions. This mode allows us to follow the complete experiment as it evolves in time and it is very valuable to investigate a possible instability of samples under microbeam irradiation or specimen damage. Storage of the energy spectra for each pixel can easily provide various cross-sections through the data set but restricts scanning frames to 64×64 pixels. Major, minor and trace elements with atomic number >15 were identified by micro-PIXE and quantification of the acquired spectra revealed elemental compositions down to absolute masses of 50 fg in the case of Ti, V or Cr. Three different aerosol types could be distinguished: particles with a marine origin including pure sea-salt particles and sea-salt combined with high contents of S, K and Ca and anthropogenic particles rich in Ti, Cr, Fe and Ni. The lateral elemental distribution within different giant North Sea aerosol particles was characterized with a constant resolution of 3 μm throughout the depth of the particles. The elemental maps showed that these particles were heterogeneous and seemed to consist of agglomerated large particles.

3.4 REMOTE AEROSOLS

The Antarctic continent is an ideal site to follow the trends in regional and global concentrations of trace elements in the atmosphere. Due to the surrounding ocean and the very low soil dust aerosol load, the Antarctic peninsula is also an ideal site to study properties of marine aerosol particles. [66]

A large diversity of particles of different origin was observed in the Antarctic troposphere; among them a significant number were heterogeneous. [67] NaCl and $MgCl_2$ particles contained small amounts of S, possibly indicating reactions of these particles with gaseous sulfur compounds. Internal mixtures of silicates and marine aerosols were observed too. Most of the S in particles larger than 0.1 μm was in the form of $CaSO_4$, and the abundance of these particles showed a seasonal variability with a maximum in summer. The S-containing particles are important because they could play a role in global climate change by influencing the radiation budget and the concentration of cloud condensation nuclei. [68]

3.5 INDOOR AEROSOLS

In an effort to evaluate the possible effects of airborne particles on paintings, a combination of different analytical techniques, including EPXMA, SEM- and STEM-EDX and SPM, was successfully applied to the characterization of individual aerosol particles collected at the Correr museum in Venice. [69] Multivariate techniques performed on the EPXMA and SEM-EDX data revealed three different aerosol sources and six to eight different particle types. To study the spatial distribution of the elements inside individual Ca-rich giant aerosol particles (>8 μm), the possibilities of a JEOL JSM 6300 SEM in the field of X-ray mapping were investigated. The optimal parameter adjustment was obtained by testing the variation of different instrumental parameters on the collection of an X-ray mapping. Processing of the collected X-ray mappings was performed by Image Math (Figure 4), a program which enables mathematical operations on images. The X-ray mapping results clearly demonstrated the heterogeneous character of this particle type. They mainly contained Ca and Si with small particles, identified as aluminosilicates, $CaSO_4$ particles or very small Fe-rich particles, adsorbed at their surface. Na and Cl were correlated as sea-salt crystals or found as a coating on the surface

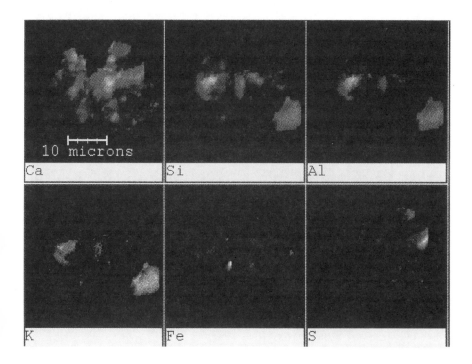

Figure 4. (a) X-ray elemental mappings by SEM-EDX on an indoor aerosol particle collected in the Correr museum in Venice, Italy.

of the particles. The occurrence of these giant particles can probably be attributed to the coagulation of giant Ca–Si-rich particles, produced by decomposition of wall plaster and cement through ageing or introduced from the outside, with small particles present in the indoor environment. Verification of these results as well as information on the distribution and semi-quantization of trace elements inside these individual giant aerosol particles was achieved by SPM measurements.

3.6 ARTIFICIALLY GENERATED AEROSOLS

Inorganic salts account for a 25–50% of the fine aerosol mass in polluted urban environments; among them, $(NH_4)_2SO_4$ and NH_4NO_3 are the most common. Hence, understanding the mechanism of crystallization of different aerosol salts is crucial to the elucidation of the atmospheric processes affecting air quality, visibility degradation and climate change. But very few studies have been

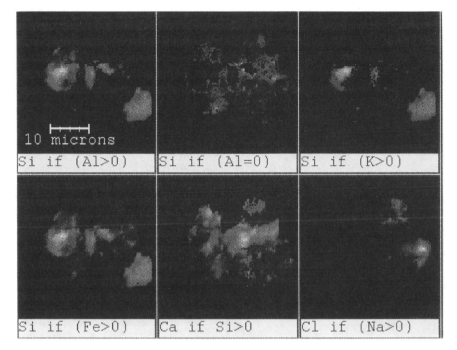

Figure 4. (b) Application of the Image Math program on the elemental X-ray mappings. Ca and Si are clearly associated and three aluminosilicate particles as well as a few NaCl particles and an Fe-rich particle appear to be adsorbed at the surface of the particle

performed in this direction, due to limitations in available thermodynamic data and experimental methods. Based on thermodynamic analysis, the work of Ge *et al.* [70] has shown that particles dried from multi-component aqueous aerosols do not have a homogeneous chemical morphology except at the eutonic point. (When two salts, at a certain RH, precipitate together they form a mixture of solid phases, and a corresponding aqueous phase composition is called the eutonic composition.) Rapid single particle mass spectrometry (RSMS) was used to examine the surface morphology and chemical composition of particles dried from KCl/NaCl, KCl/KI and $(NH_4)_2SO_4/NH_4NO_3$ mixed solutions at different mole ratios. For a simple salt mixture such as KCl/NaCl and KCl/KI, the surface composition of the dried particles is identical to the eutonic composition and is independent of the original solution mole fraction. Thermodynamic analysis showed that the composition of dried particles containing KCl/NaCl is not spatially homogeneous, but the surface composition of particles dried from solutions with different mole ratio of NaCl and KCl is the same. The same pattern was observed for particles containing

KCl/KI mixtures. One consequence of this behaviour is that the surface layer is enriched in the minor component. Particles dried from $(NH_4)_2SO_4/NH_4NO_3$ mixture solutions contained substantial amounts of $(NH_4)_2SO_4$ at or near the surface. The surface composition of the particles is in general identical to each other, it is independent of the original solution composition and is most likely equal to that of the eutonic point.

4 CONCLUSIONS

In this chapter we have demonstrated how an increasing interest in the analysis of particle surfaces and the need for three-dimensional characterization of a single aerosol particle have led to the application of many micro- and trace analytical techniques. Microanalysis has been able, in many cases, to produce interesting and almost unique information on the intra-particle structure. Since no existing microanalysis technique presently offers a panacea for all analytical demands, the most successful research programmes will be those in which several complementary techniques are to be used to provide relevant analytical information of environmental interest. As the structural information of a microparticle is contained in its exterior boundary and interior texture, there are continued demands for higher resolutions and better sensitivities. Also, the power of computer-based image processing for a variety of applications has not yet been fully realized.

In this review, mostly off-line methods for particle characterization were described. These methods collect particles on substrates or filters for an extended period of time for later compositional analysis. In such a way, the information on particle composition and possible sources is given, but the time information for each particle is missing. Additionally, a size selectivity as well as artefacts due to the sample preparation can be introduced. As a consequence, the particles from one or more events of interest are not readily distinguished from particles unassociated with the event. Also, particles smaller than roughly 0.08 μm do not usually contain sufficient mass for reliable compositional determination. At present there are only two methods described in the literature capable of single particle analysis in an on-line mode: atomic emission spectroscopy and mass spectrometry. For atomic emission spectroscopy, the particles are vaporized and ionized by either a laser spark or an inductively coupled plasma. The detection limits are element dependent and range from 100 to a few thousand ppm. The laser ionization mass spectroscopy offers new capabilities to the aerosol community like: highly efficient vaporization and ionization, general applicability generation of molecular and elemental information, ability to detect volatile species (particles spend less than 1 ms under vacuum conditions before ionization), possibility to study gas-phase reactions yielding solid products. In contrast to filters or microscopic grids retained for analysis, each particle is sampled at a well-defined time. Even in rapidly changing conditions, it is possible to go back and retroactively correlate the chemical composition with meteorological or other conditions. Particles of different sizes (0.02–$10\,\mu m$) and composition like $(NH_4)_2SO_4$, SiO_2, $RbNO_3$ and $CsNO_3$ could be detected and analized. [20]

The application of the X-ray photo electron spectroscopy (XPS or ESCA) in the environmental field is still limited to bulk analysis. XPS is, like AES, a surface analytical technique in which in this case the core-level electrons from the sample are ejected upon X-ray beam interaction. The energy analysis of the produced photoelectrons reveals information on the chemical bonding. XPS is not restricted to low Z-elemental analysis and the detection limits are comparable to AES. Up to now single particle analysis is only possible with the spatial resolution being in the order of $10\,\mu m$. Improvements in both the spatial resolution and the sensitivity are under investigation.

ACKNOWLEDGEMENTS

Lieve De Bock acknowledges financial support by the Instituut voor de Bevordering van het Wetenschappelijk-Technologisch Onderzoek in de Industrie (IWT). This work was partially sponsored by the Belgian State Prime Minister's Service-Science Policy Office, in the framework of the Impulse Programme in Marine Sciences (Contract MS/06/050).

REFERENCES

1. Jambers, W., L. De Bock, L. and Van Grieken, R. (1996) Applications of microanalysis to individual environmental particles, *Fresenius' J. Anal. Chem.*, **355**, 521.
2. Xhoffer, C. and van Grieken, R. (1993) Environmental Aerosol Characterization by Single Particle Analysis Techniques, in: *Environmental Particles*, Vol. 2. IUPAC Environmental Analytical and Physical Chemistry Series, J. Buffle and H. P. Van Leeuwen Eds. (CRC Press, Inc., Florida, USA, Lewis Publishers), 207.
3. Armstrong, T. J. (1995) CITZAF. A package of correction programs for the quantitative electron microbeam X-ray analysis of thick polished materials, thin films, and particles, *Microbeam Analysis*, **4**, 177.
4. Danilatos, D. G. (1994) Environmental scanning electron microscopy and microanalysis, *Mikrochim. Acta*, **114/115**, 143.
5. Nockolds, C. E. (1994) Low-voltage electron probe X-ray microanalysis in an in-lens field emission gun SEM: an evaluation, *Microbeam Analysis*, **3**, 185.
6. Katrinak, K., Rez, P. and Buseck, P. R. (1992) Structural variations in individual carbonaceous particles from the urban aerosol, *Envir. Sci. Technol.*, **26**, 1967.
7. Maynard, A. D. and Brown, L. M. (1992) Electron energy loss spectroscopy of ultrafine aerosol particles in scanning transmission electron microscopy, *J. Aerosol Sci.* **23**, S433.
8. Xhoffer, C., Jacob, W., Buseck, P. R. and Van Grieken, R. (1995) Problems in quantitatively analysing individual salt aerosol particles using electron energy loss spectroscopy, *Spectrochim. Acta*, **Part B 50**, 1281.

9. Shina, M. P. (1984) Laser-induced volatilization and ionisation of Microparticles, *Rev. Sci. Instrum.*, **55**, 886.

10. Marijnissen, J., Scarlett, B. and Verheijen, P. (1988) Proposed on-line aerosol analysis combining size determination, laser-induced fragmentation and time-of-flight mass spectroscopy, *J. Aerosol Sci.*, **19**, 1307.

11. McKeown, P. J., Johnston, M. V. and Murphy, D. M. (1991) On-line single-particle analysis by laser desorption mass spectrometry, *Anal. Chem.*, **63**, 2069.

12. Kievit, O., Marijnissen, J. C. M., Verheijen, P. J. T. and Scarlett, B. (1992) On-line measurements of particle size and composition, *J. Aerosol Sci.*, **23**, 301.

13. Mansoori, B. A., Johnston, M. V. and Wexler, A. S. (1994) quantitation of ionic species in single microdroplets by on-line laser desorption/ionisation, *Anal. Chem.*, **66**, 3681.

14. Murray, K. K. and Russel, D. H. (1994) Aerosol matrix-assisted laser desorption ionization mass spectrometry, *J. Am. Soc. Mass Spectrom.*, **5**, 1.

15. Prather, K. A., Nordmeyer, T. and Salt, K. (1994) Real-time characterization of individual aerosol particles using time-of-flight mass spectrometry, *Anal. Chem.*, **66**, 1403.

16. Nordmeyer, T. and Prather, K. A. (1994) Real-time measurement capabilities using aerosol time-of-flight mass spectrometry, *Anal. Chem.*, **66**, 3540.

17. Hinz, K. P., Kaufmann, R. and Spengler, B. (1994) Laser-induced mass analysis of single particles in the airborne state, *Anal. Chem.*, **66**, 2071.

18. Carson, P. G., Neubauer, K. R., Johnston, M. V. and Wexler, A. S. (1995) On-line chemical analysis of aerosols by rapid single-particle mass spectrometry, *J. Aerosol Sci.*, **4**, 535.

19. Murphy, D. M. and Thomson, D. S. (1995) Laser ionization mass spectroscopy of single aerosol particles, *Aerosol Sci. Technol.*, **22**, 237.

20. Reents, W. D., Downey, S. W., Emerson, A. B., Mujsce, A. M., Muller, A. J., Siconolfi, D. J., Sinclair, J. D. and Swanson, A. G. (1995) Single particle characterization by time-of-flight mass Spectrometry, *Aerosol Sci. Technol.*, **23**, 263.

21. Hinz, K-P., Kaufmann, R. and Spengler, B. Simultaneous detection of positive and negative ions from single airborne particles by real-time laser mass spectrometry, *Aerosol Sci. Technol.*, in press.

22. Malmqvist, K. G. (1994) Ion beam analysis for the environment, *Nucl. Instr. Methods in Physics Research*, **B 85**, 84.

23. Kondratyev, K. Y. and Binenko, V. I. (1987) Optical properties of dirty clouds, in: *Interactions between Energy Transformations and Atmospheric Phenomena*, M. Beniston and R. A. Pielke, Eds. (Reidel, Dordrecht).

24. Linton, R. W., Loh, A., Natusch, D. F. S. and Williams, P. (1975) Surface predominance of trace elements in airborne particles, *Science*, **191**, 853.

25. Linton, R. W., Williams, P., Evans, C. A. Jr and Natusch, D. F. S. (1977) Determination of the surface predominance of toxic elements in airborne particles by ion microprobe mass spectrometry and Auger electron spectrometry, *Anal. Chem.*, **49**, 1514.

26. Cox, X. B. III, Bryan, S. R., Linton, R. W. and Griffis, D. P. (1987) Microcharacterization of trace elemental distributions within coal combustion particles using secondary ion mass spectrometry and digital imaging, *Anal. Chem.*, **59**, 2018.

27. Vis, R. D., A. J. J. Boss, Valkovic, V. and Verheul, H. (1983) The analysis of fly-ash particles with a proton microbeam, *IEEA Trans. Nucl. Sci.*, **NS-30**, 1236.

28. Jakšić, M., Watt, F., Grime, G. W., Cereda, E., Marcazzan, G. M. B. and Valković, V. (1991) Proton microprobe analysis of trace element distribution in fly-ash particles, *Nucl. Instr. Methods in Physics Research*, **B 56/57**, 699.

29. Cereda, E., Marcazzan, G. M. B., Pedretti, M., Grime, G. W. and Baldacci, A. (1995) Nuclear microscopy for the study of coal combustion related phenomena, *Nucl. Instr. Methods in Physics Research*, **B 99**, 414.
30. Van Malderen, H., Van Grieken, R., Bufetov, V. and Koutzenogii, K. (1996) Chemical characterization of individual aerosol particle types in central Siberia, *Environ. Sci. Technol.*, **30**, 312.
31. Natusch, D. F. S., Wallace, J. R. and Evans, C. A. (1974) Toxic trace elements: preferential concentrations in respirable particles, *Science*, **183**, 202.
32. Mamane, Y., Miller, J. L. and Dzubay, T. G. (1986) Characterization of individual fly-ash particles emitted from coal-and oil-fired power plants, *Atmos. Environ.*, **11**, 2125.
33. Biermann, A. H. and Ondov, J. M. (1980) Application of surface-deposition models to size-fractionated coal fly-ash, *Atmos. Environ.*, **14**, 289.
34. Hock, J. L. and Lichtman, D. (1983) A comparative study of in-plume and in-stack collected individual coal fly-ash particles, *Atmos. Environ.*, **17**, 849.
35. Boni, C., Cereda, E., Braga Marcazzan, G. M. and Parmigiani, F. (1995) Bulk and surface studies of fly-ash particles, in: *Combustion Efficiency and Air Quality*, Eds. I. Hargittai and T. Vidoczy (Plenum Press, New York).
36. Parungo, F., Kopcewicz, B., Nagamoto, C., Schnell, R., Sheridan, P., Zhu, C. and Harris, J. (1992) Aerosol particles in the Kuwait oil fire plumes: their morphology, size distribution, chemical composition, transport, and potential effect on climate, *J. Geophys. Res.*, **97**, 15867.
37. Bigg, E. K., Ono, A. and Williams, Y. A. (1974) Chemical tests for individual submicron aerosol particles, *Atmos. Environ.*, **8**, 1.
38. McHugh, J. A. and Stevens, J. F. (1972) Elemental analysis of single micrometer-size airborne particulates by ion microprobe mass spectrometry, *Anal. Chem.*, **44**, 2187.
39. Hachimi, A., Poitevin, E., Krier, G., Muller, J. F., Pironon, J. and Klein, F. (1993) Extensive study of oxidation states of chromium in particles by LAMMA and Raman microprobes: application to industrial hygiene, *Analusis*, **21**, 77.
40. Goschnik, J., Schuricht, J. and Ache, H. J. (1994) Depth-structure of airborne microparticles sampled downwind from the city of Karlsruhe in the river Rhine valley, *Fresenius' J. Anal. Chem.*, **350**, 426.
41. Keyser, T. R., Natusch, D. F. S., Evans, C. A. Jr and Linton, R. W. (1978) Characterizing the surfaces of environmental particles, *Envir. Sci. Technol.*, **12**, 768.
42. Bentz, J. W. G., Fichtner, M., Goschnick, J. and Ache, H. J. (1994) Depth-resolved analysis of microparticles generated by a household fire, in: *Abstracts from Fourth International Aerosol Conference*, 29 August–2 September 1994, Los Angeles, California, R. C. Flagan, Ed. (American Association for Aerosol Research, Kemper Meadow Drive, Cincinnati, OH, USA).
43. Orlić, I., Watt, F., Loh, K. K. and Tang, S. M. (1994) Nuclear microscopy of single aerosol particles, *Nucl. Instr. Methods in Physics Research*, **B 85**, 840.
44. Mueller, P. K., Mosley, R. W. and Pierce, L. B. (1972) Carbonate and non-carbonate carbon content in photochemical aerosols as a function of size, *J. Colloid Interface Sci.*, **39**, 235.
45. Appel, B. R., Hoffer, E. M., Kothny, E. L., Wall, S. M. and Haik, M. (1979) Analysis of carbonaceous material in Southern California atmospheric aerosols, *Environ. Sci. Technol.*, **13**, 98.
46. Turpin, B. J. and Huntzicker, J. J. (1991) Secondary formation of organic aerosol in the Los Angeles Basin: a descriptive analysis of organic and elemental carbon concentrations, *Atmos. Environ.*, **25**, 207.

47. Katrinak, K., Rez, P. and Buseck, P. R. (1992) Structural variations in individual carbonaceous particles from an urban aerosol, *Environ. Sci. Technol.*, **26**, 1967.

48. Wonders, J. H. A. M., Houweling, S., De Bont, F. A. J., Van Leeuwen, H. P., Eeckhaoudt, S. M., and Van Grieken, R. (1994) Characterization of aerosol-associated lead by DPASV and LAMMA, *Intern. J. Environ. Anal. Chem.*, **56**, 193.

49. Walton, W. H., Dodgson, J., Hadden, G. G. and Jacobsen, M. (1977) The effect of quartz and other non-coal dusts in coal workers' pneumoconiosis, in: *Inhaled Particles* Vol. 4, W. H. Walton, Ed. (Oxford, Pergamon Press) pp. 669.

50. Tourmann, J. L. and Kaufmann, R. (1993) Laser microprobe mass spectrometry (LAMMS) of coal mine dusts: single particle analysis and toxicity correlation, *Intern. J. Environ. Anal. Chem.*, **52**, 215.

51. Le Bouffant, L., Daniel, H., Martin, J. C. and Bruyère, S. (1982) Effect of impurities and associated minerals on quartz toxicity, *Ann. Occup. Hyg.*, **26**, 625.

52. Tourmann, J. L. and Kaufmann, R. (1994) Laser microprobe mass spectrometric (LAMMS) study of quartz-related and non-quartz related factors of the specific harmfulness of coal mine dusts, *Ann. Occup. Hyg.*, **38**, 455.

53. Andreae, M. O., Charlson, R. J., Bruynseels, F., Storms, H., Van Grieken, R. and Maenhaut, W. (1986) Internal mixture of sea-salt, silicates and excess sulfate in marine aerosols, *Science*, **232**, 1620.

54. Dierck, I., Michaud, D., Wouters, L. and Van Grieken, R. (1992) Laser microprobe mass analysis of individual North Sea aerosol particles, *Environ. Sci. Technol.*, **26**, 802.

55. Qian, G. W. and Ishizaka, Y. J. (1993) Electron microscope studies of methane sulfonic acid in individual aerosol particles, *J. Geophys. Res. Oceans*, **98**, 8459.

56. Pardess, D., Levin, Z. and Ganor, E. (1992) A new method for measuring the mass of sulfur in single aerosol particles, *Atmos. Environ.*, **4**, 675.

57. Cheng, R. J. (1993) Shell structured marine aerosol: a laboratory observation, *Proc. of the Twelfth Annual Meeting of the American Association for Aerosol Research*, held at Oak Brook, Illinois, 11–15 October.

58. Cheng, R. J., Blanchard, D. C. and Cipriano, R. J. (1988) The formation of hollow sea salt particles from the evaporation of drops of seawater, *Atmos. Environ.*, **22**, 15.

59. Bruynseels, F., Storms, H. and Van Grieken, R. (1988) Characterisation of North Sea aerosols by individual particle analysis, *Atmos. Environ.*, **11**, 2593.

60. Ottley, C. J. and Harrison, R. M. (1992) The spatial distribution and particle size of some inorganic nitrogen, sulphur and chlorine species over the North Sea, *Atmos. Environ.*, **9**, 1689.

61. Otten, Ph., Bruynseels, F. and Van Grieken, R. (1986) Nitric acid interaction with marine aerosols sampled by impaction, *Bull. Soc. Chim. Belg.*, **95**, 447.

62. Mamane, Y. and Gottlieb, J. (1992) Nitrate formation on sea-salt and mineral particles—a single particle approach, *Atmos. Environ.*, **9**, 1763.

63. Mamane, Y. and Pueschel, R. F. (1980) A method for the detection of individual nitrate particles, *Atmos. Environ.*, **14**, 629.

64. Injuk, J., Van Malderen, H., Van Grieken, R., Swietlicki, E., Knox, J. M. and Schofield, R. (1993) EDXRF study of aerosol composition variations in air masses crossing the North Sea, *X-Ray Spectrom.*, **22**, 220.

65. Injuk, J., Breitenbach, L., Van Grieken, R. and Wätjen, U. (1994) Performance of a nuclear microprobe to study giant marine aerosol particles, *Mikrochim. Acta.*, **144/155**, 313.

66. Fitzgerald, J. (1991) Marine aerosols: a review, *Atmos. Environ.*, **25A**, 533.

67. Artaxo, P., Rabello, M. L. C., Maenhaut, W. and Van Grieken, R. (1992) Trace elements and individual particle analysis of atmospheric aerosols from the Antarctic Peninsula, *Tellus*, **44B**, 318.
68. Götz, G. (1991) Aerosols and climate, in: *Atmospheric Particles and Nuclei*, Eds. G. Götz, E. Mészàros and G. Vali (Akademiai Kiaido, Budapest). . . .
69. De Bock, L. A., Van Grieken, R. E. and Camuffo, D. (1996) Micro-analysis of museum aerosols to elucidate the soiling of paintings: case of the Correr Museum in Venice, Italy, *Environ. Sci. Technol.*, **30**, 3341.
70. Ge, Z., Wexler, A. S. and Johnston, M. V. (1996) Multicomponent aerosol crystallization, *J. Coll. Int. Sci.*, submitted.

6 Kinetics and Thermodynamics of Tropospheric Aerosols

ANTHONY S. WEXLER and SUDHAKAR POTUKUCHI

University of Delaware, Newark, USA

1 INTRODUCTION

Condensation is the primary mechanism for growth of tropospheric particles. Coagulation has been shown to be a much slower growth mechanism than condensation and heterogeneous chemical reactions do not appear to contribute substantial mass at relative humidities below 95% or so. [53] The earliest models of aerosol mass assumed gas–particle equilibrium when partitioning volatile compounds such as ammonium nitrate and ammonium chloride between the vapour and condensed phases [5, 6, 31, 43, 44] and more accurate versions have recently been developed. [24–26] This assumption was applied in Los Angeles to predict the total particulate mass of these volatile compounds. [31, 32, 38, 39, 40] Subsequently, it was observed that equilibrium may not

Atmospheric Particles, Edited by R. M. Harrison and R. Van Grieken.

always be a valid assumption in the troposphere for predicting total particle mass [1, 48]—the equilibrium assumption is better in highly polluted locations such as Los Angeles because large particle surface area facilitates rapid condensation, [18, 50] but in less polluted locations the time to reach equilibrium may be too long for equilibrium to apply.

More recently emphasis has shifted from predicting total aerosol mass to predicting the size distribution of particles in the atmosphere and how composition varies with size. This interest is motivated in part by a number of studies that suggest a correlation between particle concentrations in the atmosphere, and human morbidity and mortality, [27] along with the fact that deposition in human airways is size dependent. It has been shown that assuming gas–particle equilibrium leads to an underdetermined system of equations when predicting the size distribution of volatile compounds. [50] By considering mass transport between the gas and particle phases, the size distribution of the volatiles can be uniquely predicted.

This chapter explores two issues: aerosol kinetics and aerosol thermodynamics. Kinetics, including mass transport, places limits on the validity of thermodynamic equilibrium assumptions. By offering a more complete description of kinetic issues in tropospheric aerosols, we hope to elucidate when thermodynamic equilibrium assumptions are valid and identify research areas where theoretical and experimental exploration of aerosol kinetic issues are warranted. Thermodynamics describes the equilibrium state of a system. Understanding the thermodynamics of atmospheric aerosols is crucial to defining the end state of a kinetic process, and therefore it is essential for understanding atmospheric aerosol dynamics.

Furthermore, aerosol thermodynamics is unique in conventional solution and phase equilibrium thermodynamics because in the atmosphere the independent variable is relative humidity, whereas most texts view composition as the independent variable. Here we derive from first principles the equations and concepts that govern aerosol thermodynamics with the goal of making this field accessible to a wider audience.

2 AEROSOL KINETICS

Before we explain aerosol thermodynamics, it is worthwhile exploring common assumptions employed in aerosol modelling. Many of the assumptions are forced by limitations in the available experimental data and some are employed to simplify the situation sufficiently so that it can be readily understood or efficiently modelled mathematically. In any case, all models of systems may break down when applied inappropriately or under conditions where the underlying assumptions do not apply. As a consequence, we review the assumptions often employed by those trying to understand atmospheric aerosol systems using thermodynamic principles.

2.1 AEROSOL–VAPOUR EQUILIBRIUM

Aerosols often contain volatile components, such as water, nitric acid, ammonium and hydrochloric acid. The vapour pressure of volatile compounds is determined by the composition of the particles. Thus, it is convenient to assume that the vapour phase of these components is in equilibrium with their aerosol phase, since acquisition of the particle composition then leads to prediction of the vapour phase composition.

There are a wide variety of reasons why this assumption may not hold. These include surface tension effects and other microscale phenomena, effects of particle composition variations with size or within a particle size, and departures from phase equilibrium within the particles. These issues are discussed below—here we focus on departures from equilibrium due to rate limitations.

Consider a single aerosol particle surrounded by the atmospheric gases, including vapours which may condense or evaporate. If a vapour condenses, the composition of both the particles and the ambient gas changes. The change in mass, M_i, of compound i in a single particle can be written as

$$\frac{dM_i}{dt} = \frac{4\pi R_p D_i}{RT} \frac{(p_{i,\infty} - p_{i,s})}{1 + D_i/\alpha_i \bar{C}_i R_p} \tag{1}$$

where R_p is the particle radius assuming it is spherical, R is the gas constant, T is the ambient temperature, D_i is the molecular diffusivity of the vapour, $p_{i,\infty}$ is the partial pressure far from the particle, $p_{i,s}$ is the vapour pressure at the particle surface, α_i is the accommodation coefficient of compound i, and $\bar{C}_i$ is the root mean square velocity of the vapour molecules. [50] A number of alternate expressions are available in the literature—see Pandis $et\ al.$ [29] for a review. In the large particle and large accommodation coefficient limit, mass transport to the particle is governed by diffusion and is proportional to R_p, while in the small particle or small accommodation coefficient limit, mass transport is governed by surface accommodation and is proportional to R_p^2. A characteristic time for changing the vapour pressure of compound i over the particle can be defined as

$$\tau_{i,p}^{-1} \equiv \frac{1}{p_{i,s}} \frac{dp_{i,s}}{dt} = \frac{K}{m_w RT} \frac{4\pi R_p D_i}{1 + D_i/\alpha_i \bar{C}_i R_p} \frac{(p_{i,\infty} - p_{i,s})}{p_{i,s}} \tag{2}$$

where $p_{i,s} = KM_i/m_w$ is the equilibrium relation between particle composition and surface vapour pressure, K is the equilibrium constant and m_w is the water content of the particle. [50, 52]

When condensation occurs, it occurs on a population of particles, so that the change in ambient partial pressure during condensation is the integral of equation (1) over the size spectrum

$$\frac{\mathrm{d}p_{i,\infty}}{\mathrm{d}t} = \int_0^\infty 4\pi R_\mathrm{p} D_i n(R_\mathrm{p}) \frac{(p_{i,\infty} - p_{i,s})}{1 + D_i/\alpha_i \bar{C}_i R_\mathrm{p}} \mathrm{d}R_\mathrm{p} \tag{3}$$

where $n(R_\mathrm{p})$ is the number distribution of particles on which the vapour condenses. Once again we can define a characteristic time, now for changes in the ambient partial pressure, as

$$\tau_{i,\infty}^{-1} \equiv \frac{1}{p_{i,\infty}} \frac{\mathrm{d}p_{i,\infty}}{\mathrm{d}t} = 4\pi D_i \int_0^\infty \frac{R_\mathrm{p} n(R_\mathrm{p})}{1 + D_i/\alpha_i \bar{C}_i R_\mathrm{p}} \frac{p_{i,\infty} - p_{i,s}}{p_{i,\infty}} \mathrm{d}R_\mathrm{p}. \tag{4}$$

Many processes may cause the partial pressures $p_{i,\infty}$ and $p_{i,s}$ to change with time. The ambient partial pressure may change due to mixing with other air masses, photochemical formation or destruction, and deposition. The particle surface vapour pressure may change due to condensation or evaporation of compounds on the particle, temperature variations, or chemical reactions or phase transformations within the particle.

Equilibrium between the gas and particle phases will be a good assumption if the particle composition time-scale or vapour time-scale is short compared to the time-scales over which other processes change the ambient concentrations. If the particle composition time-scale is short, the particle surface vapour pressure adjusts rapidly due to condensation maintaining equilibrium, whereas if the vapour phase time-scale is short, the ambient partial pressures change rapidly in response to condensation to maintain equilibrium.

We can see from equation (4) that the magnitude of the time-scale is governed by both a thermodynamic driving force, $(p_{i,\infty} - p_{i,s})/p_{i,\infty}$, and a term that we refer to as the transport moment of the particle size distribution. We leave it inside the integral because the vapour pressure, $p_{i,s}$, may be a function of particle size. The driving force is unity for compounds with negligible surface partial pressure such as sulfuric acid or ammonia on acidic particles. The driving force is zero in gas–particle equilibrium.

The transport moment ranges between the first and second moments depending on the particle size distribution and transport properties of the aerosol that alter the relative size of $D_i/\alpha_i \bar{C}_i R_\mathrm{p}$ compared to unity. We have shown that $D_i/\alpha_i \bar{C}_i R_\mathrm{p} \sim Kn/\alpha_i$—transport is surface accommodation limited and proportional to R_p^2 when $R_\mathrm{p}\alpha_i \ll \lambda$, whereas it is diffusion limited when $R_\mathrm{p}\alpha_i \gg \lambda$— where Kn is the Knudsen number and λ is the mean free path of air.

In the troposphere, the partial pressure, $\tau_{i,\infty}$, and vapour pressure, $\tau_{i,s}$, time-scales can take on values ranging from seconds to hours. [21, 22, 50, 52] When either time-scale is short compared to the characteristic time over which vapour concentrations change in the atmosphere, particle–vapour equilibrium can be assumed, whereas when both are long equilibrium is not a good assumption. The particle vapour pressure time constant is infinite when the particle contains a solid phase of the condensing compound or an aqueous phase that is dominated by the condensing compound [50, 52] because in these circumstances the vapour pressure does not change with composition. This is the case for ammonium nitrate in the SoCAB (South Coast Air Basin of California) under dry conditions (the aerosol contains a solid phase of ammonium nitrate) or when the particulate aqueous concentration of ammonium nitrate is much higher than that of other solutes. When this constant is infinite, equilibration between the gas and aerosol phases can only take place by changes in the ambient partial pressure since the particle vapour pressure is constant.

The particle vapour pressure time-scale is short when the concentration of the compound in the particle is small—small amounts of condensation or evaporation lead to large relative changes in composition and therefore vapour pressure—but other circumstances also favour small values. [50]

The ambient partial pressure time-scale is generally governed by the transport moment of the particle size distribution. For a given size distribution, a

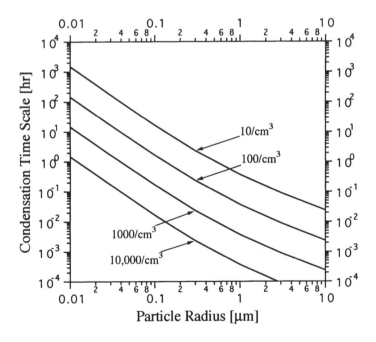

Figure 1. The time constant for equilibration between the gas and aerosol phases as a function of particle number concentration and size. The accommodation coefficient is assumed to be 0.1

larger number of particles leads to a shorter time constant value, and for a given total number of particles, larger particle sizes lead to shorter time constant values. Generally, less polluted locations have longer vapour time constant values and therefore larger departures from equilibrium than more polluted locations. The time-scale for various particle numbers and sizes is illustrated in Figure 1.

Compared to other topics governing aerosol kinetics, the rates of condensation and evaporation are relatively well understood. The primary difficulty resides with the accommodation coefficient. The accommodation coefficient is the ratio of the number of molecules sticking to a particle to the number that strike the particle. Its magnitude is a function of the condensing molecule, the phase and composition of the particle surface, and possibly other factors such as the ambient temperature and relative humidity. Accommodation coefficients are notoriously difficult to measure in bulk phases. [4] In aerosol phases, the large surface area to volume ratio makes the measurements somewhat more straightforward, but many difficulties may still arise. Nevertheless, numerous measurements of accommodation coefficients relevant to atmospheric aerosols have been performed on a limited number of aerosol systems. Most measurements are of accommodation on pure aqueous droplets which are not atmospherically relevant at relative humidities below 100%. Only a very few studies include surface coatings of organic compounds which may lower the accommodation coefficient of highly soluble compounds by an order of magnitude or more. [13, 16, 37] Lack of knowledge of accommodation coefficient values for common atmospheric constituents on actual atmospheric particles leads to significant uncertainty in predicting the size distribution of condensing compounds. [23, 30]

2.2 PHASE PARTITIONING IN PARTICLES

We now look at particle morphology and how it may limit equilibration between the particle and vapor phases. Consider a particle composed of solid phases of sodium chloride and ammonium sulfate. Such an aerosol is common in coastal urban areas due to condensation of ammonia and sulfuric acid on sea-salt particles. Solid–solid chemical reactions are notoriously slow because the diffusivity in solid phases is small. Thus these chemical compounds can coexist for long periods of time, as long as they remain solid. If they were to react under appropriate stoichiometric conditions, ammonium chloride might form and possibly evaporate. The deliquescence point of sodium chloride is about 75% and that for sodium sulfate is about 80%, so at relative humidities substantially below these values the solid phases could be unstable, but coexist. In contrast, it has been shown that even at relative humidities below the deliquescence point, water may be adsorbed on the surface of particles resulting in a small volume where reactions between solid phases may occur. [49] Thus, the atmospheric conditions and persistence of metastable aerosol solid phases is unclear.

Now consider two non-reacting compounds in a single aerosol particle, say ammonium nitrate and ammonium sulfate—a common combination in western Europe and North America. At high relative humidity, these compounds exist as dissolved ions in an aqueous droplet. Let us examine what transpires as the relative humidity is lowered. First water evaporates to maintain the water activity equal to that of the ambient relative humidity. Eventually, one of the solutes becomes saturated, and if a pre-existing nucleus is present, a solid phase of that compound forms. Continued lowering of the relative humidity causes more of the solid phase to form and also concentrates the non-precipitating compound in the remaining aqueous phase. Eventually, the second compound becomes saturated and further reductions in relative humidity result in complete efflorescence. The relative humidity of the aqueous phase that is in equilibrium with both solid phases is unique and defines the so-called mutual deliquescence point. [51] The composition of the aqueous phase at this point is also unique and is termed the eutonic. [47] For more complex mixtures, multiple deliquescence points may exist. [33, 34]

This drying process may have profound implications for the evaporation rate of particles. For the ammonium nitrate/sulfate particle just discussed, the ammonium nitrate is volatile while the ammonium sulfate is stable. As we proceeded through the drying process, at first we formed a single solid phase. To minimize the surface free energy, this solid phase is most likely to exist in the interior of the particle surrounded by the aqueous phase. At low relative humidity, when crystallization occurs, the initial solid phase will be trapped in the core surrounded by a mixture of ammonium nitrate and ammonium sulfate at the fixed mole ratio defined by the eutonic composition. The solid phases in the shell may exist as a grain structure or as a mixed salt, but this has yet to be explored experimentally. In either case, the volatility of the ammonium nitrate may be severely impaired since ammonium sulfate is mixed with the ammonium nitrate or at the very least may shield some or all the ammonium nitrate from evaporation. Reduced evaporation rates in these aerosols have been observed by Hightower and Richardson. [19]

2.3 EFFLORESCENCE VERSUS CRYSTALLIZATION

Numerous laboratory observations have demonstrated that pure particles, that is, those that do not contain a pre-existing solid phase, do not crystallize until relative humidities are reached that are much below the efflorescence point. [36] This is because the new solid phase must nucleate first and this is not likely in homogeneous aqueous solutions until substantial supersaturations are achieved. These supersaturations have not been thoroughly explored in the atmosphere so their occurrence there is not as certain. In the atmosphere, many of the aerosol particles contain pre-existing solid phases, such as soot and crustal material that provide heterogeneous nucleation sites for the formation of new phases. Heterogeneous nucleation typically does not require supersaturations as high as those required for homogeneous nucleation.

Consider now an aerosol composed of an aqueous phase of water, ammonium nitrate, and possibly other dissolved electrolytes. As we lower the relative humidity, the concentration of the ammonium nitrate in the particle increases until it saturates. If no pre-existing solid phase is present, continuing decreases in relative humidity will result in supersaturations of the ammonium nitrate in solution, which result in partial pressures of ammonia and nitric acid over the particle that are higher than their equilibrium value for the solid phase.

Now consider a population of such particles in the atmosphere. Some of the particles may have formed by binary nucleation of water and sulfuric acid, and therefore may not contain a pre-existing solid phase, whereas others may have been formed by condensation on a pre-existing solid. During transitions to low relative humidity, a solid phase of ammonium nitrate will readily nucleate in particles with a pre-existing solid phase and this solid ammonium nitrate will limit the partial pressure product of ammonia and nitric acid to its equilibrium value. But the particles without a pre-existing solid phase will supersaturate with ammonium nitrate and therefore exhibit a partial pressure product of ammonia and nitric acid that is above that for particles containing the solid phase. As a result, during transitions to low relative humidities, ammonium nitrate will evaporate from particles where it supersaturates and condense on ones where the solid phase forms. Thus relative humidity cycles, such as those that occur diurnally, may redistribute volatiles such as ammonium nitrate from particles without pre-existing solid cores to those with cores that facilitate formation of solid phases by heterogeneous nucleation.

2.4 PHASE EQUILIBRIUM FOR WATER AND THE KELVIN EFFECT

Most aerosol modelling efforts assume that the ambient relative humidity is equal to the water activity in the particles. This equilibrium may be disturbed by the Kelvin effect or if the rate of water condensation or evaporation is sufficiently large. The Kelvin effect increases the vapour pressure of volatile compounds due to the particle's surface free energy. For a given ambient relative humidity the Kelvin effect lowers the water activity in the particle. For pure water at standard temperature and pressure, a 200 nm particle has a 1% lowering in water activity due to the Kelvin effect, and particles smaller than this will experience a greater departure—a 20 nm particle has a lowering of about 10%.

Aqueous atmospheric aerosol particles must contain solutes, such as ammonium nitrate or sodium chloride, or they would not be stable at relative humidities below 100%. These solutes increase the surface tension of the droplets proportional to their concentrations, [35] but at saturation the surface tension increase is only about 10% over its value for a dilute solution. Acting to decrease the droplet surface tension are surface active organics, which have been observed in both the marine boundary layer [2, 3, 7, 15] and in urban smog. [16] The counteracting effects of increased surface tension due to solutes and decreases due to surface active compounds probably result in actual surface tensions about that for pure water. Considering the substantial uncertainty in most ambient relative humidity measurements and the small amount of mass present in the ultra-fine particles, the Kelvin effect can probably be ignored in most atmospheric aerosols and relative humidities below about 95%.

Rapid condensation or evaporation rates may also lead to departures from equilibrium between the liquid and vapour phases of water in the atmosphere. Condensation of water on particles is the result of either changes in ambient relative humidity or condensation of hygroscopic compounds. An analytical expression that describes this rate of change has been derived by Wexler and coworkers. [53] Each of these two effects can be addressed independently.

Consider the change in aerosol water due to condensation of hygroscopic compounds at constant relative humidity. Essentially, condensation of a solute, such as ammonium nitrate, must result in a concomitant condensation of water to maintain the molality of ammonium nitrate in solution constant. The general rate of condensation of any compound is proportional to its atmospheric concentration and since the concentrations of water in the atmosphere are orders of magnitude higher than that of the hygroscopic pollutants, the water will be in relative equilibrium compared to these pollutants. This paradigm holds even at high relative humidities when many moles of water must condense for each mole of pollutant to maintain water equilibrium. For instance, at 95% relative humidity the concentration of most electrolytes is around 1 molal, so over 50 moles of water must condense for each mole of pollutant. The atmospheric pollutant concentrations are typically lower than 100 ppb, whereas that of water is almost always over 1000 ppm. Thus the water is unlikely to exhibit noticeable departures from equilibrium due to condensation of hygroscopic pollutants.

Ambient changes in relative humidity may also potentially cause departures from water equilibrium. In non-precipitating air masses, the relative humidity changes primarily due to changes in temperature since otherwise the overall water content of a parcel of air changes slowly. Rapid changes in ambient temperature are usually associated with fronts, but we are considering here changes in the temperature of a parcel of air, so our reference frame moves with the air mass. Under these conditions, it is safe to assume that the temperature of an air mass will not change more than about 20 °C over 10 hours, so a temperature gradient of 2 °C hour^{-1} will be taken as the limit. This corresponds to a relative humidity change of about 7% RH hour^{-1}. So the minimum characteristic time-scale for changing relative humidity in the atmosphere is about 15 hours. The condensation time-scales are typically much shorter than this—in the clean marine boundary layer they are typically less than 10 hours and in the polluted urban environment they can be as short as a few minutes. [21, 22, 50] Thus, changes in relative humidity in the atmosphere are sufficiently slow that water vapour equilibrium is likely to be maintained.

2.5 CONCLUDING REMARKS

Mass transport limitations can result in departures from gas–particle equilibrium for volatile compounds in the atmosphere. Some of these departures have been explored theoretically and experimentally, but many are still in need of further investigation. In particular, analytical methods for exploring the morphology of ambient particles are not generally available so that surface coatings of organics and phase partitioning of crystalline solids are not amenable to measurement for at least the near future.

As a result of its high concentration, water vapour is well modelled as in equilibrium between the gas and particle phases. In highly polluted urban areas, the transport moment of aerosol particles is sufficiently high that gas–particle equilibrium is probably a good assumption for pollutant vapours such as ammonia and nitric acid. Less polluted areas are more likely to experience some departures. Under low relative humidity conditions, characteristically slow solid–solid reaction rates and diffusivities may also severely limit equilibration. In the next section, equilibrium thermodynamics will be used to predict the composition and phase state of aerosol particles by assuming that some of these limitations to equilibration are not significant.

3 AEROSOL EQUILIBRIUM THERMODYNAMICS

To use the principles of equilibrium thermodynamics to analize atmospheric aerosol particles, we must assume that the system under consideration is in equilibrium—that is, the phases under consideration have the same temperature and pressure, and the components are in chemical equilibrium. To date, all aerosol thermodynamics models have assumed that the phases in the particles are in equilibrium, as are water vapour and liquid water. Beyond this, two paths have been taken. The first assumes that the particles are also in equilibrium with the gas [5, 6, 24–26, 31, 32, 38, 39, 43, 44] and the second does not. [52, 53] The material presented here applies to both paradigms.

3.1 CHEMICAL EQUILIBRIUM BASICS

Consider a closed system. The components and phases of this system are in equilibrium with each other when the entropy of the system is maximized at constant energy and this occurs when the Gibbs free energy, G, is minimized.

[14] At minimum Gibbs free energy, $dG = 0$. By definition, $G \equiv U - TS + pV$, so a differential change in G is

$$dG = dU - T\,dS - S\,dT + p\,dV + V\,dp$$

where U is the internal energy, T the temperature, S the entropy, p the pressure and V the volume. Eliminating dU with the Gibbs relation gives

$$dG = -S\,dT + V\,dp + \sum_i \mu_i dn_i$$

where μ_i and n_i are the chemical potential and number of moles of compound i, respectively. If the minimization takes place at given (i.e. constant) temperature and pressure, $dT = 0$ and $dp = 0$, and we obtain

$$dG\,|_{T,p} = \sum_i \mu_i dn_i = 0 \tag{5}$$

Let us now restrict our attention to systems that are undergoing only one chemical reaction. Then the amount of each component in the system can be expressed as $n_i = n_i^0 + v_i \varepsilon$ where n_i^0 is the initial amount of each compound, v_i is their stoichiometric coefficients and ε is the reaction coordinate. Taking the derivative here gives $dn_i = v_i d\varepsilon$, which when substituted into equation (5) gives

$$\sum_i v_i \mu_i = 0 \tag{6}$$

The chemical potentials of each compound are often expressed in terms of their activity, a_i, by $\mu_i = \mu_i^0 + RT \ln a_i$, where μ_i^0 is the chemical potential at a standard state. Substituting this into equation (6) gives $\sum_i v_i \mu_i^0 + RT \sum_i v_i \ln a_i = 0$. The chemical potential is equal to the partial Gibbs free energy, g_i, and the sum of logarithms is equal to the logarithm of the product which, upon exponentiation of both sides, gives

$$\exp\left(\frac{-\sum_i v_i g_i^0}{RT}\right) = \prod_i a_i^{v_i} = K \tag{7}$$

where K is the equilibrium constant for the reaction.

We will be considering compounds in three phases: vapour, aqueous solution and solid crystal. In the crystalline phase, the compound is assumed to be pure and the activity is unity, $a_i = 1$. In solid solutions the activities of the compounds may be less than unity, but these cases will not be considered here. Atmospheric compounds relevant to this discussion include ammonium sulfate and ammonium nitrate solids.

All the vapours considered in air pollution modelling, including water vapour, are at sufficiently low partial pressures that they are well modelled in air as a Gibbs–Dalton mixture, that is an ideal solution of ideal gases. In this case, $a_i = p_i/p_0$, where p_i is the partial pressure of the compound and p_0 is atmospheric pressure. Typical atmospheric vapours are water, ammonia and nitric acid.

Finally, the aqueous solutions are often highly concentrated in aerosols so the activities are often expressed as $a_i = \gamma_i m_i$, where γ_i is the activity coefficient and m_i is the molality [moles/kg^{-1} water] of compound i. Common aerosol electrolytes include ammonium, nitrate and sulfate ions.

3.2 AQUEOUS PHASE PARTICLES

At sufficiently high relative humidities, those above the so- called deliquescence point, the electrolytes are present in an aqueous phase in the particles and no solid crystalline phase of water-soluble compounds is present. We are implicitly assuming here, as justified above, that the water activity in the aerosol is equal to the ambient relative humidity. We will now use basic thermodynamic principles to understand aerosol behaviour.

First, let us explore the concentration of electrolytes in the aerosol as the relative humidity changes. Our intuition is that as the relative humidity increases, the concentration of solutes in the particle should decrease, and in fact this can be proven. At constant temperature and pressure, the Gibbs–Duhem relation states

$$\sum_i n_i \mathrm{d}\mu_i = 0 \tag{8}$$

If we restrict our attention to one solute in water, equation (8) reduces to $n_1 \mathrm{d}\mu_1 + n_w \mathrm{d}\mu_w = 0$. Using $\mu_i = \mu_i^0 + RT \ln a_i$ to replace the chemical potentials with activities gives

$$\mathrm{d}a_w = -\frac{n_1}{n_w}\frac{a_w}{a_1}\mathrm{d}a_1 \tag{9}$$

and since the activities and mole numbers are positive, increases in water activity, which correspond to increases in relative humidity, result in decreases in activity, and therefore molality, of the solute. Figure 2 shows the molality of various electrolytes of atmospheric interest as a function of water activity. At zero molality the water activity is one—the activity of pure water is defined as unity.

We have shown both theoretically and empirically how the molality of a single solute changes with ambient relative humidity. Now let us explore the behaviour for a mixture of solutes. The mixing rule used in most atmospheric aerosol modelling is the so-called ZSR relation named after Zdanovskii, Stokes and Robinson. It is a purely empirical relationship and is violated to one degree or another in all mixtures. [8] Nevertheless, in most atmospheric applications it has been shown to be an excellent approximation. [11, 12, 26, 56] Zdanovskii [56] first proposed the relationship in his studies of electrolytes and Stokes and

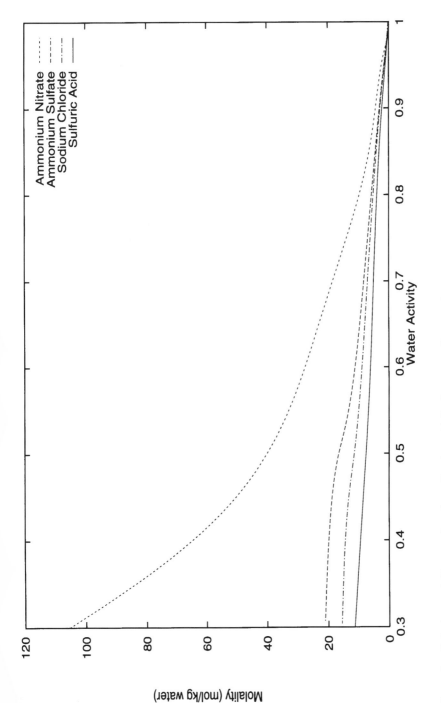

Figure 2. Solution molality as a function of water activity for pure salt aerosols

Robinson [45] appear to have found it independently for non-electrolytic solutes.

For our purposes, the basis for the ZSR relation can be summarized as follows. Consider a number of single-solute aqueous solutions, where each solution has the same water activity, say a_w. The mixture of these single-solute solutions has the same water activity as each single-solute solution alone, that is, a_w. From this principle we derive the ZSR relation.

Consider N single-solute solutions with water activity a_w, each containing n_i moles of the solute and W_i kilograms of water. The molality in the single-solute solutions is defined as $m_{i,ss} \equiv n_i/W_i$. If we mix each of the solutions together, the total amount of water in the mixture is simply $W = \sum_i^N W_i = \sum_i^N n_i/m_{i,ss}$. Now if we invoke ZSR, the water activity of the mixture is equal to the water activity of all the single-solute solutions, $m_{i,ss} = m_{i,ss}(a_w)$, and since we are assuming the ambient relative humidity is equal to the water activity, $m_{i,ss} = m_{i,ss}(\text{RH})$, we obtain

$$W = \sum_i^N \frac{n_i}{m_{i,ss}(\text{RH})} \tag{10}$$

If we know the number of moles of solute in the aerosol and the ambient relative humidity, equation (10) enables us to calculate the water content of the particles providing that all the compounds are in the aqueous phase.

We have some understanding of the basic thermodynamics of atmospheric aerosols and we will now consider gas–particle equilibrium and discuss the effect of condensation or evaporation of an electrolyte solute.

Since ammonium nitrate is both an electrolyte, volatile, and present in many urban aerosols, it is an instructive compound for understanding gas–particle equilibrium. For the aqueous particles being considered here, the equilibrium reaction is

$$\text{NH}_4^+(\text{aq}) + \text{NO}_3^-(\text{aq}) = \text{NH}_3(\text{g}) + \text{HNO}_3(\text{g})$$

The stoichiometric coefficients on the left-hand side of the reaction are each $v_i = -1$ while those on the right-hand side are both $v_i = +1$. The activity of each solute is $a_i = \gamma_i m_i$ and the activity of each vapour is $a_i = p_i/p_0$. Substituting these into the equilibrium equation (equation 7) gives

$$K_{AN} = \frac{p_{\text{NH}_3}/p_{\text{HNO}_3}/p_0^2}{\gamma_{AN}^2 m_{AN}^2} \tag{11}$$

The equilibrium constant for aqueous ammonium nitrate is available from the literature [28] as is the activity coefficient. [17]

Now we can try to understand what happens to solute-containing aqueous particles in the atmosphere. First, let us consider condensation. As a solute, such as ammonium nitrate, condenses on a particle the number of moles of ammonium nitrate in the particle increases. But as we have shown, the relative humidity fixes the single-solute molality of ammonium nitrate, so as ammonium nitrate condenses, so too must water (see equation 10).

Conversely, let us consider what transpires as we increase the relative humidity. Increasing relative humidity leads to increasing water activity in the particles, which leads to decreases in solute molality and commensurate decreases in surface partial pressure of ammonia and nitric acid. Thus, increasing relative humidity results in ammonia and nitric acid condensation.

3.3 CRYSTALLINE SALT PARTICLES

Under some circumstances, the electrolytes exist as a solid crystal phase in the aerosol particles. Thermodynamics predicts that the aerosol will be a solid at relative humidities below the deliquescence point—that is, when ions are saturated in solution. As discussed in section 2.3, crystallization has been observed to occur at substantially lower relative humidities in laboratory studies and confirming atmospheric studies have not been performed. Full crystallization of particles will not occur if they are sufficiently acidic. For example, if the particle is composed of a mixture of ammonium sulfate and sulfuric acid, different salts or combinations of salts form depending on the relative humidity and the ratio of ammonium to sulfate ion in the solution. When the ratio is 2, only ammonium sulfate forms. Between 2 and 1.5, a combination of ammonium sulfate and letovicite forms. At 1.5, only letovicite forms. Between 1.5 and 1 is a combination of letovicite and ammonium bisulfate, whereas at 1 only ammonium bisulfate forms. Finally, at ratios below 1, the particle never crystallizes completely—at low relative humidity it is composed of ammonium bisulfate crystal in equilibrium with hydrogen, ammonium, bisulfate and sulfate ions in an aqueous phase.

Let us restrict our attention to particles that are composed only of crystalline phases of salts. First consider solid–vapour equilibrium for ammonia and nitric acid:

$$NH_4NO_3(s) = NH_3(g) + HNO_3(g)$$

The activity of solid ammonium nitrate is 1, the stoichiometric coefficients are $+1$ for the vapours and their activity is p_i/p_0. Using these with equation (7) gives

$$K_{AN} = \frac{p_{NH_3} p_{HNO_3}}{p_0^2} \tag{12}$$

so that at a given temperature, the presence of ammonium nitrate solid fixes the partial pressure product of ammonia and nitric acid vapours.

Now let us consider solid–solid equilibrium within an aerosol particle. As an example, we can examine the equilibrium

$$NaCl(s) + NH_4NO_3(s) = NaNO_3(s) + NH_4Cl(s)$$

Here the activity of all the compounds is unity and equation (7) becomes $K = 1$ which is clearly not true—the equilibrium equation cannot be used in this form for solid–solid reactions—so we return to minimizing the Gibbs free energy. The Gibbs free energy for the system composed of only solid phases is $G = \sum_i n_i g_i^0$ and using the stoichiometry $n_i = n_i^0 + \nu_i \varepsilon$ gives

$$G = \sum_i n_i^0 g_i^0 + \varepsilon \sum_i \nu_i g_i^0 \tag{13}$$

Now the g_i^0 are constants, so both summations on the right-hand side of equation (13) are constants and the Gibbs free energy is minimized when ε is either a minimum or maximum depending on the sign of $\sum \nu_i g_i^0$. The extreme values of ε are determined purely by stoichiometry and occur when one of the solids in the reaction is completely depleted. Thus we conclude that reacting solid salts are in equilibrium when one of them has been completely depleted in favour of formation of the other compounds.

3.4 AEROSOLS WITH BOTH AQUEOUS AND SOLID PHASES

Single component aerosols, such as those containing just ammonium sulfate, have a sharp change in phase state at the deliquescence point. Above this point the aerosol is aqueous and below it is a solid crystal. Multi-component aerosols do not have such a simple behaviour. [54, 55] Multi-component aerosols are much more common in the atmosphere than the single component variety, so understanding their thermodynamics is essential for understanding atmospheric aerosols.

To begin our discussion of mixed phase aerosols, we first note that everything proven so far for aqueous aerosols and solid aerosols also applies to mixed phase aerosols. That is, aqueous phases are more dilute at higher relative humidity, the ZSR mixing rule can be applied to the aqueous phase to calculate the particulate mass of water, gas–particle equilibrium holds for both the solid and aqueous phases, and multiple, reacting solid phases are in equilibrium when one of the phases is depleted. Let us further explore aerosol properties when solid–aqueous solution equilibrium also holds.

Consider the reaction $NaCl(s) = Na^+(aq) + Cl^-(aq)$. The activity of the solid sodium chloride is unity, and that of the ions is $a_i = \gamma_i m_i$. Using these in the equilibrium equations gives

$$K_{NaCl} = \gamma_{Na^+} m_{Na^+} \gamma_{Cl^-} m_{Cl^-} = a_{NaCl(aq)}^2$$

from which we conclude that the activity of a solute is fixed if the solution is in equilibrium with the solid phase of the solute. This will prove useful in subsequent discussions of deliquescence.

It has been observed that atmospheric aerosols can abruptly change from a solid to an aqueous solution at a relative humidity called the deliquescence point. Equilibrium thermodynamics predicts that the aqueous to solid phase transition will occur at the same relative humidity but kinetics, discussed in section 2.3, results in this often occurring at much lower humidity levels. Let us first examine deliquescence behaviour of single-salt solutions, and for illustrative purposes we will assume that efflorescence and deliquescence occur at the same relative humidity.

If we take a single-salt aqueous aerosol at high relative humidity and dry it, the water evaporates from the particle until it becomes saturated. The point at which the salt is saturated is the deliquescence point. If the relative humidity is further lowered, the ions will come out of solution and form a solid phase, and all the water will evaporate. Thus thermodynamics predicts that a sharp relative humidity dividing line exists between particles existing in the solid phase and the aqueous phase. Only exactly at the deliquescence relative humidity can both phases exist in equilibrium.

Now consider a particle containing two salts. For example, at high relative humidity the particle might contain sodium, chloride and nitrate ions, while at low relative humidity it would be composed of sodium chloride and sodium nitrate salts. Let us examine the deliquescence properties of such a mixture.

Before we explore these binary salt solutions it is worth defining deliquescence in these cases. The deliquescence point is the relative humidity where just one salt is saturated and no solids exist in the particle. The mutual deliquescence point is the relative humidity where salts form solid phases and the particle completely dries. [52]

We begin by substituting activity for chemical potential in the Gibbs–Duhem relation (equation 8) to obtain

$$\frac{n_1}{a_1} \mathrm{d}a_1 + \frac{n_2}{a_2} \mathrm{d}a_2 + \frac{n_\mathrm{w}}{a_\mathrm{w}} \mathrm{d}a_\mathrm{w} = 0 \qquad (14)$$

where the subscripts 1 and 2 refer to the salts 1 and 2, and the subscript w refers to water, and n_i and a_i are the number of moles and activity of i in the particle, respectively. If we consider the particles starting in the aqueous state at high relative humidity, as we lower the humidity one of three things can occur: either salt 1 forms first, or salt 2 does, or they both saturate simultaneously. If salt 1 forms first, its presence fixes the activity of the dissolved components of salt 1 in the remaining aqueous phase, so $\mathrm{d}a_1 = 0$ and equation (14) becomes

$$\frac{n_2}{a_2} \mathrm{d}a_2 + \frac{n_\mathrm{w}}{a_\mathrm{w}} \mathrm{d}a_\mathrm{w} = 0 \qquad (15)$$

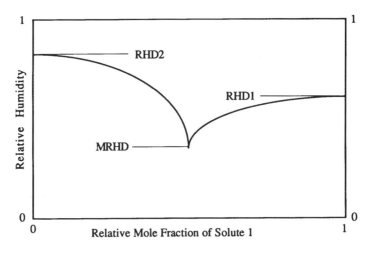

Figure 3.   Deliquescence behaviour as a function of the relative mole fraction of solutes

There are a number things to be learned from equation (15). First, as we increase the amount of salt 2 in the solution, the water activity decreases which means that increased amounts of salt 2 lower the deliquescence point of salt 1. Thus, the deliquescence point for multiple-salt solutions is lower than that of single-salt solutions. When salt 2 occurs in infinitesimal concentrations, salt 1 deliquesces at its single-salt point. As the concentration of salt 2 increases, the deliquescence point of salt 1 decreases. Clearly, there is nothing special about salt 2 in salt 1, so the same must occur for salt 1 in a solution containing salt 2.

A plot of the deliquescence point as a function of relative mole factions of the salts is given in Figure 3. The cusp where the two deliquescence curves meet is termed the mutual deliquescence point. From the figure we see that above RHD2 no solid phase is possible, regardless of composition. For relative humidities between RHD1 and RHD2, a solid phase of salt 2 can exist for compositions sufficiently dilute in salt 1. Between RHD1 and MRHD, solid phases of either compound can exist depending on the composition, and for relative humidities below MRHD the particle is composed of a solid phase of salt 1 and one of salt 2.

If the solution is composed of a deliquescent salt mixed with a non-deliquescent electrolyte such as sulfuric acid, there is no MRHD point. As the mole fraction of the acid is increased the deliquescence point of the salt decreases monotonically.

Curves such as Figure 3 are useful for understanding phase transitions in aerosols. First consider what transpires as we lower the relative humidity. Let us assume that the particle has an overall relative mole fraction of solutes that is to the left of the mutual deliquescence point. As we lower the relative humidity,

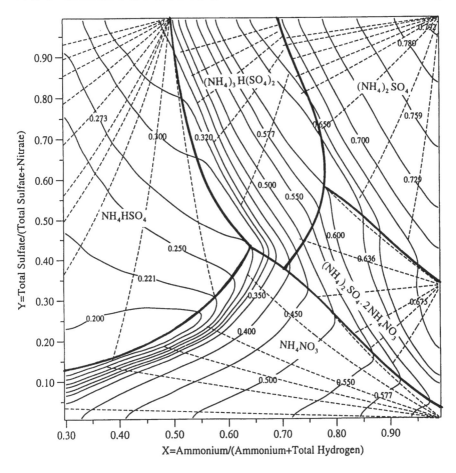

Figure 4.   Deliquescence phase diagram for acid ammonium nitrate–sulfate solutions

eventually the particle saturates with salt 2—this occurs on the deliquescence line. As the relative humidity is lowered further, salt 2 forms so the relative mole fraction of salt 1 increases. In fact, the deliquescence curve describes the relative mole fraction of the aqueous phase as a function of relative humidity.

Now let us consider increases in relative humidity. If the particle starts out as two solid salt phases at a relative humidity below the mutual deliquescence point, as we raise the relative humidity, eventually it reaches the MRHD value. At this point, water is absorbed into the particle and both salts dissolve until one of them is depleted. As the relative humidity continues to be increased, the remaining solid salt phase dissolves, moving the relative mole fraction of the salts away from that of the mutual deliquescence point. Eventually the remaining salt dissolves completely and the particle is a single aqueous phase.

Most atmospheric aerosols contain more than two electrolytes, and often they are acidic so they may not fully effloresce at low relative humidity. Nevertheless, the deliquescence/efflorescence behaviour of more complicated solutions can be visualized using the techniques just described. [33,34] Figure 4 is an illustration of the deliquescence behaviour of an acidic ammonium nitrate–sulfate aerosol. The horizontal and vertical axes are the relative mole fractions of the cations and anions in the particle, respectively. The solid contour lines represent the deliquescence relative humidity, and the bold solid lines represent phase boundaries. The dashed lines will be described shortly.

The upper and lower right-hand corners correspond to pure ammonium sulfate and pure ammonium nitrate aerosols, respectively. Their respective deliquescence points are 80 and 61%. Consider an aerosol particle with an anion mole fraction of 0.80 and a cation mole fraction of 0.93. From Figure 4 we can see that the deliquescence point is about 76% relative humidity and ammonium sulfate is the salt that forms first. The dashed lines show the change in anion and cation relative mole fraction as the salt, in this case ammonium sulfate, is precipitated out of the solution. So if we start at 76% and start drying this particle, solid ammonium sulfate forms and the composition of the remaining aqueous phase follows the dashed line down and to the left on the figure.

Eventually, at about 63% relative humidity, a phase boundary is encountered—here an additional phase of the double salt $(NH_4)_2SO_4 \cdot 2NH_4NO_3$ forms. Upon further reductions in relative humidity, the aqueous phase composition follows the phase boundary until about 60% relative humidity where letovicite $((NH_4)_3H(SO_4)_2)$ forms. Further lowerings in relative humidity cause the particle to complete effloresce. Similar diagrams have been developed for various aerosol compositions relevant to the atmosphere at 298 K. [33,34]

Now let us see how these phase diagrams can be used to calculate the surface partial pressures of volatile inorganics. For the acidic ammonium nitrate sulfate system considered above, the relevant partial pressures are P_{NH_3} and P_{HNO_3}. As shown in equation (11), to calculate the partial pressure product of ammonia and nitric acid we need to know the equilibrium constant for the reaction, K_{AN}, the composition of the solutes, m_{AN} and the activity coefficient of ammonium nitrate, γ_{AN}. Given the relative mole fractions of cations, anions and relative humidity we can verify whether the given relative humidity is above the deliquescence relative humidity or below it. If the relative humidity is below the deliquescence point, one or more solids form and so the relative mole fractions in the aqueous phase vary from their total value because the solid phases change the aqueous phase composition. To find the new relative mole fractions we can follow the dashed lines and the phase boundaries (see Figure 4), as explained previously, until we reach the given relative humidity. Using these relative mole fractions, charge balance and the given relative humidity we can

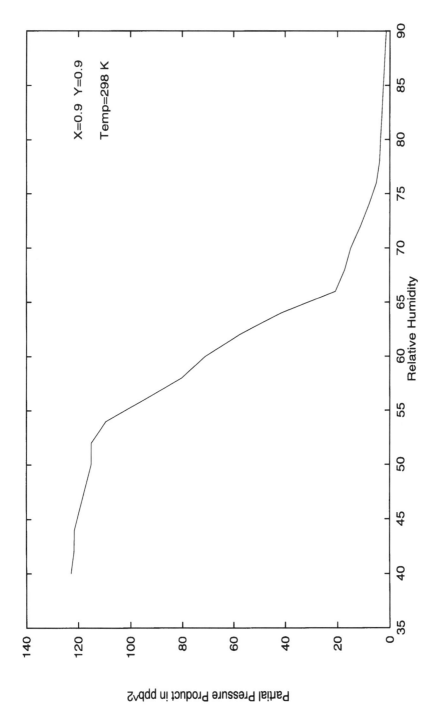

Figure 5. Partial pressure product of ammonium nitrate as a function of relative humidity

calculate the water content. Once the water content is known, molalities of various electrolytes can be calculated and using these concentrations we can evaluate the activity coefficient of ammonium nitrate in this multi-component solution mixture using activity coefficient models. [9] Plugging all these values into equation (11) we arrive at the partial pressure product.

Figure 4 is valid if we assume that the electrolytes are non-volatile. This is reasonable if the changes in ambient relative humidity are fast compared to the time-scales for evaporation or condensation of the electrolytes, but in general this is not the case. Atmospheric compositions are confined to mole factions that are near the top and left-hand axes. The top axis corresponds to situations where there is insufficient ammonia to neutralize particle sulfate—commonly the case in eastern Europe and eastern North America. The pH of the particles is low so that volatile acids may not condense. The right axis corresponds to cases where the particle sulfate is completely neutralized by ammonia, which permits volatile acids and ammonia to condense—a common occurrence in western North America and western Europe.

Figure 5 shows the variation in the partial pressure product with relative humidity when the cation and anion mole fractions are 0.9 and 0.9 respectively. In this figure we can identify the phase transitions near 76, 66 and 51%, corresponding to figure 4 (the mixed salts have been ignored in this calculation). For the reported anion and cation mole fractions the particle is completely aqueous until a relative humidity of 76%. At that point solid ammonium sulfate forms and as we further dry the particle letovicite forms at 66% and finally at a relative humidity of 51% ammonium nitrate also forms. After this the particle completely effloresces. Before concluding let us explore how the deliquescence behaviour is affected by temperature.

Deliquescence of many compounds of atmospheric interest depends on the ambient temperature. This dependence has been investigated for both single-salt [51] and multiple-salt solutions, [47] both of which will be reviewed here. Relative humidity is defined as the ratio of p_w, the ambient partial pressure of water, to p_s, its saturation vapour pressure, $RH = p_w/p_s$. Taking the natural logarithm of both sides and differentiating with respect to temperature gives

$$\frac{d \ln RH}{dT} = \frac{d \ln p_w}{dT} - \frac{d \ln p_s}{dT} \tag{16}$$

The RH on the left-hand side of equation (16) is the deliquescence point if the aerosol particles contain a solid phase in equilibrium with an aqueous phase, and this fact will be used in subsequent development. The terms on the right-hand side of equation (16) can be evaluated using the Clausius–Clapeyron equation:

$$\frac{d \ln p}{dT} = \frac{\Delta H}{RT^2} \tag{17}$$

where the latent heat of vaporization, ΔH, is dependent on the composition of the solution.

For pure water, the heat of vaporization is equal to the latent heat of water, $\Delta H = L_{pure}$. For the aqueous solution phase of the aerosol particle, a small amount of water evaporation coincides with a small amount of solute precipitation to maintain the solution molality constant so the appropriate heat of vaporization is

$$\Delta H = L_{solvent} + (M_w/1000)\Sigma_i m_{solute,i} L_{solute,i}$$

where M_w is the molar mass of water, $m_{solute,i}$ is the molality of electrolyte i that is deliquescing, and $L_{solvent}$ and $L_{solute,i}$ are the latent heat of vaporization of water in the saturated solution and the latent heat of fusion of salt i, respectively. [14,47]

Combining these expressions with those of equations (16) and (17) gives

$$\frac{d \ln \text{RHD}}{dT} = \frac{L_{solvent} + (M_w/1000)\Sigma_i m_{solute,i} L_{solute,i} - L_{pure}}{RT^2} \tag{18}$$

We now assume that the heat of vaporization of water is not altered substantially if it is pure or in solution, *i.e.*

$$m_w/1000\Sigma_i m_{solute,i} L_{solute,i} >> L_{solvent} - L_{pure}$$

so we can simplify equation (18) to

$$\frac{d \ln \text{RHD}}{dT} = \frac{M_w}{1000}\frac{\Sigma_i m_{solute,i} L_{solute,i}}{RT^2} \tag{19}$$

For the modest temperature excursions typical of the atmosphere, the molality and latent heats can be assumed to be constant and equation (19) can be integrated to obtain

$$\ln\frac{\text{RHD}(T)}{\text{RHD}(T_0)} = -\frac{M_w}{1000}\frac{\Sigma_i m_{solute,i} L_{solute,i}}{R}\left(\frac{1}{T} - \frac{1}{T_0}\right) \tag{20}$$

where $\text{RHD}(T_0)$ is the deliquescence relative humidity at a known temperature. Equation (20) agrees with experimental data over typical excursions in ambient temperature. [47,51]

3.5 CONCLUDING REMARKS

If phase equilibrium is assumed within atmospheric aerosol particles, thermodynamics can be used to predict the phase state of the particles, their water content, and the vapour pressure of the volatile components. Equilibrium thermodynamics predicts that solid phases will form at lower relative humidities when the particles contain mixtures of compounds—the deliquescence relative humidity is lowered. Deliquescence is also a function of ambient temperature and again equilibrium thermodynamics can be used to successfully predict this dependence. Finally, much of what is known about the thermo-dynamics of electrolytes and highly concentrated electrolyte solutions can be used to generate graphical representations of the deliquescence behaviour and the vapour pressures of the volatile compounds in moderately complicated mixtures. For more complicated mixtures computer codes can be used to predict these quantities.

4. SUMMARY

All systems move towards thermodynamic equilibrium and the rate of this motion is governed by kinetics. Limits to phase equilibrium in atmospheric aerosols are caused by mass transfer, chemical kinetics and nucleation of new phases. Many of these areas have not received sufficient attention in either field or laboratory studies, and as a result our current understanding of the kinetics of atmospheric aerosols is somewhat limited. Although some of the mechanisms discussed here are amenable to measurement with current laboratory techniques, exploration of a number of these mechanisms may require development of new techniques and instruments so that concentration and composition measurements can be made in the minute samples typified by atmospheric particles. Efforts to develop new instruments capable of these measurements are under way in a number of groups. [20]

Substantial work also remains in studies of the thermodynamics of atmospheric particles. These particles are often highly concentrated and may be supersaturated due to kinetic limitations, so predicting the activity of the solvent (water) and solutes in the aerosols is often fraught with uncertainty, especially at lower relative humidities. Recently, solution thermodynamics models have been developed that remove the distinction between solute and solvent, which is appropriate for atmospheric aerosols, and as a result are more successful at predicting solute and solvent activities at low relative humidity. [9,10]

This review has focused almost exclusively on the water soluble, inorganic electrolyte portion of the atmospheric aerosol. These compounds, along with associated water, often comprise more than half the mass of atmospheric aerosols. Another substantial portion is organic, and the volatility, water solubility, phase partitioning and mixture characteristics of these compounds in atmospheric aerosol are just now beginning to be explored. [41,42,46] When these studies have matured a more comprehensive view of the kinetics and thermodynamics of the organic portion of the aerosol will emerge, as it did for the inorganics.

NOMENCLATURE

a_i	Activity of species i
a_{NaCl}	Activity of sodium chloride
a_w	Water activity
$\bar{C}_i$	root mean square velocity of vapour molecule i
D_i	Molecular diffusivity of species i
g_i	Partial Gibbs free energy of species i
g_i^0	Partial Gibbs free energy of species i at a standard state
G	Gibbs free energy
K	Equilibrium constant
K_{AN}	Equilibrium constant of ammonium nitrate
K_{NaCl}	Equilibrium constant of sodium chloride
Kn	Knudsen number
L_{pure}	Latent heat of vaporization of water
$L_{solute,i}$	Latent heat of fusion of salt i
$L_{solvent}$	Latent heat of vaporization of water in the saturated solution
m_i	Molality (moles / kg^{-1} water) of species i
m_{AN}	Molality of ammonium nitrate
m_{cl^-}	Molality of chloride ion
m_{Na^+}	Molality of sodium ion
m_w	Water content of the particle
$m_{i,ss}$	Molality of single-solute aqueous solution of solute i
$m_{solute,i}$	Molality of electrolyte i that is deliquescing
M_i	Mass of species i
M_w	Molar mass of water
MRHD	Mutual deliquescence relative humidity
n_i	Number of moles of species i
n_i^0	Initial number of moles of species i
n_w	Number of moles of water
$n(R_p)$	Number distribution of the particles
P	Pressure
P_i	Partial pressure of species i
P_0	Atmospheric pressure
P_s	Ambient saturation vapour pressure of water
P_w	Ambient partial pressure of water
$P_{i,s}$	Surface vapour pressure of species i
$P_{i,\infty}$	Ambient partial pressure of species i
P_{HNO_3}	Vapour pressure of nitric acid
P_{NH_3}	Vapour pressure of ammonia
R	Gas constant
R_P	Particle radius
RH	Relative humidity

RHD	Deliquescence relative humidity
RHD1	Deliquescence relative humidity of salt 1
RHD2	Deliquescence relative humidity of salt 2
S	Entropy
t	Time
T	Ambient temperature
T_0	Known temperature
U	Internal energy
V	Volume
W	Total amount of water in the mixture
W_i	Amount of water corresponding to single solute i
α_i	Accommodation coefficient of species i
ΔH	Latent heat of vaporization
ε	Reaction coordinate
γ_i	Activity coefficient of species i
γ_{AN}	Activity coefficient of ammonium nitrate
γ_{Cl^-}	Activity coefficient of chloride ion
γ_{Na^+}	Activity coefficient of sodium ion
λ	Mean free path of air
μ_i	Chemical potential of species i
μ_i^0	Chemical potential of species i at a standard state
μ_w	Chemical potential of water
ν_i	Stoichiometric coefficient of species i
$\tau_{i,p}$	Characteristic time for the change in the vapour pressure of species i over a particle
$\tau_{i,\infty}$	Characteristic time for the change in the ambient partial pressure of species i

REFERENCES

1. Allen A. G., Harrison R. M., and Erisman J. Field measurements of the dissociation of ammonium nitrate and ammonium chloride aerosols. *Atmos. Environ.* **23**: 1591–1599, 1989.
2. Barger W. R. and Garrett W. D. Surface active organic material in the marine atmosphere. *J. Geophys. Res.*, **75**: 4561–4566, 1970.
3. Barger W. R. and Garrett W. D. Surface active organic material in air over the Mediterranean and over the Easter equatorial Pacific. *J. Geophys. Res.*, **81**: 3151–3157, 1976.
4. Barnes G. T. The effects of monolayers on the evaporation of liquids. *Advances in Colloid and Interface Sci.*, **25**: 89–200, 1986.
5. Bassett M. and Seinfeld J. H. Atmospheric equilibrium model of sulfate and nitrate aerosols. *Atmos. Environ.* **17**: 2237–2252, 1983.
6. Bassett M. and Seinfeld J. H. Atmospheric equilibrium model of sulfate and nitrate aerosols—II. Particle size analysis. *Atmos. Environ.* **18**: 1163–1170, 1984.

7. Blanchard D. C. Sea-to-air transport of surface active material. *Science*, **146**: 396–397, 1964.
8. Chen H., Sangster J., Teng T. T., and Lenzi F. A general method for predicting the water activity of ternary aqueous solutions from binary data. *Canadian J. of Chem. Eng.* **51**: 234–241, 1973.
9. Clegg S. L. and Pitzer K. S. Thermodynamics of multicomponent, miscible, ionic solutions: Generalised equations for symmetrical electrolytes. *J. Phys. Chem*, **96**: 3513–3520, 1992.
10. Clegg S. L., Pitzer K. S. and Brimblecombe, P. Thermodynamics of multicomponent, miscible, ionic solutions: 2. Mixtures including unsymmetrical electrolytes. *J. Phys. Chem*, **96**: 9470–9479, 1992.
11. Cohen M. D., Flagan R. C., and Seinfeld J. H. Studies of concentrated electrolyte solutions using the electrodynamic balance. 1. Water activities for single electrolyte solutions. *J. Phys. Chem.*, **91**: 4563–4574, 1987a.
12. Cohen M. D., Flagan R. C., and Seinfeld J. H. Studies of concentrated electrolyte solutions using the electrodynamic balance. 2. Water activities for mixed electrolyte solutions. *J. Phys. Chem.*, **91**: 4575–4582, 1987b.
13. Daumer B., Niessner R., and Klockow D. Laboratory studies of the influence of thin organic films on the neutralization reaction of H_2SO_4 aerosol with ammonia. *J. Aero. Sci.*, **23**: 315–325, 1992.
14. Denbigh K. *The Principles of Chemical Equilibrium*. Cambridge University Press, Cambridge, 1981.
15. Garrett W. D. Retardation of water drop evaporation with monomolecular surface films. *J. Atmos. Sci.*, **28**: 816–819, 1971.
16. Gill P. S., Graedel T. E., and Weschler C. J. Organic films on atmospheric aerosol particles, fog droplets, cloud droplets, raindrops, and snowflakes. *Rev. Geophys. Space Phys.*, **21**: 903–920, 1983.
17. Hamer W. J. and Wu Y.-C. Osmotic coefficients and mean activity coefficients of uni-univalent electrolytes in water at 25C. *J. Phys. Chem. Ref. Data* **1**: 1047–1099, 1972.
18. Harrison R. M. and MacKenzie A. R. Numerical simulation of kinetic constraints upon achievement of the ammonium nitrate dissociation equilibrium in the troposphere. *Atmos. Environ.* **24A**: 91–102, 1990.
19. Hightower R. L. and Richardson C. B. Evaporation of ammonium nitrate particles containing ammonium sulfate, *Atmos. Environ.*, **22**: 2587–2591, 1988.
20. Johnston M. V. and Wexler A. S. Mass spectrometry of individual aerosol particles, *Anal. Chem.* **67**:721A–726A, 1995.
21. Kerminen V.-M. and Wexler A. S. Post-fog nucleation of H_2SO_4–H_2O particles in smog. *Atmos. Environ.*, **28**: 2399–2406, 1994a.
22. Kerminen V.-M. and Wexler A. S. Particle formation due to SO_2 oxidation and high relative humidity in the remote marine boundary layer. *J. Geophys. Res.* **99**: 25, 607–25, 614, 1994b.
23. Kerminen V.-M. and Wexler A. S. Enhanced formation, and development of sulfate particles due to marine boundary layer circulation. *J. Geophys. Res.* **100**: 23,051–23,062, 1995.
24. Kim Y. P. and Seinfeld J. H. Atmospheric gas–aerosol equilibrium: III. Thermodynamics of crustal elements Ca^{2+}, K^+, and Mg^{2+}. *Aerosol Sci. Tech.*, **22**: 93–110, 1995.
25. Kim Y. P., Seinfeld J. H., and Saxena P. Atmospheric gas–aerosol equilibrium: I. Thermodynamic model. *Aerosol Sci. Tech.*, **19**: 157–181, 1993a.
26. Kim Y. P., Seinfeld J. H., and Saxena P. Atmospheric gas–aerosol equilibrium: II. Analysis of common approximations and activity coefficient calculation methods. *Aerosol Sci. Tech.*, **19**: 182–198, 1993b.

27. Lipfert F. W. *Air Pollution and Community Health*. Van Nostrand Reinhold, New York, 1994.
28. Mozurkewich M. The dissociation constant of ammonium nitrate and its dependence on temperature, relative humidity and particle size. *Atmos. Environ.* **27A**: 261–270, 1993.
29. Pandis S. N., Wexler A. S., and Seinfeld J. H. Dynamics of tropospheric aerosols. *J. Phys. Chem.* **99**: 9646–9659, 1995.
30. Pandis S. N., Wexler A. S., and Seinfeld J. H. Secondary organic aerosol formation and transport—II. Predicting the ambient secondary organic aerosol size distribution. *Atmos. Environ.*, 27A: 2403–2416, 1993.
31. Pilinis C. and Seinfeld J. H. Continued development of a general equilibrium model for inorganic multicomponent atmospheric aerosols. *Atmos. Environ.* **21**: 2453–2466, 1987.
32. Pilinis C. and Seinfeld J. H. Development and evaluation of an Eulerian photochemical gas–aerosol model. *Atmos. Environ.*, **22**: 1985–2001, 1988.
33. Potukuchi S. and Wexler A. S. Identifying solid–aqueous phase transitions in atmospheric aerosols. I. Neutral acidity solutions. *Atmos. Environ.* **29**: 1663–1676, 1995a.
34. Potukuchi S. and Wexler A. S. Identifying solid–aqueous phase transitions in atmospheric aerosols. II. Acidic solutions. *Atmos. Environ.* **29**: 3357–3364, 1995b.
35. Pruppacher H. R. and Klett J. D. *Microphysics of Clouds and Precipitation*. D. Reidel Publishing Co., Dordrecht, Holland, 1980.
36. Rood M. J., Shaw M. A., Larson T. V., and Covert D. S. Ubiquitous nature of ambient metastable aerosol. *Nature* **337**: 537–539, 1989.
37. Rubel G. O. and Gentry J. W. Measurement of water and ammonia accommodation coefficients at surfaces with adsorbed monolayers of hexadecanol. *J. Aerosol Sci.*, **16**: 571–574, 1985.
38. Russell A. G., McRae G. J., and Cass G. R. Mathematical modeling of the formation and transport of ammonium nitrate aerosol. *Atmos. Environ.* **17**: 949–964, 1983.
39. Russell A. G. and Cass G. R. Verification of a mathematical model for aerosol nitrate and nitric acid formation and its use for control measure evaluation. *Atmos. Environ.* **20**: 2011–2025, 1986.
40. Russell A. G., McCue K. F., and Cass G. R. Mathematical modeling of the formation of nitrogen-containing air pollutants. 1. Evaluation of an Eulerian photochemical model. *Environ. Sci. Technol.* **22**: 263–271, 1988.
41. Saxena P., Hildemann L. M., McMurry P. H., and Seinfeld, J. H. Organics alter the hygroscopic behavior of atmospheric particles. *J. Geophys. Res.* **100**: 18 755–18 770, 1995.
42. Saxena P. and Hildemann L. M. Water-soluble organics in atmospheric particles: A critical review of the literature and application of thermodynamics to identify candidate compounds. *J. Atmos. Chem.* **24**: 57–109, 1996.
43. Saxena P., Seigneur C., and Peterson T. W. Modeling of multiphase atmospheric aerosols. *Atmos. Environ.* **17**: 1315–1329, 1983.
44. Saxena P., Hudischewskyj A. B., Seigneur C., and Seinfeld J. H. A comparative study of equilibrium approaches to the chemical characterization of secondary aerosols. *Atmos. Environ.* **20**: 1471–1483, 1986.
45. Stokes R. H. and Robinson R. A. Interactions in aqueous nonelectrolyte solutions. I. Solute–solvent equilibria. *J. Phys. Chem.*, **70**: 2126–2130, 1966.
46. Storey J. M. E., Luo W., Isabelle L. M., and Pankow J. F. Gas/solid partitioning of semivolatile organic compounds to model atmospheric solid surfaces as a function of relative humidity. 1. Clean quartz. *Environ. Sci. Technol.* **29**: 2420–2428, 1995.

47. Tang I. N. and Munkelwitz H. R. Composition and temperature dependence of the deliquescence properties of hygroscopic aerosols. *Atmos. Environ.* **27A**: 467–473, 1993.

48. Tanner R. L. An ambient experimental study of phase equilibrium in the atmospheric system: aerosol H^+, NH_4^+, SO_4^{2-}, NO_3^-–$NH_3(g)$, $HNO_3(g)$. *Atmos. Environ.* **16**: 2935–2942, 1982.

49. Vogt R. and Finalyason-Pitts B. J. A diffuse reflectance infrared Fourier transform spectroscopic (DRIFTS) study of the surface reaction of NaCl with gaseous NO_2 and HNO_3. *J. Phys. Chem.* **98**: 3747–3755, 1994.

50. Wexler A. S. and Seinfeld J. H. The distribution of ammonium salts among a size and composition dispersed aerosol. *Atmos. Environ.*, **24A**: 1231–1246, 1990.

51. Wexler A. S. and Seinfeld J. H. Second generation inorganic aerosol model. *Atmos. Environ.*, **25A**: 2731–2748, 1991.

52. Wexler A. S. and Seinfeld J. H. Analysis of aerosol ammonium nitrate: departures from equilibrium during SCAQS. *Atmos. Environ.*, **26A**: 579–591, 1992.

53. Wexler A. S., Lurmann F. W., and Seinfeld J. H., Modelling urban and regional aerosols—I. Model development. *Atmos. Environ.*, **28**: 531–546, 1994.

54. Winkler, P. The growth of atmospheric aerosol particles as a function of the relative humidity—II. An improved concept of mixed nuclei. *Aerosol Sci.* **4**: 373–387, 1973.

55. Winkler, P. The growth of aerosol particles with relative humidity. *Physica Scripta* **37**: 223–230, 1988.

56. Zdanovskii A. B. New methods of calculating solubilities of electrolytes in multicomponent systems. *Zhur. Fiz. Khim.*, **22**: 1475–1485, 1948.

7 Dioxins, Dibenzofurans and PCBs in Atmospheric Aerosols

STUART J. HARRAD

University of Birmingham, UK

1 INTRODUCTION

This chapter is concerned with the role played by atmospheric aerosols, in the environmental fate and behaviour of polychlorinated dibenzo-*p*-dioxins and polychlorinated dibenzofurans (collectively referred to as PCDD/Fs) and polychlorinated biphenyls (PCBs). Research into these compounds has been driven by concern over their potential adverse immunological, carcinogenic, developmental and reproductive effects. [*inter alia* 1–5] Such adverse effects are compounded by the tendency of these compounds to bioaccumulate, and their environmental persistence—PCDD/F half-lives in soil are typically reported at *ca* 10 years, while Panshin and Hites [6] report the atmospheric half-life of PCBs to exceed 23 years. Typical daily adult non-occupational human exposure

Atmospheric Particles, Edited by R. M. Harrison and R. Van Grieken.
© 1998 John Wiley & Sons Ltd.

in industrialized Western societies is around $100\,pg\sum i - TE^*$ for PCDD/Fs [*inter alia* 7, 8] and $1\,\mu g\sum PCB$. [9] Direct human exposure to these compounds via inhalation is negligible [9,10], and budget calculations show the atmosphere to represent only a very minor reservoir for PCDD/Fs and PCBs. [11,12] Despite this, the atmosphere plays a vital role in the environmental fate and behaviour of these compounds, not least as an important vector for their long-range transport, via which they have established a ubiquitous environmental presence. [*inter alia* 13,14] Given the overriding importance of the diet—especially beef and dairy products—as a source of both PCDD/Fs and PCBs, this chapter will also address the influence of the atmospheric behaviour of these compounds on the air-to-cattle-to-human food chain.

2 STRUCTURES/NOMENCLATURE

Figures 1 and 2 illustrate the basic structures and nomenclature of PCDD/Fs and PCBs respectively. Scientific interest has centred on those PCDD/Fs con-

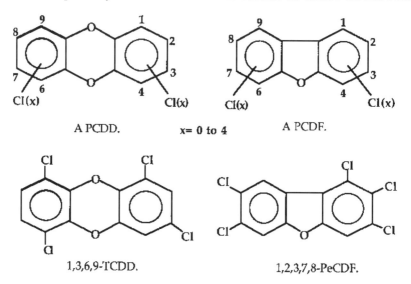

A PCDD.   x= 0 to 4   A PCDF.

1,3,6,9-TCDD.   1,2,3,7,8-PeCDF.

Figure 1. Basic structures and nomenclature of PCDD/Fs

---

* i-TE is an acronym for 2,3,78-TCDD equivalents. It is a means of expressing the toxicity of a complex mixture of different PCDD/F congeners in terms of an equivalent quantity of 2,3,7,8-TCDD. Each of the seventeen 2,3,7,8-substituted congeners are assigned a toxic equivalency factor (TEF) based on its toxicity relative to that of 2,3,7,8-TCDD, which is universally assigned a TEF of 1. Multiplication of the concentration of a PCDD/F by its assigned TEF gives its concentration in terms of TE and the toxicity of a mixture is the sum of the TEs calculated for all congeners. While several different weighting schemes exist, the one referred to in this chapter is that devised by the

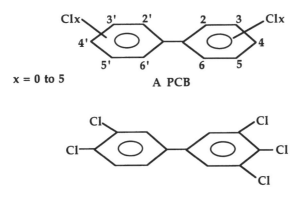

x = 0 to 5   A PCB

3,3',4,4',5-pentachlorobiphenyl - PCB # 126

Figure 2. Basic structures and nomenclature of PCBs

taining four or more chlorines—and especially those 17 substituted at the 2,3,7 and 8-positions—and those PCBs containing three to seven chlorines, particularly those unchlorinated at the 2, 2', 6 and 6'-positions (the so-called non-*o*-congeners). This concentration of interest rests predominantly with the enhanced toxicity of compounds displaying such chlorination patterns, [15] although some workers have recognized the importance of the remaining compounds in source apportionment. [*inter alia* 16,17] It is relevant at this point to define the following terms often used in the literature relating to PCDD/Fs and PCBs: viz. isomer, congener and homologue. Individual PCDD/Fs and PCBs are referred to as *congeners*, while PCDD/F and PCB congeners possessing identical chemical formulae are *isomers*. As a further subdivision, a *homologue* group is defined as a set of PCDD/F or PCB isomers. To illustrate, while 2,3,7,8-tetrachlorodibenzo-*p*-dioxin (2,3,7,8-TCDD) is an isomer of the same homologue group as e.g. 1,2,6,7-TCDD, neither are isomers of the same homologue group as 2,3,7,8-tetrachlorodibenzofuran (2,3,7,8-TCDF), which in turn is a member of the TCDF homologue group and an isomer of e.g. 1,2,7,8-TCDF. All four compounds are, however, congeners of each other and all other PCDD/Fs. Such distinctions are important when considering the environmental fate and behaviour of PCDD/Fs and PCBs.

NATO Committee on Challenges to Modern Society (18). Under this system, 2,3,4,7,8-pentachlorodibenzofuran and ocatachlorodibenzo-*p*-dioxin for example, possess TEFs of 0.5 and 0.001 respectively. Note that although tentative TEFs have been assigned to some PCBs–*inter alia* Safe (15) – these have yet to be accepted by any government body.

3 SOURCES

Although the origins of both are predominantly anthropogenic, a fundamental distinction between PCBs and PCDD/Fs is that while the principal source of the former has been their deliberate anthropogenic production, the latter have never been intentionally manufactured, other than for use as analytical standards. Industrial production of PCBs commenced in 1929 in the USA, spreading to Germany by 1930, and the UK by 1954. By the time that concerns over their potential adverse effects caused the cessation of Western production in the late 1970s, the myriad uses of these compounds as *inter alia* dielectric fluids in transformers and capacitors, had led to a total Western production of around 1.2 Mt. Less is known about production in the former Soviet bloc nations, although it is known that such manufacture continued well into the 1980s. It is more difficult to assess the extent to which this volume of PCBs has escaped into the environment, but interpretation of archived soil data reported for various UK sites [19] leads to the conclusion that a significant proportion—perhaps up to 50% or more—of UK PCB sales (*ca* 40 000 t), were released into the environment. It is this factor—combined with the comparable environmental recalcitrance of both PCDD/Fs and PCBs—that accounts for the fact that environmental levels of the latter are approximately two to three orders of magnitude greater than the former.

Following the cessation of production and restrictions on their use, environmental contamination by PCBs has declined dramatically. [*inter alia* 19–23] Consequently, source inventories conducted for the UK since the introduction of such controls. [12,24] now indicate that the principal source of atmospheric PCBs is the revolatilization of the existing PCB burden, with other sources including leaks from electrical transformers and capacitors remaining in use.

Identification of PCDD/F sources is more difficult, although the low levels detected in archived materials predating the twentieth century indicate that they are predominantly anthropogenic. [25] Indeed, comparisons of atmospheric releases with depositional inputs conducted as part of several source inventories, [11,26–28] have only managed to account for 10–20% of depositional input. Although a more recent inventory—based on the most comprehensive database to date—has explained some of this discrepancy for the UK, [29] there remains the possibility that even after 20 years of intensive investigation, some significant sources of PCDD/Fs remain undiscovered. Uncertainty over emission factors for different sources means that rankings carry a significant degree of uncertainty, and the relative importance of different sources will vary between individual nations. In summary, however, the present atmospheric burden of PCDD/Fs arises as a result of a wide range of combustion activities (such as waste incineration and fossil fuel combustion), and industrial practices (such as metal smelting, and the production and use of organochlorine chemicals, especially chlorophenol-based products). Natural sources such as forest fires and volcanic activity may be of significance in some areas, and the possibility that recirculation of the existing burden may be important cannot be ruled out.

4 AMBIENT ATMOSPHERIC CONCENTRATIONS

Owing to the extremely low levels present, levels of PCDD/Fs in ambient air have only become measurable in the last decade, with the development of sufficiently sophisticated analytical instrumentation. To illustrate, even in major cities, concentrations of 2,3,7,8-TCDD are only around 5–10 fg m^{-3}. Table 1 lists typical PCDD/F concentrations detected in Belfast air [30], with Table 2 providing concentrations of selected PCBs in air sampled in the centre of Manchester, UK. [31] Global distribution of airborne PCBs essentially reflects the extent of local and regional use and disposal of these compounds. As a result, although PCBs are globally ubiquitous and concentrations in "unpolluted" air are similar in both hemispheres, the highest concentrations are found close to the major sources of these compounds; viz. the temperate industrialized areas of the northern hemisphere between 40° and 60° N latitude. [32]

Table 1. Mean and range of PCDD/F levels* in Belfast air

Congener	Mean	Minimum	Maximum
2,3,7,8-TCDD	6.1	5.1	7.7
1,2,3,7,8-PeCDD	20	9.0	32
1,2,3,4,7,8-HxCDD	18	6.1	30
1,2,3,6,7,8-HxCDD	19	6.3	37
1,2,3,7,8,9-HxCDD	30	10	47
1,2,3,4,6,7,8-HpCDD	310	120	460
OCDD	1000	430	1300
2,3,7,8-TCDF	33	14	62
1,2,3,7,8-PeCDF	42	20	85
2,3,4,7,8-PeCDF	22	22	150
1,2,3,4,7,8-HxCDF	73	25	150
1,2,3,6,7,8-HxCDF	63	20	130
1,2,3,7,8,9-HxCDF	27	9.1	54
2,3,4,6,7,8-HxCDF	85	26	160
1,2,3,4,6,7,8-HpCDF	230	94	430
1,2,3,4,7,8,9-HpCDF	41	21	72
OCDF	240	140	340
Σi-TE	96	37	178

* Levels are the sum of vapour and particle-bound phases in fg m^{-3} reported by ERM/HMIP. [30]

Table 2. Mean, median and range of PCB levels* in UK urban air

Congener No.	Mean	Median	Minimum	Maximum
28	120	120	74	170
52	110	110	68	150
101	68	65	19	120
118	18	22	2.3	27
138	23	23	12	34
153	30	29	15	52
180	20	18	8.2	49
ΣPCB	1000	1000	610	1600

* Values reported are the sum of vapour and particle-bound phases in pg m^{-3} for Manchester. [31] ΣPCB represents the sum of the 31 congeners measured.

Hermanson and Hites [33] reported a clear linear correlation between average ambient temperature and atmospheric concentrations of $\sum$PCB at three separate sites in Bloomington, Indiana, USA, with a similar correlation observed by Panshin and Hites. [34] Similarly, seasonal variations in UK atmospheric concentrations of PCBs are marked, with a clear maximum occurring in the warmer summer months. [35] These observations are consistent with source inventories showing revolatilization of previously deposited material to be the principal source of the contemporary atmospheric loading of PCBs. [12,24] These findings are supported by the observation that this "summer peak" in concentrations is especially marked for the more volatile lower chlorinated congeners. [35]

Such clear seasonal variations in atmospheric concentrations are not evident for PCDD/Fs [36], although there are indications of higher concentrations in winter. This may indicate that revolatilization of the existing PCDD/F burden is not a major source to the atmosphere, although there is a possibility that any "summer peak" in PCDD/F concentrations due to volatilization from soil, may be masked by a decrease in atmospheric emissions arising from the combustion of coal, oil and wood—all identified as PCDD/F sources [16,37,38]—during the summer months.

5 ATMOSPHERIC BEHAVIOUR

The role that the atmosphere plays in transforming the congener profile of airborne PCDD/Fs and PCBs in rural areas is of great interest, not least because that it is in such locations that these compounds enter the human food chain. In particular, the toxicity (measured in terms of i-TE) of a given environmental mixture of PCDD/Fs and PCBs can vary considerably according to the exact congener profile, hence any processes that significantly modify this profile are important.

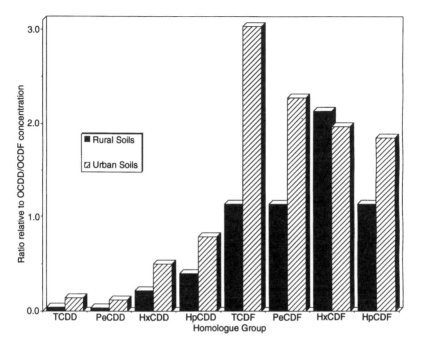

Figure 3. PCDD/F homologue profiles in urban and rural UK soils [39,40]. Ratios are median PCDD and PCDF homologue group concentrations relative to OCDD and OCDF respectively

It is therefore significant that the relative concentrations of PCDD/F homologues are noticeably different in urban as opposed to rural UK soils. [39,40] Specifically, there is a relative enrichment of the lower chlorinated homologues in urban areas. This homologue profile modification—which has been reported by other workers in other countries [*inter alia* 14]—is illustrated in Figure 3 (which shows median PCDD and PCDF homologue concentrations ratioed to levels of OCDD and OCDF respectively), and is important as it results in a less toxic PCDD/F profile prevailing in rural agricultural areas. Conversely, Atlas and Giam [41] reported air samples at the remote Enewetak Atoll site to be dominated by the less chlorinated PCBs, and suggested that this was due to photolytic dechlorination. However, in a review of data produced by his laboratory, Ballschmiter [13] was unable to confirm enrichment of any PCB congeners/homologues in air samples taken during a cruise from Bremerhaven in Germany to Punta Arenas in Chile.

The following sections will address the processes influencing the atmospheric behaviour of PCDD/Fs and PCBs, and in particular, the relative importance of such processes, and how they may affect the homologue profiles of PCDD/Fs and PCBs.

5.1 VAPOUR–PARTICULATE PHASE PARTITIONING

Understanding the vapour/particle phase partitioning of airborne PCDD/Fs and PCBs is crucial to understanding their atmospheric fate/residence time, since the vapour and particle phases will display different atmospheric lifetimes. Both PCDD/Fs and PCBs may be considered as semi-volatile organic compounds (SVOCs), with vapour pressures lying between around 10^{-4} and 10^{-11} atm, and generally decreasing with increasing degree of chlorination within a given compound class. Along with the atmospheric burden of particulate matter, a given congener's vapour pressure will influence its relative partitioning between the vapour and particle-bound phases in the atmosphere, with the more volatile lower chlorinated congeners displaying a markedly greater tendency to exist in the vapour form than the higher chlorinated congeners. This results in a wide congener-specific variation in the relative proportions of a given PCDD/F or PCB that exists in the vapour phase. For example, while the majority (typically *ca* 95%) of airborne PCB No. 28 exists in the vapour phase in the UK, [31] under similar urban conditions, OCDD has been shown to exist almost exclusively (*ca* 96%) in the particle-bound form. [42]

Vapour/particulate (V/P) ratios may be predicted for PCDD/Fs and PCBs using the following algorithm developed by Junge [43] to describe exchangeable adsorption of SVOCs to aerosols:

$$f_p = (c^* S_t)/(P + (c^* S_t)) \tag{1}$$

where f_p is the fraction of SVOC adsorbed to particulates, c the Junge constant (atm cm), dependent on molecular mass, heat of condensation and a compound-specific adsorption constant, [13] assumed here to be 1.7×10^{-4} atm cm [42], S_t the available surface area per unit volume of atmospheric particles (cm^2 cm^{-3}), assumed for the purposes of comparison—citing Lorber *et al.* [42]—to be 3.5×10^{-6} cm^2 cm^{-3} for areas defined as "background with local sources" and P = SVOC vapour pressure (atm) at ambient temperature, assumed here to be 298 K.

It is vital to distinguish between the vapour pressure of the sub-cooled liquid and solid compound in the above equation, and hence the V/P partitioning of PCDD/Fs and PCBs. Using P_s—vapour pressure of the solid—gives an extremely low forecasted V/P ratio. By comparison, use of the sub-cooled liquid vapour pressure (P_l), results in a higher predicted value for V/P, which as demonstrated by Tables 3 and 4, correlates more closely with observed ratios for selected PCDD/Fs and PCBs in ambient air. Note that P_l is calculated as follows:

$$P_l = P_s^* \exp[6.79^*((T_m - 298)/298)] \tag{2}$$

Table 3. Comparison of "typical" observed [42] vapour/particulate (V/P) ratios for selected PCDD/Fs with those predicted using equations (1) and (2)

Congener	Observed V/P	Predicted V/P
2,3,7,8-TCDD	6.7	0.95
1,2,3,7,8-PeCDD	2.2	0.29
1,2,3,4,7,8-HxCDD	0.43	0.063
1,2,3,6,7,8-HxCDD	0.43	0.029
1,2,3,7,8,9-HxCDD	0.43	0.016
1,2,3,4,6,7,8-HpCDD	0.11	0.017
OCDD	0.042	0.0017
2,3,7,8-TCDF	4.88	2.00
1,2,3,7,8-PeCDF	1.86	0.59
2,3,4,7,8-PeCDF	1.86	0.35
1,2,3,4,7,8-HxCDF	0.54	0.052
1,2,3,6,7,8-HxCDF	0.54	0.055
1,2,3,7,8,9-HxCDF	0.54	0.097
2,3,4,6,7,8-HxCDF	0.54	0.057
1,2,3,4,6,7,8-HpCDF	0.12	0.037
1,2,3,4,7,8,9-HpCDF	0.12	0.021
OCDF	0.02	0.0017

where P_s is the contaminant crystalline solid vapour pressure (atm), T_m the contaminant melting point (K) and 6.79 is an average value for S_f/R, where S_f is the entropy of fusion and R the universal gas constant. [13] Note that the values used here for both P_s and T_m are those cited by Lorber et al. [42] for PCDD/Fs, and by Ballschmiter and Wittlinger [32] for PCBs.

Despite the improved correlation with forecasted V/P values obtained using P_1, Tables 3 and 4 show that the "typical" observed V/P ratios for PCDD/Fs and PCBs are greater and lower respectively than would be predicted using

Table 4. Comparison of "typical" observed [31] vapour/particulate (V/P) ratios for selected PCBs with those predicted using equations (1) and (2)

Congener No.	Observed V/P	Predicted V/P
28	51.7	890
52	42.9	880
101	46.4	140
118	10.1	87.5
138	14.2	18.0
153	11.8	54.6
180	4.0	43.6

equations (1) and (2). This discrepancy is at least partly due to the fact that present "Hi-Vol" experimental sampling methods may fail to accurately reflect "true" V/P ratios for PCDD/Fs and PCBs. Specifically, they may overestimate due to volatilization of particle-bound compounds during sampling, this is likely to be significant owing to the high flow rates (up to $1\,m^3\,min^{-1}$) required to obtain adequate sample volumes and especially so for the more volatile lower chlorinated congeners. Alternatively, underestimation may occur as a result of vapour-phase congeners undergoing irreversible adsorption on particles entrained on the filter. Unfortunately, definitive assessments of the exact influence of such sampling artefacts on reported V/P ratios have yet to be conducted for PCDD/Fs and PCBs.

Also of relevance is the correlation between the V/P ratio and the ambient atmospheric temperature that is observed for both PCDD/Fs [44] and PCBs. [33] To illustrate, the experimental data of Eitzer and Hites [44] shows excellent correlation ($r = 0.81$ for 1,2,3,4,6,7,9-heptachlorodibenzo-p-dioxin) when substituted into equation (3), [45] thus confirming the relationship between V/P and temperature:

$$\log[V(\text{TSP})/P] = a'_0 + a_1/T \qquad (3)$$

where V is the fraction of compound present in vapour phase, TSP the concentration of total suspended particulates in ambient air, P the fraction of compound adsorbed to particulates, T the ambient air temperature and a_0' and a_1 are both determined experimentally. Interestingly, Hermanson and Hites [33] found TSP concentrations to exert only a negligible influence on the relationship between V/P values for individual PCBs and ambient atmospheric temperature.

5.2 CONGENER/HOMOLOGUE/ISOMER DISTRIBUTION WITH RESPECT TO PARTICLE SIZE

One possible explanation for the observed relative enrichment of the less chlorinated PCDD/F homologues in urban locations, could be the preferential deposition in urban areas of larger particles containing proportionately greater levels of the lower chlorinated homologues. This hypothesis is lent some credence by the observations of Eduljee and Townsend [46] who observed that homologue profiles of PCDD/Fs varied according to the size fraction of fly-ash particles from a municipal waste incinerator. They found the relative proportion of lower chlorinated homologues to be greater on the larger size fractions, hypothesizing that this was linked to the fact that such larger particles were more able to absorb "thermal stress", with the result that the less thermodynamically stable lower chlorinated PCDD/Fs were enriched on the larger particle size fractions. However, the particle size distribution of PCDD/Fs in ambient air was investigated by Kaupp et al., [47] who reported no significant variation in congener/homologue/isomer pattern between different size fractions, thus indicating that particle mediated transport and deposition should be similar for all PCDD/Fs. It must be noted, however, that their findings were based on a limited number of samples, and that no tetrachlorinated dioxins and furans, as well as pentachlorinated dioxins, were detected in any of the samples, and it therefore seems that further work is required to validate their important conclusions. They also found around 90% of detected tetra-through octachlorinated PCDD/Fs to be associated with accumulation mode particles of aerodynamic diameter < 1.35μ m. The predominant association of PCDD/Fs with such particles—which display the longest atmospheric residence times—owing to their resistance to rapid gravitational settling and relatively slow removal by precipitation and dry deposition—helps to explain the ubiquitous distribution of PCDD/Fs. It also has implications for human health, as such small particles can more easily penetrate the lower regions of the respiratory tract.

Holsen et al. [48] reported on the distribution of individual PCBs with respect to particle size, splitting the total airborne particle loading into five size fractions between < 1 and 36.5 μ m aerodynamic diameter. Although they reported the presence of PCBs on all of these fractions, there was a decrease in the particle mass normalized PCB concentration with increasing particle size—e.g. from ca 50 μg g^{-1} for particles of < 1μm to ca 25 μg^{-1} for particles of between 25 and 36.5 μ m. They also reported the finer particles to contain relatively higher particle mass normalized concentrations of the higher chlorinated PCBs. Although further evidence appears to be required in order to confirm the existence of a relationship of this type, the data of Holsen et al. [48] suggest that—contrary to PCDD/Fs—the homologue pattern of PCBs is to some extent dependent on particle size.

5.3 ATMOSPHERIC REACTIVITY

Koester and Hites [49] stated that atmospheric photolysis of PCDD/Fs is confined to the vapour phase fraction. Both PCDD/Fs and PCBs undergo photolytic reactions in the vapour phase—yielding inter alia dechlorinated and hydroxylated products. [50, 51] Furthermore, Atkinson [52] has demonstrated gas phase reaction with the hydroxyl (OH) radical to occur for both PCBs and PCDD/Fs. Although Atkinson suggested that gas phase reaction via OH radical addition to the benzene rings of PCDD/Fs and PCBs was the more important of these two removal mechanisms, Bunce et al. [51] considered that for the monochlorobiphenyls, direct photolysis and hydroxyl radical reactions were approximately equally likely to occur with half- lives of around 1 week. More recently, however, Anderson and Hites [53] estimated the tropospheric lifetime of a range of mono-to pentachloro-PCBs due to their gas-phase reaction with the OH radical to range between 2 and 34 days, with values increasing with increasing degree of chlorination. Atkinson and co-workers [54, 55] also estimated atmospheric lifetimes due to reaction with OH radicals to be between 2 and 312 days for vapour-phase PCDD/Fs, again increasing with increased degree of chlorination. With regard to the environmental significance of these removal processes, Bunce *et al.* [51] considered that for a given removal process to play a significant rôle, the associated lifetime must not exceed 1–2 weeks. Given that atmospheric half-lives of these compounds due to deposition processes—which act on both vapour and particle phase pollutants—are around 1 week [13], it may be supposed that both OH radical and photolytic reactions of PCDD/Fs and PCBs—both of which are restricted to the gas phase—are of relatively minor importance, and will become less so with increasing chlorination. With respect to photolysis, this conclusion is supported by Ballschmiter and Wittlinger, [32] who considered the evidence for atmospheric photodegradation of those PCBs shown to be photolabile under laboratory conditions, and concluded that there was no clear evidence that such photodegradation was significant under environmental conditions. Furthermore, the fact that atmospheric transport appears to effect homologue but not isomer-specific changes in PCDD/F profile during atmospheric transport, [56, 57] indicates that photolysis and reaction with OH radicals—which are isomer-specific—will exert only a minor influence on the atmospheric behaviour of PCDD/Fs. In marked contrast, however, the recent findings of Anderson and Hites [53] indicate that PCB reaction with the OH radical is perhaps the most important *permanent* loss mechanism of PCBs from the atmosphere, and it is clear that this remains a fertile research topic.

5.4 WET DEPOSITION

The scavenging of PCDD/Fs and PCBs via wet gaseous and particle deposition (i.e. precipitation and fog) is dependent on their water solubilities and vapour/particle partitioning ratios. More specifically, the dissolution into precipitation and fog of those compounds that are predominantly associated with the vapour phase, is dependent on the compound's Henry's law constant (H—effectively the ratio of a compound's vapour pressure and its water solubility, expressed in atm m^3 mol^{-1}). By comparison, those compounds predominantly associated with particles are removed via particle wash-out. It can thus be seen that the predominant rain-scavenging process for a given compound is dependent on the relative magnitudes of W_v and W_p (the unitless vapour and particle wash-out ratios respectively), and the V/P ratio. Both W_v and W_p may be calculated as shown in equations (4) and (5):

$$W_v = \frac{\text{the dissolved phase concentration of a compound in rain } (w/v - e.g.\ pg\ dm^{-3})}{\text{the vapour phase concentration of that compound in air } (w/v - e.g.\ fg\ dm^{-3})}$$

(4)

$$W_p = \frac{\text{the particulate} - \text{bound concentration of a compound in rain}}{\text{the particulate phase concentration of that compound in air}} \quad \frac{(w/v - e.g.\ pg\ dm^{-3})}{(w/v - e.g.\ fg\ m^{-3})} \quad (5)$$

Alternatively, W_v may be expressed as RT/H, where $R =$ the gas constant (atm m^3 mol^{-1} K^{-1}) and $T =$ the ambient temperature (K). In order to determine the relative importance of W_v and W_p, one must also calculate the total wash-out ratio (W) using equation (6):

$$W = W_v(1 - f_p) + W_p f_p \quad (6)$$

where f_p is the particle-bound fraction of the pollutant in question.

Standley and Hites [58] calculated values of W_v and W_p from experimental data obtained in the mid-western USA for different PCDD/F homologue groups. Although limited by the fact that their air measurements were conducted over significantly longer time periods than those for rainfall—wash-out ratios are ideally calculated from simultaneous pollutant measurements in both air and rain—some broad conclusions can be made. For most homologues, particle wash-out was more important—accounting for 92% of total wash-out for heptachlorinated dioxins—with W_v and W_p about equal for pentachlorinated dioxins and furans, and W_v more important for tetrachlorinated furans. Standley and Hites [58] also reported a significant inverse correlation between PCDD/F vapour pressure and W_v, but none with W_p, and a general increase in overall scavenging efficiency with increasing degree of chlorination due to increasing association with the particulate phase.

For PCBs, Duinker and Bouchertall [59] reported particle scavenging to be the dominant wet deposition process, despite the high V/P ratios reported for PCBs measured in simultaneously collected air samples. The same authors also reported the efficiency of such wet particle deposition of PCBs to increase with increasing chlorine number. Conversely, Eisenreich *et al.* [60] considered vapour phase wash-out to dominate wet deposition of PCBs to Lake Superior.

5.5 DRY DEPOSITION

Particle flux (F_p — in pg cm^{-2}s^{-1}) and particle concentration in air (C_p — in pg cm^{-3}) are related to the particle dry deposition velocity (V_d — in cm s^{-1}) by equation (7):

$$V_d = F_p/C_p \quad (7)$$

V_d is strongly dependent on particle size, and is comparatively low for the accumulation mode particles known to carry the majority of the particle-bound fraction of PCDD/Fs. Koester and Hites [61] reported values of V_d for all tetra-through octa-PCDD/F homologue groups—except for TCDD. These values—from two sites—ranged between 0.064 cm s^{-1} for PeCDFs and 0.60 cm

s^{-1} for OCDF, with an overall average of $0.19\,cm\ s^{-1}$, and a standard deviation of $0.13\,cm\ s^{-1}$. From this they concluded that the efficiency with which particle-bound PCDD/Fs are removed from the atmosphere is approximately the same for all congeners, although there was a discernible enhancement of dry deposition removal efficiency of PCDD/Fs with increased chlorination.

Holsen et al. [48] measured congener and homologue- specific dry deposition fluxes for PCBs, concluding that the overall PCB dry deposition velocity ranged between 0.4 and $0.6\,cm\ s^{-1}$, with an average of $0.5\,cm\ s^{-1}$. These measurements agreed well with the range of 0.04–$3\,cm\ s^{-1}$ reported by Eisenreich et al. [62] By size fractionating atmospheric particle-bound PCBs, Holsen et al. [48] were also able to estimate the relative contribution of gaseous, fine particle and coarse particle deposition to the overall deposition velocities quoted above. Their findings confirmed dry deposition of particles—especially coarse particles—to be the most significant process—accounting for up to 92% of total dry deposition of PCBs. Given that coarse particles contain roughly equal concentrations of all PCB homologue groups, it may be construed that—unlike PCDD/Fs—no significant increase in dry deposition removal efficiency will occur with increasing PCB chlorination.

For PCDD/Fs, Koester and Hites [61] concluded that wet and dry deposition are roughly equal in magnitude, and that both wet and dry deposition of particle-bound PCDD/Fs far outweigh that of vapour-phase pollutants. Given that f_p increases with increasing chlorination, it follows that soils will be enriched with the higher chlorinated homologues, with this enrichment enhanced with increasing distance from urban sources, owing to the more facile atmospheric degradation of the relatively volatile lower chlorinated homologues.

As observed for PCDD/Fs, the evidence is that both wet and dry deposition of particle-bound PCBs far outweigh that of the vapour-phase fraction. With regard to the relative significance of wet and dry deposition of PCBs, Holsen et al. [48] reported the mean $\sum$ PCB dry deposition flux to Chicago in 1989/90 to be $4.5\,\mu g\ m^{-2}\ day^{-1}$, with a range of $2.8 - 9.7\,\mu g\ m^{-2}\ day^{-1}$. Unfortunately, the lack of simultaneous measurement of total depositional flux prohibits direct assessment of the relative contribution made by wet and dry deposition to the total depositional flux. However, the fact that during 1991, $\sum$ PCB deposition to both Manchester and Cardiff in the UK ranged between 0.2 and $3.1\,\mu g\ m^{-2}\ day^{-1}$ [31], indicates that dry deposition is at least as important as wet deposition as an atmospheric removal mechanism for PCBs. This view is supported by the findings of Eisenreich et al. [60] and Murphy et al. [63] who respectively reported dry deposition to account for 60–80% of the PCB flux to Lake Superior, and wet deposition to be the most significant atmospheric source of PCBs to Lake Huron.

6 ATMOSPHERIC UPTAKE PATHWAYS BY PLANTS

Given the minor role that direct inhalation normally plays in the exposure of both humans and food animals to both PCDD/Fs and PCBs, the extent and mechanisms of uptake of these compounds by plants is of considerable interest. Above-ground plants—in particular, grass—can play an especially important role via which these compounds can enter the human food chain. Uptake of organic chemicals into plant foliage can occur via several processes: root uptake and subsequent translocation via the transpiration stream, together with foliar uptake via dry particulate deposition, wet deposition of both particle-bound and dissolved material, and dry gaseous deposition.

Recent work by Welsch-Pausch et al. [64] and Umlauf et al. [65] indicates dry gaseous deposition to be the key process governing the accumulation of PCDD/Fs in grass and PCBs in spruce needles respectively. An important implication of these findings is that current efforts to reduce the atmospheric load of particulates may inadvertently result in an increase in human exposure to such compounds by shifting their vapour/particle equilibrium to the gas phase, thereby leading to more efficient foliar uptake, and entry into food chains like the air–grass–cattle–human chain, responsible for the majority of human exposure to both PCDD/Fs [10] and PCBs [9].

7 CONCLUSION

Recent advances in analytical instrumentation have greatly facilitated studies of the atmospheric behaviour of PCDD/Fs and PCBs. Despite this, there remains much work to be done if the processes governing the atmospheric fate of these compounds are to be fully understood. Although the atmosphere is a minor reservoir for PCDD/Fs and PCBs, such an understanding is vital if we are to comprehend the means by which they distribute throughout the environment and enter the human food chain. There are many areas of interest, but among the most important is a concerted move towards separate measurements of dry, wet, vapour and particle deposition fluxes and velocities, not least because of the dominant rôle played by dry gaseous deposition in the uptake of PCDD/Fs and PCBs by above-ground vegetation. More detailed examination of the relative distribution of these compounds on different atmospheric particle size fractions is also a priority, as are studies of the relative significance of different atmospheric removal mechanisms.

REFERENCES

1. B. G. J. Heinzow and H. R. Tinneberg, Effects of polychlorinated biphenyls on T-lymphocytes using a modified erythrocyte rosette inhibition test, *Chemosphere*, **19**, 823 (1989).

2. E. Schaeffer, H. Greim and W. Goessner, Pathology of chronic polychlorinated biphenyl (PCB) feeding in rats, *Toxicol. Appl. Pharmacol.*, **75**, 278 (1984).
3. Y. L. Guo, T. J. Lai, S. H. Ju, Y. C. Chen and C. C. Hsu, Sexual developments and biological findings in Yucheng children, *Organohalogen Compounds*, **14**, 235 (1993).
4. J.-E. Kihlström, M. Olsso, S. Jensen, Å. Johansson, J. Ahlbom and Å. Bergman, Effects of PCB and different fractions of PCB on the reproduction of the mink (*Mustela vison*), *Ambio*, **21**, 563 (1992).
5. USEPA, Health assessment document for 2,3,7,8- tetrachlorodibenzo-*p*-dioxin (TCDD) and related compounds (External Review Draft Document Ref: EPA/600/BP-92/01a-c) (1994).
6. S. Y. Panshin and R. A. Hites, Atmospheric concentrations of polychlorinated biphenyls at Bermuda, *Environ. Sci. Technol.*, **28**, 2001 (1994).
7. L. S. Birnbaum, P. M. Cook, W. Farland, P. Preuss and J. C. Schaum, Update: The U. S. EPA's scientific reassessment of the risks of exposure to dioxin, *Organohalogen Compounds*, **14**, 1 (1993).
8. MAFF (Ministry of Agriculture, Fisheries and Food), *Dioxins in Food: UK Dietary Intakes*, Food Surveillance Information Sheet No. 71, HMSO, London (1995).
9. R. Duarte-Davidson and K. C. Jones, Polychlorinated biphenyls (PCBs) in the UK population: estimated intake, exposure and body burden, *Sci. Tot. Environ.*, **151**, 131 (1994).
10. S. R. Wild, S. J. Harrad and K. C. Jones, The influence of sewage sludge applications to agricultural land on human exposure to polychlorinated dibenzo-*p*-dioxins (PCDDs) and- furans (PCDFs), *Environ. Poll.*, **83**, 357 (1994).
11. S. J. Harrad and K. C. Jones, A source inventory and budget for chlorinated dioxins and furans in the United Kingdom environment, *Sci. Tot. Environ.*, **126**, 89 (1992).
12. S. J. Harrad, A. P. Sewart, R. Alcock, R. Boumphrey, V. Burnett, R. Duarte-Davidson, C. Halsall, G. Sanders, K. Waterhouse, S. R. Wild and K. C. Jones, Polychlorinated biphenyls (PCBs) in the British environment: sinks, sources and temporal trends, *Environ. Poll.*, **85**, 131 (1994).
13. K. Ballschmiter, Transport and fate of organic compounds in the global environment, *Angew. Chem. Int. Ed. Engl.*, **31**, 487 (1992).
14. J. M. Czucwa and R. A. Hites, Airborne dioxins and dibenzofurans: sources and fates, *Environ. Sci. Technol.*, **20**, 195 (1986).
15. S. Safe, Polychlorinated biphenyls (PCBs), dibenzo-*p*-dioxins (PCDDs), dibenzofurans (PCDFs) and related compounds: environmental and mechanistic considerations which support the development of toxic equivalency factors (TEFs), *Crit. Rev. Toxicol.*, **21**, 51 (1990).
16. R. Bacher, M. Swerev and K. Ballschmiter, Profile and pattern of monochloro-through octachlorodibenzodioxins and -dibenzofurans in chimney deposits from Wood Burning, *Environ. Sci. Technol.*, **26**, 1649 (1992).
17. S.J. Harrad, T.A. Malloy, M.A. Khan and T.D. Goldfarb, Levels and Sources of PCDDs, PCDFs, Chlorophenols (CPs) and Chlorobenzenes (CBzs) in Composts from a Municipal Yard Waste Compositing Facility, *Chemosphere*, **23**, 181 (1991).
18. NATO/CCMS, *International Toxicity Equivalent Factor (I-TEF) Method of Risk Assessment for Complex Mixtures of Dioxins and Related Compounds. Pilot Study on International Information Exchange on Dioxins and Related Compounds. Report No. 176, NATO/CCMS (1988)*.
19. R. E. Alcock, A. E. Johnston, S. P. McGrath, M. L. Berrow and K. C. Jones, Long-term changes in the polychlorinated biphenyl (PCB) content of UK soils, *Environ. Sci. Technol.*, **27**, 1918 (1993).

20. K. C. Jones, G. Sanders, S. R. Wild, V. Burnett and A. E. Johnston, Evidence for a decline of PCBs and PAHs in rural vegetation and air in the United Kingdom, *Nature* **356**, 137 (1992).

21. G. Sanders, K.C. Jones, J. Hamilton-Taylor and H. Dörr, Historical inputs of polychlorinated biphenyls and other organochlorines to a dated lacustrine sediment core in rural England, *Environ Sci. Technol.*, **26**, 1815 (1992).

22. R. M. Smith, P. W. O'Keefe, D. R. Hilker, B. Bush, S. Connor, R. Donnelly, R. Storm and M. Liddle, The historical record of PCDDs, PCDFs, PAHs, PCBs and lead in Green Lake, New York—1860 to 1990, *Organohalogen Compounds*, **12**, 215 (1993).

23. K. C. Jones, R. Duarte-Davidson and P. A. Cawse, Changes in the PCB concentration of United Kingdom air between 1972 and 1992, *Environ. Sci. Technol.*, **29**, 272 (1995).

24. G. H. Eduljee, PCBs in the environment, *Chem. Br.*, **24**, 241 (1988).

25. L.-O. Kjeller, K. C. Jones, A. E. Johnston, and C. Rappe, Increases in the polychlorinated dibenzo-*p*-dioxin and -furan (PCDD/F) contents of soil and vegetation since the 1840s, *Environ. Sci. Technol.*, **25**, 1619 (1991).

26. C. Rappe, Dioxins, *Organohalogen Compounds*, **4**, 33 (1990).

27. H. Hagenmaier, Contributions of diesel-powered vehicles and wood burning to overall PCDD/PCDF emissions, *Organohalogen Compounds*, **20**, 267 (1994).

28. C. C. Travis and H. A. Hattemer-Frey, Human exposure to dioxin, *Sci. Tot. Environ.*, **104**, 97 (1991).

29. G. H. Eduljee and P. Dyke, An updated inventory of potential PCDD and PCDF emission sources in the UK, *Sci. Tot. Environ.*, **177**, 303 (1996).

30. ERM/HMIP, *Risk Assessment of Dioxin Releases from Municipal Waste Incineration Processes*, HMIP, London, UK (1996).

31. P. Clayton, B. J. Davis, K. C. Jones and P. Jones, *Toxic Organic Micropollutants in Urban Air*, Warren Spring Laboratory Report No. LR904, Warren Spring Laboratory, Stevenage, UK (1992).

32. K. Ballschmiter and R. Wittlinger, Interhemisphere exchange of hexachlorocyclohexanes, hexachlorobenzene, polychlorobiphenyls, and 1,1,1-trichloro-2,2-bis(*p*-chlorophenyl)ethane in the lower troposphere, *Environ. Sci. Technol.*, **25**, 1103 (1991).

33. M. H. Hermanson and R. A. Hites, Long-term measurements of atmospheric polychlorinated biphenyls in the vicinity of superfund dumps, *Environ. Sci. Technol.*, **23**, 1253 (1989).

34. S. Y. Panshin and R. A. Hites, Atmospheric concentrations of polychlorinated biphenyls at Bloomington, Indiana, *Environ. Sci. Technol.*, **28**, 2008 (1994).

35. C. Halsall, V. Burnett, B. Davis, P. Jones, C. Pettit and K. C. Jones, PCBs and PAHs in U.K. urban air, *Chemosphere*, **26**, 2185 (1993).

36. R. Duarte-Davidson, P. Clayton, P. Coleman, B. J. Davis, C. J. Halsall, P. Harding-Jones, K. Pettit, M. J. Woodfield and K. C. Jones, Polychlorinated dibenzo-*p*-dioxins (PCDDs) and furans (PCDFs) in urban air and deposition, *Environ. Sci. & Pollut. Res.*, **1**, 262 (1994).

37. S. J. Harrad, A. R. Fernandes, C. S. Creaser and E. A. Cox, Domestic coal combustion as a source of PCDDs and PCDFs in the British environment, *Chemosphere*, **23**, 255 (1991).

38. H. Thoma, PCDD/F concentrations in chimney soot from house heating systems, *Chemosphere*, **17**, 1369 (1990).

39. C. S. Creaser, A. R. Fernandes, A. Al-Haddad, S. J. Harrad, R. B. Homer, P. W. Skett and E. A. Cox, Survey of background levels of PCDDs and PCDFs in UK soils, *Chemosphere*, **18**, 767 (1989).

40. C. S. Creaser, A. R. Fernandes, S. J. Harrad and E. A. Cox, Levels and sources of PCDDs and PCDFs in urban British soils, *Chemosphere*, **21**, 931 (1990).
41. E. Atlas and C. S. Giam, Global transport of organic pollutants: ambient concentrations in the remote marine atmosphere, *Science*, **211**, 163 (1981).
42. M. Lorber, D. Cleverly, J. Schaum, L. Phillips, G. Schweer and T. Leighton, Development and validation of an air-to-beef food chain model for dioxin-like compounds, *Sci. Tot. Environ.*, **156**, 39 (1994).
43. C. E. Junge, Basic considerations about trace constituents in the atmosphere as related to the fate of global pollutants, In: *Fate of Pollutants in the Air and Water Environments*, Part I; Suffet, I. H. ed., Wiley: New York, 1977, pp. 7–26.
44. B. D. Eitzer and R. A. Hites, Polychlorinated dibenzo-*p*-dioxins and dibenzofurans in the ambient atmosphere of Bloomington, Indiana, *Environ. Sci. Technol.*, **23**, 1389 (1989).
45. H. Yamasaki, K. Kuwata and J. Miyamato, Effects of ambient temperature on aspects of airborne polycyclic aromatic hydrocarbons, *Environ. Sci. Technol.*, **16**, 189 (1982).
46. G. H. Eduljee and D. I. Townsend, Simulation and evaluation of potential physical and mass transfer phenomena governing congener group profiles in soils near combustion sources, *Chemosphere*, **16**, 1841 (1987).
47. H. Kaupp, J. Towara and M. S. McLachlan, Distribution of polychlorinated dibenzo-*p*-dioxins and dibenzofurans in atmospheric particulate matter with respect to particle size, *Atmos. Environ.*, **28**, 585 (1994).
48. T. M. Holsen, K. E. Noll, S-P. Liu and W-J. Lee, Dry deposition of polychlorinated biphenyls in urban areas, *Environ. Sci. Technol.*, **25**, 1075 (1991).
49. C. J. Koester and R. A. Hites, Photodegradation of polychlorinated dioxins and dibenzofurans adsorbed to fly ash, *Environ. Sci. Technol.*, **26**, 502 (1992).
50. H. R. Buser, Preparation of qualitative standard mixtures of polychlorinated dibenzo-*p*-dioxins and dibenzofurans by ultraviolet and γ-irradiation of the octachloro compounds, *J. Chromat.*, **129**, 303 (1976).
51. N. J. Bunce, J. P. Landers, J-A. Langshaw and J. S. Nakai, An assessment of the importance of direct solar degradation of some simple chlorinated benzenes and biphenyls in the vapor phase, *Environ. Sci. Technol.*, **23**, 213 (1989).
52. R. Atkinson, Estimation of OH radical reaction rate constants and atmospheric lifetimes for polychlorobiphenyls, dibenzo-*p*-dioxins, and dibenzofurans, *Environ. Sci. Technol.*, **21**, 305 (1987).
53. P. N. Anderson and R. A. Hites, OH radical reactions: the major removal pathway for polychlorinated biphenyls from the atmosphere, *Environ. Sci. Technol.*, **30**, 1756 (1996).
54. R. Atkinson, Atmospheric lifetimes of dibenzo-*p*-dioxins and dibenzofurans, *Sci. Tot. Environ.*, **104**, 17 (1991).
55. E. C. S. Kwok, J. Arey and R. Atkinson, Gas-phase atmospheric chemistry of dibenzo-*p*-dioxin and dibenzofuran, *Environ. Sci. Technol.*, **28**, 528 (1994).
56. S. J. Harrad, Levels and sources of PCDDs and PCDFs in UK soils, PhD thesis, UEA, Norwich, UK (1989).
57. H. Hagenmaeir, C. Lindig and J. She, Correlation of environmental occurrence of polychlorinated dibenzo-*p*- dioxins and dibenzofurans with possible sources, *Organohalogen Compounds*, **12**, 271 (1993).
58. L. J. Standley and R. A. Hites, Chlorinated organic contaminants in the atmosphere, In: *Organic Contaminants in the Environment, Environmental Pathways and Effects*, Jones, K. C. ed., Elsevier Applied Science: London, 1991, pp. 1–32.

59. J. C. Duinker and F. Bouchertall, On the distribution of atmospheric polychlorinated biphenyl congeners between vapor phase, aerosols, and rain, *Environ. Sci. Technol.*, **23**, 57 (1989).
60. S. J. Eisenreich, S. J. Hollod and T. C. Johnson, Atmospheric concentrations and deposition of polychlorinated biphenyls to Lake Superior, In: *Atmospheric Pollutants in Natural Waters*, Eisenreich, S. J. ed., Ann Arbor Science: Ann Arbor, MI, 1981, pp. 425–444.
61. C. J. Koester and R. A. Hites, Wet and dry deposition of chlorinated dioxins and furans, *Environ. Sci. Technol.*, **26**, 1375 (1992).
62. S. J. Eisenreich, B. B. Looney and J. D. Thornton, Airborne organic contaminants in the Great Lakes ecosystem, *Environ. Sci. Technol.*, **15**, 30 (1981).
63. T. J. Murphy, A. Schinsky, G. Paolucci and C. P. Rzeszutko, Inputs of polychlorinated biphenyls from the atmosphere to Lakes Huron and Michigan, In: *Atmospheric Pollutants in Natural Waters*, Eisenreich, S. J. ed., Ann Arbor Science: Ann Arbor, MI, 1981, pp. 445–458.
64. K. Welsch-Pausch, M. S. McLachlan and G. Umlauf, Determination of the principal pathways of polychlorinated dibenzo-*p*-dioxins and dibenzofurans to *Lolium multiflorum* (Welsh rye grass), *Environ. Sci. Technol.*, **29**, 1090 (1995).
65. G. Umlauf, H. Hauk and M. Reissinger, Deposition of semivolatile organic compounds to spruce needles II. Experimental evaluation of the relative importance of different pathways, *Environ. Sci. & Poll. Res.*, **1**, 209 (1994).

8 Polycyclic Aromatic Hydrocarbons in Atmospheric Particles

D. J. T. SMITH AND R. M. HARRISON

University of Birmingham, UK

1 INTRODUCTION

Air pollution caused by human activities has been significant for 2000 years or more, as man has strived to discover new ways to apply the conversion of fuel into energy. It was over 200 years ago that British surgeon Sir Percival Pott noticed a high incidence of scrotal cancer among chimney sweeps in London; however, not until 1949 was the highly active carcinogen benzo(a)pyrene identified in domestic soot. Today it is motor vehicle exhaust emissions that are the major concern in urban and suburban areas—diesel emissions in particular

Atmospheric Particles, Edited by R. M. Harrison and R. Van Grieken.
© 1998 John Wiley & Sons Ltd.

contain large amounts of particulate matter on to which are adsorbed relatively high concentrations of a large class of semi-volatile organic compounds (SOCs), many of which are suspected as being carcinogenic and mutagenic [1]—a group known as polycyclic aromatic hydrocarbons (PAHs).

Organic compounds, mainly in the form of partially or unburned combustion products, are known to be a major constituent of atmospheric aerosol mass. [2] Due to their widespread presence and persistence in the urban environment, human exposure to PAH occurs principally by direct inhalation of polluted air or tobacco smoke, ingestion of contaminated food and water, or by dermal contact with soot, tar and oils [3–6]. Health concerns have focused on their oxygenated metabolites (i.e. diol epoxides) which have the potential to bind to and disrupt DNA and RNA, thus behaving as cancer-initiating agents. [1]

1.1 SOURCES OF PAHs

PAHs contain only carbon and hydrogen, and consist of two or more fixed benzene rings in linear, angular or cluster arrangements (Tables 1 and 2). The broader class of polycyclic aromatic compounds (PACs) incorporate a range of substituent groups and/or heteroatoms (N, O, S) in the ring structure. Their formation arises primarily as a result of incomplete combustion of organic fuels. The formation mechanism comprises two sequential stages of pyrolysis followed by pyrosynthesis, with an optimum formation temperature of 660–740 °C. [6] Once formed, simple PACs may undergo further pyrosynthetic reactions to yield higher condensed ring structures. [7] Natural sources of PAH include forest fires, biosynthesis by algae, bacteria and plants [8] and synthesis from degraded biological matter (e.g. fossil fuels). [9]

Atmospheric emissions of the sum of parent PAH by source type have been estimated for the USA and throughout much of Europe. [10–13] Anthropogenic combustion processes are the more easily quantifiable and studies have typically considered "mobile" sources (principally motor vehicles) as well as various "stationary" industries (general chemical-based operations and controlled combustion) processes, e.g. petroleum industry, coke manufacture, refuse incineration, power generation and domestic heating). Naturally PAH production and consequent release by each source will depend on combustion efficiency, and fuel composition as well as pollution control measures in operation. [10] The debate over the accuracy of this atmospheric emission data is fuelled by uncertainties in emission factors used, lack of standardized procedures for data collection and analysis, as well as doubts over the accuracy of available emission inventories—nevertheless estimated emissions have been included for reference in Tables 3 and 4. The estimates for the USA [11–14] appear to indicate that stationary sources accounted for between 80 and 90% of total PAH emissions prior to 1980. However, there have been growing concerns

Table 1.   Structure of the 16 parent PAH species described in USEPA Method 610

EPA No.	Nomenclature	Structure	MP (°C)	BP (°C)
55	Naphthalene		80	218
77	Acenaphthylene		92	270
80	Fluorene		116	294
1	Acenaphthene		96	279
81	Phenanthrene		100	340
78	Anthracene		218	340
39	Fluoranthene		110	375
84	Pyrene		156	399
72	Benz(a)anthracene		158	400
76	Chrysene		255	448
74	Benzo(b)fluoranthene		168	481
75	Benzo(k)fluoranthene		217	480
73	Benzo(a)pyrene		177	495
82	Dibenz(a,h)anthracene		262	524
79	Benzo(ghi)perylene		273	542
83	Indeno[1,2,3,c,d]pyrene		163	534

Table 2. PAH species often investigated in airborne pariculates

PAH EPA No.	Compound	Carcin- ogenic activity	Formula	No. of rings	MW
55	Naphthalene	0	$C_{10}H_8$	2	128
77	Acenaphthylene	0	$C_{12}H_8$	3	152
80	Fluorene	0	$C_{13}H_{10}$	3	166
1	Acenaphthene	0	$C_{12}H_{10}$	3	154
81	Phenanthrene	0	$C_{14}H_{10}$	3	178
78	Anthracene	0	$C_{14}H_{10}$	3	178
39	Fluoranthene	+	$C_{16}H_{10}$	4	202
84	Pyrene	0	$C_{16}H_{10}$	4	202
72	Benz(a)anthracene	+	$C_{18}H_{12}$	4	228
76	Chrysene	+	$C_{18}H_{12}$	4	228
–	Benzo(b)naphtho[2,1-d]thiophene	0	$C_{16}H_{10}S$	4	234
74	Benzo(b)fluoranthene	++	$C_{20}H_{12}$	5	252
75	Benzo(k)fluoranthene	++	$C_{20}H_{12}$	5	252
73	Benzo(a)pyrene	+++	$C_{20}H_{12}$	5	252
82	Dibenz(a,h)anthracene	+	$C_{22}H_{14}$	5	278
79	Benzo(ghi)perylene	+	$C_{22}H_{12}$	6	276
83	Indeno[1,2,3-cd]pyrene	+	$C_{22}H_{12}$	6	276
–	Coronene	+	$C_{24}H_{12}$	7	300

Note: With regards to carcinogenic activity, 0 represents inactive and ++ represents appreciable carcinogenic activity. Carcinogenic activity relates to the percentage of treated animals which develop tumours (cited in Lee *et al.* 1981): [7]
none (non-carcinogen) (0)
up to 33% (weakly carcinogenic) (+)
above 33% (strongly carcinogenic) (++)

over the accuracy of emission factors used in these particular studies, with uncertainty in emission data as high as two orders of magnitude. [15] More recently, it has been suggested that "local" mobile sources are more likely to be the major PAH contributors especially in an urban and suburban environment. [16–21] Motor vehicles have been shown to account for at least 36% of total yearly emissions in the USA. [22] Harkov and Greenberg [14] estimated that motor vehicles accounted for 98% of non-heating season benzo(a)pyrene emissions throughout the state of New Jersey. Recent receptor modelling indicates that mobile sources account for up to 72% of PAHs in Paris [23]—this is similar to that found at an an urban location in the UK, where 88% of ambient benzo(a)pyrene was estimated as having a vehicular origin. [24]

Exhaust emissions of PAHs from motor vehicles can originate by three distinct mechanisms. These are synthesis from simpler molecules in the fuel (particularly from aromatic compounds), storage in engine deposits and subsequent emission of PAHs already in the fuel, and by pyrolysis of lubricant. [17,

Table 3. Estimated atmospheric emissions of benzo(a)pyrene by source type

Source	World-wide Suess (1976) (t yr⁻¹)	World-wide Edwards (1983) (t yr⁻¹)	USA Suess (1976) (t yr⁻¹)	USA Edwards (1983) (t yr⁻¹)	NJ, USA (winter) Harkov and Greenberg (1985) (kg yr⁻¹)	NJ, USA (summer) Harkov and Greenberg (1985) (kg yr⁻¹)	Germany Ahland (1981) (t yr⁻¹)	UK Wild and Jones (1995) (kg yr⁻¹)
Residential heating (coal, wood, oil and gas)	2604 (51%)	2360 (51.6%)	475 (37%)	430 (36.8%)	6134 (98%)		10.1 (56%)	28300 (91.3%)
Power generation (power plants and industrial boilers)					0.3 (< 1%)	0.1 (< 1%)	< 0.1 (< 0.5%)	20.8 (< 0.1%)
Industrial processes (coke, asphalt, carbon black, Al, etc.)	1045 (21%)	950 (20.8%)	198 (15%)	180 (15.4%)	2 (< 1%)	3 (2%)	5.4 (30%)	1.5 (< 0.1%)
Incineration (municipal and commercial)	102 (2%)	90 (2.0%)	34 (3%)	30 (2.6%)				0.3 (< 0.1%)
Open burning (coal refuse, agricultural, forest, etc.)	1248 (25%)	1130 (24.7%)	554 (43%)	510 (43.6%)				< 500 (< 1.6%)
Mobile sources (petrol and diesel engines)	45 (1%)	40 (0.9%)	22 (2%)	20 (1.6%)	130 (2%)	183 (98%)	2.3 (13%)	2100 (6.8%)

Note: blank space indicates no data available.

Table 4. Estimated atmospheric emissions of total PAH by source type (t yr^{-1})

Source	USA Peters et al. (1981)	Ramdahl et al. (1983) USA	Sweden Ramdahl et al. (1983)	Norway Ramdahl et al. (1983)	Wild and Jones (1995) UK
Residential heating (coal, wood, oil and gas)	3956 (36%)	1380 (16%)	132 (26%)	63 (21%)	604 (84.8%)
Power generation (power plants and industrial boilers)	88 (<1%)	401 (5%)	7 (1%)	1 (< 1%)	5.8 (< 0.8%)
Industrial processes (coke, asphalt, carbon black, Al, etc.)	640 (6%)	3497 (41%)	312 (62%)	203 (69%)	19.1 (2.7%)
Incineration (municipal and commercial)	56 (<1%)	50 (<1%)	2 (< 1%)	1 (< 1%)	0.056 (< 0.1%)
Open burning (coal refuse, agricultural, forest, etc.)	4025 (36%)	1100 (13%)	2 (1%)	7 (2%)	6.3 (< 0.8%)
Mobile sources (petrol and diesel)	2266 (21%)	2170 (25%)	47 (9%)	20 (7%)	80.2 (11.3%)

25, 26] The rate of exhaust emission is dependent on factors such as engine type (including design and build quality of combustion chamber, temperature in combustion chamber, air to fuel ratio), operating conditions (i.e. emissions will increase with load, cold starts, acceleration and a decreasing air to fuel ratio), as well as the aromatic constituents of fuel and lubricating oil (e.g. Candeli *et al.* [27] showed benzene derivatives such as ethyl benzene and xylene to be better precursors for PAH formation than benzene). [28–30]

1.2 SIZE DISTRIBUTION

Comparatively few studies have investigated the size distribution of PAH-containing particles due to the large air volumes required to yield sufficient material for analysis and the sampling errors introduced by the greater pressure gradient inherent with sampling devices. [30] Reported size distributions may be biased due to sampling artefacts resulting from the use of particle size fractionating samplers which have been designed mainly for non-volatile species. These artefacts include the adsorption of low molecular weight organic gases on to the collection substrates during sampling, and volatilization of particle-associated species under the subatmospheric pressures within the samplers. [31, 32] These effects may be especially apparent for samples collected using high-volume cascade impactors—for example, Miguel and de Andrade [33] noticed a tendency for PAH concentrations to be reduced as sample duration increased, although Baek *et al.* [25] and van Vaeck and van Cauwenberghe [34] reported the two methods of filtration and impaction to yield total particulate PAH levels within 20% of each other. The use of Berner and Hering low-pressure impactors minimizes adsorption artefacts. [35]

   The greater fraction of airborne organic particulate matter is undoubtedly present in the respirable particle size range. [36] In Los Angeles, Miguel and Friedlander [37] reported that 75 and 85% of benzo(a)pyrene and coronene respectively, to be associated with particles with an aerodynamic mean diameter less than $0.26\,\mu m$; a similar distribution for particle-associated benzo(a)pyrene was measured in Rio de Janeiro. [38] In the UK, 95% of total particulate PAHs sampled in London was found to be associated with aerosols less than $3.3\,\mu m$ in diameter, while during winter up to 82% was less than $1.1\,\mu m$. [25, 39] Van Vaeck and van Cauwenberghe [40] found about 80% of the atmospheric particulate PAHs to be associated with particles less than $1\,\mu m$. Similar observations were made in Brussels and Paris, with at least two-thirds of particle-associated PAHs having a diameter less than $1\,\mu m$. [41] Although it is accepted that atmospheric aerosols generally follow a bimodal size distribution, samples collected by both Baek [42] and Miguel and Friedlander [37] indicate the existence of a unimodal distribution for particulate PAHs, the peak being between 0.1 and $1\,\mu m$. In addition, PAHs and elemental carbon size distributions measured in a road tunnel in Los Angeles were found to be unimodal with over 85% of the mass in particles less than 0.12 μm. [43] However, other reports do indicate the existence of two distinct size bands in ambient air [23, 35]—firstly, particles less than 1 μm, corresponding to relatively non-volatile PAHs, formed by adsorption, and secondly, particles greater than 1 μm, corresponding to volatile PAHs attaching to particles by condensation [34, 40]. The exception appears to be fluoranthene—evidence provided by Pistikopoulos *et al.* [23] suggests that this volatile species has anomalous behaviour. Figure 1(a) illustrates the unimodal size distribution measured at the Caldecott road tunnel, California, while Figure 1(b) shows the bimodal distribution found in ambient air at Los Angeles. [35, 43]

1.3 ATMOSPHERIC TRANSPORT, RESIDENCE TIME AND REACTIONS

Atmospheric residence time with respect to airborne transport is primarily related to the behaviour of the carrier particles. Most particulate phase PAH is adsorbed on to particles in the accumulation mode (0.2–2 μm). [37] This can

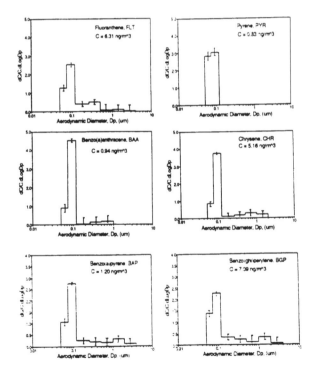

Figure 1.(a) Size distribution of 4,5 and 6-ring PAHs measured at the Caldecott Tunnel, Berkeley, California (taken from ref. 35).

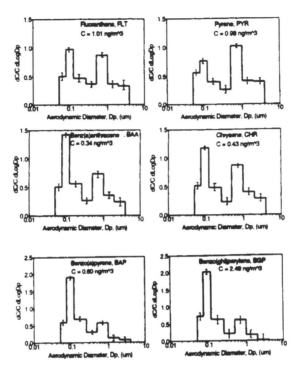

Figure 1.(b) Size distribution of 4,5 and 6-ring PAHs measured at Paco Rivera, Los Angeles, California (taken from ref. 43)

be explained by the condensation formation mechanism, i.e. larger specific surface areas are associated with this particle range. [40] These particles deposit only slowly from the atmosphere; depending on atmospheric conditions, they may be airborne for days or even weeks, being transported over long distances—perhaps in excess of 1000 km. [44, 45]. Indeed PAHs have been identified in the world's most remote parts such as the Arctic and marine atmospheres. [46] Wet and dry deposition can be collected using deposit gauges—the TOMPS UK network is one the few long-term studies that has employed this sampling technique for the investigation of PAHs (cited in section 4.3). As a result of the surface area and associated lipids of plants, it has been speculated that vegetation may well be a significant sink for lipophilic organic compounds such as PAHs [47]—indeed it has been estimated from studies undertaken at Bloomington, Indiana that $44 \pm 18\%$ of local atmospheric PAH emissions are removed by vegetation. [48] Particle scavenging rather than gas scavenging is believed to be the dominant removal mechanism for PAHs from the atmosphere by precipitation. [49] The transport distance for both particulate and vapour phase fractions is limited ultimately by chemical reactions. [50, 51]

Atmospheric PAHs are susceptible to both chemical oxidation and photochemical alterations. [52–54] Photochemical degradation of PAHs results in oxygenated compounds such as endoperoxides or diones. It is generally recognized that photochemical degradation plays a dominant role in the atmosphere, with the rate of decay influenced by the nature of the substrate on which they are absorbed [55]—aerosol age, origin, colour and carbon content appear to affect the extent of degradation. [56] Estimates by Kamens *et al.* [52] suggest that at moderate humidities and temperature, half-lives of PAHs on atmospheric soot particles in chamber atmospheres were of the order of 1 hour. Under conditions of low sunlight, low water vapour concentration or low temperature, the daytime half-life on airborne soot particles was found to increase for many PAH compounds to a period of several days.

PAHs have been found to react readily with atmospheric ozone (forming diacids, quinones and epoxides), NO_x (typically undergoing electrophilic substitution to form nitro-PAH), SO_x (producing sulfonic acids) and OH radicals— often the products formed have stronger carcinogenic and mutagenic properties than the parent compounds. [57–59]. Results from laboratory experiments are, however, difficult to extrapolate to ambient conditions—this has led to difficulties in predicting physicochemical reactions and transformations occurring in the atmosphere. Published half-lives appear to vary from hours to weeks. [10] Miguel [60] found half-life times of 12.4, 8.2 and 3.0 days for perylene, chrysene and benzo(a)pyrene respectively, exposed to 1 ppm NO_2, while Butler and Crossley [61] reported values of 6.9 and 29 days for benzo(a)pyrene and coronene exposed to 10 ppm NO_2. Reactions involving nitrogen oxides with PAHs appear to vary strongly as a function of adsorption media, [15] thus helping to explain the variation in reported half-lives of benzo(a)pyrene from several hours to several days. Atmospheric chemical reactions would prove to be a major loss pathway if PAH half-lives were only several hours in the presence of NO_2. On the other hand, if the time-scale of the reaction were several days, this loss mechanism would play only an insignificant role as this time period would exceed the lifetimes of the carrier particles. Lane and Katz [62] found the decomposition rate of selected PAH species on a filter to be first order with respect to time exposed to light and dependent on the ozone concentration. Very significant differences between the half-lives of different PAHs under exposure was noted. Half-lives were calculated as 52.7 and 29 hours for benzo(a)pyrene, and 34.9 and 3.3 hours for benzo(k)fluoranthene for exposures in darkness and simulated light respectively; in the light, half-lives were drastically reduced. Ambient levels of SO_2 appear to have no significant reaction with PAHs. [61] The hydroxyl radical has been recognized as a dominant scavenger of many airborne hydrocarbons. [19] The atmospheric reactions of vapour phase PAHs with the hydroxyl radical have been shown to result in the formation of products such as nitroarenes. [63]

2 SAMPLING

Traditionally, most atmospheric sampling has been performed by drawing large volumes of ambient air using conventional mass flow controlled $(1.1\,m^3\,min^{-1})$ "Hi-vol" samplers, thus enabling the PAH-laden airborne particles to be trapped on filters. [64] This method originated in the USA and Canada in the 1970s where Hi-vol samplers were routinely employed to measure TSP. There appeared to be no consistency in sample duration (typically 24–72 hours depending on analytical technique)—or type of filter medium (including glass fibre, Teflon, Teflon-coated glass fibre, cellulose acetate and silver membrane). Glass fibre filters have been a popular choice due to their good mechanical strength and low cost [65], while cellulose filters were identified as being less suitable due to having a higher content of soluble organic material [40]. Observations made by Pupp in 1974 suggested that 3- and 4-ring PAH were not fully filter retained due to their volatility. [66] Yamasaki confirmed these observations in the early 1980s, thus bringing about renewed research interest into ambient PAH levels. [67] Thus many of the particulate only samples taken previously may well have grossly underestimated PAH concentrations, especially for the compounds with a molecular weight less than 252 Daltons. [68] Quantification of the higher molecular weight species, which are generally associated with the higher carcinogenic activity, will not, however, be significantly affected.

Numerous sampling strategies have been adopted in order to collect total airborne PAHs—these include impregnating the filter medium, using a back-up filter, or alternatively employing a liquid impinger, cryogenic trap or solid adsorbent behind the filter. [64] This latter method has proved to be one of the more practicable techniques for collecting gaseous phase PAHs using a Hi-vol sampling system. [69, 70] Examples of solid adsorbents include open pore polyurethane foam (PUF) plugs, Tenax GC and XAD-2 resin. [71–73] With all adsorbent traps care must be taken not to exceed breakthrough volumes— which are a function of the vapour pressure for each particular PAH species— retention efficiency decreases with increasing air temperature. [69, 74, 75] PUF cartridges are currently favoured for use with Hi-vols as a cheap, practical, easy-to-use option, giving low blank and a low pressure drop. Polymeric resins have higher resultant back pressures and are therefore favoured for use with low-volume sampling systems. [68, 76] Recent research recommends that these adsorbents be used in combination with Teflon filters—this filter medium is preferred owing to its chemical inertness and low level of impurity. [77] Grosjean et al. (58) found particle-associated concentrations on glass filters to be only 85–90% of those collected by Teflon filters. There are, however, inconsistencies between the results of comparisons between these filter types. [25]

Regardless of sampling location and method, there are two major potential sources of error associated with particulate phase PAHs collected on filter media. They include, firstly, volatilization, and secondly, chemical reaction/filter artefact formation. The effects of volatilization, i.e. losses of particulate phase PAHs due to the pressure drop across the filter, known as "blow-off" losses, may be minimized by keeping sampling time as short as possible bearing in mind analytical detection limits. [32, 66, 78] In addition "adsorption gains" of volatile PAHs to the particles on the filter may take place. Foreman and Bidleman [79] found that compounds with vapour pressures near 10^{-9} atm adsorb to filters in significant quantities under Hi-vol sampling conditions. This contribution has been shown to be substantial for glass and quartz fibre filters, where as much as half of the total organic carbon has been attributed to adsorption. [80] Ligocki and Pankow [69] report that adsorption of low molecular weight PAHs to filter surfaces exposed to ambient air average 30% of the measured particle-bound PAH concentration for vapour pressures in the range of 10^{-7}–10^{-9} atm. Bidleman has suggested that thin liquid films may accumulate on filters during sampling, thereby enhancing adsorption of vapours. [49]

Artefact formation, i.e. the reaction of PAHs on the filter paper with gaseous pollutants such as NO_2, O_3 and SO_2, may be minimized by using inert Teflon filter media. [58] However, use of these filters will mean a high pressure drop which will in turn increase the volatilization effect. Diffusion denuders have recently been designed to collect PAHs, thus enabling further investigation into vapour to particle partition and the volatilization artefact inherent in Hi-vol sampling. [81, 82] Ambient air is drawn through concentric tubes that have walls coated with an adsorbent; vapour-phase PAHs diffuse to and are adsorbed by the denuder walls. Particles, however, have a much slower rate of diffusion and consequently pass through the denuder to be collected on a filter paper. A back-up adsorbent trap behind the filter ensures collection of any volatilized PAHs. [74] As yet, diffusion denuders have achieved only limited success for the collection of PAHs and other SOCs, due in part to the low flow rates inherent in most denuder designs. The task of finding a suitable denuder coating capable of complete collection of vapour-phase PAHs has proved difficult. Approaches have included employment of organic liquids (requires a high boiling point) and solid adsorbents such as polymeric resins (found to degrade with use). [83–85]

It was during the early 1950s that Kotin reported that filter-deposited PAHs may be modified by photochemical and thermal reactions with major pollutant gases, such as O_3, NO_2 and SO_2. [86] However, the complex issue of filter-promoted reactions is still not yet fully understood. There have been many discrepancies over results, with the atmospheric chemical reactions being difficult to re-create in laboratory conditions. Chemical reactions of PAHs with nitrogen oxides are particularly important because the nitrogen compounds are usually emitted together with PAHs from combustion sources. A number of studies have demonstrated that several PAHs are readily reacted in the dark with nitrogen oxides (particularly nitrogen dioxide) to form nitro-derivative PAHs. As well as depending on the chemical structure of individual PAHs, the rate of reaction was found to be highly dependent upon the nature of the substrate on which the PAHs are adsorbed. [87, 88] Reactions have been shown to be facilitated by the presence of nitric acid. [58] Atmospheric particulates have been found to contain nitro-PAHs, many of which are strong direct-acting mutagens; [53, 54, 89, 90] also some nitro-PAHs are known to possess strong carcinogenic activity. [1, 91]

The analysis of nitro-PAHs in airborne particulates is difficult and few data have, as yet, been published. 2- Nitrofluoranthene has been observed as the most abundant nitro- PAH in extracts of atmospheric particulates from the USA [53, 92]; however this compound has not been reported to be directly emitted from combustion sources. [54, 86, 93] This compound is known not to be artifactually generated during the collection of particulate organic matter on a filter; this is suggestive of atmospheric chemical transformation of parent fluoranthene to be largely responsible for the formation of 2-nitrofluoranthene. [53, 92]

Although circumstantial evidence has been presented for atmospheric nitration of PAHs, [59, 94, 95] the occurrence of nitro-PAHs in airborne particulate matter still appears to be controversial, since their identification in the particulate matter might also be the result of artefact formation during sampling. [85]

2.1 PARTICLE/VAPOUR PARTITION

Only relatively recently has it been established that the lower molecular weight PAH species are present in air mainly as vapours, with higher molecular weight species being physically adsorbed on particulate surfaces. [66] This distribution between particle and gaseous phase for individual compounds has been shown to be dependent upon temperature and TSP concentration. [67] Partitioning is in fact dependent on a number of factors, including vapour pressure of the PAH (itself a function of temperature), amount of fine particles available (in terms of available surface area for adsorption of PAHs), ambient temperature, sampling method and sample duration, reactivity and stability of each PAH species, state at emission source, and the affinity of individual PAH for the organic matrix of the particles. [39, 41] The exchange between phases may be related to the aerosol age. For example, the gas–particle distribution may be shifted by dilution of the gaseous and particulate phase emissions during transport in the atmosphere, resulting in a desorption of some constituents in the particulate matter to the gas phase—known as the aerodynamic dilution effect. [96]

It was Junge [97] who presented the first fundamental equation of reversible SOC adsorption to aerosols. Junge used a linear Langmuir isotherm to describe gas–particle partitioning in ambient air, whereby

$$\phi = c_J S_T / (p^0 + C_J S_T)$$

where ϕ is the adsorbed fraction, c_J is a parameter dependent on sorbate molecular weight, surface concentration for monolayer coverage and the difference between the heat of desorption from the particle surface (ΔH_d) and the heat of vaporization of the liquid sorbate ($\Delta H_{v,1}$) [98] (atm cm), S_T is Whitby's average total surface area ($cm^2 cm^{-3}$) and p^0 the solute saturation vapour pressure (atm).

The Langmuir adsorption model is simplified when the surface covered by PAHs adsorbed on particulate matter is assumed to be very small compared to the total surface area of the particulate matter in air, thus

$$\theta_x = \frac{b_x P}{1 + b_x P} \approx b_x P$$

where θ_x is the ratio of the particulate surface covered by compound x to the total particulate surface, b_x the ratio of the rate of adsorption to the rate of desorption for compound x and P the gas phase partial pressure of compound x.

Yamasaki *et al.* [67] postulated that an equilibrium would be established between the vapour and particulate phases of PAH species at a given temperature in the ambient air, the vapour to particle distribution being operationally defined by the ratio of the vapour phase concentration (A) to the concentration in the particulate phase (F/TSP):

$$K = A / (F/TSP) = A(TSP)/F$$

where K is the sorption equilibrium ratio, A the gaseous phase PAH concentration, determined using adsorbent ($ng\,m^{-3}$), F the particulate phase PAH concentration, determined on filter ($ng\,m^{-3}$) and TSP the total suspended particulate matter ($\mu g\,m^{-3}$). Yamasaki *et al.* [67] demonstrated that over a small range of temperatures this ratio for PAHs can be described by a simple linear regression equation derived from a Langmuir isotherm relationship (equivalent to Junge's equation if $\Delta H_d = \Delta H_{v,1}$), whereby

$$\log[A(TSP)F^{-1}] = -mT^{-1} + c$$

where m and c are compound dependent empirical constants (i.e. gradient and intercept in straight-line relationship), A is absorbent-retained PAH concentration ($ng\,m^{-3}$), F is filter-retained PAH concentration ($ng\,m^{-3}$), TSP is

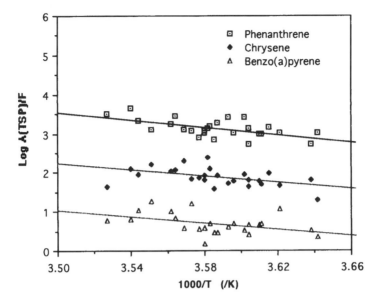

Figure 2. Yamasaki plots of log[A(TSP)/F] versus $1/T$ for selected PAHs. Sampling location: Birmingham, UK: winter 1/2 to 28/2/92 [169]

suspended particle concentration (μg m^{-3}) and T is average sampling temperature (K).

The adherence of experimental data to this relationship has been demonstrated using results from several urban air studies. [16, 39, 67, 99] Typically, only data for 3- to 5-ring PAH species have been utilized due to there being sufficiently detectable quantities in both vapour phase and particulate phase components (Figure 2). It is apparent that simple linear regression parameters (i.e. m and c) do vary quite considerably throughout the world (Table 5). These gradient and intercept constants appear to vary according to climate, sampling method and conditions employed, [39] as well as location (e.g. distance from source and characteristics of carrier particles). Theory predicts that similar compounds sorbing on the same particulate matter should possess very similar y-intercepts (i.e. values of c). [100] The Yamasaki *et al.* [67] data set was fitted using a special common y-intercept regression (CYIR) giving a value of 18.48 [100]—this was deemed to be close to the type of value that can be predicted for PAH compounds. [98] All of these studies used conventional sampling techniques (i.e. Hi-vol filters with back-up solid adsorbents) which may have resulted in volatilization artefact altering the gas/particle distribution during sampling. [82] Deviations from this Langmuir adsorption model may also be explained by relatively large temperature changes occurring during a single sampling period, or even inaccuracies in extraction and analysis. For

Table 5. Selected published parameters for "Yamasaki" plots of log AP/F versus T^{-1}

EPA No.	Compound	Yamasaki et al. (1982)			Baek et al. (1991c)			Smith and Harrison (1996)		
		m	c	r	m	c	r	m	c	r
81	Phenanthrene	4117	21.45	0.88	3805	19.98	0.81	4809	23.07	0.63
78	Anthracene				4081	20.45	0.80	4157	20.19	0.43
39	Fluoranthene	4421	18.46	0.89	4463	21.20	0.85	4618	21.72	0.63
72	Benz(a)anthracene	5826	24.89	0.94	7155	29.82	0.72	5980	26.72	0.68
76	Chrysene				4376	19.93	0.86	3359	16.88	0.63
74	Benzo(b)fluoranthene	5693	23.24	0.89	4639	20.17	0.73	4527	19.64	0.69
75	Benzo(k)fluoranthene				5143	21.92	0.67	6522	26.91	0.61
73	Benzo(a)pyrene	4864	19.99	0.83	4179	18.66	0.63	3662	16.35	0.63

example, it is believed that some PAH compounds are partially bound inside carbonaceous particles, making complete recovery of this fraction by solvent extraction difficult. [98]

In addition to the above relationship, SOC adsorption to aerosols is controlled by vapour pressure. [74] As an adjunct to the linear sorption isotherm relationship derived by Yamasaki et al., [67] it has been shown that within a given class of compounds the equilibrium ratio (K) is related to saturation vapour pressure of the sorbate (p^0) in the following form: [69, 98]

$$\log K = \log[A(\text{TSP})/F] = \log p^0 + \log k$$

where k is a temperature-dependent constant.

Foreman and Bidleman [101] described a preferential fit using the sub-cooled liquid-phase vapour pressure (p_{L^0}) as opposed to the vapour pressure of the crystalline solid (p_S^0), where

$$\log K = \log p_L^0 + \log k'$$

For samples collected in ambient air in Denver, Colorado, regression parameters were established as gradient 0.760 (p_L^0) and intercept 6.713 (k') for the class of PAH compounds at 5 °C.

2.2 AIR QUALITY STANDARDS

No USA national ambient air quality standard currently exists for PAHs. However, Title III of the Clean Air Act Amendments of 1990 identifies polycyclic organic matter (POM)—which would contain higher molecular weight PAHs—as a hazardous air pollutant for future legislation (cited in ref. 43). WHO [102] has estimated a unit lifetime risk from benzo(a)pyrene of respiratory cancer of 8.7×10^{-5} per ng m^{-3}. In Europe the Netherlands have introduced an interim goal of reducing the annual average benzo(a)pyrene to 5 ng m^{-3} (Tweede Kamer der Staten-General, 1984/5, cited in ref. 17), while a guideline of 10 ng m^{-3} for the annual average benzo(a)pyrene has been proposed by the German Federal Environmental Agency (Unweltsbudesamt, 1979, cited in ref. 17). Sampling studies have indicated that the annual average concentrations of benzo(a)pyrene are substantially below these guideline values throughout the majority of Europe. In the UK over recent years, emissions caused by open coal burning have decreased since control and enforcement of the Clean Air Acts 1956 and 1968. The recent trend to gas for domestic heating has undoubtedly contributed in helping to lower PAH emissions. An air quality limit value for smoke emissions has been introduced by the EC Directive (80/779/EEC); control of smoke indirectly controls PAHs in ambient air.

Legislation now requires that all new petrol cars sold in the USA and Europe (EC Directive 91/441/EEC) must be fitted with catalytic converters. There is substantial evidence to suggest that the use of catalytic converters with vehicles drastically restricts PAH emissions. [103–105]

3 EXTRACTION AND ANALYSIS

3.1 PARENT PAHS

Extraction of filter papers and back-up vapour phase traps can be carried out by means of Soxhlet extraction or ultrasonic vibration using organic solvents such as dichloromethane or hexane. This is usually followed by clean-up with the use of silica gel or alumina chromatography columns, with Sephadex, or more recently the employment of solid phase extraction cartridges. [106]

Analysis of PAH mixtures has traditionally been accomplished by means of chromatographic techniques (e.g. TLC, GC-FID, GC-MS, GC-LC and LC-MS). One particularly favourable application is high performance liquid chromatography (HPLC) with the use of UV–visible and/or fluorescence detectors. [107, 108] With the vast improvement and availability of specialized PAH columns, quantification of picogram amounts has become feasible. [109, 110] More recent instrumental techniques for measurement of PAHs in occupational environments include synchronous luminescence (SL) (for liquid samples, a constant differential wavelength is maintained between excitation and emission monitoring wavelengths), room temperature phosphorescence (RTP) (heavy atom doped filters are spiked with sample to enhance sensitivity), and second derivative UV absorption spectrometry (DUVAS) (second derivative of absorption spectra in the UV can be obtained for vapours and liquids). In addition real-time PAH particulate monitors have recently become available, with detection by RTP or UV absorption. [56, 111]

Typically, quantitative analysis and method calibration are completed for the 16 parent PAH species specified in the priority list of the USEPA (Method 610). These 16 species are certified in Standard Reference Material (SRM) 1647 (*Priority Pollutant Polynuclear Aromatic Hydrocarbons in Acetonitrile*); this SRM was prepared specifically for use in the calibration of chromatographic systems. Additionally, coronene and benzo(b)naphtho[2,1-d]thiophene (BNT) are often identified due to their ubiquitous presence in urban air, originating principally from vehicle emissions. [16, 112] NIST SRM can be employed to check extraction efficiencies. *Urban Particulate Matter* SRM 1648 and SRM 1649 are certified for selected PAH species from USEPA Method 610. [112, 113]

3.2 NITRO-PAHS

The determination of nitro-PAHs has become an important concern since the mutagenicity of several of these compounds was demonstrated. Several methods for detecting nitro-PAHs have been published—these are complicated by the fact that nitro-PAHs are typically found at two to three orders of magnitude lower levels than their parent compounds (i.e concentrations of picograms per cubic metre). [115] Tejeda *et al.* [116] reported a method for determining PAHs in automotive emissions using catalytic reduction of the nitro group to the amino group followed by fluorescence detection, while MacCrehan *et al.* [117] discussed a similar though less cumbersome method using a post-column packed with a zinc/silica mixture which quantitatively reduces the nitro-PAH to the corresponding and highly fluorescent amino-PAH.

Nitro-PAH species heavier than 1-nitropyrene have been found to be present in ambient air predominantly in the particulate phase. [115]

4 AMBIENT CONCENTRATIONS

4.1 INTRODUCTION

Atmospheric concentrations of PAHs have been measured on an *ad hoc* basis throughout much of the world since the mid-1970s. Earlier efforts focused on particle-associated PAHs only. Benzo(a)pyrene was generally selected for quantification; occasionally, levels of other of the more prevalent species were investigated (e.g. pyrene, chrysene and benzo(ghi)perylene). [118] Since the mid-1980s both particle and vapour phase contributions have been identified, typically for parent species selected from those listed in USEPA Method 610. In addition to these compounds it is becoming more frequent for researchers to take an interest in other PAH compounds, especially "source markers"—examples of which include coronene and thiophenes, which are indicative of petrol and diesel exhaust emissions respectively. [112]

4.2 NORTH AMERICA

Comparative data on recent ambient PAH levels found throughout the USA and Canada are illustrated in Tables 6 and 7. Comparisons of measurements should be carried out with extreme caution due to the unique characteristics of each sampling site, differing procedures for sample collection (e.g. differing sampling flow rate and duration) as well as the variety of methods used for clean-up/analysis and the varying analytical uncertainties contained within. [16] Nevertheless, it appears that urban levels for most species are roughly similar throughout the whole of North America, with the concentration of benzo(a)pyrene typically around $1 \, \text{ng} \, \text{m}^{-3}$. Winter levels were found to

Table 6. Comparison of total PAH concentrations obtained at urban centres in non-European locations (ng m^{-3})

PAH EPA No.	Compound	Keller and Bidleman (1984) Ohio, USA	Ligocki and Pankow (1989) Oregon, USA	Hoff and Chan (1984) Ontario, USA	Yamasaki et al. (1982) Osaka, Japan
80	Fluorene	na	11.1	na	na
1	Acenaphthene	na	na	na	na
81	Phenanthrene	23/140	26.3	na/13.8	100/129
78	Anthracene	1/4.2	3.44	na/1.04	100/129
39	Fluoranthene	4/23	7.90	5.96/5.10	37/36
84	Pyrene	9/27	7.32	0.37/5.2	25/30
72	Benz(a)anthracene	na	1.52	na/2.80	8.6/15
76	Chrysene	na	1.99	na/3.90	
74	Benzo(b)fluoranthene	na	3.71	na	6.4/14
75	Benzo(k)fluoranthene	0.04/0.3		na/1.10	na
73	Benzo(a)pyrene	0.3/2.0	na	na/2.30	5.7/12
79	Benzo(ghi)perylene	0.6/5.0	na	na/0.53	3.3/7.6
83	Indeno[1,2,3-cd]pyrene	na	na	na/0.39	na
–	Coronene	0.6/0.8	na	na	na

Note: na = not analysed; / = summer / winter concentrations.

be higher than those found in summer by a factor of 2–10 (Table 9). [119–122] These observations are probably explained by the increase in emissions from domestic heating (coal, oil and wood) and traffic (due to congestion and greater emissions from cold starts) during winter, and meteorological conditions which are less favourable for dispersion. Also the annual summer industrial holiday fortnight may have some contributory effect. One would expect the summer aerosol to be more chemically active than that in winter; it has been suggested that photochemical decay may play a more significant role in summer. [15] Few simultaneous urban and rural sampling studies have been reported; however, it appears that urban levels are greater by a factor of 2–10 (Table 10)—this difference is probably mainly a result of proximity to sources. [15, 123]

4.3 UK AND EUROPE

Comparative data on recent ambient PAH levels found throughout Europe are illustrated in Tables 7, 8 and 9. Extensive studies have been undertaken throughout most of the European continent including a variety of differing ambient environments in Germany, [124, 125] Austria, [126] the Netherlands, [127] Belgium [127] and Scandinavia. [68, 129] These studies revealed levels to

Table 7. Comparison of mean particulate phase PAH concentrations at urban sites throughout the world (ng m^{-3})

PAH EPA No.	Compound	De Raat et al. (1987) Kralingen, Netherlands	Baek (1988) London, UK	Harrison et al. (1996) Birmingham, UK	Greenberg et al. (1996) New Jersey, USA	Grosjean et al. (1983) Los Angeles, USA	Katz et al. (1978) Toronto, Canada	Cretney et al. (1985) Christchurch, New Zealand	Miguel et al. (1986) Rio de Janeiro, Brazil	Smith et al. (1996) Lahore, Pakistan	Chakraborti et al. (1988) Calcutta*, India
80	Fluorene	na	na	0.21/1.06	na	na	na	na	na	0.98	na
1	Acenaphthene	na	na	0.29/1.60	na	na	na	na	na	2.78	na
81	Phenanthrene	na	0.17/0.34	0.25/1.08	na	na	na	na	na	0.97	11.2
78	Anthracene	0.1	0.22/0.41	0.16/0.39	na	<0.1/0.8	na	5.3	na	4.99	2.5
39	Fluoranthene	2.2	1.01/1.53	0.35/1.17	na	0.8/1.0	na	na	1.8	2.81	10.6
84	Pyrene	1.7	0.71/1.31	0.55/2.36	0.35/2.77	1.5/1.7	na	na	1.0	2.93	24.2
72	Benz(a)anthracene	1.4	0.41/1.02	0.13/1.48	0.15/0.88	0.2/0.6	na	15.0	0.8	5.39	30.2
76	Chrysene	2.7	0.79/1.72	0.21/2.21	0.52/2.42	0.6/1.2	na	7.1	1.5	8.64	32.4
–	BNT	na	0.37/1.35	0.15/0.62	na	na	na	na	na	2.02	na
74	Benzo(b)fluoranthene	2.5	0.92/1.93	0.34/1.87	0.34/1.09	0.4/1.2	0.8/1.4	23.3	1.6	9.80	111.7
75	Benzo(k)fluoranthene	1.0	0.37/0.91	0.14/1.12	0.15/0.63	0.2/0.4	0.6/0.9	na	0.8	4.61	22.4
73	Benzo(a)pyrene	1.1	0.74/1.87	0.230.73	0.21/1.06	0.2/0.6	1.0/1.7	17.1	1.7	9.32	43.2
82	Dibenz(a,h)anthracene	na	0.06/0.18	0.07/0.78	na	na	na	na	na	3.85	12.1
79	Benzo(ghi)perylene	0.2	2.38/4.04	0.76/1.91	0.62/1.44	0.3/4.5	7.1/10.5	19.7	2.5	14.64	57.8
83	Indeno[1,2,3-cd]pyrene	1.4	1.02/1.67	0.42/1.95	0.37/0.98	na	na	12.8	4.3	12.31	na
–	Coronene	na	1.39/2.65	0.27/1.03	0.37/0.52	1.4/4.7	na	8.7	na	5.40	na

Note: na = not analysed; / = summer/winter concentrations; * = winter concentrations.

Table 8. Comparison of total PAH concentrations obtained at urban centres throughout Europe (ng m^{-3})

PAH EPA No.	Compound	Baek (1988) London, UK	Clayton et al. (1992) London, UK	Brown et al. (1996) London, UK	Harrison et al. (1996) Birmingham, UK	Clayton et al. (1992) Stevenage, UK	Clayton et al. (1992) Cardiff, UK	Clayton et al. (1992) Manchester, UK	Jaklin et al. (1985) Vienna, Austria	Jaklin et al. (1988) Linz, Austria	Thrane et al. (1987) Norway
80	Fluorene	na	28.2	na	7.00/13.7	19.1	18.3	21.3	na	na	na
1	Acenaphthene	na	3.85	na	4.23/13.5	3.60	4.60	2.48	na	na	na
81	Phenanthrene	5.12	75.1	26.06	3.84/24.1	42.9	52.2	62.3	112/197	40/91	376–888/195–1760
78	Anthracene	2.84	5.30	2.80	0.61/4.49	3.20	4.18	4.28	23/47	3.3/9.2	8–60/23–55
39	Fluoranthene	3.46	11.4	20.58	2.11/12.4	6.38	18.8	18.0	52/94	17/37	145–297/94–812
84	Pyrene	3.79	10.2	18.98	3.33/38.0	5.28	11.5	10.9	55/95	10/23	71–234/68–491
72	Benz(a)anthracene	1.41	1.20	3.83	0.34/5.59	0.78	2.88	2.25	5.1/16	2.5/6.7	13–44/19–158
76	Chrysene	1.26	2.28	6.61	0.61/6.49	1.60	4.23	2.70	8.7/22	5.3/11	33–101/26–265
74	Benzo(b)fluoranthene	1.78	1.10	2.53	0.38/2.15	0.85	2.90	1.43			
75	Benzo(k)fluoranthene	0.74	1.25	2.27	0.16/1.20	0.93	3.30	1.48	10/30	6.9/13	28–125/28–183
73	Benzo(a)pyrene	1.44	0.70	1.83	0.25/0.81	0.45	2.55	0.85	3.9/15	2.5/5.3	5–24/6–61
79	Benzo(ghi)perylene	3.30	1.93	3.64	0.76/0.83	0.98	2.85	1.93	8.5/20	2.4/4.4	4–30/9–45
83	Indeno[1,2,3-cd]pyrene	1.57	na	2.93	0.42/1.96	na	na	na	1.0/4.0	2.2/4.1	4–23/7–38
–	Coronene	1.67	na	na	0.27/1.03	na	1.05	1.10	6.3/15	1.4/2.0	1–8/4–9

Note: na = not analysed; / = summer/winter concentrations.

Table 9. Comparison of particulate concentrations obtained during summer at rural locations throughout the world (ng m^{-3})

PAH EPA No.	Compound	Baek, (1988) Folkstone, Kent, UK	Grimmer et al. (1981) N. Rhine-land, Germany	Broddin et al. (1980) Botrange, Belgium	Smith and Harrison (1996) Wasthills, Birming-ham, UK
80	Fluorene	na	na	na	0.06
1	Acenaphthene	na	na	na	0.06
81	Phenanthrene	0.02	na	0.8/2.4	0.06
78	Anthracene	0.03	na		0.03
39	Fluoranthene	0.21	na	0.8/5.3	0.07
84	Pyrene	0.21	na	0.4/3.2	0.10
72	Benz(a)anthracene	0.28	1.9/2.3	1.4/5.8	0.04
76	Chrysene	0.16	4.7/7.4		0.07
74	Benzo(b)fluoranthene	0.52	3.4/5.8	2.1/6.5	0.12
75	Benzo(k)fluoranthene	0.21	2.6/4.5		0.06
73	Benzo(a)pyrene	0.43	1.2/2.3	1.8/5.2	0.06
79	Benzo(ghi)perylene	1.16	1.8/3.1	na	0.21
83	Indeno[1,2,3-cd]pyrene	0.54	1.5/2.4	na	0.11
–	Coronene	0.21	1.6/1.3	na	0.06

Note: na = not analysed; / = summer/winter concentrations.

be higher in winter than summer by a factor of 2–5, with urban levels around two to three times greater than rural locations (Tables 9 and 10). [124, 127]

Recent data for the UK rely on the findings of three basic studies, the first of which includes the Toxic Organic Micropollutants Survey (TOMPS) network; this is a long-term monitoring programme currently investigating PAH concentrations in ambient air at four urban locations throughout the UK. [130, 131] As can be seen from Table 8, the annual mean concentrations at the three major city sites (i.e. London, Manchester and Cardiff) were similar—within a factor of 2 for each PAH species—with lower levels being obtained at the residential town of Stevenage. In addition TOMPS revealed total PAH deposition rates at the four sites also to be within a twofold factor. For example, total deposition of PAHs in London was found to range from 0.3 to 7.1 μg m^{-2}day^{-1} with a mean of 3.5 μg m^{-2}day^{-1}, whereas in Manchester values ranged from 2.7 to 17.3 μg m^{-2}day^{-1} with a mean of 6.9 μg m^{-2}day^{-1}.

Secondly, Baek *et al.* [17] investigated atmospheric PAH concentrations at South Kensington, London during the period 1985–1987 (Table 8). Results indicated an increase in PAH concentrations in winter time by a factor of approximately 4, while a study at a rural site in Kent found PAH levels to be lower than at an urban location also by a factor of 4. [42] Thirdly, seasonal intensive sampling campaigns have been undertaken at an urban site in Bir-

Table 10. Approximate ratios of winter to summer levels of PAHs in urban air and urban to rural levels of atmospheric PAH at rural locations throughout the world

Country	Location	Reference	Winter to summer ratio	Urban to rural ratio
USA	New Jersey	Harkov and Greenberg (1985)	10	2
USA	New Jersey	Greenberg et al. (1985)	4–6	3–5
USA	Los Angeles	Grosjean (1983)	2–5	
USA	Los Angeles	Gordon (1976)	4–10	
USA	Various	Faoro and Manning (1981)	2–3	10
Canada	Toronto	Katz et al. (1978)	2–3	
Japan	Osaka	Yamasaki et al. (1982)	2	
Australia	Brisbane	Yang et al. (1991)	4	5
New Zealand	Christchurch	Cretney et al. (1985)		
Germany	Essen	Grimmer et al. (1981)	2–3	2–4
Netherlands	Rijnmond	De Raat et al. (1987)	2–4	
Austria	Vienna	Jaklin and Krenmayr (1985)	2–4	
Belgium	Antwerp	Broddin et al. (1980)	10	2–10
Sweden	Stockholm	Colmsjo et al. (1985)	2	
UK	London	Baek et al. (1991b)	4	4
UK	Birmingham	Smith and Harrison (1996)	5	4
Pakistan	Lahore	Smith et al. (1996)	1–2	

Note: blank space indicates data not available.

mingham and simultaneously at a nearby rural location in the West Midlands conurbation. [24] PAHs at the urban site were found to be present in greater concentrations than at the rural site by a factor of approximately 4, with concentrations in the winter campaign exceeding those in the summer by a factor of 5 (Table 10). [16]

These findings for the UK agree well with other published European data, although values are perhaps marginally higher in comparison with those found in the USA (Figure 3). The ratios of urban to rural concentrations appear to have a lower differential in Europe—perhaps this is due to European urban areas being relatively closer in proximity to one another compared to the expanse of North America. Another observation is that the differential between winter and summer concentrations appears larger in the USA (Table 10). This may be associated with the different sources of domestic heating employed. Since the introduction of the Clean Air Acts and establishment of smokeless zones by many UK local authorities, there has been a marked reduction in coal combustion and a trend towards oil and gas. In the USA, coal and wood combustion still contribute significantly as residential heating sources.

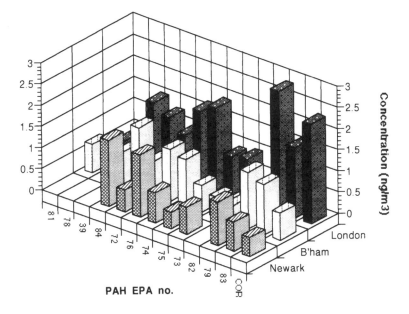

Figure 3. Comparison of yearly filter PAH concentrations at Birmingham, UK with those identified at London, UK and Newark, New Jersey. Note: A gap in the above figure indicates that no data are available. Sampling periods: Edgbaston, Birmingham: winter 1/2 to 28/2/92 and summer 27/7 to 23/8/92 (Smith and Harrison, 1996); Newark, New Jersey, USA: winter 17/1 to 25/2/83 and summer 6/7 to 14/8/82 (Greenberg *et al*, 1985); Exhibition Road, London SW7: weekly throughout 1985/6 (Baek *et al*. 1991c)

4.4 OTHER COUNTRIES OF THE WORLD

Japanese [67] and Australasian [3, 132] data appear to be reasonably similar to those found in Europe. However, the evidence from developing countries in Asia and South America presents some much higher concentrations. [133, 135]

During a global air quality monitoring programme, WHO identified TSP concentrations in Lahore, Pakistan to be among the highest in the world. This is due to a high intensity of polluting sources (for example, from numerous small industrial activities and a diversity of poorly controlled combustion sources). Also, the arid climate aggravates the high pollution levels, resulting in enormous levels of suspended particulate matter. Particle-associated PAH concentrations at three sites in Lahore have recently been investigated. [16] Results are comparable with particulate phase concentrations obtained in India (Table 7). [133] The Lahore urban yearly mean concentration for benzo(a)pyr-ene (9.32 $ng\,m^{-3}$) was found to be of the same order of magnitude as that identified in the Indian cities of Ahmedabad (3.7 $ng\,m^{-3}$) and Bombay (5.6 $ng\,m^{-3}$), [136–138] as well as the Chinese city of Beijing (6 $ng\,m^{-3}$). [102]

Concentrations of particle-associated PAHs throughout India and Pakistan are at least one order of magnitude higher than UK samples. Furthermore, given the difference in ambient temperature, vapour to particle ratios of PAHs are expected to be far higher in the hotter climate of Lahore.

It appears that current airborne particulate phase PAH concentrations in Asian cities are comparable to levels that were found in European capitals throughout the 1950s and 1960s, [17, 25, 39, 139–141] i.e. prior to the implementation of remediative legislation such as the Clean Air Acts, introduction and enforcement of pollution control measures, and the trend away from coal towards cleaner fuels. Unlike many developed countries where PAH concentrations are on the decline, it appears that emissions of PAHs in large Asian cities such as Lahore may still be on the increase.

4.5 ROAD TUNNEL STUDIES

Studies carried out in the Queensway road tunnel in Birmingham, UK revealed concentrations of PAHs and TSP approximately one order of magnitude greater than levels found in ambient air samples from the urban location during winter. [16] Good agreement was established between mean particle phase PAH

Table 11. Summary of mean total PAH concentrations identified in various road tunnel studies around the world (ng m^{-3})

PAH EPA No.	Compound	Gordon et al. (1989) Baltimore, USA	Smith and Harrison, (1996) Birmingham Queensway, UK	Khalili et al. (1995) Chicago, USA
80	Fluorene	na	168	406
1	Acenaphthene	na	114	168
81	Phenanthrene	209	333	300
78	Anthracene	37.6	51.1	177
39	Fluoranthene	56.4	47.5	117
84	Pyrene	57.5	55.3	193
72	Benz(a)anthracene	7.6	14.0	90.2
76	Chrysene	na	25.8	77.9
74	Benzo(b)fluoranthene	na	11.6	43.6
75	Benzo(k)fluoranthene	na	5.4	41.2
73	Benzo(a)pyrene	5.8	12.7	62.6
82	Dibenz(a,h)anthracene	na	4.4	14.7
79	Benzo(ghi)perylene	8.0	35.2	17.0
83	Indeno[1,2,3-cd]pyrene	4.6	21.5	20.0
–	Coronene	4.7	11.8	bd

Note: bd = below detection limit; na = not analysed.

Table 12. Comparison of particulate phase PAH concentrations obtained in world-wide road tunnel studies (ng m^{-3})

PAH EPA No.	Compound	Grimmer et al. (1981) Essen, Germany	Smith and Harrison (1996) Birmingham Queensway, UK	Gordon et al. (1989) Baltimore, USA	Fox and Staley (1976) Baltimore, USA	Colmsjo et al. (1985) Stockholm, Sweden	Miguel et al. (1989) Santa Barbara, Brazil
81	Phenanthrene	na	25.6	18.0	na	8.1	96.9
78	Anthracene	na	9.41	2.9	na	2.1	4.19
39	Fluoranthene	na	21.1	20.0	93.0	35.2	69.6
84	Pyrene	na	29.9	27.0	120.0	47.1	76.3
72	Benz(a)anthracene	31.6/48.6	11.6	7.6	102.0	16.4	51.3
76	Chrysene	44.4/66.6	18.5	na	na	24.7	69.5
74	Benzo(b)fluoranthene	19.9/41.6	11.6	na	na	na	88.2
75	Benzo(k)fluoranthene	19.2/35.7	5.4	na	na	17.9	36.7
73	Benzo(a)pyrene	20.0/38.9	12.7	5.8	66.0	16.1	90.7
79	Benzo(ghi)perylene	12.8/20.5	35.2	8.0	85.0	44.0	162.0
83	Indeno[1,2,3-cd]pyrene	12.7/27.1	21.5	4.6	na	16.3	84.2
—	Coronene	16.9/35.5	11.8	4.7	na	28.8	na

Note: na = not analysed;/=summer/winter concentrations.

concentrations obtained in this study and those in other road tunnels throughout the world. [16]

There is a fair degree of similarity between absolute PAH concentrations obtained in road tunnels throughout Europe. [16, 125] It appears that absolute concentrations identified in Europe are higher than those found over recent years in the Baltimore road tunnel (Table 11); this may well be a direct result of the use of catalytic converters, which were introduced in the USA in the mid-1970s but became mandatory in the UK only in 1993. Also the average aromatic content of petrol in the USA is considerably lower than that typically found in Europe. [17] The aromatic content of the fuel and lubricant (especially the heavier aromatics > C_9) has been shown to have a strong effect on particle-bound PAH exhaust emissions [142–144]—Candeli *et al.* [27] found 95% of PAHs present in petrol exhaust emissions to be combustion derived. There are also comparatively fewer diesel vehicles in the USA. PAHs present in diesel exhaust emissions have been found to originate from the sources of fuel survival and pyrosynthesis. [145] Radioactive tracer experiments have revealed the relative contributions from these two sources to vary considerably for individual PAHs. For example, fuel survival provided 80% of benzo(a)pyrene while pyrosynthesis contributed 70% of pyrene to diesel emissions: [146, 147]

Particle-associated PAH concentrations determined in the Santa Barbara road tunnel, Rio de Janeiro, Brazil were observed to be considerably higher than those concentrations measured in Europe and the USA, particularly for the higher molecular weight species (Table 12). This may be a consequence of the high traffic volumes and frequent congestion experienced in Rio, combined with the low proportion of vehicles with catalytic converters. Fuel produced from sugar-cane is used in the operation of locally produced motor vehicles as either hydrous alcohol or gasohol. The aromatic content of these Brazilian alcohol blended fuels may well be lower than in European fuels. [148]

Levels of nitro-PAHs in tunnel samples from Baltimore, USA were found to be similar to those found in ambient urban samples. [22] Motor vehicles did not appear to be an important source of most atmospheric nitro-PAHs, the one notable exception being 1- nitropyrene. This conclusion is consistent with the belief that much of the nitro-PAHs measured in air particulate matter is formed during atmospheric reactions of the parent PAH with hydroxyl radicals and nitrogen oxides. Diesel vehicle hydrocarbon exhaust emissions are considerably greater per kilometre than from petrol engines. [112, 149] Emissions of particulate matter from diesel engines are typically an order of magnitude higher than their petrol counterparts. Diesel particulate matter is more mutagenic and probably more carcinogenic than that from petrol, based on the results from Ames tests. [150] This may well be due to the presence of nitro-derivatives of PAHs, several species of which have been positively identified in diesel exhaust particulates. These nitroaromatic compounds are formed by reaction of PAHs

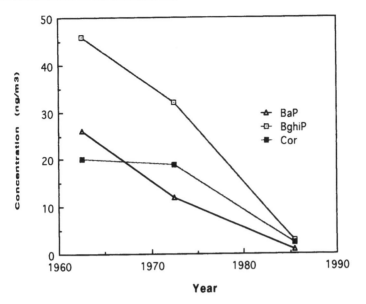

Figure 4. Concentrations of selected particle-associated PAHs in central London, 1962–1986. Sampling: Fleet Street and St Bartholomew's Hospital, near St Paul's (Commins and Hampton, 1976); Imperial College, Exhibition Road, South Kensington (Baek *et al*, 1992)

with NO_x—diesel engine exhaust is rich in NO_x, especially under heavy load and high speed conditions.

5 TRENDS

The majority of sampling campaigns carried out before the mid- 1980s utilized only filter papers, thus collecting only the particulate phase of SOCs. This means that in the majority of cases it is only the particle-associated PAH concentrations—which are arguably of greater concern for carcinogenicity—that can be compared.

In a study of 26 towns in the USA, Faoro and Manning [120] found benzo(a)pyrene concentrations in 1977 to be less than $1 ng m^{-3}$—only one-fifth of those obtained in 1966 (cited in ref. 118), while in 1958 the annual average benzo(a)pyrene level was recorded as $7 ng m^{-3}$ (cited in ref. 151). Recent declines could be explained by the reduction in coal consumption and restrictions on domestic burning, as well as introduction of vehicle catalysts. This is supported by Gordon *et al*. [152] who found PAH concentrations to be between 5 and 10 times lower than those levels measured by Fox and Staley [153] some 15 years earlier in the Baltimore Harbour road tunnel in the USA—

Table 13. Concentrations of selected particulate phase PAH in Birmingham, 1977/78 and 1992 (ng m^{-3})

PAH EPA No.	Compound	Aston University 1977/8	Birmingham University 1992
84	Pyrene	2.6	1.5
76	Chrysene	4.6	1.2
73	Benzo(a)pyrene	2.9	0.5
–	Coronene	0.9	0.7

Sampling periods:
* Edgbaston, Birmingham, UK: [daily samples] winter 1/2 to 28/2/92 and summer 27/7 to 23/8/92 (Smith and Harrison, 1996).
* Aston, Birmingham, UK: monthly throughout 1977/78 (Butler and Crossley, 1981).

this difference was attributed mainly to the introduction of catalytic converters in the USA around the mid-1970s. Catalytic converters have been shown to reduce total gasoline vehicular PAH emissions by approximately 90%. [103–105]

In several European cities in the 1960s, it was not uncommon for benzo(a)pyrene concentrations to be as high as 100 ng m^{-3} [108]—recent findings are two orders of magnitude lower. [16] Brasser reported a dramatic decrease in atmospheric levels of PAH in the Netherlands between 1964 and 1979, and Pott [154] has described a substantial reduction of at least one order of magnitude between 1969 and 1984 in benzo(a)pyrene in Germany's Ruhr district. [118]

In central London during the period 1949–1951, an annual average benzo(a)pyrene concentration was recorded as 46 ng m^{-3}. [155] Between 1949 and 1973 Lawther and Waller observed a 90% reduction in London's annual average benzo(a)pyrene concentration. [156] Baek et al. [157] found London samples taken in 1985 to be only approximately a fifth of those levels reported in the early 1970s. Coronene has shown the lowest reduction over recent years in the UK, probably due to its presence in petrol vehicle exhausts emissions and the growth in road traffic (Figure 4).

Birmingham is the UK's second city and is widely recognized as being the country's industrial heart—it has been tentatively suggested that particulate phase PAH concentrations in the Birmingham area have decreased dramatically over the past 15 years by a factor of between 2 and 10 (Table 13). [16, 20] This observation corresponds with the previously described marked decrease in metal concentrations found in the UK over the past 20 years. [36, 158] Additionally, TSP concentrations were found to be far greater in the late 1970s. The declining trend may be explained by the introduction of the Clean Air Acts of 1956 and 1968, and establishment of smokeless zones by many local authorities both within and surrounding conurbations. There has been a marked reduction in coal combustion and a trend towards oil and natural gas, as well as improved combustion technology. This has resulted in the combustion of oil-based vehicular fuels being the major source of airborne PAHs. Indeed, coronene has shown the lowest reduction over recent years in Birmingham and the rest of the UK due to its indicative presence in petrol vehicle exhausts (Table 13).

In addition, evidence for a decline of PAHs in rural vegetation and air in the UK has been put forward by Jones et al., [159] who utilized plant foliage, a reliable monitor of ambient levels of particulate and vapour phase compounds in air. Archived herbage samples (1860–1986) from Rothamsted Experimental Station, Hertfordshire were analyzed to show that PAH levels have decreased six- to sevenfold since the 1920s. [160, 161]

6 SOURCE EMISSION PROFILES AND SOURCE APPORTIONMENT MODELLING

Standardized PAH profiles are constructed by establishing the ratio between the amount of each individual PAH species relative to one specific compound— usually selected as benzo(e)pyrene or benzo(a)pyrene due to their being photo-chemically stable and found in ambient air predominantly in the particulate phase. [102] Sawicki [65] discussed the viability of using profiles or "finger-prints" of individual PAH compounds to distinguish the impact of different sources on airborne concentrations. Recent research has reinforced evidence that shows significant variation in the composition of PAH emissions for different combustion sources—known as "signatures"—with the potential for each source to contain one or more unique "tracer" compound. [112]

Unfortunately, deficiencies in existing data on the organic composition of source emissions have led to an absence of reliable fingerprints. [161, 162] Under certain circumstances the use of tracers may be hampered due to short atmospheric half-lives and potential for evaporative and reaction losses during sampling. In many cases PAH data have been collected for the determination of emission rates and are consequently inadequate for source profile/receptor modelling purposes. Deficiencies reported by Daisey [163] include the follow-ing: PAH data rarely exist with other composition variables; data are seldom representative of the average emissions for a source type and are often repres-entative of a single set of operating conditions; sampling methods used for source emissions are frequently incompatible with those used for ambient samples collected at a receptor site; differences in particle size fraction of samples may affect composition; and analytical methods and quality assurance practices used to collect existing source composition data vary widely.

Notwithstanding the above, Daisey [163] reiterated the belief that differences among emission sources were sufficient to develop fingerprints for use in receptor modelling, enabling sources of pollution to be both identified and quantified at a particular site—namely, source apportionment at a receptor. Recent studies have indeed illustrated that each source of PAHs provides an individual signature, i.e. the chemical composition patterns obtained for each categorical pollutant were different. [23, 112, 161, 164, 165]

The PAH content of a sample of urban air can be considered to be a composite of the various local sources combined with a contribution from outside that area. [25] Source emission profiles are widely variable, with relative proportions of some PAHs from a given source varying by up to several orders of magnitude [25] (Table 14); however, it appears that PAH profiles for ambient air display similar characteristics throughout many cities. [102] Indeed, it has been shown that urban air profiles are relatively uniform throughout much of Europe. [154] The fall in coal burning (source marker, benzo(a)pyrene) has resulted in a shift in PAH profile towards vehicle emissions (source markers, benzo(ghi)perylene and coronene). Obviously in some locations, profiles may be dominated by a local emission source that can "mask" background contributions. [166, 167] At the very least, profile analysis may be a preliminary method of identifying the presence of any dominant PAH indicators which are potentially useful for source discrimination. [112, 168]

Road tunnel traffic source profiles recently found in Birmingham appeared to be similar to those found at road tunnels in Europe and the USA. [169] Fingerprints agreed with those established by Daisey et al. [112] whereby the volatile species fluorene and the higher molecular weight species indeno[1,2,3-cd]pyrene featured strongly, probably occurring through the combustion of lubricating oil. [168] Petrol markers appeared to be the higher molecular weight species such as benzo(ghi)perylene, while diesel markers comprise the more volatile species phenanthrene, fluoranthene, pyrene and BNT. [48, 149, 170]

Stocks et al. [141] established particulate organic samples collected in traffic tunnels to be enriched in benzo(ghi)perylene and coronene, while Sawicki [65] found relatively high levels of benzo(a)pyrene to be present. Hering et al. [171] extended this work, showing benzo(ghi)perylene emissions to be independent of diesel population in the traffic mix. Greenberg et al. [172, 173] found a good correlation between coronene and lead data taken in New Jersey and indicated that benzo(ghi)perylene and coronene could potentially be used as vehicular emission tracers. Inspired by these studies, Duval and Friedlander [149] collected PAH emission data from selected sources in Los Angeles—the following source fingerprints were identified which enabling chemical mass balance (CMB) modelling to be completed: coal combustion (markers include anthracene, phenanthrene, fluoranthene, pyrene, benz(a)anthracene, and chrysene); coke production (anthracene, phenanthrene, benzo(a)pyrene and benzo(ghi)-perylene); incineration (phenanthrene, fluoranthene and especially pyrene); wood combustion (anthracene, phenanthrene, fluoranthene and pyrene); oil burning (fluoranthene and pyrene); petrol-powered cars (fluoranthene, pyrene

Table 14. Typical emission characteristics of PAH relative to benzo(e)pyrene (cited in Daisey et al. [112])

PAH EPA No.	Compound	Road Tunnel (Grimmer et al., 1980)	Diesel engine (Spindt, 1974)	Petrol engine (water-cooled) (Grimmer et al., 1977)	Petrol engine (air-cooled) (Grimmer et al., 1977)	Residential bituminous coal burning (Sanborn et al., 1983)	Residential wood (oak) burning (Cooke et al, 1982)	Residential oil burning (IARC, 1983)	Coal-fired power plant (Cuffe et al., 1964)	Oil-fired power plant (Bennett et al., 1979)
81	Phenanthrene		1602	64	50	3.9	17		3.2	15.6
78	Anthracene	2.2		917	9	1.9	4.8		0.42	3.2
39	Fluoranthene		63	45	18	3.2	5.5	30.4	6.5	3.9
84	Pyrene		87	78	36	2.2	3.0	12.4	3.4	0.7
72	Benz(a)anthracene	1.2	8.8	1.4	1.4	2.0	1.8	2.1		
76	Chrysene	1.8	1.2	2.3	2.1	1.6	0.8	4.2		
74	Benzo(b)fluoranthene	1.2		0.5	0.8	3.3	4.0	3.1		1.6
75	Benzo(k)fluoranthene	1.0		0.2	0.30					
73	Benzo(a)pyrene	1.2	5.5	1.4	1.4	1.4	1.4	0.1	1.7	0.4
79	Benzo(ghi)perylene	1.8	1.8	3.1	5.6	0.8	5.9	0.4	0.65	0.08
83	Indeno[1,2,3-cd]pyrene	0.8		0.9	1.5	1.8	4.1	0.6		0.04
—	Coronene	0.9		2.9	4.6	0.5	2.7	0.2	0.02	0.05

Note: blank space indicates no data available.

and especially benzo(ghi)perylene and coronene); diesel-powered vehicles (similar to petrol with perhaps higher ratios of benzo(b)fluoranthene and benzo(k)-fluoranthene, plus thiophene compounds). These markers indicate some degree of similarity and overlap between profiles from different source categories. This was also evident—especially for the lower molecular weight compounds—in fingerprints presented by Khalili et al. [161], which were used in preliminary CMB modelling calculations in Chicago. [174]

In recent years interest in the individual organic constituents of aerosols has grown, in part because of their potential for use as tracers in source apportionment models. Daisey surmised that PAHs as well as other organic compounds (such as alkanes) could be used to assist in distinguishing emissions from particular pollutant sources, [112, 175, 176] and that these compounds might be used to best advantage in combination with trace element data by simultaneously determining the organic and inorganic compositions.

The purpose of a receptor source apportionment model is to estimate the contributions of specific types of source to pollutant levels in the atmosphere at a sampling or receptor site. The contributions of the sources are distinguished through differences in their physical and chemical properties. [177–181] Multivariate source apportionment methods such as factor analysis extract information about a source's contribution on the basis of the variability of elements measured on a large number of samples (usually in excess of 50). The potential of multivariate factor analysis has been clearly demonstrated [182–186]. Factor analysis has been used generally as an exploratory tool to identify the major sources of aerosol emissions and to select statistically independent source tracers. [175, 182, 186] It can also be used to develop source emission composition profiles. [184, 185] The factor analysis/multiple regression (FA/MR) modelling method developed by Daisey and Kneip [46] utilizes stepwise multiple regression. The disadvantages of this multivariate approach are that large numbers of ambient samples must be collected and measured, and that statistically independent source tracers are required for each major source type. It has been observed that in practice multivariate techniques can only usually identify between five and eight sources. [187] Typical inorganic marker species have been identified as follows (Gordon, 1988): oil burning (V, S, Ni); marine (Na, Cl); refuse burning (Zn, Cu); soil (Mn, Fe, Ti, TSP); coal (As, Se); vehicular (Pb, Br, elemental carbon); secondary aerosol (SO_4, NO_3, NH_4); metallurgical (Fe, Mn, Cr).

A recent study based in Birmingham, UK, investigated a suite of inorganic pollutant and PAH concentrations at an urban location over a period of 2 months, and the combined data were used to conduct multivariate source apportionment for the Birmingham aerosol load. [24] The chemical source apportionment methodology took the form of principal component analysis (PCA) followed by multi-linear regression analysis (MLRA). This multivariate technique enabled major air pollution source categories to be identified (namely vehicular/road dust, coal combustion, secondary aerosol, oil combustion, incineration, direct vehicular emissions and marine/road salt), along with the quantitative contributions of pollutant species to each source group. The results demonstrated that a combination of measurements of PAHs and inorganic pollutants is a far more powerful tracer of emission sources than PAH data alone. PAHs were found to be associated predominantly with emissions from road traffic, although other sources such as fuel oil, coal combustion and incineration also contribute.

It would appear from this study that certain individual PAH species can also be used as source markers, as follows:

1. High "factor loadings" for chrysene and benzo(k)fluoranthene, along with moderate loadings for BNT, indicate coal combustion.
2. Benzo(ghi)perylene and coronene along with phenanthrene and BNT indicate direct vehicular emissions; the former two species are indicative of gasoline engine emissions, with the latter two compounds being of diesel origin. [171, 173] Mobile sources are clearly the major PAH contributors to ambient air in Birmingham. Benzo(a)pyrene is often looked upon as the most toxic of the PAHs. In past years, coal combustion was largely responsible for elevated concentrations of the compound. [6, 139] Nowadays, however, the two road traffic-related factors together account for 88% of ambient concentrations at the Birmingham site, typical of a large UK urban area.
3. It would appear that road salt particles may be adsorbing volatile vehicular PAH emissions, namely phenanthrene and other semi-volatile four-ringed PAH species (fluoranthene and pyrene).
4. Pyrene along with fluoranthene and phenanthrene appear to be markers for incineration. [188]
5. It would appear that oil combustion is associated with high loadings for some of the more volatile PAH species (fluorene, fluoranthene and pyrene), along with moderate loadings for the higher molecular weight compounds (benzo(b)fluoranthene and indeno(1,2,3-cd)pyrene).

These findings appear to agree favourably with the PAH source profiles observed by Daisey et al. [112] as well as the PAH source fingerprints used by Duval and Friedlander's [149] CMB model. This study also arrived at the conclusion that vehicular emissions are the predominant source of airborne PAHs. Due to the similarity of PAH profiles from different source types, it would appear that individual PAH species can be employed as source markers only when used in combination with inorganic pollutant data. At certain receptor sites, however, the use of supplementary organic concentrations may possibly identify additional source groups that could not be differentiated when using traditional trace element data alone. Another important element of this study was the use of total (i.e. particulate plus gaseous) PAH concentrations reather than simply the particulate phase. This overcomes problems associated with temperature- dependent phase transfers which confound the data analysis if only particle-associated PAH concentrations are used.

REFERENCES

1. IARC in *Polynuclear Aromatic Compounds*, Part 1, *Chemical, Environmental and Experimental Data*. Monographs on the Evaluation of the Carcinogenic Risk of Chemicals to Humans 32, 33–91. International Agency for Research on Cancer, Lyon, 1983.
2. Lioy P. J. and Daisey J. M. in *Toxic Air Pollutants*. Lewis Publishers, Michigan, 1987.
3. Yang S. Y., Cornell D. W., Hawker D. W. and Kayal S. I., *Sci. Tot. Environ.* 1991 **102**, 229–240.
4. Lioy P. J., Waldman J. M., Greenberg A., Harkov R. and Pietarinen C. *Archives of Environ. Health* 1988, **43**, 304–312.
5. Dennis A. J. in *Polynuclear Aromatic Hydrocarbons: Formation, Metabolism and Measurement. Seventh International Symposium on Polynuclear Aromatic Hydrocarbons*, Cooke M. and Dennis A. J., Ed., Batelle Press, Columbus, 1983, 405–412.
6. Commins B. T., *Atmos. Environ.* 1969, **3**, 565–572.
7. Lee M. L., Novotny M. V. and Bartle K. D. in *Analytical Chemistry of Polycyclic Aromatic Compounds*. Academic Press, New York, 1981.
8. Laflamme R. E. and Hites R. A., *Geochimica et Cosmochimica Acta* 1978, **42**, 289–303.
9. Blumer M. and Youngblood W. W., *Science* 1975, **18**, 53–55.
10. Wild S. R. and Jones K. C., *Environ. Poll.* 1995, **88**, 91–108.
11. Ramdahl T., Alfheim I. and Bjorseth A., in *Mobile Source Emissions Including Polycyclic Organic Species*, Rondia D., Cooke M. and Harkov R., Ed., D. Reidel Publishing Company, Dordrecht, Netherlands, 1983, 277–298.
12. Peters J. A., Deangelis D. G. and Hughes T. W., in *Chemical Analysis and Biological Fate: Polycyclic Aromatic Hydrocarbons*, Cooke M. and Dennis A. J., Ed., Batelle Press, Columbus, 1981, 571–582.
13. Suess M. J., *Sci. Tot. Environ.* 1976, **6**, 239–250.
14. Harkov R. and Greenberg A., *JAPCA* 1985, **35**, 238–243.
15. Greenberg A., Darack F., Harkov R., Lioy P. J. and Daisey J. M., *Atmos. Environ.* 1985, 19, 287–291.
16. Smith D. J. T. and Harrison R. M., *Atmos. Environ.* 1996, **30**, 2513–2525.
17. Baek S. O., Field R. A., Goldstone M. E., Kirk P. W. W., Lester J. N. and Perry R., *Wat. Air Soil Pollut.* 1991a, **60**, 279–299.
18. Harkov R. and Greenberg A. in *77th A. APCC Meeting*, San Francisco, California, 1984.
19. US. National Academy of Sciences in *Committee on Biological Effects of Atmospheric Pollution: Particulate Polycyclic Organic Matter*. Washington, 1983.
20. Butler J. D. and Crossley P. in *Polycyclic Aromatic Hydrocarbons in the Atmosphere in Birmingham*. TRRL Supplementary Report No. 656, 1981.
21. Bridbord K. in *Carcinogenesis* Volume 1, Freudenthal R. I. and Jones P. W., Ed., Raven Press, New York, 1976, 319.
22. Benner B. A. in Mobile Sources of Polycyclic Aromatic Hydrocarbons (PAH) and Nitro-PAH : A Roadway Tunnel Study. Ph.D Thesis, University of Maryland, MD, 1988.

23. Pistikopoulos P., Masclet P. and Mouvier G., *Atmos. Environ.* 1990, **24**, 1189–1197.
24. Harrison R. M., Smith D. J. T. and Luhana L. M., *Environ. Sci. Technol.* 1996, **30**, 825–832.
25. Baek S. O., Goldstone M. E., Kirk P. W. W., Lester J. N. and Perry R., *Environ. Technol.* 1991b, **12**, 107–129.
26. Gross G. P., *Soc. Automotive Engineers*, 1974, 740654.
27. Candeli A., Mastrandrea V., Morozzi G., Toccacelli S., *Atmos. Environ.* 1974, **8**, 693.
28. Candeli A., Morozzi G. and Shapiro M. A., in *Mobile Source Emissions Including Polycyclic Organic Species*, Rondia D., Cooke M. and Harkov R., Ed., D. Reidel Publishing Company, Dordrecht, Netherlands, 1983, 29–47.
29. Pedersen P. S., Ingwersen J., Nielsen T. and Larsen E., *Environ. Sci. Technol.* 1980, **14**, 71.
30. Poster D. L., Hoff R. M. and Baker J. L., *Environ. Sci. Technol.* 1995, **29**, 1990–1997.
31. Cotham W. E. and Bidleman T. F., *Environ. Sci. Technol.* 1992, **26**, 469.
32. Zhang X. and McMurry P. H., *Environ. Sci. Technol.* 1991, **25**, 456–459.
33. Miguel A. H. and de Andrade J. B., *Int. J. Environ. Anal. Chem.* 1989, **35**, 35–41.
34. Van Vaeck I. and van Cauenberghe K., *Atmos. Environ.* 1978, **12**, 2229–2239.
35. Venkataraman C. and Friedlander S. *Environ. Sci. Technol.* 1994, **28**, 563–572.
36. QUARG, *Urban Air Quality in the UK. The First Report of the Urban Air Review Group*. Department of the Environment, London, 1993.
37. Miguel A. H. and Friedlander S. K., *Atmos. Environ.* 1978, **12**, 2407–2413.
38. Miguel A. H., *Intern. J. Environ. Analyt. Chem.* 1982, **12**, 17.
39. Baek S. O., Goldstone M. E., Kirk P. W. W., Lester J. N. and Perry R., *Chemosphere* 1991c, **22**, 503–520.
40. Van Vaeck I. and van Cauenberghe K., *Environ. Sci. Technol.* 1985, **19**, 707–716.
41. Van Vaeck I., van Cauenberghe K. and Janssens J., *Atmos. Environ.* 1984, **18**, 417–430.
42. Baek S. O. in Significance and Behaviour of Polycyclic Aromatic Hydrocarbons in Urban Ambient Air. Ph.D Thesis, Imperial College of Science and Technology, University of London, 1988.
43. Venkataraman C., Lyons J. M., and Friedlander S., *Environ. Sci. Technol.* 1994, **28**, 555–562.
44. Bjorseth A. in *Polynuclear Aromatic Hydrocarbons—Chemistry and Biological Effects*. M. Dekker, New York, 1979.
45. Lunde G. and Bjorseth A., *Nature* 1977, **268**, 518.
46. Daisey J. M. and Kneip T. J. in *Atmospheric Particulate Organic Matter—Multivariate Models for Identifying Sources and Estimating their Contributions to the Ambient Aerosol*. American Chemical Society, Symposium Series No. 167, 1981 . . .
47. Simonich S. L. and Hites R. A., *Environ. Sci. Technol.* 1995, **29**, 2905–2913.
48. Simonich S. L. and Hites R. A., *Nature* 1994, **370**, 49–51.
49. Poster D. L. and Baker J. E., *Environ. Sci. Technol.* 1996, **30**, 349–354.
50. Tuominen J., Salomaa S., Sorsa M. and Himberg K., *Environ. Sci. Technol.* 1988, **22**, 1228–1234.
51. Schroeder W. H. and Lane D. A., *Environ. Sci. Technol.* 1988, **22**, 240–247.
52. Kamens R. M., Gou Z., Fulcher J. N. and Bell D. A., *Environ. Sci. Technol.* 1988, **22**, 103.
53. Pitts J. N., Sweetman J. A., Zielinska B., Winer A. M. and Atkinson R., *Atmos. Environ.* 1985, **19**, 1601.
54. Nielsen T., *Environ. Sci. Technol.* 1984, **22**, 2249.

55. Korfmacher W. A., Wehry E. L., Mamantov G. and Natusch D. F. S., *Environ. Sci. Technol.* 1980, **14**, 1094.
56. Davis C. S., Fellin P. and Otson R., *JAPCA* 1987, **37**, 1397–1408.
57. Brorström E., Grennfelt P., Lindskog A., Sjödin A. and Nielsen T. in *Polynuclear Aromatic Hydrocarbons: Formation, Metabolism and Measurement. Seventh International Symposium on Polynuclear Aromatic Hydrocarbons*, Cooke M. and Dennis A. J., Ed., Batelle Press, Columbus, 1983, 201–210.
58. Grosjean D., Fung K. and Harrison J., *Environ. Sci. Technol.* 1983, **17**, 673–679.
59. Pitts J. N., van Cauwenberghe K., Grosjean D., Schmid J. P., Fitz D. R., Belser W. L., Knudson G. B. and Hynds P. M., *Science* 1978, **202**, 515–518.
60. Miguel A. H. in *Polynuclear Aromatic Hydrocarbons: Formation, Metabolism and Measurement. Seventh International Symposium on Polynuclear Aromatic Hydrocarbons*, Cooke M. and Dennis A. J., Ed. Batelle Press, Columbus, 1983, 897–904.
61. Butler J. D. and Crossley P., *Atmos. Environ.* 1981, **15**, 91–94.
62. Lane D. A. and Katz M. in *Fate of Pollutants in the Air and Water Environment: Part 2*, Suffet I. H. Ed., Wiley- Interscience, New York, 1977, 137–154.
63. Arey J., Atkinson R., Zielinska B. and McElroy P. A., *Environ. Sci. Technol.* 1989, **23**, 321–327.
64. Lodge J. P. in *Methods of Air Sampling and Analysis*. Lewis Publishers, Michigan, 1988.
65. Sawicki E., *J. Amer. Indust. Hygiene Assoc.* 1962, **23**, 137–142.
66. Pupp C., Lao R. C., Murray J. J. and Pottie R. F., *Atmos. Environ.* 1974, **8**, 915–925.
67. Yamasaki H., Kuwata K. and Miyamoto H., *Environ. Sci. Technol.* 1982, **16**, 189–194.
68. Thrane K. E. and Mikalsen A., *Atmos. Environ.* 1981, **15**, 909–918.
69. Ligocki M. P. and Pankow J. F., *Environ. Sci. Technol.* 1989, **23**, 75–83.
70. Billings W. N. and Bidleman T., *Environ. Sci. Technol.* 1980, **14**, 679–683.
71. Chuang J. C., Hannan S. W. and Wilson N. K., *Environ. Sci. Technol.* 1987, **21**, 798–804.
72. Eckerman H., *Chemosphere* 1990, **21**, 889–904.
73. Billings W. N. and Bidleman T., *Atmos. Environ.* 1983, **17**, 383–391.
74. Bidleman T. F., *Environ. Sci. Technol.* 1988, **22**, 361–367.
75. Bidleman T. F. and You F., *Environ. Sci. Technol.* 1984, **18**, 330–333.
76. Hart K. M., Isabelle L. M. and Pankow J. F., *Environ. Sci. Technol.* 1992, **21**, 377–380.
77. Hart K. M. and Pankow J. F., *J. Aerosol Sci.* 1990, **21**, 377–380.
78. De Wiest F., *Atmos. Environ.* 1978, **8**, 915–925.
79. Foreman W. T. and Bidleman T. F., *Atmos. Environ.* 1990, **24**, 2405–2416.
80. McDow S. R. and Huntzicker J. J., *Abstracts of Papers of the American Chemical Society* 1990, **200**, 106.
81. Eatough D. J., Wadsworth A., Eatough D. A., Crawford J. W., Hansen L. D. and Lewis E. A., *Atmos. Environ.* 1993, **27**, 1213–1219.
82. Lane D. A., Johnson N. D., Barton S. C., Thomas G. H. S and Schroeder W. H., *Environ. Sci. Technol.* 1988, **22**, 941–947.
83. Kreiger M. S. and Hites R. A., *Environ. Sci. Technol.* 1992, **26**, 1551–1555.
84. Coutant R. W., Callahan P. J., Kuhlman M. R. and Lewis G. R., *Atmos. Environ.* 1989, **23**, 2205–2211.
85. Coutant R. W., Brown L., Chuang J. C., Riggin R. M. and Lewis G. R., *Atmos. Environ.* 1988, **22**, 403–409.
86. Ciccioli P., *Aerosol Sci. Technol.* 1989, **10**, 296–310.
87. Lucas S. V., Lee K. W. and Melton C. W., *Aerosol Sci. Technol.* 1991, **14**, 210–223.
88. Nielson T., *Environmental Health Perspectives* 1983, **47**, 103–114.

89. Ramdahl T., Becher G. and Bjorseth A., *Environ. Sci. Technol.* 1982, **16**, 861.
90. Campbell R. M. and Lee M. L., *Analyt. Chem.* 1984, **56**, 1026.
91. Rosenkranz H. S. and Mermelstein R. in *Nitrated Polycyclic Aromatic Hydrocarbons*, White C. M., Ed., Huethig, Heidelberg, 267–292, 1985.
92. Ramdahl T., Zielinska B., Arey J., Atkinson R., Winer A. M. and Pitts J. N., *Nature* 1986, **321**, 425.
93. Paputa-Peck M. C., Marano R. S., Schuetzle D., Riley T. L., Hampton C. V., Prater T. J., Skewes L. M., Jensen T. E., Ruehle P. H., Bosch L. C. and Duncan W. P., *Anal. Chem.* 1983, **55**, 1946.
94. Gibson T. L., *JAPCA* 1986, **36**, 1022–1025.
95. Arey J., *Journal of Resources of the National Bureau of Standards* 1978, **93**, 279.
96. De Wiest F. and Fiorentina H., *Atmos. Environ.* 1975, **8**, 915–925.
97. Junge C. E. in *Fate of Pollutants in the Air and Water Environment: Part 1* Suffet I. H., Ed., Wiley, New York 1977, 7–26.
98. Pankow J. F., *Atmos. Environ.* 1987, **21**, 2275–2283.
99. Keller C. D. and Bidleman T. F., *Atmos. Environ.* 1984, **18**, 837–845.
100. Pankow J. F., *Atmos. Environ.* 1991, **25**, 2229–2239.
101. Foreman W. T. and Bidleman T. F., *Environ. Sci. Technol.* 1987, **21**, 869–875.
102. WHO in *Air Quality Guidelines for Europe.* WHO Regional Publications, European Series 23, 105–117, 1987.
103. Rogge W. F., Hildemann L. M., Mazurek M. A. and Cass G. R., *Environ. Sci. Technol.* 1993, **27**, 636–651.
104. Westerholm R. N., Alsberg T. E., Stenberg U. and Strandell M. E., *Environ. Sci. Technol.* 1985, **19**, 43–50.
105. Tejeda S., Lang J. M., Snow L., Carlson R., and Black F., *SAE Technical Paper. Series No. 801371*, 1981.
106. Adamek S., Kettrup A. and Kicinski H., *Chromatographia* 1989, **28**, 203–208.
107. Bartle K. D., Lee M. N. and Wise S. A., *Chemical Society Review* 1981, **10**, 113–158.
108. Sawicki E., *INSERM* 1976, **52**, 297–354.
109. Miguel A. H., *Intern. J. Environ. Analyt. Chem.* 1988, **35**, 35–41.
110. May W. E. and Wise S. A., *Analyt. Chem.* 1984, **56**, 225–232.
111. Gammage R. B., Hawthorne A. R., Vo-Dinh T. and Schuresko D. D., *Alternative Energy Sources* 1983, **3**, 209.
112. Daisey J. M., Cheney J. L. and Lioy P. J., *JAPCA* 1986, **36**, 17–33.
113. Wise S. A., Hilpert L. R., Chestler S. N. and May W. E., *Fresenius J. Analyt. Chem.* 1988, **332**, 573–582.
114. Wise S. A., Benner B. A., Chestler S. N. and May W. E., *Analyt. Chem.* 1986, **58**, 3067–3077.
115. Dimashki M., Smith D. J. T. and Harrison R. M., *Journal of PAC* 1996 (in press).
116. Tejeda S. B., Zweidinger R. B. and Sigsby J. E., *Anal Chem.* 1986, **58**, 1827.
117. MacCrehan W. A., May W. E., Yang S. D. and Benner B. A., *Anal Chem.* 1988, **60**, 284.
118. Menichini E., *Sci. Tot. Environ.* 1992, **116**, 109–135.
119. Harkov R. and Greenberg A., *JAPCA* 1985, **35**, 238–243.
120. Faoro R. B. and Manning J. A., *JAPCA* 1981, **31**, 62–64.
121. Katz M., Sakuma T. and Ho A., *Environ. Sci. Technol.* 1978, **12**, 909–915.
122. Gordon R. J., *Environ. Sci. Technol.* 1976, **10**, 370–373.
123. Faoro R. B., *JAPCA* 1975, **25**, 638–640.
124. Grimmer G., Naujack K. W. and Schneider D., *Inter. J. Environ. Analyt. Chem.* 1982, **10**, 265–276.

125. Grimmer G., Naujack K. W. and Schneider D. in *Polycyclic Aromatic Hydrocarbons: Chemistry and Biological Effects*. Bjorseth A. and Dennis A. J., Eds., Batelle Press, Columbus, Ohio, 110–115, 1981.
126. Jaklin J. and Krenmayr P., *Inter. J. Environ. Analyt. Chem.* 1985, **21**, 33–42.
127. De Raat W. K. and de Meijere F. A., *Sci. Tot. Environ.* 1987, **103**, 1–17.
128. Broddin G., Cautreels W. and van Cauwenberghe K., *Atmos. Environ.* 1980, **14**, 895.
129. Colmsjö A. L., Zebühr Y. U. and Ostman C. E., *Chemosphere* 1986, **15**, 169–182.
130. Jones K. C., Halsall C., Burnett V., Davis B., Jones P. and Pettit C., *Chemosphere* 1993, **26**, 2185–2197.
131. Clayton P., Davis B. J., Jones K. and Jones P. in *Toxic Organic Micropollutants in Urban Air*. Warren Spring Laboratory Report No. LR 904, 1992.
132. Cretney J. R., Lee H. K. and Wright G. J., *Environ. Sci. Technol.* 1985, **19**, 397–404.
133. Chakraborti D., van Vaeck L. and van Espen P., *Inter. J. Environ. Analyt. Chem.* 1988, **32**, 109–120.
134. Miguel A. H., de Andrade J. B. and Hering S. V., *Intern. J. Environ. Analyt. Chem.* 1986, **26**, 265–275.
135. Smith D.J.T., Harrison R. M., Luhana L., Pio C. A., Castro L. M., Nawaz-Tariq M., Hayat S. and Quraishi T., *Atmos. Environ.* 1996, **30**, 4031–4040.
136. Negi B. S., *Atmos. Environ.* 1987, **21**, 1259–1266.
137. Aggarwal A. L., Raiyani C. V., Patel P. D., Shah P. G. and Chatterjee S. K., *Atmos. Environ.* 1982, **16**, 867–870.
138. Mohan-Rao A. M. and Vohra K. G., *Atmos. Environ.* 1975, **9**, 403–408.
139. Commins B. T. and Hampton I., *Atmos. Environ.* 1976, **10**, 561–562.
140. Waller R. E. and Commins B. T., *Environ. Res.* 1967, **1**, 295–306.
141. Stocks P., Commins B. T. and Aubrey K. V., *Inter. J. Air Water Pollut.* 1961, **4**, 141–153.
142. Williams P. T., Abbass M. K., Andrews G. E. and Bartle K. D., *Combustion and Flame* 1989, **75**, 1–24.
143. Williams P. T., Bartle K. D. and Andrews G. E., *Fuel* 1986, **65**, 1150–1158.
144. Pedersen P. S., Ingwersen J., Nielson J. and Larsen E., *Environ. Sci. Technol.* 1980, **14**, 71–79.
145. Tancell P. J. in The Origin of Polycyclic Aromatic Hydrocarbons in Diesel Exhaust Emissions. Ph.D Thesis, University of Plymouth, 1995.
146. Tancell P. J., Rhead M. M., Trier C. J., Bell M. A. and Fussey D. E., *Sci. Tot. Environ.* 1995a, **162**, 179–186.
147. Tancell P. J., Rhead M. M., Trier C. J., Bell M. A. and Fussey D. E., *Environ. Sci. Technol.* 1995b, **29**, 2871–2876.
148. Homewood B., *New Scientist* 1993, 9 January, 22–23.
149. Duval M. M. and Friedlander S. K. in *Source Resolution of Polycyclic Aromatic Hydrocarbons in the Los Angeles Atmosphere—Application of a CMB with First Order Decay*. USEPA Report No. EPA-600/2-81-161, 1981.
150. Naujack K. W., Schneider D. and Grimmer G., *Fresen. J. Analyt. Chem.* 1982, **311**, 475–484.
151. Finlayson-Pitts B. J. and Pitts J. N. in *Atmospheric Chemistry: Fundamentals and Experimental Techniques*. Wiley-Interscience, New York, 870–960, 1985.
152. Gordon G. J., Benner B. and Wise S., *Environ. Sci. Technol.* 1989, **23**, 1269–1278.
153. Fox M. and Staley S., *Analyt. Chem.* 1976, **48**, 992–998.
154. Pott F., *Staub, Reinhaltung der Luft* 1985, **45**, 369–379.
155. Stocks P. and Campbell J. M., *British Medical Journal* 1955, 15 October, 923–929.

156. Lawther P. J. and Waller R. E., *IARC Scientific Publications* 1976, **13**, 27–40.
157. Baek S. O., Goldstone M. E., Kirk P. W. W., Lester J. N. and Perry R., *Sci. Tot. Environ.* 1992, **111**, 169–199.
158. McInnes G. in *Multi-element Survey: Summary and Trend Analysis 1976/7 to 1988/9*. Warren Spring Laboratory Report No. LR 771, 1992.
159. Jones K. C., Stratford J. A., Tidridge P., Waterhouse K. S. and Johnston A. E., *Environ. Poll.* 1989, **56**, 337–351.
160. Jones K. C., Sanders G., Wild S. R., Burnett V. and Johnston A. E., *Nature* 1992, **356**, 137–140.
161. Khalili N. R., Scheff P. A. and Holsen T. M., *Atmos. Environ.* 1995, **29**, 533–542.
162. Masclet P. and Mouvier G., *Atmos. Environ.* 1990, **24**, 1187–1197.
163. Daisey J. M., *Environ. Inter.* 1985, **11**, 285–291.
164. Li C. K. and Kamens R. M., *Atmos. Environ.* 1993, **27**, 523–532.
165. De Raat W. K. and de Meijere F. A., *Sci. Tot. Environ.* 1991, **103**, 1–17.
166. Hopke P. K. and Glover D. M., *J. Air Waste Man.* 1991, **41**, 294–305.
167. Grimmer G., Naujack K. W. and Schneider D. in *Polycyclic Aromatic Hydrocarbons—Fourth International Symposium.* Bjorseth A. and Dennis A. J., Eds., Batelle Press, Columbus, Ohio, 107–125, 1980.
168. Hangebrauck R. P., von Lehmden D. J. and Meeker J. E. in *Sources of PAH in the Atmosphere.* Public Health Service Publication 993–AP–33. Cincinnati, US Department of Health, Education and Welfare, 1967.
169. Smith D. J. T. in Source Apportionment of Atmospheric Particles in the UK, Portugal and Pakistan. Ph.D. Thesis, University of Birmingham, 1995.
170. Grimmer G. in *Analysis of Automobile Condensates. Pollution and Cancer in Man.* Mohr U., Schmahl D. and Tomalis L., Ed., IARC Publication No. 16, Lyon, France, 1977.
171. Hering S., John W. and Goren S., *Atmos. Environ.* 1983, **17**, 115–119.
172. Greenberg A., Harkov R., Darack F., Daisey J. M. and Lioy P. J., *Environ. Sci. Technol.* 1984, **18**, 287–291.
173. Greenberg A., Bozelli J. W., Stout D., and Yokoyama R., *Environ. Sci. Technol.* 1981, **16**, 566–570.
174. Khalili N. R. in Atmospheric Polycyclic Aromatic Hydrocarbons in Chicago: Characteristics and Receptor Modelling. Ph.D Thesis, Illinois Institute of Technology, Illinois.
175. Daisey J. M., Morandi M. T. and Lioy P. J. *Atmos. Environ.* 1987, **21**, 1821–1831.
176. Lioy P. J., Kneip T. J. and Daisey J. M. in *Receptor Model Technical Series,* Volume 4: *Factor Analysis and Multiple Regression (FA/MR) Techniques in Source Apportionment.* US EPA Report No. EPA-450/4-85-007, 1985.
177. Friedlander S. K., *ACS Symposium Series* 1981, **167**, 1–19.
178. Cooper J., *JAPCA* 1980, **30**, 1116–1125.
179. Gordon G. E., *Environ. Sci. Technol.* 1980, **14**, 792–800.
180. Friedlander S. K., *Environ. Sci. Technol.* 1973, **7**, 235–240.
181. Thurston G. D. and Spengler J. D., *Atmos. Environ.* 1985, **19**, 9–25.
182. Kleinman M. T. and Kneip T. J., *Environ. Sci. Technol.* 1980, **14**, 62–65.
183. Thurston G. D. and Spengler J. D. in *Receptor Models Applied to Contemporary Air Pollution Problems,* Dattner S. L. and Hopke P. A. Eds. Pollution Control Association, Pittsburgh, 175–187, 1984.
184. Alpert D. J. and Hopke P. K., *Atmos. Environ.* 1981a, **15**, 675–687.
185. Alpert D. J. and Hopke P. K., *Atmos. Environ.* 1981b, **14**, 1137–1146.

186. Kleinman M. T. in The Apportionment of Sources of Airborne Particulate Matter. Ph.D Thesis, New York University, 1977.
187. Henry R. C., Lewis C. W., Hopke P. K. and Wiliamson H. J., *Atmos. Environ.* 1984, **18**, 1507–1515.
188. Chiang P. C., You J. H., Chang S. C. and Wei Y. H., *J. Hazardous Materials* 1982, **31**, 29–37.

9 Carbonaceous Combustion Aerosols

Hélène CACHIER

Centre des Faibles Radioactivités, Gif sur Yvette, France

Atmospheric Particles, Edited by R. M. Harrison and R. Van Grieken.
© 1998 John Wiley & Sons Ltd.

1 INTRODUCTION

Combustion results in direct emissions of reactive or unreactive gases and aerosols. The composition and relative abundance of products emitted are primarily related to the fuel composition itself and to a significant extent to the physical characteristics of the combustion processes. Fuels used in combustion are fossil fuels or living or dead vegetation. They are mainly of a carbonaceous nature.

Combustion produces primarily CO_2 and H_2O:

$$(C_xH_yO) + O_2 \longrightarrow CO_2 + H_2O \tag{1}$$

However, the chemical composition of combustion plumes is more complex. A principal important parameter is the presence in the fuel matrix of other elements such as S, N or trace elements. It is also important to consider the physical conditions of the combustion process which determine deviations from the "ideal" combustion case and hence formation of more reduced species. There is a vast array of reduced carbonaceous species which may be produced in the atmosphere by combustion, either as gases (CO, CH_4, hydrocarbons, organic gases—volatile organic compounds, VOCs) or aerosols.

Following Goldberg, [1] formation of aerosols may be explained by the dismutation of carbon monoxide through the reaction:

$$2CO \longleftrightarrow C + CO_2 \tag{2}$$

or by polymerization/dehydrogenation of the fuel (noted as Ri H) under the oxidative action of the hydroxyl radical or any oxidant:
action of the hydroxyl radical or any oxidant:

$$R_iH + OH^{\bullet} \longrightarrow Ri^{\bullet} + H_2O \tag{3}$$

$$2R_i^{\bullet} + OH^{\bullet} \longrightarrow R_iCO + R_h^{\bullet} \tag{4}$$

$$CO + OH^{\bullet} \longleftrightarrow H^{\bullet} + CO_2 \tag{5}$$

$$H_2 + OH^{\bullet} \longrightarrow H^{\bullet} + H_2O \tag{6}$$

$$R_h + R_iH \longrightarrow R_{h+i}H \quad \text{(polymerization)} \tag{7}$$

Fresh combustion-derived particles may be presented as a carbonaceous matrix with inclusions of various trace elements. The relative importance of non-carbon atoms in combustion aerosols is related to several important determining factors. These include: the initial composition of the fuel matrix, the volatilization capability of a given element in derived products and also the mode of production for particles (primary vs secondary production). Microscopic studies of biomass burning aerosols have shown, for example, that carbonaceous cores ubiquitously possess a potassium signal. This may be explained by the fact that potassium cannot escape into the atmosphere in any gaseous form. However, the potassium signal is not found in fossil fuel derived aerosols because this type of fuel is very impoverished in potassium by comparison with vegetation.

Finally, aerosols evolve in the atmosphere by interactions with other species, and upon entering cloud systems they undergo major coagulation and chemical reactions. Ageing phenomena severely affect the particle surface and thus "old" combustion particles are probably different from "fresh" ones. On average, old combustion particles are heterogeneous particles consisting of a carbonaceous core covered by a coating of organics and electrolytes among which sulfate is often the most important component.

In this chapter, emphasis is placed on analysis of the carbonaceous fraction of combustion aerosols, keeping in mind the possible heterogeneous nature of the particles which critically determines their atmospheric behaviour. Their increasing importance in the tropospheric reservoir will also be pointed out.

2 CHEMICAL AND STRUCTURAL COMPLEXITY OF CARBONACEOUS AEROSOLS: DEFINITIONS AND VOCABULARY

It may be assumed that two main sources contribute to the burden of ambient carbon particles: primary production by direct combustion emissions and secondary production of particles by oxidative conversion of organic gases through emission by living vegetation or through production in combustion plumes. At most sites, combustion activities overwhelmingly account for atmospheric carbonaceous aerosols, [2] which provides the main reason why the composition of the aerosol carbonaceous component will be discussed in terms of combustion sources only.

2.1 CHEMICAL COMPOSITION

The total carbonaceous fraction of aerosols (TC), which is related to the carbon remaining after an acidic treatment (carbonates are thus excluded) is classically divided into two groups: the organic component (OC) and the black carbon component (BC).

2.1.1 Organic and Black Carbon Particle Structure

Thermal analysis of these aerosols is shown in Figure 1(a) which points out the presence of various components with different volatility properties. Peaks are attributed to volatile, primary and secondary organics, following increasing temperature. The last peak (at around 350 °C under oxygen) corresponds to the evolution of a dark (grey) component and is attributed to BC (also referred to as soot). Carbonates, if present, would appear at higher temperatures.

Peak shape and even peak position on the temperature axis are not constant (Fig 1(b)) and it may easily be deduced that ambient carbonaceous aerosols form a complex and variable mixture in which each component is not chemically defined. Moreover, there is not a clear-cut distinction between the OC and BC components.

Using several analytical techniques, Smith and co-workers [5] made extensive studies on hexane soot and proposed a structure for particles shown in Figure 2. This structure is a "weighted" picture of results obtained with elemental analyses (H/C/O), spectroscopy (FTIR, UV, Raman) and other techniques such as ^{13}C-NMR, EPR, GC/MS or ESCA used on bulk or solvent-extracted material. Up to now, this work on hexane soot is the most exhaustive and coherent approach to elucidating the chemical composition of combustion particles and will be used here as a reference to support further assessments.

Combustion carbonaceous aerosols appear to consist of a backbone of graphitic structure associated with C–O functions and hydrocarbon segments at the particle surface. The type of association comprises a vast set of loose or strong chemical or physical bondings with the core, the most important bondings being chemical or hydrogen-type bondings, Π-electron or dipole–dipole attractions.

Structure and properties of both OC and BC aerosols may be deduced from the schematic in Figure 2. *The gradual relative importance of the graphitic backbone in the whole structure determines a shift in aerosol properties* (optical absorption, resistance to different types of agression) with an apparent cut-off point which will be used in the separation and definition of the OC and BC fractions given in Table 1. The crucial problem is to verify that the different cuts provide components with satisfactory overlap.

Black carbon (soot) is the most polymerized and refractory fraction of combustion aerosols. Although the graphitic core is overwhelmingly important in particles it must be underlined that organic functionalities are still present on the surface. For example, Smith *et al.* [5] found for hexane soot, that the amount of neutral C was of the order of 90% and that graphitic core surface coverage by organic groups was of the order of 50%. Elemental or graphitic carbon (EC) may be considered as the ultimate stage in the aromatization process and forms only a minor part of soot. [6]

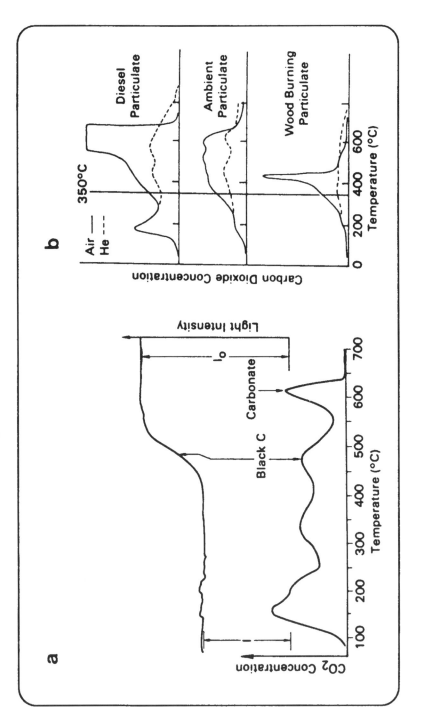

Figure 1. Thermoanalysis of various ambient and combustion aerosols: (a) ambient samples (in oxygen) (from Novakov [3]); (b) various samples (in air and helium) (from Cadle and Groblicki [4])

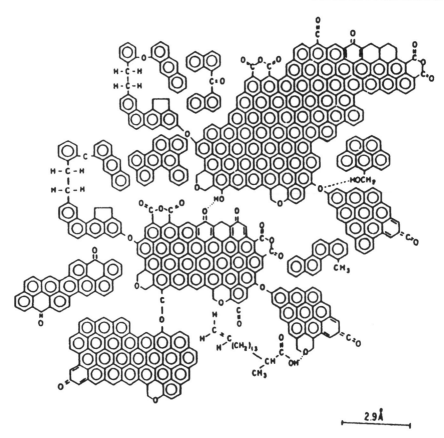

Figure 2. Proposed structure for hexane soot (from Smith *et al.* [5])

Table 1. Covariation of physicochemical properties and composition of combustion carbonaceous aerosols

	Total carbon (TC)					
	Organic carbon (OC)			Black carbon (BC)		
	VOC	$\longrightarrow$ POC	$\longrightarrow$	BC	$\longrightarrow$	EC
H/C ratio	2	1.5	0.2	0.1	0.05	0
Aromaticity	0	0.1	0.8	0.8	0.9	1
Colour	No	$\longrightarrow$	yellowish	grey or black		
Imaginary refractive index: 0.0035				~ 0.8		
Resistance to thermal destruction: poor				important		
Resistance to chemical oxidation: poor				important		
Resistance to bacterial digestion: poor				important		

2.1.2 Organic Speciation

On average, *organic aerosols* contain organic material which may be found in many different chemical forms, depending on a considerable variety of factors linked to the nature of the combustion fuel, the quality of combustion and age of the aerosol. The extractable part of the organic component (EOM) may be visualized as the outer part of the particles. In a pioneering effort, Clain [7] describes an attempt to obtain a comprehensive EOM speciation in combustion aerosols. About 3700 compounds are unambiguously identified by GC/MS, representing more than 90% of the extractable organic matter but only 10–30% of the total carbonaceous aerosol mass. More than 70% of the organic aerosol is thus present as unidentifiable polymerized material, pointing out the relative importance of the particle core. However although a relatively minor component, the EOM situated at the outer part of the particles will be of major importance in determining the aerosol properties which will rely primarily on the abundance of reactive molecules, as is the organization of these molecules in surface layers. [8] It may be seen in Figure 3 that most of the important chemical families are significantly present in aerosol extracts but in different proportions. This figure also illustrates the importance of the nature of the combustion source, and that of ageing processes on the chemical composition of the particle EOM. A salient feature of Clain's data also is that very light (C-4 to C-8) compounds which theoretically are in gaseous form are actually found to be attached to aerosols, confirming the hypothesis of important interactions occurring at the particle surface.

2.1.3 Organic Markers

Organic molecular markers found in aerosols encompass a wide range of compounds. These molecules originate directly from vegetation or fossil fuels by distillation-like processes or are derived products as a result of more or less oxidation processes and thermal maturation (Figure 4). PAHs (polycyclic aromatic hydrocarbons) and the graphitic core of particles may be considered as an advanced state of these transformations.

Biomass burning aerosols have a predominant biogenic signature and lesser amounts of PAHs [9] probably due to average relatively low temperature combustion processes. The plant wax signature is often found in lipidic compounds such as long-chain (C-19, C-31) alkanes with odd-to-even carbon number predominance, *n*-alkanoic acids or alkanols. Other individual biopolymers or biomarkers from the pyrolytic breakdown of resins (di- or triterpenoids) or lignin precursors (phenolic products) may serve as tracers of specific vegetation taxa. [10–13]

Pyrene and benzo(a)pyrene are relatively abundant in savanna and bush fires and owing to their particular chemical stability could be used as tracers for

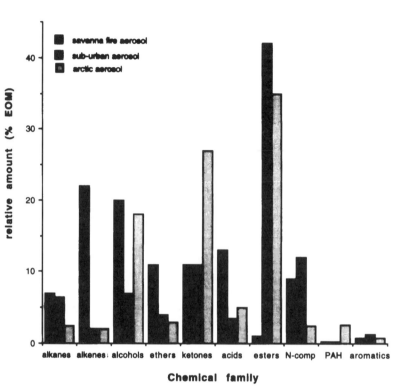

Figure 3. Extractable organic matter (EOM) speciation in biomass-burning aerosols (adapted from Clain [7])

these types of combustions. Retene (1-methyl, 7-isopropyl phenanthrene) resulting from the thermal transformation of abietic acid present in the resin of wood (as seen in Figure 4) could serve as an unambiguous marker and tracer of wood burning. [14, 15]

In industrial aerosols, molecules used to "fingerprint" the fossil fuel combustion sources are not numerous. The set of "unresolved" alkanes (with no odd-to-even carbon number preference) is a satisfactory indication of their presence. Although many other organic compounds may be detected in these aerosols [7] none are actually specific and the best tracers remain PAH compounds. [16, 17] The use of benzanthracene has been proposed to trace automotive exhausts, whereas benzofluoranthenes could indicate influence of diesel exhausts [18] (see also Chapter 8).

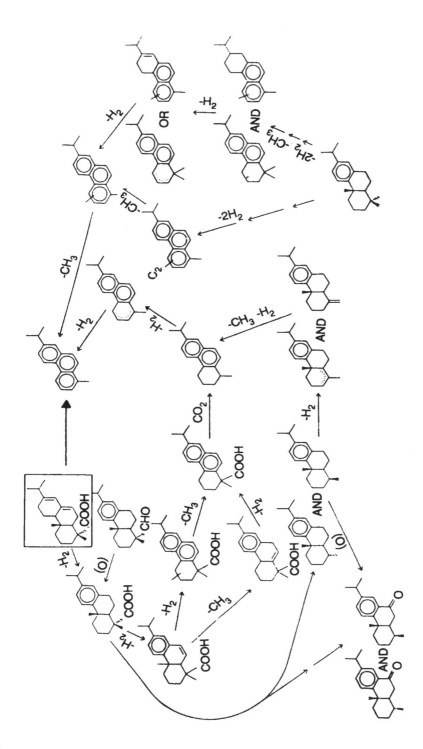

Figure 4. Scheme of pyrolytic maturation of abietic acid or abietane diterpenoids to retene. Several of these compounds may be found in conifer biomass-burning aerosols (from Standley and Simoneit [11])

Much attention has been paid to the PAH content of combustion aerosols owing to their carcinogenic properties and their use as tracers due to their satisfactory resistance to chemical agression. The major part of PAHs found in biomass-burning aerosols is also present in fossil fuel aerosols but their abundance histogram is different and could help source apportionment. [19, 20] Industrial aerosols bear on average more unbranched PAHs than biomass-burning particles or non-pyrolytic residues. Noteworthy is the fact that a significant amount of PAHs attached to aerosols should be found as gases, and that partition between the two phases is reversible and sensitive to temperature. [21] As experiments conducted on various 3- to 5-ring PAH compounds proved that most species retain their isotopic integrity, $\partial^{13}C$ isotopic composition determinations on the bulk fraction or on individual particles have also been proposed to apportion sources in aerosols or in sediment BC particles. [22] It may also be noted that PAHs with a highly pericondensed structure such as coronene or benzo(e)pyrene are found in very old sediments that are associated with charred woody remains and BC, and they trace past fires. [23]

2.1.4 Recommendations

In light of previous considerations, the important following recommendations are proposed for the sake of clarity in the use of vocabulary for combustion aerosols:

- Whatever the type of combustion (industrial combustion or biomass burning) aerosols emitted contain both OC and BC particles. Organic carbon is always the predominant form (50–95% in mass) which has often been ignored especially for industrial aerosols.
- The use of "soot" is ambiguous, as is the term "smoke". Some authors call "soot" industrial aerosols, others call "smoke" biomass-burning aerosols in such a way that it is preferable to use "total particulate carbon". Use of elemental carbon or graphitic carbon instead of BC is erroneous too and should be abandoned.
- Combustion processes produce both primary and secondary particles. Primary particles are formed through ejection of liquid or solid fuel and this material retains structural and chemical information on the fuel matrix. Charcoals or industrial fly ashes are found in this first particle category. Secondary particles may appear rapidly in the initial stages of the plume as a result of reactions such as those presented in the introduction, and others may appear initiated by oxidative attack of gaseous organic material. As a matter of clarity, either primary or secondary particles which are both present in combustion plumes are considered as "primary aerosols". Further attachment of organic material at the soot particle surface may also induce important

chemical transformations, but up to now this has not been quantified. Moreover the fate of pyrolytic organic gases (VOC) has never been taken into consideration although this sort of species is likely to undergo oxidation and photochemical attacks. These reactions are likely to favour the formation of secondary particles as least at easily as presumed for natural hydrocarbons.

Major anthropogenic aerosol types as presented by the last IPCC report [24] on climate change (1994–95) are, in our opinion, artificially divided due to a misappreciation of the chemical structure of combustion aerosols and their reactivity in the atmosphere. The following clarification to the particle categorization is proposed as shown in Table 2.

Due to the polymeric and variable nature of combustion particles, isotopic tracers seem the most promising tools to trace the bulk carbon of the aerosols. Enrichment in stable (^{13}C) or radioactive (^{14}C) has been tentatively used. ^{14}C is adapted in source regions to apportion wood burning and fossil fuel-based combustion. [15, 21] ^{13}C isotopic signatures have been used in urban regions [25, 26] and in the tropics to discriminate C-3 and C-4 biomass-burning particles. [27, 28] In the marine atmosphere of the northern hemisphere, this tracer gave evidence of overwhelming combustion-derived inputs. [29]

2.2 PARTICLE MORPHOLOGY

Electron microscopy is a powerful technique for observations of combustion particles [30]. The procedure generally includes a transfer of particles on to a metallic grid after partial dissolution in chloroform of the sample filter such as a Nuclepore-type membrane.

Table 2. Categorization of the major anthropogenic and carbonaceous aerosols

IPCC report (1994–95)	Our categorization
Primary	
Industrial soot (except soot)	Industrial BC
Soot (includes biomass burning)	Industrial OC
Biomass burning (except soot)*	Biomass-burning BC
	Biomass-burning OC
Secondary	
Sulfates from SO_2	Sulfates from SO_2
Nitrates from NO_x	Nitrates from NO_x
Organic matter from biogenic VOC	Organic matter from biogenic VOC
	Organic matter from pyrolytic VOC

*Also forms secondary particles.

2.2.1 Electron Microscopic Techniques

Scanning electron microscopy (SEM) investigations are generally conducted on primary particles (charcoal and fly ashes) which are relatively coarse particles (diameter from 1 to 50 µm on average) with interesting structures as shown in Figure 5. These particles are mostly found in source region atmospheres although occasionally they can be transported over a long range. [1] Their accumulation in various matrices such as lake sediments or monumental black crusts may constitute historical records of combustion, and source identification will be possible using microscopic morphological characteristics. [31–33] As an example, charcoals retain the taxon inner-cell structure which may allow vegetation reconstruction. [32] Fly-ash particles are found as smooth, porous or spongy spheres with sometimes smaller spheres embedded in bigger structures, with morphology depending on several factors, primarily the fossil

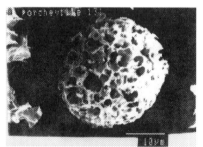
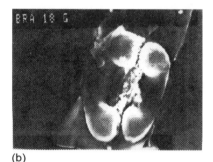

(a)   (b)

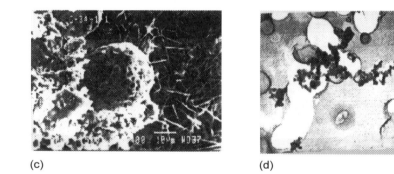

(c)   (d)

Figure 5. Microscopic studies of combustion carbonaceous particles: (a) SEM study of a spongy (carbonaceous) fly-ash (courtesy of R. Lefèvre); (b) SEM study of a charcoal from Amazonian fires (courtesy of A. Gaudichet); (c) SEM detail of a monumental black crust showing a porous fly ash embedded in gypsum filaments (courtesy of P. Ausset); (d) TEM study of atmospheric microsoots from Arles (France) showing biomass burning and diesel clusters (diesel spheres are the smallest) (courtesy of F. Bannery)

fuel used (coal or oil) and combustion temperature. [34, 35] Occasionally, some industrial processes may release carbonaceous particles exhibiting different morphologies, many of them being non-spherical. [36] In smelter effluents Bradley and Buseck (1983) [37] found filaments consisting of complex inter-growth of carbon and metals with important tracer capability.

Transmission electron microscopy (TEM) is used for smaller particles. These "micro-soots" (up to a few microns) are actually agglomerates of much smaller spheres. As for fly ashes, spherical shape of particles is the signature of high-temperature transformations. Biomass burning or diesel microsoot morpho-logy are identical which is striking at first glance. However, close observations of elemental spheres in clusters show that on average diesel spheres are a little smaller (10–20 nm diameter) than biomass-burning ones (15–30 nm). Also, the presence of larger isolated spheres (~50 nm) could be an indicator of vegetation fires at their smouldering stage. [30, 38] X-ray analysis of individual microsoot particles allows detection of trace elements which may constitute unambiguous evidence for a source tracer: diesel- or oil-burning particles always contain sulfur (sometimes associated with vanadium) whereas potassium is a marker of biomass-burning particles emitted during flaming fires. [30,39]

Using TEM with electron energy loss spectroscopy (EELS) Katrinak *et al.* [40] could investigate the inner structure of elemental carbon spheres within aggregates. The studies revealed a range of carbon structures from amorphous to graphitic carbon. Such variations occur among individual spheres, a result which could be interpreted as the confirmation of the "reference structure" (see Figure 2) described earlier.

2.2.2 Particle Transformation Study

Microscopic studies can also lead to further information on carbon particle morphology and other properties, which up to now have received little attention. During their atmospheric lifetime, particles are probably shaped by coagulation and more likely by heterogeneous interactions with cloud droplets: if young particles are primarily elongated branched chains, aged particles are expected to be found as more spherical clusters. Indeed, this evolution has been in evidence during laboratory experiments [41] and on natural aerosols. [42] Some aged soot particles may even appear as dense spheres up to a few microns, a feature never encountered in young aerosols. [43, 44] Statistical studies per-formed on bulk samples may help determine real particle size distributions; studies on size-segregated samples also have the potential to give information on the aerodynamical behaviour of particles (related to both particle mean shape and density).

Finally, some microscopic studies using barium chloride as a detector of sulfates, [45] have given evidence of rapid formation of a sulfate coating at the soot particle surface, giving a picture of heterogeneous particles found in a so-called internal aerosol. Surface coatings of sulfate and nitrate may also be detected by the TEM/EELS method.

Internal aerosols are possibly formed also by intrusion of suspended dust or marine salts during particle formation in combustion plumes, or by electrostatic attraction. Some mixed particles of this type were found in Kuwait fire air masses (BC with inorganic aerosols [42]) and in savanna plumes (BC in organic droplets, [2]).

3 ANALYSIS OF ORGANIC (OC) AND BLACK (BC) CARBON AEROSOLS

3.1 TOTAL CARBON

The TC content of aerosols is determined after removal of carbonates, by burning particles accumulated on carbon-free collection surfaces (glass, quartz or silver fibre filters). The amount of carbon is obtained by a carbon analizer using infrared, chromatographic or coulometric detection of CO_2. The limiting factor of the analytical protocol is due to the filter blank especially for OC analyses which is of the order of $1.5 \pm 1\,\mu g\,C\,cm^{-2}$.

3.2 BLACK CARBON

The BC content of aerosols is determined by a wide variety of methods, [46,47] all techniques being based on an operational definition of BC. The main concern is to ensure a sufficient overlap of results using different methods which remains difficult even for methods of the same type. [48]

3.2.1 Optical Methods

In optical methods, BC is considered to be the most effective absorbent of the aerosol for visible atmospheric radiation. Methods are based on the degree of attenuation of a reflected or transmitted light beam. Particles are analysed as a filter deposit by instruments such as the aethalometer [49] or the integrating plate or sandwich method. [50] In such devices, reflection on the filter surface or multi-scattering by the particle deposit [51] are artefacts that should be taken into account. If a fibre filter is used, multi-reflections inside the fibrous matrix enhances the measurement sensitivity. A more sophisticated technique, the "integrating sphere", has been described by Heintzenberg. [52] In this procedure, aerosols are dispersed in a liquid and are supposed to be closer to their atmospheric state than when deposited on a filter and the multiple reflections on the sphere walls also ensure a satisfactory sensitivity. Furthermore, all the light scattered is integrated, ensuring more reliable measurements than with other methods such as the integrating plate method. [53]

A disadvantage of optical methods is the need for calibration generally by using artificial carbon, a material which poorly imitates atmospheric BC. As quoted by Liousse *et al.* [44] calibration is also likely to vary with type and age of BC particles but this consideration has never been taken into account: a striking result would be to multiply by a factor of 3–5 concentration values found in remote atmospheres where particles are expected to be bigger and with a thicker coating than in source regions. Nevertheless, optical determinations of BC contents are very sensitive and may permit on-line measurements, [54,55] yielding important information either in source regions or in remote atmospheres as shown in Figure 6 for the South Pole.

A minor problem comes from interferences which may occur with crustal dust. This non-BC absorbent fraction has such a considerable range of its absorption cross-section due to its variable size distribution and mineralogy (constituents such as limonite, haematite or magnetite have very different imaginary refractive indices, [56]) that in some situations it appears reasonable to make additional measurements. Indeed, the fraction of absorption attributable to aerosol dust can be derived from firing the quartz sample filter with the aerosol deposit to burn off BC and then remeasuring the absorption. Results reveal, however, that

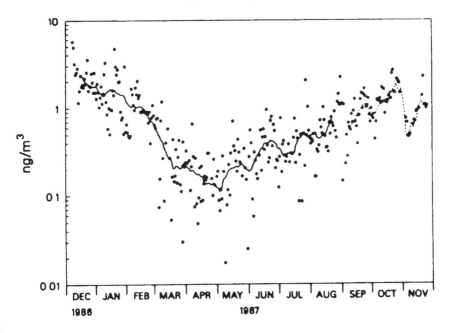

Figure 6. Optical measurements (aethalometer) of BC concentrations at the South Pole (from Hansen *et al.* [54])

on average absorption by non-BC particles is negligible or not of major importance (5–20%) even at sites significantly influenced by crustal dust. [57–59]

3.2.2 Chemical Methods

Chemical methods are used to determine BC content. They are based on the resistance of BC to an oxidative attack which is optimized to eliminate all types of organics. They are conducted either under acidic conditions, HNO_3 or H_2SO_4, [60,61] or under alkaline conditions, KOH/H_2O_2 or NaOH/solvent. [62,63] The analytical protocols are demanding and time-consuming, and so these methods are generally used for complex matrices such as sediments when other methods are not available. They have proved to be successful even in very old sediments such as Cretaceous–Tertiary clays. [61] When combined with a thermal treatment, chemical extraction (HNO_3) is also promising for analysis of ash residues and recently led to interesting results by Kuhlbusch and Crutzen [64] on the sink of carbon formed by vegetation fires.

3.2.3 Thermal Methods

Thermal methods are based on the thermal resistance of BC. They include a thermal pretreatment which allows escape of any organic material. These methods have to avoid two contradictory artefacts: the untimely removal of BC and the charring of the organic matter which may lead to an overestimate of BC content. This is generally achieved under an oxidative gas flow at a temperature of around 340 °C. [4,65] Thermo-optical techniques use continuous measurements of the filter deposit darkness in parallel with thermoanalysis. [66,67] Charring may thus be visualized as an increase in the filter darkness before evolution of black carbon. Quantification of charring assumes, however, that the light extinction per unit mass of pyrolitically produced BC is the same as that of original BC aerosols. Another problem has been recently shown by Novakov and Corrigan [68] who have stated that alkaline metals could catalyze thermal destruction of the BC aerosols at temperatures significantly lower than that ordinarily expected, thereby giving a severe underestimate of the BC component. This has been shown to occur with biomass-burning aerosols which are commonly enriched in potassium but could also occur with marine aerosols.

3.3 ANALYSIS OF ORGANIC AEROSOLS WITH THE THERMO-OPTICAL METHOD

When focus is put on the different components of organic aerosols, it is possible to enhance the clarity of thermograms by using an inert gas carrier (helium or nitrogen). At present, the most accomplished procedure is probably that developed at DRI-Reno. [48] The method consists of using successive steps in temperature under firstly pure helium up to 550 °C and then under a mixture of oxygen (2%) and helium. Main peaks attributed to the different organic aerosol components are separated satisfactorily, and BC itself appears as different components as seen in Figure 7. The darkening of the filter by charring (which may represent 20% of the initial attenuation) is followed by laser reflectance. This procedure appears to have the potential to apportion combustion sources (diesel versus wood burning) by observation of differences in the apparent make-up between OC peak populations. [69,70]

However, independent measurements of aerosol light absorption and BC concentrations have pointed out some potential failures of this thermo-optical method for recognizing OC and BC peaks. This problem might be linked to variable scattering effects of internally mixed BC and other atmospheric species such as sulfates and/or modification of BC particle absorption by evolution of

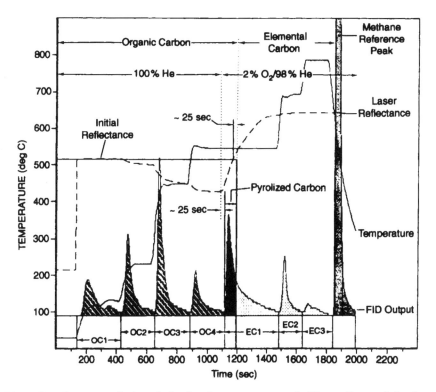

Figure 7. Thermo-optical analysis of carbonaceous aerosols (Desert Research Institute design) (from Chow *et al.*, [48]). The sample is submitted to volatilization at temperature of 120, 250, 450 and 550 °C in a pure helium atmosphere for the OC component analysis then to combustion at 550, 700 and 800 °C in a 2% oxygen and 98% helium atmosphere for BC analysis. Charring of the organic material is quantitatively followed by the reflectance of the deposit and corrected

its coating. [69,70] It could be also inferred that a portion of the OC aerosol is light adsorbing but not thermally refractory, revealing a discrepancy in the two operational thermal and optical definitions of BC.

4 REACTIVITY AND ATMOSPHERIC BEHAVIOUR OF COMBUSTION PARTICLES

The majority of combustion particles can be considered to have been shaped and chemically transformed during the combustion and ageing processes and can be satisfactorily represented schematically in Figure 8(a) for young elongated aerosols and in Figure 8(b) for more spherical aged aerosols with a surface coating. There is evidence, however, that freshly emitted BC elongated particles may be already internally mixed with VOCs in the combustion plume. [71]

Therefore, combustion particles are not inert species and display an interesting reactivity leading to atmospheric properties which are now progressively taken into consideration. This interest is attested by the quality and diversity of studies reported in the special issues edited by Novakov, [72] Puxbaum and Novakov [73] and Penner and Novakov. [74]

4.1 PARTICLE SURFACE REACTIVITY

Combustion particles have an important surface reactivity due the favourable combination of two main factors. The first is that particles have a relatively

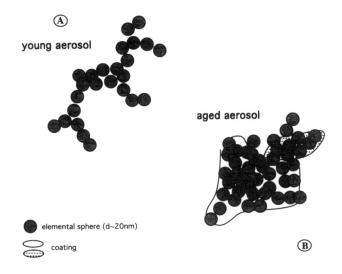

Figure 8. Schematic diagrams for young and aged BC aerosols

large surface to mass ratio: particle surface area is found to be of the order of $100 \, m^2 \, g^{-1}$ for gas fixation [5,75] and about half this value for liquid fixation. [76] This property may be primarily explained by the particle structure (sphere clusters) and to a lesser degree by their surface roughness (degree of pore structure). Haynes *et al.* [77] showed that in high-temperature combustion of coal, submicron particles ($d \sim 0.03 \, \mu m$) provide most of the active ash surface although representing only 1% of the total mass.

The second factor to take into account is a significant surface reactivity due to the presence of organic functions and aromatic systems which will favour attachment of atmospheric gaseous species and water vapour and induce further chemical reactions. Oxidative processes will then rapidly transform part of the particle surface. The coating acquired will enhance the particle hydrophilic character. It may be assessed that on average for tropical biomass-burning fresh aerosols, the coating is primarily of an organic nature, whereas in atmospheres influenced by industrial combustions (using coal, oil or diesel fuels), sulfates are probably the dominant species at the particle surface. [76] The coating is also likely to be modified during long-range transport in the atmosphere and when particles enter could droplets due to cloud nucleation processes and by deposition of solutes when droplets evaporate. From polar aerosol studies, Clain [7] found that the organic coating could turn into a quite unreactive material (with relatively a lot of ketones and esters). Conversely, the inorganic coating of sulfate or nitrate is expected to be more important. However, this picture of an internal population for combustion-derived aerosols, even off source regions, remains controversial for some authors (e.g. Heintzenberg and Covert [78]). The possible existence of a coating is illustrated by Novakov, [79] showing a remarkably constant S/C ratio value (0.22 in mass) in urban aerosols whereas older aerosols generally display higher values (0.5–1). These results suggest too that the build-up of sulfate coating begins rapidly near the source. [80] Figures 9(a) and (b) illustrate among several others, the significant correlation frequently found between particulate BC and (excess) sulfate concentrations. This strongly suggests similar atmospheric pathways for the two anthropogenic aerosol components and/or a sulfur coating formation on to the BC particle surface. In Figure 9(a) (obtained at Finokalia, a remote site in Crete), BC/S concentration ratios are low due primarily to exceedingly high S concentrations, confirming model predictions for S peaks in this region where prevailing winds bring "old" atmospheric material from western and central Europe. [81] Figure 9(b) shows evidence of sulfate formation in source regions by two different processes, probably by heterogeneous interactions directly on to carbonaceous particles in winter and through aqueous film interactions in summer when average atmospheric humidity is high. [82]

Particle surface and reactivity are affected by the build-up of a coating, but also by the physical accumulation and coagulation processes which produce a shift in number, surface and mass distribution of particles. Following a smoke

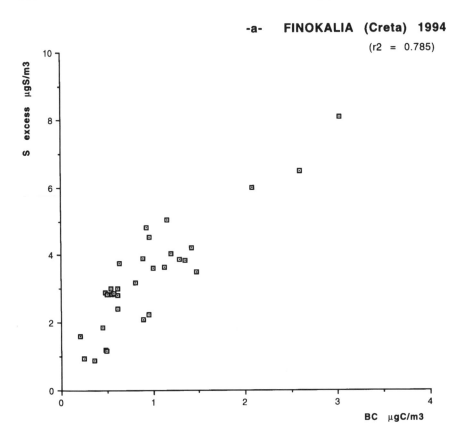

Figure 9. Correlation between sulfate and carbon concentrations in aerosols: (a) at Finokalia, a remote site in Crete (Cachier and Mihalopoulos, unpublished data);

plume parcel in a "Lagrangian" type experiment may be an opportunity to gain some insights into such phenomena. Radke *et al.* [83] were able to sample a smoke plume from Californian fires transported some 1000 km over the ocean which represented a transport time of up to 3 days. Ageing effects were in evidence but with an unexpected significant increase of the particle coarse mode. Such a transfer from accumulation to coarse mode is not predicted by coagulation theory for spherical particles and could be primarily related to particle shape. An additional complexity which could account for this result might reside in the numerous gas–particle interactions in the smoke, a driver for heterogeneous particle growth. If this result reflects a general behaviour, sedimentation of particles could be a more important phenomenon than generally expected.

(b) BEIJING 1983-1984

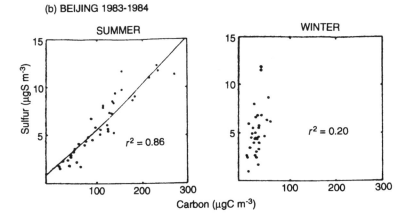

Figure 9. (b) in the urban area of Beijing (from Dod *et al.* [82])

All these considerations on shape and coating appear critical to a greater understanding of the geochemical behaviour of particles and of their impact in the atmosphere.

4.2 PARTICLE OPTICAL PROPERTIES

Optical properties of the combustion carbonaceous particles are closely dependent on their shape, size, chemical composition and their state of mixing whether external or internal.

Ageing through accretion processes shifts the particle size distribution and is an important factor to take into account to estimate aerosol extinction which is shown by calculation to reach a maximum for particles in the accumulation mode. [83–85] Sensitivity to particle size is illustrated by changes in the light-scattering coefficient of a smoke plume which is positively correlated to the mass of the accumulation mode through ageing. [83]

Light absorption and scattering properties of the total aerosol phase is also very sensitive to the presence of BC, an efficient light absorber, whereas organic particles have the light-scattering properties typical of atmospheric particles generally. Patterson *et al.* [86] observed that various combustion conditions (with different types of fuel and during intense or smouldering combustion conditions) produce aerosol mixtures with significantly different absorption properties. In temperate regions, the contribution of carbonaceous particles to the degradation of visibility may be comparable to anthropogenic sulfates. [87] Occurrences of severe visibility reduction which are observed during pollution episodes such as "brown haze" in Denver, Colorado [88] or over regional basins favouring smog production [89] are striking examples of the rapid secondary production of particles from combustion gases. A concentrated smoke plume on average, due to backscattering and absorption in the smoke layer, produces a surface cooling which may reach several degrees as observed over the Amazon basin during intense regional-scale forest fires [90] or near the Kuwait fires [91].

BC is by far the most effective component in determining the absorption cross-section of atmospheric particulate matter in the visible wavelength range: [56] BC may induce radiative heating [92] and even in small quantities may dramatically affect the aerosol cooling properties especially above surfaces of high reflectivity or when incorporated in clouds. [93–96] Concern has been raised over the last decade especially by Clarke [97] of the potential effect of BC particles for radiative transfer in remote environments. In Arctic regions, it is quoted that reduction of snow albedo by BC deposition on snow is of the order of 10% which approximately doubles that of Arctic haze. [98]

Different models may be used to describe optical properties of the aggregates. The simplest and most popular is the Mie theory in which particles are assumed to be spherical. As discussed earlier, this theory could reasonably be used for aged particles. For more elongated particles, theoretical calculations tend to show that Mie theory poorly depicts particles which behave more or less as fractal clusters. [99,100] The situation is even more complex as some works have shown that under "resonance" frequency conditions (i.e. when the effective length of the agglomerate is of the same order as the incident radiation wavelength) error on calculated optical properties may be important. [101]

Optical properties are also affected by the presence of a solid or an aqueous coating and vary with its chemical nature and size. [44,102] Studies on this particular facet are just in their infancy. The formation of heterogeneous aerosols is particularly crucial for the light absorption of BC which has been calculated to be larger in a transparent sphere than when it is in air. [103,104] Haywood and Shine [105] have shown that adsorption could be enhanced by a factor of 4 when BC is incorporated in cloud droplets. Pinnick et al. [59] have shown that a similar phenomenon could prevail in heterogeneous particles when BC is present as a host in sulfate or organic particles. However, if the volume fraction of BC is too small (less than 5%), the sulfate particle behaves as a pure homogeneous particle. [95]

4.3 PARTICLE SENSITIVITY TO ATMOSPHERIC HUMIDITY

This sensitivity has numerous consequences. Fixation of water changes the aerosol size distribution and the aerosol weight and the phenomenon is dependent on the nature of the coating and on the size of particles. The coarser particles are the most effective water scavengers. [106] For example, it has been shown that both fresh wood smoke and diesel particles increase notably in weight by adsorption of water (2–10%) but the diesel particles proved to be three to five times less effectively scavenged by water. [107]

Thereby, combustion particles have the capability to serve as efficient *cloud condensation nuclei* (CCN) even near sources as shown by the significant CCN/CN (condensation nuclei) ratio values obtained on field plumes or laboratory experiments as shown in Table 3. An illustration is also given by Hudson's observations [108] in the Californian basin showing that fresh combustion particles are instantaneously activated as CCN. Although there is some discrepancy in the available data set, there is a consensus that young biomass-burning aerosol particles are on average more easily activated than industrial particles (Table 3). The CCN activity of combustion particles is attributed to their association with hygroscopic inorganic [41] or surfactant species, [109] or may be an intrinsic property due to the presence of polar organic compounds at the particle surface. [7,110]

Combustion sources are the main CCN producers with the exception of remote forested areas, over continents. On a global scale, comparison of different

Table 3. Activated fraction (%) of young particles for various combustion sources at 1% supersaturation (partially adapted from Lammel and Novakov[111])

Fuel type	Combustion type	CCN/CN (%)	Soluble ions (meq/g^{-1})	References
Aviation (JP4)	Flaming	1	nd	Hallett *et al.* [41]
Crude oil (low sulfur)	Ignition, flaming	20	1.25	Rogers *et al.* [38]
Crude oil (Kuwait fire)	Flaming	70	nd	Hudson and Clarke [112]
Light crude (high sulfur)	Ignition, flaming	58	1.22	Rogers *et al.* [38]
Acetylene gas	Welding torch	52	3.13	Hallett *et al.* [41]
Diesel	Diffusion flame	42	0.009	Lammel and Novakov [111]
Diesel	Engine motor, idling	80	0.99	Lammel and Novakov [111]
Wood smoke	Smouldering	24	2.7	Hallett *et al.* [41]
Forest and brush fire	Flaming	90	2.7	Hallett *et al.* [41]
Savanna grass	Smouldering	16*	nd	Dinh *et al.* [114]
Savanna grass	Flaming	30*	nd	Dinh *et al.* [114]
Pure cellulose	Smouldering	97†	0.003	Novakov and Corrigan [110]

*At ambient (80%) RH.
†At 3% supersaturation.
nd: not determined.

sources shows that CCN emissions by industrial or vegetation fire sources are of the same order of magnitude: Squires [113] estimated that the northern hemispheric CCN production (at 0.5% supersaturation) is $5 \times 20^{20} CCN/s^{-1}$ for natural sources and $2.5 \times 10^{20} CCN/s^{-1}$ for industrial sources. Dinh *et al.* [114] estimated a flux of $5 \times 10^{19} CCN/s^{-1}$ for African savanna fires only and for more severe conditions (lower supersaturation value: 0.1%).

In source regions of high combustion aerosol loads, particles may markedly affect the water vapour cycle. They may contribute to the formation of fog droplets. At a given site, incorporation of combustion aerosol particles in fog droplets may vary over a large range (up to 80%) owing to both water vapour availability and age of the aerosol; this behaviour was observed in California by Hansen and Novakov[115] and is illustrated in Figure 10. On a larger scale, since the acceleration of deforestation during the last decade, a marked alteration of the cloud cover has been observed from space over the Amazon basin during the burning season. [116] Indeed the increase in the number of CCN enhances droplet number concentration and thereby decreases the mean droplet radius which has been predicted to affect both the cloud optical properties and precipitation capability. This has also been experimentally verified from AVHRR satellite images for ship tracks in marine stratus clouds [117] and over the Amazon basin by comparison of "off" and "during" fire situations ([93] and is shown in Figure 11).

A crucial question must be addressed here concerning the role of ageing in modifying the chemical composition of the combustion particle coating and hence its behaviour on a more global scale. A few workers give direct evidence of the CCN role of combustion particles. At the marine but probably anthropogenically influenced, site of Puerto Rico, Novakov and Penner [118] using CCN, CN, sulfate and organic measurements suggested that organic aerosols could account for the major part (64%) of both the total aerosol number and the CCN fraction (Figure 12). Recalling that in such marine areas organic aerosols are overwhelmingly long-range transported particles of combustion origin, [29] this experiment could constitute a strong argument for the CCN role of aged combustion particles. The role of the organic aerosols in the atmospheric water cycle is not however so simple or unequivocal. Organics may augment the CCN population but in some cases, surface organic molecules are expected to behave as micelles and to prevent droplet growth. [119]

Finally, combustion particles are expected to be easily incorporated in clouds and be scavenged by in- and out-cloud processes; wet deposition thus constitutes their major sink.

4.4 ATMOSPHERIC CHEMICAL IMPACT

Combustion aerosols probably also have an *atmospheric chemical impact*. Oxidative reactions occurring at the particle surface might result in

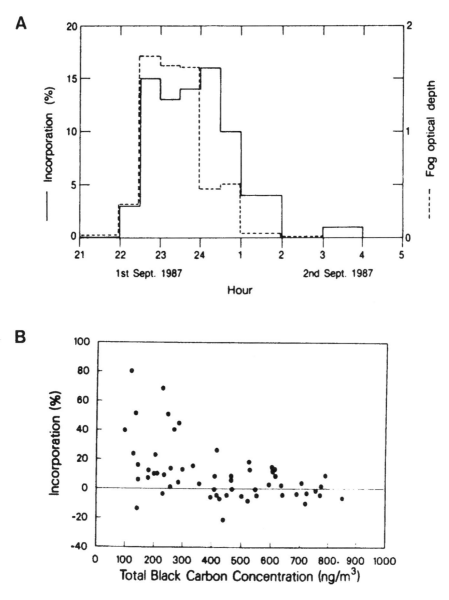

Figure 10. Incorporation of BC particles into fog droplets (a) during a maritime fog event; (b) trends with increasing concentrations (from Hansen and Novakov [115])

significant oxidant (for example: ozone, OH radical) consumption. In this respect, the presence of carbonaceous particles in the lower stratosphere could be a significant sink of ozone. [120] However, although tested in laboratory experiments [120, 121] such assumptions now need validation by field measurements.

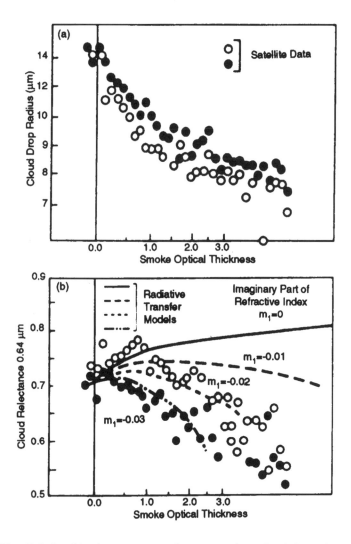

Figure 11. Relationships between aerosol concentration, cloud drop size and cloud reflectance in smoke plumes from the Amazon basin (plain and open circles correspond to near nadir observations for solar and anti-solar directions); (a) AVHRR measurements of cloud drop size as a function of smoke optical thickness; (b) simulation and AVHRR measurements of cloud reflectance at 0.64 μm for different abundances of BC (increasing concentrations of BC augment the imaginary part of the smoke refractive index) (from Kaufman [93])

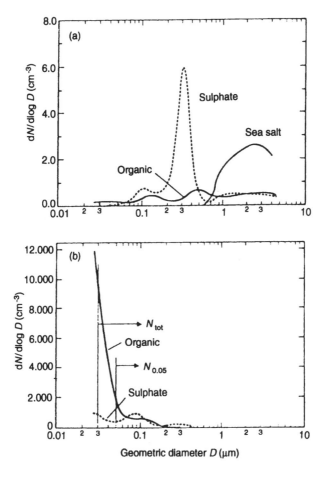

Figure 12. Mass size and number distributions of sea-salt, sulfate and organic marine aerosols (Puerto Rico), showing the prevalence of organic aerosols in the Aitken nuclei population (from Novakov and Penner [118])

Simulation experiments have also shown that heterogeneous oxidative reactions of gases on carbonaceous material may occur with water vapour, [122] producing sulfate from SO_2 or interacting in the NO_x-NO_y cycles. [123–125] Such reactions may also take place with liquid water either at the BC particle surface [126] or in aqueous droplets. [127, 128] The catalytic effect of soot particles in an aqueous medium is also enhanced by the presence of dissolved metal ions (such as iron) leached from the particles. [129]

Finally, adsorption of some species on soot particles has been shown to ensure a protection against photo-oxidative degradation. As an example, the preservation of highly pericondensed PAHs is attributed to a rapid quenching by adsorption on to BC particles. [130]

5 GEOCHEMICAL HIGHLIGHTS AND TROPOSPHERIC IMPACTS OF COMBUSTION PARTICLES

Through the presence of their BC component, combustion aerosols are ubiquitously present in the global troposphere where they form one of the main components of the atmospheric accumulation mode. [29, 131] Therefore, they are likely to have a significant and increasing global impact.

5.1 SOURCE STRENGTHS

Source strength determinations rely on several steps, and each step has the potential of being a weak point of the source strength estimation chain.

5.1.1 Source Identification

Source identification is not as easy as it may seem. The first step involves the identification of the fuel which is used in the combustion process. This is relatively simple for fossil fuels, identified as coal, oil (sometimes divided into other categories), peat, lignite or natural gas. Industrial and domestic waste burning is still not taken into account. For biomass burning, there is a general agreement to separate sources into five groups: boreal and temperate forest fires, savanna fires, agricultural waste, fuelwood and charcoal. But the situation is more complex since in developing countries, agricultural wastes such as rice straw or forest wastes after fires and even animal dungs are used as domestic fuel. Overlaps in the different budgets are also inevitable.

Identifying the way fuel is burnt is the second step of major importance. At the beginning of the industrial era, burning efficiency of coal was on average 10 times less than now in developed countries [132] which was probably accompanied by particle emissions about 10 times higher than now. Similar differences prevail today between developed and developing countries and are also found between use of coal or oil fuel in domestic settings and industrial units such as power plants. [133] This is why in exhaustive source inventories such as that from Cooke and Cooke and Wilson, [134, 135] for a given fuel, its usage is often subdivided into two or three categories. Similar considerations also apply to automobile sources for which particulate production for the same fuel may span over one order of magnitude. For example, during a short experiment conducted near a street in the USA, Hansen and Rosen [136] estimated that 20% of the vehicles accounted for 65% of emissions. For biomass burning, it is also very important to know if combustion occurred under smouldering or flaming conditions. Particulate emissions usually peak in parallel with lack of oxygen. [137] Cachier *et al.* [138–140] have shown that for this reason too, for savanna fires under flaming conditions it is important to know whether the fire spread quickly or not. A great variability is found for the global production of particulates and the relative amount of BC in aerosols for all combustion types.

5.1.2 Amount of Fuel Submitted to Combustion

Knowledge of the amount of fuel burnt is quite reliable for fossil fuels and extensive deforestation initiated by concern for the CO_2 greenhouse effect. Estimates for other sources remain very questionable. Estimates for the amount of fuel burnt in savanna fires (the most important source) vary by a factor of 2 due to lack of knowledge of the savanna fire return interval and vegetation mean density, and any agreement between different estimates is considered to be purely coincidental as assessed by Lacaux *et al.* [141] The use of charcoal is gaining more and more in importance compared to the direct use of wood for domestic purposes, due to increasing urbanization in developing countries. Particulate production by charcoal production and burning is important but the shift in sources (from wood to charcoal) has not attracted much attention up to now.

5.1.3 Emission Factors

Amounts of fuel burnt give only a curtailment of the source efficiency for the production of particles. This efficiency is known from the *emissions factors* (EF) which are generally presented as the fraction of carbon found in the aerosol and expressed in gC(aerosol) / kgC(fuel) or in gC(aerosol) / kg (dry fuel). EF values span over a large range showing inter- and intra-source variability. "Best guess" estimates for mean emission factors for the principal sources are reported in Table 4 but it is recalled that there is a need to study EF distribution from real sources under actual operating conditions [135, 136, 138].

Emission factors may be experimentally obtained in a smoke plume by measuring in the same air parcel excess concentration (relative to background concentrations) of carbonaceous aerosols and all C-containing gaseous species, [138, 140, 143] using the relation

$$EF = \Delta[C](\text{aerosol})/\Delta CO_2 + \Delta CH_4 + \Delta NMHC + \Delta[C](\text{ashes}) + \Delta VOC \quad (8)$$

This approach may be simplified considering that 90% of the fuel carbon burns as CO_2. [144] Then

$$EF = \Delta[C](\text{aerosol})/0.9\Delta CO_2 \quad (9)$$

EF is more accurately determined considering that nearly all the fuel carbon burns as $CO_2 + CO$. [144] Then

Table 4. Estimated BC and TC emission factors for different combustion sources (OC represents 70% of TPM; C% in vegetation fuel is 44%. Values are means for industrial/domestic uses in developed/developing countries) (adapted from Cachier[2], Cooke and Wilson[135], Liousse et al.[96])

Fuel source	BC emission factor EF (BC) in gC(aerosol) / kgC(fuel)	OC emission factor EF(OC) in gC(aerosol) / kgC(fuel)
Fossil fuels		
natural gas	0.0003	0.0012
heavy fuel	0.02	0.08
motor gasoline	0.1	0.4
kerosene	0.3	1.2
jet fuel	1	1
diesel	5	5
liquefied petroleum gas	0.06	0.25
coal	3	3
lignite	4	15
peat	1	6
coke	3	3
Biomass burning		
charcoal production	3.5	45
charcoal consumption	2	7
dung consumption	2	25
fuelwood	2.5	11
tree burning	1.5	13
grass burning	0.8	6

$$EF = \Delta[C](aerosol)/0.95(\Delta CO_2 + \Delta CO) \tag{10}$$

Figure 13 is an example of an experimental determination for EF of savanna burning aerosols by means of real-time simultaneous measurements of total aerosol mass and $CO_2 + CO$ concentrations.

EF may be also retrieved in remote atmospheres when satisfactorily differentiated air parcels show simultaneous increase of CO_2 and aerosols. This has been shown by Hansen et al. [145] in Alaska during Arctic haze events and at Amsterdam Island (in the Indian Ocean) during episodes of rapid transport of biomass burning pollution off South Africa. [144]

5.1.4 Source Fluxes

In pioneering work, Duce [146] presented a possible global budget for atmospheric carbonaceous particles showing a discrepancy between inputs and outputs, with sources being considerably underestimated. A striking feature of the global carbonaceous budget, though based on very few data, was the quasi-absence of particulate production by biomass burning.

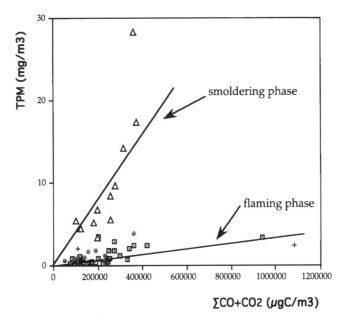

Figure 13. Determination of emission factors (EF) for total particulate matter (total aerosol) emitted during savanna fires using real-time data of TPM and excess CO_2 and CO concentrations. EF is calculated as the line slope. Smouldering fires are more efficient emitters of particles than flaming fires (from Cachier *et al.* [139])

Since that period, concern about and knowledge of particulate production by combustion processes have progressed a lot although important gaps still remain.

There is now a general agreement to estimate particulate production by combustion through the use of an accurate source inventory including numerous different types of processes and based on accurate estimates of mass of fuel burnt and emission factors.

Annual flux F is calculated according to

$$F = \sum F_i \tag{11}$$

and

$$F_i = M_i \times EF_i \tag{12}$$

It must be emphasized here that EF which are generally obtained in combustion plumes include the production of supermicron particles. However, they are generally less abundant (less than 20%) and are likely to be preferentially scavenged in source regions.

Source–receptor modelling applied on a global scale may be another independent approach which is infrequently used as measurements of particulate carbon as atmospheric aerosols or in precipitation in the remote atmosphere are scarce. Table 5 is a summary of estimate attempts for global determinations of source strength of combustion aerosols, which up to now, are not so numerous.

'The example of rice straw burning is a good case study to illustrate precautions to take for the development of accurate source climatology. Rice straw may be burned either in the field or as domestic fuel for cooking or heating. As well as emission chemical composition, 3D spatial and temporal distributions of aerosols depend on parameters displayed in Table 6.

At present, the most accomplished efforts in source strength determination are certainly in the inventories from Cooke and Wilson [135] (ISPRA) and Liousse et al. [96, 142] (resulting from collaboration between Lawrence Livermore Laboratory and Centre des Faibles Radioactivités). The very good agreement between both total BC estimates is, however, partly coincidental as some minor sources (though not identical) are missing in both inventories.

Cooke and Wilson [135] present a very detailed inventory for the use of fossil fuel whereas the Liousse et al. [96, 142] inventory has much more focus on

Table 5. Global source strength of combustion aerosols (in MT carbon = 10^{12} gC)

Authors	Annual flux (MT carbon)	Study type	Study motivation
Duce*[146]	TC: 1 (BB) 16 (IND)	Source receptor	Secondary sources of aerosols
Turco et al.[147]	BC: 7.5 BB 5 (IND)	Source inventory	Nuclear winter
Cachier et al.*[29]	TC: 15 (BB) 16 (IND)	Source receptor	Importance of tropical biomass burning
IPCC[24]	BC + OC : 130 (BB + IND)	"Best guess"	Global radiative impact
Penner et al.[148]	BC (BB + IND): 24 13	SO$_2$-based inventory fuel-based inventory	Global radiative impact
Cooke and Wilson[135]	BC: 8 (BB) 6 (IND)	Source inventory	Global radiative impact
Liousse et al.[96]	BC: 6 (BB) 7 (IND) OC: 35 (BB) 22 (IND) (natural organics: 6)	Source inventory	Global radiative impact

IND: industrial, mainly based on fossil fuel.
BB: biomass burning.
*fine aerosols only.

Table 6. Annual particulate emissions by rice straw burning in developed and developing countries (adapted from Liousse *et al.* [142])

	Developed countries	Developing countries
Straw biomass (Tg dry matter)	39	506
Fraction submitted to fire	10%	50%
Combustion efficiency	85%	85%
Burnt biomass as open fires	100%	40%
Burnt biomass as domestic fuel	0%	60%
Seasonality	After harvest only	After harvest and all the year
Height injection	0–2000 m	0–2000 m and 0–1000 m
Weighed emission factor (EF_{BC})	0.6	0.86
OM/BC	1.7	3.0
BC flux (TgC)	0.002	0.186
OC flux (TgC)	0.003	0.554

biomass-burning emissions. For the latter inventory, it must be emphasized that both OC and BC particles are studied which give a global picture of important OC inputs from all types of combustion.

Figure 14 shows the partition of particulate BC emissions between the main regions of the world, as obtained in different models. In spite of some discrepancies, some key features are strikingly representative of the global situation.

- *"Industrial combustion"* referred to in this chapter actually includes all types of combustion using fossil fuels. They comprise both stationary and mobile sources. Of fossil fuel consumption, 90% occurs in the northern hemisphere [149] and this situation is globally found for BC emissions too as shown in Figure 15. In developed countries, drastic control of industrial effluents combined with optimized combustion efficiencies has reduced particulate emissions dramatically; conversely mobile sources have increased a lot, with almost no pollution control. This is particularly crucial for diesel vehicles especially in western Europe where they have intensively penetrated the automobile market. [150] Source–receptor models have shown that in most cities, more than 50% of atmospheric BC could now be caused by mobile sources [151] and the situation is still worse in European cities where this percentage could reach 90% [152–154] and references therein). Even in rural areas, the mobile sources could be very important for air quality and visibility. [155]
- Some 90% of *biomass-burning* events are anthropogenically caused. The major part of biomass burning (80%) occurs in the intertropical zone with 45% as savanna fires. [116, 144] Thus, 50% of the biomass-burning activity is actually located in Africa due to the important savanna area in both hemi-

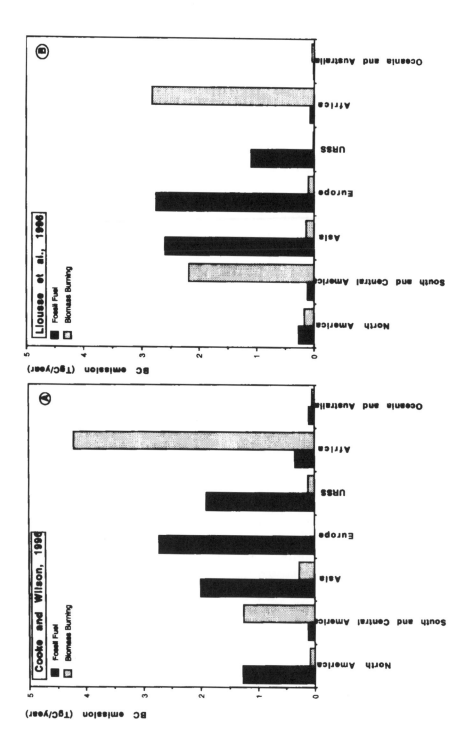

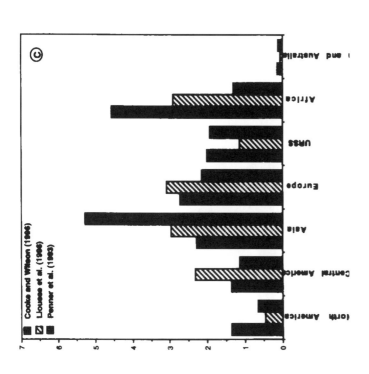

Figure 14. Comparison of different scenarios of the regional distribution of BC emissions: (a) and (b) fossil fuel and biomass burning emissions from Cooke and Wilson [135] inventory and Liousse et al. [96] inventory; (c) total combustion emissions from Cooke and Wilson [135], Liousse et al. [96] and Penner et al. [148]. The satisfactory agreement between the three inventories for the total amount of BC emitted each year (13–14 Tg carbon) could be coincidental as shown by the numerous discrepancies shown in this diagram. More than 90% of fossil fuel BC is emitted in the northern hemisphere whereas biomass burning BC is equally divided between both hemispheres

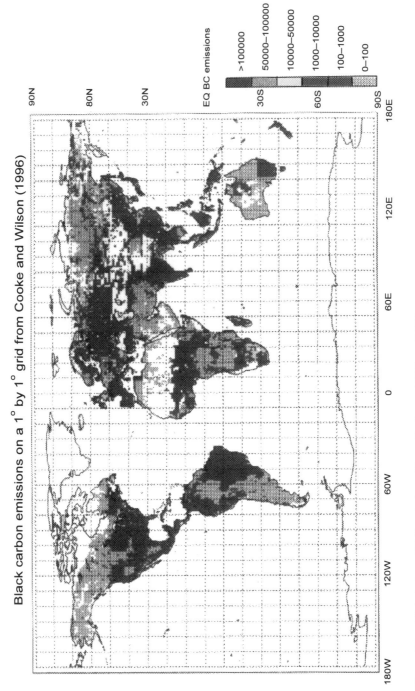

Figure 15. Global map of BC emissions (from Cooke and Wilson [135])

spheres. Following a repetitive pattern, tropical forest and savanna fires develop during the dry season and severely influence the atmospheric regional burden of particles and gases. [27, 156–159] The regional particulate pollution by extensive fires is clearly marked either in aerosols or in rain [160] and may be observed either in source regions (as seen in Figure 16) or at very remote locations downwind of source regions (Figure 17 illustrates this phenomenon for Amsterdam Island). All tropical biomass-burning sources are likely to increase in the near future due to the pressure of an expanding population. Boreal forest fires represent a few per cent only of the global biomass burning. They are driven by meteorological conditions and may display important interannual or interdecadal variability of one order of magnitude or so. [161]

It may be inferred from this discussion that Arctic latitudes are sporadically influenced by fires in summer whereas tropical latitudes are repetitively influenced by fires from November to March in the northern hemisphere and from June to September in the southern hemisphere (as attested by the fine potassium tracer of biomass-burning aeosols. [162]) Industrial emissions are active all the year long with some peaks during cold months. They are mostly located at temperature latitudes and will mostly affect this geographical band north of 30 ° N. However, during the meteorological phenomenon of Arctic haze [163] the Arctic atmosphere may also receive significant pollution from more temperate regions.

It must be kept in mind that source inventories are evolving pictures. They depend firstly on the evolution of fuel consumption and shift between the usages. For example for the last two decades, the trend in Europe was the lowering of combustion particulate emissions due to significant abatement policies for industrial plants, whereas Asia doubled fuel consumption and emissions. [134] Secondly, the accuracy of the source inventory relies primarily on available data which for some countries are difficult to obtain for several geographical or political reasons. This kind of difficulty is illustrated in Figure 18 with the evolution of global BC emissions from fossil fuel combustion: as quoted by Cook, although a global increase of emissions is visible, detailed variations such as the 1978 peak or the 4-year sudden jump from 1986 to 1989 followed by a fall in 1990, cannot be easily explained. Furthermore, this chaotic profile is also reflected significantly in the global emission profile which points out the weight of China in the global BC emission inventory and the overwhelming role this country will play in the future.

It has been shown recently that some impurities in Greenland ice may be related to the hemispheric background of combustion. [164, 165] Carbonaceous particles are also incorporated satisfactorily in polar ice and firn. [2, 166, 167] In Figure 19 which shows a continuous record profile of BC concentration since the beginning of the nineteenth century, it may be seen that both biomass

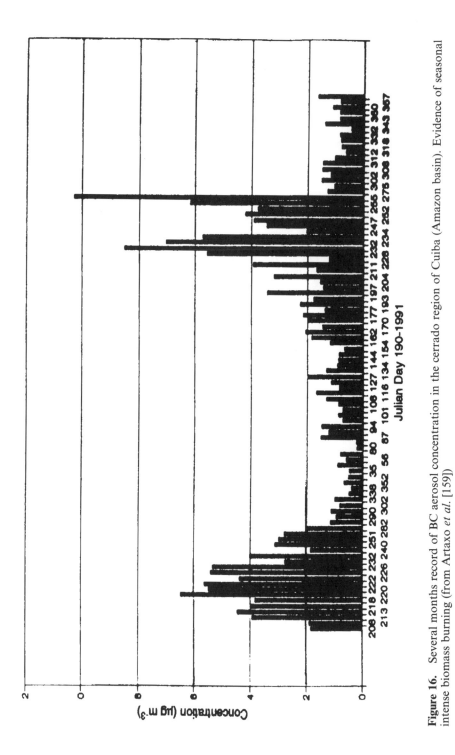

Figure 16. Several months record of BC aerosol concentration in the cerrado region of Cuiba (Amazon basin). Evidence of seasonal intense biomass burning (from Artaxo *et al.* [159])

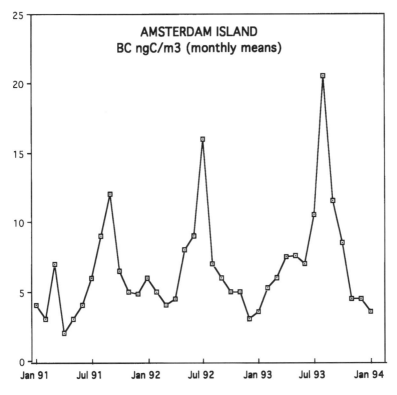

Figure 17. Three-year aethalometer data of BC concentration in the atmosphere of the remote site of Amsterdam Island in the Indian Ocean (77°E, 38°S). Important biomass-burning inputs follow the repetitive pattern of the dry season in the southern hemisphere (June to September) (from Cachier *et al.*, [140])

burning (before the industrial era) and fossil fuel combustion inputs are recorded in the ice-core. Furthermore, the well-known trends of increasing fossil fuel consumption [168] are not reflected in the profile which displays the lowest concentrations in the 1950s. Such a study points out the major importance of the search for knowledge on emission factors for each usage of fuel, but also suggests that ice-cores could serve as surrogates for past source inventories.

5.2 PARTICLE DEPOSITION

Combustion aerosols have no significant chemical sinks and accumulate in the $0.1-1\,\mu m$ size range. They are assumed to have a dry deposition of $0.1\,cm\ s^{-1}$ but their predominant sink is wet deposition. Mean residence time has been estimated by Ogren and Charlson [169] to be of the order of 7 days, allowing for

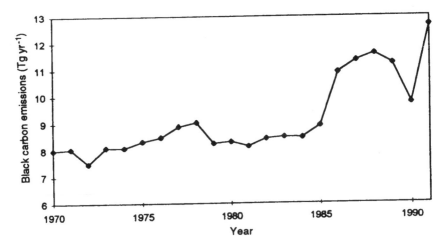

Figure 18. Trend in the emissions of BC from fossil fuel sources for the period 1970–1991 (from Cooke [134]). Abrupt changes are more likely related to the poor reliability of China data than to actual phenomena.

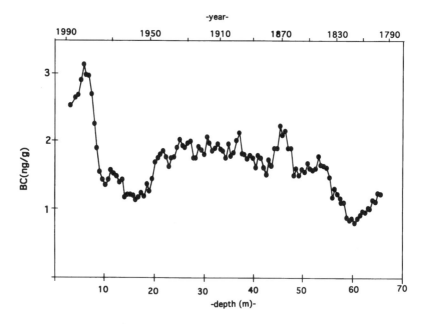

Figure 19. A 200-year continuous BC record in a Summit (Greenland) ice-core (5-year smoothed average concentrations) (from Pertuisot et al. [167])

synoptic influence of the particles. Residence time, however, is likely to span over a vast range of values from a few days to a month. Such a variability is caused by different possible precipitation patterns in source regions but also in height of injection. Another parameter to take into account is the formation of a hydrophilic coating which is likely to facilitate incorporation of particles in cloud droplets. Indeed, some experiments have shown that BC particles in source regions are incorporated in droplets, but not as easily as sulfate aerosols. [170] This is the reason why in their global transport model, Cooke and Wilson [135] divided BC into two categories: hydrophobic BC particles and hydrophilic BC particles, hydrophobic turning into hydrophilic with an appropriate transformation rate tested to be 2.5% per hour.

Data on carbonaceous particles in wet precipitation are very scarce, probably due to specific problems of collection for carbonaceous material (such as bacterial contamination and the necessary sample volume). Although combustion aerosols are ubiquitously found in all types of wet precipitation, data are mostly restricted to BC in snow (Table 7). The influence of ageing could be confirmed by the prudent use of scavenging ratios whose values obtained in various precipitation (rain and snow) show a significant increase in remote areas from 100 to 500. [160] The probable easier incorporation of organic

Table 7. Black (BC) and total (TC) particulate carbon in rain and snow ($\mu gC/kg^{-1}$) (from Cachier[2] and references therein)

(Number of samples)		BC	TC	Reference
Urban				
Gif sur Yvette, France	(54)	336	1175	Ducret and Cachier (1992)
Detroit, USA	(40)	160	920	Dasch and Cadle (1989)
Rural				
Michigan, USA	(25)	72	822	Cadle and Dasch (1988)
Sweden	(5)	100		Ogren *et al.* (1984)
Tropical				
Lamto, Ivory Coast	(62)	143	487	Ducret and Cachier (1992)
Enyele, Congo forest	(46)	73	292	Ducret and Cachier (1992)
Remote				
Mace Head, Ireland	(23)	29	97	Ducret and Cachier (1992)
Arctic				
Barrow (Alaska)	(13)	23		Clarke and Noone (1985)
Spitzbergen	(12)	31		Clarke and Noone (1985)
Abisko (Norway)	(8)	33		Clarke and Noone (1985)
Camp Century (Green)	(2)	2.5		Chylek *et al.* (1987)
Dye3 (Greenland)	(9)	6		Clarke and Noone (1985)
Dye3 (Greenland)	(22)	9	57	Ducret (1993)
Antarctica				
South Pole	(1)	2.5		Chylek *et al.* (1987)
Amundsen Scott	(nd)	0.35		Warren and Clarke (1990)

particles is more difficult to determine as there is evidence of some partial dissolution of organic material in cloud or rain water. [160]

The scavenging of combustion particles remains a very crucial open-ended question especially for concentrations at remote places. For example, Liousse *et al.* [96] have shown with their global model that for such areas, a 15% shift in the BC particle scavenging ratio could lead to a 100% change of concentration. This result clearly points out the need for further experimental effort in this topic.

5.3 TRANSPORT IN THE GLOBAL TROPOSPHERE

Combustion aerosol concentrations are found in a range of three orders of magnitude. They are likely to influence the global atmosphere as shown by the significant particulate carbon concentrations found in the remote marine or the polar atmosphere. [29, 58, 97, 131, 140, 145, 162, 171]

Due to the mean time required for interhemispheric mixing ($\sim$ 6 months) concentrations reflect the hemispheric distribution of sources on land. On average, combustion aerosols are five times more abundant in the remote atmosphere of the northern hemisphere than in the southern hemisphere as seen in Figure 20.

Source inventories have been tested through implementation in dynamic models with satisfactory results on average and with a good reproducilbility of temporal trends (Cooke [134], Cooke and Wilson: [135] Moguntia model; Penner *et al.* [148, 172] and Liousse *et al.*: [96, 142] Grantour model). The spatial agreement between modelled and experimental concentration values is illustrated in Tables 8(a) and (b) for BC and OC results. Temporal variability of BC concentrations in aerosol and rain at representative sites is shown in Figures 21 and 22.

5.4 EXPECTED RADIATIVE IMPACT

Carbonaceous combustion particles have been shown to be found mainly in the accumulation mode and dispersed in the global troposphere with significant concentrations of the order of those quoted for industrial sulfates. Although patchy, the presence of these combustion aerosols is expected to have a significant radiative atmospheric impact. In the light of previous discussions, this impact is primarily related to the interaction of incoming solar radiation with particles in the atmosphere or in clouds.

OC particles have a direct effect due to scattering which produces a perturbation to the energy balance of the earth–atmosphere system. The radiative forcing is expressed negatively and of the order of that quoted for industrial sulfates (mean value in the range: -1 to 0.5 W m^{-2}).

It is clear that the presence in the aerosol mix of BC which is a very effective absorber of light may drastically change this direct radiative effect of combus-

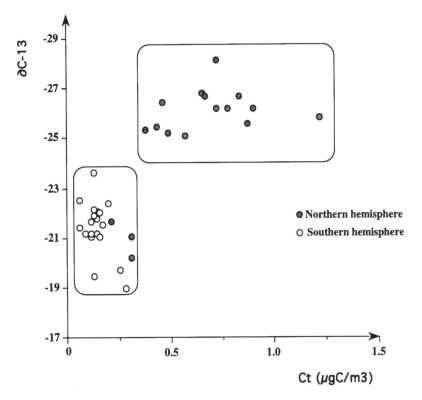

Figure 20. Isotopic composition and concentration of various bulk carbonaceous aerosols sampled in the marine atmosphere (from Cachier *et al.* [29])

tion aerosols. This is illustrated by Figure 23 adapted from the work of Penner [119] where global averages of estimated changes in reflected solar radiation are reported as a function of total OC emission source rate. The relationship is shown to be highly dependent on the presence of BC and for a realistic BC/OC ratio value of 20%, a shift in 50% of the scattering effect of particles may be observed. It is thus crucially important to gain accurate knowledge of the relative amounts of BC and OC in aerosols. Liousse *et al.* [96] have shown that in some areas situated in industrial regions of the northern hemisphere, the presence of BC aerosols could overwhelm the scattering effect of both industrial sulfate and organic particles. This phenomenon is, however, dependent on the albedo of the underlying surface [173] and the height of particle injection.

OC and BC particles may also act as CCN agents and modify the cloud cover which, to a first approximation, increases reflectivity of clouds and their residence time in the atmosphere. This indirect cooling effect is still more difficult to

Table 8. Comparison of concentration simulated values given by the Grantour model and observed surface concentrations (from Liousse et al.[96], references therein)

(-a-) Black carbon concentrations (ngC/m^{-3})

Locations	Time period	Observed	Simulated	References
Northern Hemisphere				
Locations				
10 to 30°N, 160°W (Pacific Ocean)	*Oct. to July*	7	6.6	Andreae et al. (1984); Clarke (1989)
11.3°N, 162.1°E	*April*	50	20	Cachier et al. (1990)
10°N, 165°E	*July*	25	9	Parungo et al. (1994)
10°N, 113°E (China Sea)	*June*	33	87	Parungo et al. (1994)
0°, 150°E (Western Pacific Ocean)	*July*	9	10	Parungo et al. (1994)
0°, 160°W (Pacific Ocean)	*Year*	3	2.95	Andreae et al. (1984); Clarke (1989)
0°, 120°W (Equatorial Pacific)	*June*	10	14	Andreae et al. (1984)
Southern hemisphere				
Locations				
2°S, 77.3°W (Ecuador)	*Annual*	100–520	289	Andreae et al. (1984)
10 to 30°S 160°W (Pacific Ocean)	*Oct. to July*	8	4	Andreae et al. (1984); Clarke (1989)
10°S, 55°W (Brazil)	*Year*	200–620	536	Andreae et al. (1984)
10°S, 76°W (Peru)	*April*	24	25.5	Cachier et al. (1986)
13.6°S, 172°W (Samoa)	*July*	19	14	Cachier et al. (1986)
20°S, 40°W (Atlantic Ocean)	*October*	20	74	Andreae et al. (1984)
37.5°S, 77.3°E (Amsterdam Island)	*Year*	5.5	4.5	Cachier et al. (1994)
40.7°S, 144.4°E (Cape Grim)	*Year*	3	8.9	Heintzenberg and Bigg (1990)
41°S, 174°E (New Zealand)	*Year*	22	15.2	Cachier et al. (1986)
75.4°S, 27°W (Halley Bay)	*Year*	0.3–3	0.34	Cachier et al. (1986); Hansen et al. (1988)
87°S, 102°W (South Pole)	*Year*	0.3	0.16	Hansen et al. (1988)
87°S, 102°W (South Pole)	*Year*	1.3	0.16	Bodhaine et al. (1995)

Table 8. Comparison of concentration simulated values given by the Grantour model and observed surface concentrations (from Liousse et al.[96], references therein)

(-b-)Organic carbon concentrations ($ngC\,m^{-3}$)	Time period	Observed	Simulated	References
Northern hemisphere				
Locations				
11.3°N, 162.1°W (Enewetak)	*Apr.–May*	700	70	Chesselet et al. (1981)
18.2°N, 62.5°W (Puerto Rico)	*Mar.*	660	500	Novakov and Penner (1993)
15°N, 27°W (North Atlantic)	*Nov.*	330–1600	384	Ketsederis et al. (1976); Andreae (1983)
1.6 to 5.5°N, 81°W (Pacific)	*Feb.*	425	417	Wolff et al. (1986)
Southern Hemisphere				
Locations				
37.5°S, 77.3°E (Amsterdam Island)	*Year*	22	27.4	Cachier et al. (1994)
41°S, 174°E (New Zealand)	*Year*	130	81	Cachier et al. (1986)
2°S, 77.3°W (Ecuador)	*Year*	510	406–5000	Andreae et al. (1984)
10°S, 76°W (Peru)	*Mar.–Apr.*	160	385	Cachier et al. (1986)
13.6°S, 172°W (Samoa)	*Aug.*	59	52	Cachier et al. (1986)
40.7°S, 144.7°E (Cape Grim)	*Year*	230	50–127	Andreae (1983)
20°S, 40°W (South Atlantic)	*Oct.–Nov.*	280	538	Andreae et al. (1984)
2.4°S to 0°, 81°W (Pacific)	*Feb.*	152	242	Wolff et al. (1986)

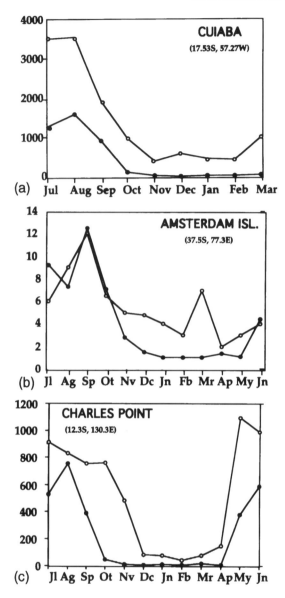

Figure 21. Annual pattern of BC surface concentrations: observed and simulated data with the Grantour model; (a) Mace Head (Ireland); (b) Amsterdam Island (Indian Ocean); (c) Charles Point (savanna, Australia) (from Liousse *et al.* [142])

quantify and rough estimates presented for mean perturbation, are also of the order of -1 W m^{-2}. However, the crucial role of BC particles embedded in cloud

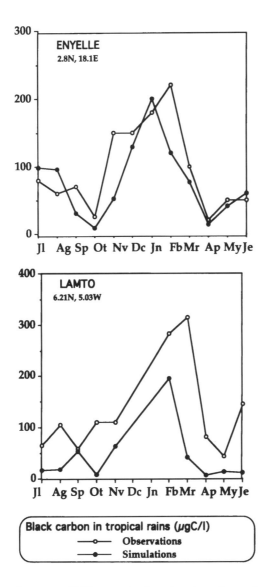

Figure 22. Annual pattern of BC concentrations in rains: observed and simulated data with the Grantour model: (a) Enyelle (equatorial forest of northern Congo); (b) Lamto (savanna site, Ivory Coast) (from Liousse *et al.* [142])

droplets must be taken into account. Indeed, these particulate light absorbers have the potential to offset the cooling caused by the decrease in average cloud droplet size [93] due to amplification of the absorption effect in the droplet medium.

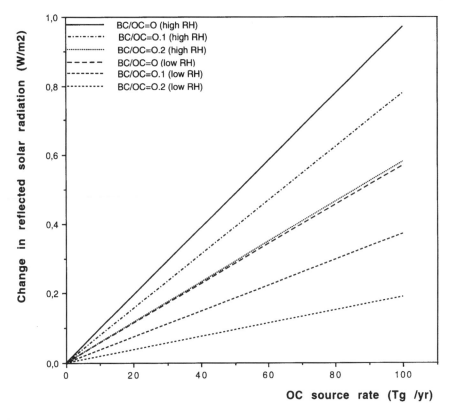

Figure 23. Influence of the abundance of BC in aerosols for the estimated average change in reflected solar radiation (W/m^2) by combustion organic aerosols. Change is estimated for low (40%) and high (85%) relative atmospheric humidity (from Penner [119])

As debated in the IPCC report, [24] combustion aerosols have the potential, at some locations, to counteract the additive radiative effects of anthropogenic greenhouse gases. As quoted in this chapter, carbonaceous particles containing primarily OC and/or BC material form with sulfur the major fraction of the combustion aerosols. These anthropogenic particles could be one "gremlin" in the greenhouse. [174] They could be one of the causes for regional anomalies observed for the past few decades for an increase in global surface temperature [175–177] or for the interhemispheric asymmetry in cloud cover changes. [178] Studies on the aerosol perturbations are, however, just beginning and available preliminary estimates of the mean radiative forcing of particles suggest that they could completely obscure the greenhouse warming which is of course unrealistic. It is now considered to be of vital importance to investigate in particular the internal mixture of combustion particles in atmospheric aerosols. This type of mixing is expected to change the particle optical properties but more dramatically induce a global decrease of their number in the tropospheric reservoir and thus to reduce their global radiative impact.

ACKNOWLEDGEMENTS

This work is CFR contribution No. 1896 and was performed under the auspices of CNRS and CEA in France. The author gratefully acknowledges S.G. Jennings and C. Liousse for a thorough review of the chapter.

REFERENCES

1. E. D. Goldberg, *Black Carbon in the Environment*, Wiley, New York (1985)
2. H. Cachier in *Ice-core Studies of Biogeochemical Cycles*, NATO-ASI Series 30, edited by R. Delmas, pp 313–346, Springer-Verlag, Berlin (1995).
3. T. Novakov in *Particulate Carbon: Atmospheric Life Cycle*, edited by G. T. Wolff and R. L. Klimisch, pp 19–37, Plenum Press, New York, London (1982).
4. H. Cadle and P. J. Groblicki in *Particulate Carbon: Atmospheric Life Cycle*, edited by G. T. Wolff and R. L. Klimisch, pp 89–108, Plenum Press, New York, London (1982).
5. D. M. Smith, M. S. Akhter, J. A. Jassim, C. A. Sergides, W. F. Welch and A. R. Chughtai, *Aerosol Sci. Technol.* **10**, 311 (1989).
6. S. G. Jennings, C. D. O'Dowd, W. F. Cooke, P. J. Sheridan and H. Cachier, *Geophys. Res. Lett.* **21**, 1719 (1994).
7. M. P. Clain, PhD dissertation (in French), Université de Savoie, 158 pp (1995).
8. P. S. Gill, T. E. Graedel and C. J. Weschler, *Rev. Geophys. Space Phys.* **21**, 903 (1983).
9. L. J. Standley and B. R. T. Simoneit, *Environ. Sci. Technol.* **21**, 163 (1987).
10. L. J. Standley and B. R. T. Simoneit, *Atmos. Environ.* **24**, 67 (1990).
11. L. J. Standley and B. R. T. Simoneit, *J. Atmos. Chem.* **18**, 1 (1994).
12. B. R. T. Simoneit, W. F. Rogge, M. A. Mazurek, L. J. Standley, L. M. Hildemann and G. R. Cass, *Environ Sci. Technol.* **27**, 2533 (1993).
13. B. R. T. Simoneit, M. Radzi bin Abas, G. R. Cass, W. F. Rogge, M. A. Mazurek, L. J. Standley and L. M. Hildemann in *Biomass Burning and the Global Cycles*, Edited by J. S. Levine, 509–518 MIT Press, Cambridge (1996).
14. T. Ramdahl, *Nature* **306**, 580 (1983).
15. T. Ramdahl, J. Schjoldager, L. A. Currie, J. E. Hanssen, M. Moller, G. A. Klouda and I. Alfheim, *Sci. Total Environ.* **36**, 81 (1984).
16. Y. L. Tan, J. F. Quanci, R. D. Borys and M. J. Quanci, *Atmos. Environ.* **26A**, 1177 (1992).
17. D. H. Lowenthal, B. Zielinska, J. C. Chow, J. G. Watson, M. Gautam, D. H. Ferguson, G. R. Neuroth and K. D. Stevens, *Atmos. Environ.* **28**, 731 (1994).
18. S. V. Hering, A. H. Miguel and R. L. Dod, *Sci. Tot. Environ.* **36**, 39 (1984).
19. D. J. Freeman and F. C. R. Cattell, *Environ. Sci. Technol.* **24**, 1581 (1990).
20. P. Masclet, H. Cachier, H. Wortham and C Liousse, *J. Atmos. Chem.* **22**, 41 (1995).

21. A. E. Sheffield, G. E. Gordon, L. A. Currie and G. E. Riederer, *Atmos. Environ.* **28**, 1371 (1994).
22. V. P. O'Malley, T. A. Abrajano and J. Hellou, *Org. Geochem.* **21**, 809 (1994).
23. S. D. Killops and M. S. Massoud, *Environ. Intern.* **18**, 1 (1992).
24. Intergovernmental Panel on Climate Change (IPCC), edited by J. T. Houghton *et al.*, Cambridge University Press, 339 pp, 1995.
25. J. D. Court, R. J. Goldsack, L. M. Ferrari and M. A. Polack, *Clean Air* **6**, (1981).
26. R. L. Tanner and A. H. Miguel, *Aer. Sci. Technol.* **10**, 213 (1989).
27. H. Cachier, P. Buat-Ménard, M. Fontugne and J. Rancher, *J. Atmos. Chem.* **3**, 469 (1985).
28. H. Cachier, M. P. Brémond and P. Buat-Ménard, *Nature* **340**, 371 (1989).
29. H. Cachier, P. Buat-Ménard, M. Fontugne and R. Chesselet, *Tellus* **38B**, 161 (1986).
30. A. Gaudichet, F. Echalar, B. Chatenet, J. P. Quisefit, G. Malingre, H. Cachier, P. Buat-Ménard, P. Artaxo and W. Maenhaut, *J. Atmos. Chem.* **22**, 19 (1995).
31. M. J. Cope and W. G. Chaloner, *Nature* **283**, 647 (1980).
32. W. A. Patterson, K. J. Edwards and D. J. Maguire, *Quaternary Sci. Rev.* **6**, 3 (1987).
33. P. Ausset, R. Lefèvre J. Philippon and C. Vernet, *C. R. Acad. Sci. Paris* **318**, 493 (1994).
34. C. Sabbioni and G. Zappia, *Atmos. Environ.* **27**, 1331 (1993).
35. N. L. Rose and S. Juggins, *Atmos. Environ.* **28**, 177 (1994).
36. P. J. Sheridan, R. C. Schnell, J. D. Kahl, J. F. Boatman and D. M. Garvey, *Atmos. Environ.* **27**, 1169 (1993).
37. J. P. Bradley and P. R. Buseck, *Nature* **306**, 770 (1983).
38. C. F. Rogers, J. G. Hudson, J. Hallett and J. E. Penner, *Atmos. Environ.* **25**, 2571 (1991).
39. F. Echalar, A. Gaudichet, H. Cachier and P. Artaxo, *Geophys. Res. Lett.* **22**, 3039 (1995).
40. K. A. Katrinak, P. Rez and P. R. Buseck, *Environ. Sci. Technol.* **26**, 1967 (1992).
41. J. Hallett, J. G. Hudson and C. F. Rogers, *Aerosol Sci. Technol.* **10**, 70 (1989).
42. P. J. Sheridan, R. C. Schnell, D. J. Hofmann, J. M. Harris and T. Deshler, *Geophys. Res. Lett.* **19**, 389 (1992).
43. D. W. Johnson, C. G. Kilsby, D. S. McKenna, R. W. Saunders, G. J. Jenkins, F. B. Smith and J. S. Foot, *Nature* **353**, 617 (1991).
44. C. Liousse, H. Cachier and S. G. Jennings, *Atmos. Environ.* **27**, 1203 (1993).
45. F. Parungo, B. Kopcewicz, C. Nagamoto, R. Schnell, P. Sheridan, C. Zhu and J. Harris, *J. Geophys. Res.* **97**, 15 867 (1992).
46. G. T. Wolff and R. L. Klimisch, *Particulate Carbon: Atmospheric Life Cycle*, Plenum Press, New York, London (1982).
47. A. Petzold and R. Niessner, *Mikrochim. Acta* **117**, 215 (1995).
48. J. C. Chow, J. G. Watson, L. C. Pritchett, W. R. Pierson, C. A. Frazier and R. G. Purcell, *Atmos. Environ.* **27**, 1185 (1993).
49. A. D. A. Hansen, H. Rosen and T. Novakov, *Sci. Tot. Environ.* **36**, 191 (1984).
50. A. D. Clarke, K. J. Noone, J. Heintzenberg, S. G. Warren and D. S. Covert, *Atmos. Environ.* **21**, 1455 (1987).
51. K. Ruoss, R. Dlugi, C. Weigl and G. Hänel, *Atmos. Environ.* **26**, 3161 (1992).
52. J. Heintzenberg, *Atmos. Environ.* **16**, 2461 (1982).
53. H. Horvath, *Atmos. Environ.* **27**, 319 (1993).
54. A. D. A. Hansen, B. A. Bodhaine, E. G. Dutton and R. C. Schnell, *Geophys. Res. Lett.* **15**, 1193 (1988).
55. S. G. Jennings, F. M. McGovern and W. F. Cooke, *Atmos. Environ.* **27**, 1229 (1993).
56. J. D. Lindberg, R. E. Douglass and D. M. Garvey, *Appl. Opt.* **32**, 6077 (1993).

57. R. L. Gunter, A. D. A. Hansen, J. F. Boatman, B. A. Bodhaine, R. C. Schnell and D. M. Garvey, *Atmos. Environ.* **27**, 1363 (1993).
58. A. D. A. Hansen, V. N. Kapustin, V. N. Kopeikin, D. A. Gillette and B. A. Bodhaine, *Atmos. Environ.* **27**, 2527 (1993).
59. R. G. Pinnick, G. Fernandez, E. Martinez-Andaloza, B. D. Hinds, A. D. A. Hansen and K. Fuller, *J. Geophys. Res.* **98**, 2651 (1993).
60. M. G. Winkler, *Quat. Res.* **23**, 313 (1985).
61. W. S. Wolbach and E. Anders, *Geochem. Cosmochim. Acta* **53**, 1637 (1989).
62. J. J. Griffin and E. D. Goldberg, *Limnol. and Ocean.* **20**, 456 (1975).
63. W. S. Wolbach, R. S. Lewis and E. Anders, *Science* **230**, 167 (1985).
64. T. A. Kuhlbusch and P. J. Crutzen, *Global Biogeochem. Cycles* **9**, 491 (1995).
65. H. Cachier, M. P. Brémond and P. Buat-Ménard, *Tellus* **41B**, 379 (1989b).
66. J. J. Huntzicker, R. L. Johnson, J. J. Shah and R. A. Cary in *Particulate Carbon: Atmospheric Life Cycle*, edited by G. T. Wolff and R. L. Klimisch, pp 79–85, Plenum Press, New York, London (1982).
67. E. C. Ellis, T. Novakov and M. D. Zeldin, *Sci. Tot. Environ.* **36**, 261 (1984).
68. T. Novakov and C. E. Corrigan, *Mikrochemica Acta* **119**, 157 (1995).
69. H. D. Huffman, *Atmos. Environ.* **30**, 73 (1996).
70. H. D. Huffman, *Atmos. Environ.* **30**, 85 (1996).
71. H. Burtscher, S. Künzel and C. Hüglin, *J. Aerosol Sci.* **26** (suppl 1), S129 (1995).
72. T. Novakov (guest editor), *Aerosol Sci. and Technol.* **10** (1 and 2), special issue on *Carbonaceous Particles in the Atmosphere*, 445 pp (1989).
73. H. Puxbaum and T. Novakov (guest editors), *Atmos. Environ.* **27A**, (8) special issue on *Carbonaceous Particles in the Atmosphere* (1993).
74. J. E. Penner and T. Novakov (guest editors), *J. Geophys. Res.* **101**, special issue on *Carbonaceous Particles in the Atmosphere* (1996).
75. M. R. Schure, P. A. Soltys, D. F. Natusch and T. Mauney, *Environ Sci. Technol.* **19**, 82 (1985).
76. T. Novakov *Sci. Tot. Environ.* **36**, 1 (1984)...
77. B. S. Haynes, M. Neville, R. J. Quann and A. F. Sarofim, *J. Colloid Interface Sci.* **87**, 266 (1982).
78. J. Heintzenberg and D. S. Covert, *Tellus* **39B**, 374 (1987).
79. T. Novakov, *Sci. Tot. Environ.* **36**, 1 (1984).
80. M. Bizjak, R. Cigler, A. D. A. Hansen and V. Hudnik, *Atmos. Environ.* **27**, 1347 (1993).
81. J. Langner and H. Rodhe, *J. Atmos. Chem.* **13**, 225 (1991).
82. R. L. Dod, R. D. Giauque, T. Novakov, S. Weihan, Z. Quipeng and S. Wenzhi, *Atmos. Environ.* **20**, 2271 (1986).
83. L. F. Radke, A. S. Ackermann, D. A. Hegg, J. H. Lyons, P. V. Hobbs and J. E. Penner, *Atmos. Res.* Weickmann memorial volume (in press).
84. F. R. Faxvog and D. M. Roessler, *Applied Optics* **17**, 2612 (1978).
85. S. G. Jennings and R. G. Pinnick, *Atmos. Environ.* **14**, 1123 (1980).
86. E. M. Patterson, R. M. Duckworth, C. M. Wyman, E. A. Powell and J. W. Gooch, *Atmos. Environ.* **25**, 2539 (1991).
87. W. M. White and E. S. Macias, *Aerosol Sci. Technol.* **10**, 111 (1989).
88. P. J. Groblicki, G. T. Wolff and R. J. Countess, *Atmos. Environ.* **15**, 2473 (1981).
89. S. M. Larson, G. R. Cass and H. A. Gray, *Aerosol Sci. Technol.* **10**, 118 (1989).
90. A. Robock in *Global Biomass Burning*, edited by J. S. Levine, pp 463–476, MIT Press, Cambridge (1991).
91. K. A. Browning, R. J. Allam, S. P. Ballard, R. T. H. Barnes, D. A. Bennetts, R. H. Maryon, P. J. Mason, D. McKenna, J. F. B. Mitchell, C. A. Senior, A. Slingo and F. B. Smith, *Nature* **351**, 363 (1991).

92. J. P. Blanchet, J. Heintzenberg and P. Winkler, *Beitr. Phys. Atmos.* **59**, 359 (1986).
93. Y. J. Kaufman in *Aerosol Forcing and Climate*, Dahlem Worshop Report ES **17**, edited by R. J. Charlson and J. Heintzenberg, pp 297–332, Wiley, Chichester (1995).
94. B. A. Bodhaine, *J. Geophys. Res.* **100**, 8967 (1995).
95. P. Chylek, G. Videen, D. Ngo, R. G. Pinnick and J. D. Klett, *J. Geophys. Res.* **100**, 16 325 (1995).
96. C. Liousse, J. E. Penner, C. Chuang, J. J. Walton, H. Eddleman and H. Cachier, *J. Geophys. Res.* **101**,19411 (1996).
97. A. D. Clarke, *Aerosol Sci. and Technol.* **10**, 161 (1989).
98. A. D. Clarke and K. J. Noone, *Atmos. Environ.* **19**, 2045 (1985).
99. J. Nelson, *Nature* **339**, 611 (1989).
100. R. A. Dobbins, G. W. Mulholland and N. P. Bryner, *Atmos. Environ.* **28**, 889 (1994).
101. M. F. Iskander, H. Y. Chen and J. E. Penner, *Atmos. Environ.* **25**, 2563 (1991).
102. T. P. Ackerman and O. B. Toon, *Appl. Opt.* **20**, 3661 (1981).
103. R. Bhandari, *Appl. Opt.* **25**, 3331 (1986).
104. P. Chylek and J. Hallett, *Q. J. Res. Meteorol. Soc.* **118**, 167 (1992).
105. J. M. Haywood and K. P. Shine, *Geophys. Res. Lett.* **22**, 603 (1995).
106. F. Parungo, C. Nagamoto, M. Y. Zhou, A. D. A. Hansen and J. Harris, *Atmos. Environ.* **28**, 3251 (1994).
107. S. R. McDow, M. Vartiainen, Q. Sun, Y. Hong, Y. Yao and R. M. Kamens, *Atmos. Environ.* **29**, 791 (1995).
108. J. G. Hudson, *Atmos. Environ.* **25**, 2449 (1991).
109. E. Andrews and S. M. Larson, *Environ. Sci. Technol.* **27**, 857 (1993).
110. T. Novakov and C. E. Corrigan, *Geophys. Res. Lett.* **23**, 2141 (1996).
111. G. Lammel and T. Novakov, *Atmos. Environ.* **29**, 813 (1995).
112. J. G. Hudson and A. D. Clarke, *J. Geophys. Res.* **97**, 14 533 (1992).
113. P. Squires, *J. Res. Atmos.* **2**, 297 (1966).
114. P. V. Dinh, J. P. Lacaux and R. Serpolay, *Atmos. Res.* **31**, 41 (1994).
115. A. D. A. Hansen and T. Novakov, *Aerosol Sci. Technol.* **10**, 106 (1989).
116. J. S. Levine, *EOS* **71**, 37 (1990).
117. J. A. Coakley, R. L. Bernstein and P. A. Durkee, *Science* **237**, 953 (1987).
118. T. Novakov and J. E. Penner, *Nature* **365**, 823 (1993).
119. J. E. Penner in *Aerosol Forcing of Climate*, Dahlem Workshop Report ES **17**, edited by R. J. Charlson and J. Heintzenberg, pp 91–108, Wiley, Chichester (1995).
120. W. Fendel, D. Matter, H. Burtscher and A. Schmidt-Ott, *Atmos. Environ.* **29**, 967 (1995).
121. S. Stephens, M. J. Rossi and D. M. Golden, *Int. J. Chem. Kinetics* **18**, 1133 (1986)
122. J. C. Petit and Y. Bahaddi, *Carbon* **5**, 821 (1993).
123. L. A. Gundel, N. S. Guyot-Sionnest and T. Novakov, *Aer. Sci. Technol.* **10**, 343 (1989).
124. F. de Santis and I. Allegrini, *Atmos. Environ.* **26**, 3061 (1992).
125. K. Tabor, L. Gutzwiller and M. J. Rossi, *Geophys. Res. Lett.* **20**, 1431 (1993).
126. T. Novakov, S. G. Chang and A. B. Harker, *Science* **186**, 259 (1974).
127. S. G. Chang, R. Toossi and T. Novakov, *Atmos. Environ.* **15**, 1287 (1981).
128. G. Santachiara, F. Prodi and F. Vivarelli, *Atmos. Environ.* **23**, 1775 (1989).
129. I. Grgìc, V. Hudnik and M. Bizjak, *Atmos. Environ.* **27**, 1409 (1993).
130. T. D. Behymer and R. A. Hites, *Environ. Sci. Technol.* **19**, 1004 (1985).
131. J. Heintzenberg, *Tellus* **41B**, 149 (1989).
132. B. Etemad and J. Luciani in *World Energy Production, 1800–1985*, Université de Genève, Droz, Geneva (1991).

133. S. S. Butcher and M. J. Ellenbecker, *J. Air Pol. Cont. Ass.* **32**, 380 (1982).
134. W. F. Cooke, *PhD Dissertation*, Galway University College, National University of Ireland, 190 pp (1996).
135. W. F. Cooke and J. J. N. Wilson, *J. Geophys. Res.* **101**, 19395 (1996)..
136. A. D. A. Hansen and H. Rosen, *J. Air Waste Manage. Assoc.* **40**, 1654 (1990).
137. L. F. Radke, D. A. Hegg, J. H. Lyons, C. A. Brock, P. V. Hobbs, R. Weiss and R. Rasmussen in *Aerosols and Climate*, edited by P. V. Hobbs and M. P. MacCormick, pp 411–422, A. Deepak (1988)...
138. H. Cachier, C. Liousse, P. Buat-Ménard and A. Gaudichet, *J. Atmos. Chem.* **22**, 123 (1995).
139. H. Cachier, C. Liousse, A. Gaudichet, F. Echalar, J. P. Lacaux and T. Kuhlbusch, *J. Geophys. Res.* (special SAFARI experiment issue) (in press).
140. H. Cachier, C. Liousse, M. H. Pertuisot, A. Gaudichet, F. Echalar and J. P. Lacaux in *Biomass Burning and Global Change*, edited by J. S. Levine, MIT Press, Cambridge pp 428–440 (1996).
141. J. P. Lacaux, H. Cachier and R. Delmas, in *Fire in the Environment*, Dahlem Worshop Report ES 13, edited by P. J. Crutzen and J. G. Goldammer, pp. 150–191, Wiley, Chichester (1993).
142. C. Liousse, J. E. Penner, J. J. Walton, H. Eddleman, C. Chuang and H. Cachier in *Biomass Burning and the Global Cycles*, edited by J. S. Levine, MIT Press, Cambridge pp 492–508 (1996).
143. L. F. Radke, D. A. Hegg, P. V. Hobbs, J. D. Nance, J. H. Lyons, K. K. Laursen, R. E. Weiss, P. J. Riggan and D. E. Ward in *Global Biomass Burning*, edited by J. S. Levine, pp 209–224, MIT Press, Cambridge (1991).
144. H. Cachier, 1992 in *Encyclopedia of the Earth Science System*, edited by N. Nierenberg, Vol. 1, pp 377–385 (1992), Academic Press, San Diego.
145. A. D. A. Hansen, T. J. Conway, L. P. Steele, B. A. Bodhaine, K. W. Thoning, P. Tans and T. Novakov, *J. Atmos. Chem.* **9**, 283 (1989).
146. R. A. Duce, *Pageophys.* **116**, 244 (1978).
147. R. P. Turco, O. B. Toon, R. C. Whitten, J. B. Pollack and P. Hamill, in *Precipitation Scavenging, Dry Deposition and Resuspension*, edited by H. R. Pruppacher, R. G. Semonium and W. G. N. Slinn, pp 1337–1351, Elsevier Science, New York (1983).
148. J. E. Penner, H. Eddleman and T. Novakov, *Atmos. Environ.* **27**, 1277 (1993).
149. E. Robinson and E. C. Robbins, *American Petroleum Inst.* 4076 (1971).
150. D. J. Ball, in *Proc. Inst. Mech. Eng. Conf., Vehicle Emissions and their Impact on European Air Quality*, C337, pp 83–88, C. Phys. London Scientific Services (1987).
151. J. Trijonis, *Sci. Total Environ.* **36**, 131 (1984).
152. J. Heintzenberg and P. Winkler, *Sci. Tot. Environ.* **36**, 27 (1984).
153. M. P. Brémond, H. Cachier and P. Buat-Ménard, *Environ. Technol. Lett.* **10**, 339 (1989).
154. R. S. Hamilton and T. A. Mansfield, *Atmos. Environ.* **25**, 715 (1991).
155. G. J. Keeler, S. M. Japar, W. W. Brachaczek, R. A. Gorse, J. M. Norbeck and W. R. Pierson, *Atmos. Environ.* **24**, 2795 (1990).
156. P. J. Crutzen and M. O. Andreae, *Science* **250**, 1669 (1990).
157. P. J. Crutzen, A. C. Delany, J. Greenberg, P. Haagenson, L. Heidt, R. Lueb, W. Pollock, W. Seiler, A. Wartburg and P. Zimmerman, *J. Atmos. Chem.* **2**, 233 (1985).
158. V. W. Kirchhoff, A. W. Setzer and M. C. Pereira, *Geophys. Res. Lett.* **16**, 469 (1989).
159. P. Artaxo, F. Gerab, M. A. Yamasoe and J. V. Martins, *J. Geophys. Res.* **99**, 22857 (1994).

160. J. Ducret and H. Cachier, *J. Atmos. Chem.* **15**, 55 (1992).
161. B. J. Stocks in *Global Biomass Burning*, edited by J. S. Levine, pp. 197–202, MIT Press, Cambridge (1991).
162. M. O. Andreae, *Science* **220**, 1148 (1983).
163. H. Rosen and A. D. A. Hansen, *Atmos. Environ.* **19**, 2203 (1985).
164. J. Savarino, PhD dissertation (in French), Université Joseph Fourier-Grenoble 1, LGGE publication No 925, 390 pp (1996).
165. J. Savarino and M. Legrand, *J. Geophys. Res.* (in press).
166. P. Chylek, B. Johnson and H. Wu, *Ann. Geophys.* **10**, 625 (1992).
167. M. H. Pertuisot, H. Cachier, F. Maupetit and M. Legrand, Paper presented at the *5th International Conference on Particulate Carbon in the Atmosphere*, Berkeley (Aug. 1994).
168. R. M. Rotty and G. Marland, *Report NDP-006*, Carbon Dioxide Information Center, Oak Ridge National Laboratory (USA) (1986).
169. J. A. Ogren and R. J. Charlson, *Tellus* **35B**, 241 (1983).
170. R. Dlugi, *Aerosol Sci. Technol.* **10**, 93 (1989).
171. K. Okada, M. Ikegami, O. Uchino, Y. Nikaidou, Y. Zaizen, Y. Tsutsumi and Y Makino, *Geophys. Res. Lett.* **19**, 921 (1992).
172. J. E. Penner, R. E. Dickinson and C. A. O'Neil, *Science* **256**, 1432 (1992).
173. M. D. King, Y. J. Kaufman, W. P. Menzel and D. Tanré, *Geosci. Remote Sens.* **30**, 2 (1992).
174. W. G. N. Slinn, *Atmos. Environ.* **25**, 2473 (1991).
175. J. K. Angell, *J. Clim.* **1**, 1296 (1988).
176. J. D. Kahl, D. J. Charlevoix, N. A. Zaitseva, R. C. Schnell and M. C. Serreze, *Nature* **361**, 335 (1993).
177. J. D. Kahl, M. C. Serreze, R. S. Stone, S. Shiotani, M. Kisley and R. C. Schnell, *J. Geophys. Res.* **98**, 12 825 (1993).
178. F. Parungo, J. F. Boatman, H. Sievering, S. W. Wilkison and B. B. Hicks, *J. of Climate* **7**, 434 (1994).

10 Primary Biological Aerosol Particles: Their Significance, Sources, Sampling Methods and Size Distribution in the Atmosphere

SABINE MATTHIAS-MASER

University of Mainz, Germany

1 INTRODUCTION

The atmospheric aerosol consists of a quantity of solid and liquid components of all kinds of substances. A not negligible part of this are the biological aerosol particles or better named *primary biological aerosol particles* (PBAP), which should not be underestimated. In an international workshop held in Geneva in June 1993 as past of the International Global Aerosol Programme (IGAP) a science team consisting of international scientists worked out a definition for this important component of the atmospheric

Atmospheric Particles, Edited by R. M. Harrison and R. Van Grieken.

aerosol particles as follows: "Primary biological aerosol particles describe airborne solid particles (dead or alive) that are or were derived from living organisms, including microorganisms and fragments of all varieties of living things." This definition includes a wide spectrum of biological particles spread over a large size range. The following gives a short list of the PBAP including their size range. [1] The smallest PBAP are viruses ($0.005 \, \mu m < r <\sim 0.25 \, \mu m$). The larger particle range include bacteria ($r >\sim 0.2 \, \mu m$), algae, spores of lichen mosses, ferns and fungi ($r >\sim 0.5 \, \mu m$), pollen ($r >\sim 5 \, \mu m$), plant debris like leaf litter, parts of insects, human and animal epithelial cells (supposed $r > 1 \, \mu m$).

In this chapter the expression PBAP is purposely used in contrast to "biogenic particles". Biogenic particles like those deriving from DMS (dimethylsulfide) or phytholyths, which occur in the atmosphere in a non-biological or mineralized state, will not be considered here.

At this point the distinction should be made between PBAP and organic carbon (OC). Important emitters of OC are plants. [2] Embedded and growing from the cuticular membranes of leaf surfaces of vascular plants, wax-like lipids form crystalline structures of micron and submicron dimensions. Due to wind-induced mechanical shear and leaves rubbing against each other, such epicuticular wax protrusions and leaf deposits become airborne and have been identified in urban and rural aerosols. The waxes consist for instance of n-alkanes, n-alkanols, n-alkanoid esters and wax esters. N-alkanes are found not only on plant leaves but also on seeds, fruits, stems, pollen, fungi and on insects. They reveal a characteristic distribution for each plant species, but they could not be compared with the PBAP as mentioned here.

Most investigations according to OC refer to particles with $r < 1 \, \mu m$. PBAP of that size range do not consist exclusively of OC. Investigations of the protoplasm of some species of fungi mostly show the following 10 elements: [3] C (50%), O (29%), N (14%), H (8%), P (3%), S (1%), K (1%), Ca (0.5%), MG (0.5%) and Fe (0.2%). The values in brackets refer to the elemental composition of bacteria. [4] Heintzenberg [5] compiled the chemical composition for the aerosol particles smaller than $1 \mu m$ in radius (Figure 1). In addition to the major ions he has listed the elemental carbon and the organic carbon as constituents, but no biological material. The biological particles added from Figure 7 (see later) to Figure 1 of course contain carbon. Therefore portions of this are included in the elemental and organic carbon of Heintzenberg but also parts of the non-determined fraction could switch over to the biological particles.

Another point is that the most OC investigations neglect the giant PBAP which, while minor in number concentration, are more important in mass and volume concentration.

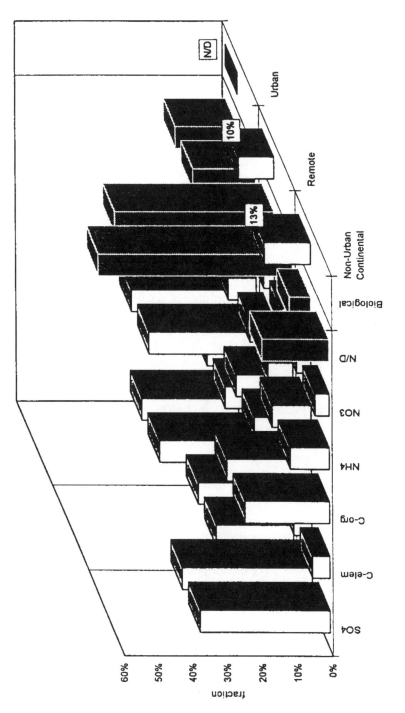

Figure 1. Mass fractions of ions and carbon (elemental and organic) of the fine atmospheric aerosol particles[5] N/D means "non determined". Adding the biological content (Figure 7), the second row of each model aerosol has been modified to the fact that the N/D portion has been adjusted

2 CHARACTERISTICS OF PBAP

The following describes some characteristics like shape and composition of PBAP, [4] starting with the smallest.

Viruses (0.005 μm < *r* < 0.25 μm) consist of nucleinacids (DNA or RNA) which are surrounded by a protein cover called *capsid*. Most of them have symmetric (geometrical, spherical or oval) shapes. They are non-cellular particles which are not able to increase on their own, but need living cells for reproduction.

Bacteria (*r* >∼ 0.2 μm) consist of DNA or RNA and proteins, lipids and phospholipids as essential compounds. Cellulose and the peptidoglycan *murein* are main components of the cell. The shape of the bacteria varies from rods, spheres to spirals, coccus, etc. For reproduction they need specific nutrients, temperatures and humidities.

The propagation of some bacteria, algae, fungi, lichen, mosses and ferns is effected by *spores* (*r* >∼ 0.5 μm). Their inner body, consisting of protoplasm, DNA, RNA and cell material, is surrounded by a thick wall, the *sporoderm*. This consists of two layers, the *intine* consisting of cellulose and protein, pectin and callose [6] and the *exine* which is mainly built up of sporopollenins. The latter are chemical-resistant, high-polymer esters or carotinoids. [7, 8] Their shape varies greatly: spherical, oval, fusiform, needle shaped, elongated and helical. For the predominantly asexual reproduction the spores only need to land on suitable ground.

The spermatophytes reproduce themselves by *pollen* (*r* >∼ 5 μm). They consist of protein 20%, carbohydrates 37%, lipids 40% and mineral compounds 3%. [8] They are surrounded by the resistant *sporoderm* (see spores), too. The shape of pollen shows a great variety. Pores and furrows within the exine lead to different designs and forms which are characteristic for many species. Some pollen are equipped with airbags (*conifers*) which support their distribution by wind.

Plant debris, parts of insects, human and animal epithelial cells or hair have a proposed radius of *r* > 1 μm. Only a few investigations have been performed with this part of PBAP, so the information is rather limited. Their constituents extend from cellulose and starch (plant debris) to skleroproteins (hair) and chitin (insect parts). They are irregular shaped and not easy to identify.

3 SIGNIFICANCE OF PBAP

PBAP in the atmosphere play an important role in air hygiene indoors as well as outdoors. Viruses, bacteria, spores, pollen and insect parts (i.e. house mite dust) are causes of diseases or allergenic reactions of human, animals and plants.

Another aspect for the investigation of PBAP in the atmosphere is the role they play in cloud physics. Atmospheric aerosol particles act as *cloud nuclei*. Therefore most particles need temperatures under $-10\,°C$. The pecularity of some PBAP is that they have freezing capability even at temperatures about $-4\,°C$. That means even if the surroundings are still supercooled the cloud-forming processes can be induced by these ice nucleation active PBAP.

The hypothesis that bacteria were causing ice nucleation was tested by heating (to $95\,°C$ for a few minutes) a suspension of faecal material to destroy proteins (see Chapter 5) and measuring the ice-nucleating activity. This procedure reduced the number of active nucleators at $-5\,°C$ from 1.3×10^6 to zero. [9] Schnell and Vali [10] reported that a portion of atmospheric freezing nuclei was of biogenic origin. The sources of these nuclei include decaying vegetation, marine plankton [11] and bacteria. [12] Sun *et al.* [13] discovered 17 species or varieties of 3 genera of ice nucleation active bacteria on plants; 78% belong to the *Erwinia herbicola* group (Gram negative facultative anaerobe plant pathogen) and the *Pseudomonas syringae* group (Gram negative aerobe plant epiphyte, occasionally pathogen). But even dead bacteria may be active in producing freezing nuclei. [14]

Ice nucleation activity of the free-living fungus *Fusarium* was mentioned by Pouleur *et al.* [15] He compared the ice nucleation activity (INA) of some species of *Fusarium* with the INA of bacteria and lichens. The warmest temperatures at which ice nucleation occurs for *F. avenaceum* was $-2.5\,°C$ which is close to that of lichen INA ($-2\,°C$) and slightly lower than that of *Pseudomonas* ($-1\,°C$). As a result of his investigations he found out that lichen INA and *Fusarium* INA could have a common origin. He even suggested that ice nucleation of *F. acuminatum* and of *F. avenaceum* could be the source of leaf-derived nucleation (LDN) observed in decomposing leaf litter. [10] They suggested that the microflora is responsible for the decomposition of organic matter. *Fusarium acuminatum* and *F. avenaceum* are both soil-borne fungi and are more prevalent in temperate and cold climates, where LDN concentration is highest. [10]

Investigations of Warren and Wolber [16] show that a single gene is responsible for the INA phenotype of the different microorganisms. This is better known for bacteria than for ice nucleation in lichens.

Even pollen can contribute to condensation processes. Durham [17] worked with 12 species of pollen to determine their response to air of various humidity levels. He found that all were hygroscopic in the sense that they acquired water from the vapour below 100% relative humidity. The species studied include six weeds, two grasses and four deciduous trees. Dingle [18] mentioned the hygroscopicity of the protoplasm of ragweed (*Ambrosia artimesiifolia*) at humidities higher than 51%.

There is no question about the fact that these INA particles are present in the atmosphere and initiate ice nucleation for cloud formation.

4 SOURCES OF PBAP IN THE ATMOSPHERE

The most important *source of PBAP* are plants. Pollen and spores are actively or passively released by different kinds of plants. Most airborne pollen are anemophilous (they are distributed by wind), although only about 30 of the over 300 families of spermatophytes (grasses, almost all conifers, deciduous trees) belong to this group. These plants can produce over a billion pollen during a season. This occurs mainly in spring and summer, depending on the *phenology*. But the studies of Markgraf [19] and Whitehead [20] show a resumption of pollen flight during autumn and winter. This "reflotation" occurs when pollen and spores which have been impacted on plant surfaces in spring and summer resuspend back into the atmosphere in autumn and winter by wind or by decaying processes.

The decaying processes of leaves produce first of all leaf litter, but also the bacteria initiating the decaying processes are carried out into the atmosphere.

Agricultural activities such as ploughing, threshing, disc harrowing, etc. can throw plant debris and microbes, often pathogens, into the air.

Industrial and municipal facilities like abattoirs, sewage treatment plants and textile mills are often a strong source of microorganisms, which were then, depending on meteorological conditions, distributed in the atmosphere. [21] Matthias-Maser and Jaenicke [22] found high concentration of bacteria and spores when measuring next to an area where old buildings were being pulled down.

A lot of bacteria and viruses are released into the atmosphere by coughing and sneezing. The microbes on human or animal skins are distributed into the atmosphere by activities like dressing, airing, bed-making, shaking, etc. [23] By this means about 10 000 bacteria per m^3 can become airborne. [24]

Over half of the earth is covered with water, so the marine atmosphere supplies an important contribution to the atmospheric aerosol. Although the major component is sea-salt, the ocean is also an important source for microorganisms. The microbes are collected in a microlayer around the rising bubble and are injected within the jet drops into the atmosphere ("bubble-burst mechanism"). [25]

Generally the *source strength* depends on different factors. First of all the season plays an important role. In spring and summer the plants are very active in producing all different kinds of pollen and spores whereas in autumn and winter the decaying processes are predominant. This seasonal dependence corresponds to the different meteorological conditions. Some plants produce more pollen on sunny days whereas some need raindrops to release pollen or spores. Lindemann [26] found an increase in bacterial concentration after rain, but generally most pollen and spores will be released into the atmosphere in dry weather and low relative humidity.

Closely coupled with the temperature and the relative humidity is the time of day. Anemophilous trees release their main pollen at noon (high temperature and low humidity) when desiccation causes the bursting of the anthers. This is also true for some fungi (*fungi imperfecti*). [27, 28] Most grasses release their pollen in the early morning as do many *ascomycetes*, which need a high humidity to eject spores.

The wind velocity could be another factor influencing the release of microorganisms. Some plants need a minimum wind to release pollen, [29] others (*Alternaria*) show a decrease of concentration with increasing wind velocity.

5 LIFETIME OF PBAP

Before mentioning the different sampling methods, there should be a short reference to the *lifetime* of PBAP in the atmosphere. The most common sampling methods are based on cultivating the samples. In this way only viable microorganisms can be investigated.

Pollen, spores and microorganisms have to travel for a long time in the atmosphere before finding a receptor. Their thick walls make them resistant to many environmental influences such as radiation, high or low temperatures, etc. Many plants have a so-called "drying tolerance"; that means those bacteria, algae, lichens, some fungi, different mosses and most pollen and spores can stay dry for a long time without losing their germination capability. The vegetative cells of bacteria will be killed at temperatures of about 60 °C within 5–10 minutes. Yeast and fungal spores die at temperatures above 80 °C and a bacterial spore above 120 °C. Natural UV-radiation can have a lethal effect, whereby bacteria are killed faster than fungal spores. The cell death of a microorganism means the irreversible loss of growing or germinating capability, but some cell damage which would normally lead to death is repairable for instance with special UV-radiation. [4]

Even the composition of the other atmospheric aerosol particles can influence the lifetime of microorganisms. Rüden *et al.* [30] found that metal components NH_4^+ and SO_2^{2-} have an antimicrobial effect.

6 SAMPLING METHODS

The choice of a *bioaerosol sampler* depends on the purpose of the investigation. Is the scientist interested in evaluating special microorganisms, dead or alive, or is the total concentration of PBAP wanted? Then in selecting a bioaerosol sampler one should consider that some samplers affect the viability of a bioaerosol more than others.

The sizes of PBAP spread over many orders of magnitude. This accounts for the variety of different sampling methods, which are adapted to the different sizes and different kinds of particles.

A very common method is the sampling of PBAP using impactors. Impaction can take place on nutrient media or solid surfaces. For particles with radius $r > 0.1 \, \mu m$ single or multi-stage slit-to-agar samplers are used. A common instrument is the Andersen sampler [31] which collects six size-grated aerosol fractions in petridishes by a system of multi-jet cascade impaction. Based on this scheme there exists a variety of slit-to-agar samplers (i.e. Mattson–Garvin) which differ in number of stages, sampling time, sampling rate and cut-off radii.

Investigating the variation of pollen and spores within the time the Hirst-or the Burkard spore trap is widely used. The particles are sucked into an orifice beneath a rain shield and are impacted on an adhesive-coated microscope slide. This is pulled upward at 2 mm per hour by clockwork. A wind vane keeps the orifice facing the wind for quasi-isokinetic sampling conditions which are necessary for particles larger than 1 μm. [32]

For particles of the size range $0.2 \, \mu m < r < \sim 50 \, \mu m$ Matthias-Maser and Jaenicke [33] used an isokinetic two-stage impactor (ITSI) and a wing impactor (WI). The instruments were placed on wind vanes and the size of the inlet of the ITSI was adjusted to the actual wind speed. The particles are sampled on adhesive-coated glass slides for evaluation in a *light microscope* (LM) and on graphitic foil for evaluation in a *scanning electron microscope* (SEM).

Impactors are easy to handle, some are commercial available and they have well-defined cut-off characteristics.

Air hygienists often use impingers. [34,35] The particles are projected into a liquid, where they are broken up into their individual component cells. The liquid is serially diluted, plated out and incubated to give a colony count and an estimate of the total viable cells in the original sample. They permit counts of several different organisms from the same sample by means of differential counts or selective media. But the impinger is not good for hygrophobic fungal spores and if a long time sample is taken, evaporation of the sampling fluid will occur. Impingers are simple, cheap and convenient, can be easily sterilized and by an addition of a pre-impinger may become an approximate simulant of the upper and lower part of the respiratory system. They exist in single and multi-stage form.

The simplest method for sampling microorganisms is the exposure of petridishes, which are filled with nutrient medium. This sedimentation principle gives a selective estimate of the concentration of microbes in the atmosphere but it is only appropriate to the giant aerosol particles ($r > 5 \, \mu m$). [36]

Finally, particle collection by filtration should be mentioned. The investigators typically use membrane filters to collect biological aerosol particles. The filters are evaluated either under a microscope or they are placed face up directly on to agar or an absorbent pad saturated with culture medium.

The previous list does not claim completeness; more details are given in Nevaleinen *et al.*, [37] Griffiths *et al* [38] and Macher [1].

7 ANALYSIS OF THE SAMPLES

When sampling microorganisms by using *culture methods* (i.e. Anderson sampler, petri-dishes, impinger) first of all a proper sampling medium must be chosen. Selective media are available for many types of microbes. The results are given in colony-forming units per cm^3 (cfu). In most cases it is difficult to count the colonies exactly, because the airborne particles may have been aggregates of two or more cells or species. The identification of the colonies is based mainly on the *morphology* and staining properties. The investigator should have specific experience in bacteriology, mycology and palynology.

This is also required for the evaluation of PBAP sampled on glass slides under LM. It is helpful to use dyes, which are often selective for special kinds or constituent of microorganisms. The *staining* capability of different biological particles is dependent on the pH-value of the stain. Normally the thicker the cell wall the smaller the pH-value. [39] Basic dyes stain cell walls while acid dyes stain cytoplasm and cellulose walls.

"Basic fuchsin" is often used for staining airborne pollen. [32,40] "Acridin orange" is used for staining nuclein acid (evaluation in epifluoresence microscope). "Giemsa" consisting of the basic "Azur B" and the acid "Eosin Y" is also used as a nuclein dye. It stains the cell nucleus red and the cytoplasm acquires a bluish colour. [41] An unselective stain was used by Matthias-Master. [42] This protein dye "Coomassie blue" reacts with the carboxyl group in the protein. Under LM (for particles with $r > 2\,\mu m$) most microorganisms including plant debris and insect parts show a blue colour. For investigation in SEM (for particles with $r < 2\,\mu m$) equipped with an energy dispersive X-ray detector (EDX) the characteristic morphology (special forms like spheres, rods, characteristic forms, etc.), the special elemental composition (P, S, K, Ca sometimes with Cl on a high background spectrum) and the behaviour during EDX (sometimes unstable due to the electron beam, shrinking) of the PBAP must be considered. [33]

8 OCCURRENCE IN THE ATMOSPHERE

The immense number of different sources of PBAP and the use of sampling and analysis methods, which are often designed for the investigation of special kinds of PBAP, lead to numerous measurements representative only of the sampling location, environmental conditions and phenology. But these measurements can give an idea of the number and concentration of PBAP in the atmosphere. Most investigators specialize in one kind of PBAP (i.e. bacteria or pollen or spores). For example, Spendlove [21] and Bovallius [43] found bacterial concentrations of 10^4–10^5 bacteria per cm^3 above sewage plants, Gregory [44] told of 290 bacteria per cm^3 of air in a park and more than 7500 bacteria per cm^3 of air in a busy street.

The concentration of pollen shows a great variability due to the phenology. For instance, the anemophilous *Ambrosia artemisiifolia*... ejects over a billion pollen per plant within a season. Even in winter (November–March) Janzon [45] found pollen concentrations of mainly birch (*Betula*) and pine (*Pinus*) of 0.1–15 pollen per m^3. Their concentration can reach 1000 and 400 per m^3 air respectively in spring. Hamilton [28] found that the counts of insect fragments (hair or scales) were five times greater in the country than in the city. These particles were most numerous in late August and September with a maximum daily count occurring between late afternoon and dusk. The numbers of insect parts show strong positive correlation between both temperature and dewpoint and the counts are inversely related to the wind velocity. Unfortunately no absolute values were given.

Even at heights of 200 m a strong accumulation of spores and pollen in and/or above the inversion layer can be found. [46] The concentration could reach values of over 100 microorganisms per m^3 of air space volume. [47]

9 SIZE DISTRIBUTION OF PBAP

In atmospheric science it is usual to determine and to discuss *size distributions* of the atmospheric particles. A first attempt to examine PBAP in their entirety was made by Matthias-Maser and Jaenicke. [22] They investigated the PBAP within the size range 0.2 μm $< r <\sim$ 50 μm using an isokinetic two-stage impactor and a wing impactor for sampling the total aerosol particles. The analysis of the samples was performed using the unselective protein dye "Coomassie blue" to recognize them under LM. In SEM/EDX the above-mentioned criteria were used for differentiation between PBAP and non-biological aerosol particles. As a result the size distribution dN/d log r for the total and for the primary biological aerosol particles were obtained. The measurements took place primarily in spring, the early summer, and some were done in autumn. Peters [48] completed the measurements with samples in autumn, winter and early spring to get an annual variation of PBAP.

The following figures show the size distributions measured in March, June, October and December, each showing characteristic features for the season. In each figure the size distributions (with error bars) for PBAP and for the total aerosol are plotted together with the percentage in each size class. Additionally the total concentrations N_{tot}, $Nbio_{tot}$... and the total volumes V_{tot}, $Vbio_{tot}$ were calculated. Figure 2 presents the results from March. Here the beginning of the pollen season is obvious, starting with birch (*Betula*) and hazel (*Corylus*), which are responsible for 20 and 42% respectively of the total aerosol in the sixth and seventh radii intervals. This results in a high volume concentration $Vbio_{tot}$. The next figure (Figure 3) shows the result of an early summer mea-

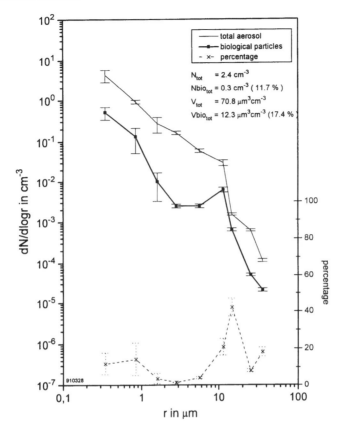

Figure 2.   Size distributions measured in March

surement in June. Here grass pollen are predominant in the fraction of PBAP (almost 70%). The measurement from October (Figure 4) shows a high percentage of PBAP in the submicron range caused by bacterial aerosol (decaying processes) and a maximum at a radius of about 3 μm in the percentage curve as a result of the enlarged occurrence of spores. The latter is more distinct in December (Figure 5). In addition to the spores the reflotation (see Chapter 4) causes an increase in the percentage of pollen. Within the largest radius interval plant debris were almost exclusively dominant.

Considering all measurements the mean concentration of $\overline{Nbio}_{tot} = 3.11\,\mathrm{cm}^{-3}$ and $\overline{Vbio}_{tot} = 6.5\,\mathrm{\mu m}^3\,\mathrm{cm}^{-3}$ were determined. That amounts to a mean percentage of 23.7% in number concentration and 22.3% in volume of the total aerosol particles. Calculating a mean density of PBAP of $1\,\mathrm{g\,cm}^{-3}$ [44] a mass of PBAP came to an amount of $6.5\,\mathrm{\mu g\,m}^{-3}$.

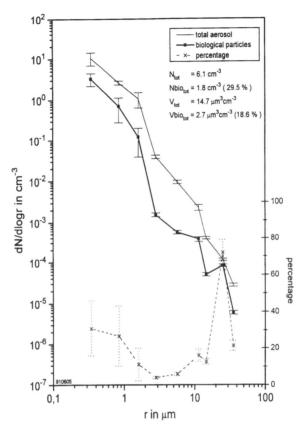

Figure 3. Size distributions measured in June

The strong dependence of the PBAP on the phenology and the annual variation of the phenology could indicate an *annual variation* of PBAP. Therefore all measurements were combined in Figure 6 which shows the number concentration $Nbio_{tot}$ and the volume $Vbio_{tot}$ respectively plotted versus time together with their percentages of total aerosol.

Obviously no characteristic annual variation of PBAP either in number or in volume can be seen. The presumption that in winter the concentration of PBAP would decrease was not confirmed. The production mechanisms of PBAP in winter (reflotation, decaying processes) are stronger than expected. The composition of the PBAP merely changed within the year as follows:

- spring: microorganism, pollen, some spores, a few fragments
- summer: microorganism, pollen, spores, a few fragments

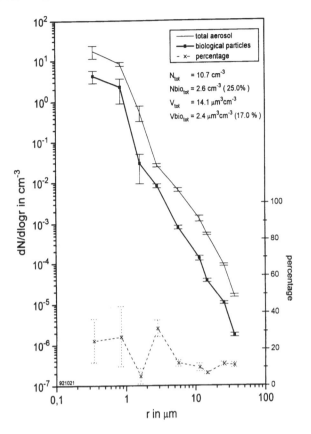

Figure 4. Size distributions measured in October

- autumn: microorganism, fragments, spores, a little pollen
- winter: microorganism, fragments, spores, some pollen

The great variability of $Nbio_{tot}$ in spring as a consequence of the different florescence of the plants in combination with the different meteorological conditions can be allusively seen. But more measurements should be taken to confirm these results.

The theoretical considerations require *model size distributions* of the aerosol particles. Mathematical expressions which are selected to describe aerosol size distributions are lognormal distributions which not only fit a great deal of data but also prossess unique mathematical properties which make them easy to handle. Willeke and Whitby [49] propose the multimodal nature of the size distribution, described by a sum of two lognormal frequency

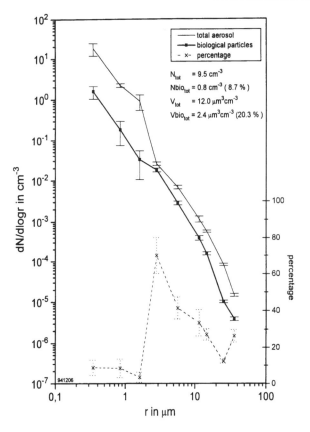

Figure 5. Size distributions measured in December

distributions. Aerosols like biological particles show a lognormal distribution and therefore should be described in that way. [50] In this chapter the following functions were used: [51]

$$\frac{\mathrm{d}N(r)}{\mathrm{d}\log r} = \sum_{i=1}^{2} n_i \frac{1}{\log \sigma_i \sqrt{2\pi}} \exp\left[\frac{-\left(\log \frac{r}{R_i}\right)^2}{2(\log \sigma_i)^2}\right] \tag{1}$$

where n_i is the particle number concentration as an integral above the individual lognormal distributions, R_i the mean geometric radius, and σ_i the polydispersity.

The parameters n_i, R_i and σ_i have been calculated for the different aerosol types (urban, rural, marine, background, etc.). [51] In 1995, Matthias-Maser and Jaenicke [22] determined these parameters for a first model size distribution

Table 1.   Parameters for the model distribution of PBAP

	n_i	R_i	σ_i
$i = 2$	5.93×10^0	1.08×10^{-1}	2.40
$i = 3$	2.75×10^{-2}	1.31×10^0	2.16

for the biological aerosol by performing a nonlinear fitting procedure. [52] These have been updated with the data from Peters. [48] The new parameters for the PBAP are listed in Table 1.

In Figure 7 the new model size distribution for PBAP together with the former one from Matthias-Maser and Jaenicke [22] and the model size distribution for rural aerosol are shown. The saddle point in the old distribution at the middle

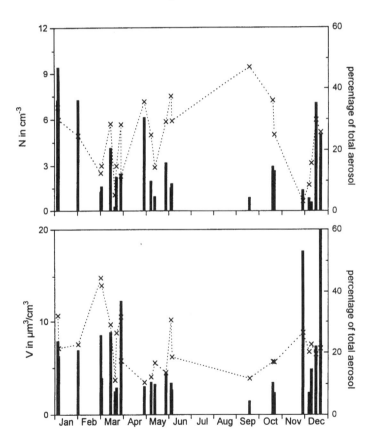

Figure 6.   Total concentration N and volume V of PBAP and their percentages (dashed line) of the total aerosol versus time

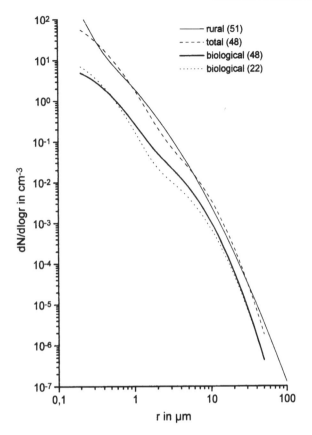

Figure 7. Model size distributions for rural, [51] total[48] and biological [22],[48] aerosol particles

radii intervals is turned into a slight dip by adding the results of the winter measurements, as expected in Matthias-Maser and Jaenicke. [22] The new model size distribution differs by about one order of magnitude from the rural size distribution in the submicron range. For the larger particles the biological distribution approaches the rural up to less than half an order of magnitude, i.e. in the giant particles range the rural aerosol mainly consists of PBAP.

10 CONCLUSIONS

PBAP are a ubiquitous component of the atmospheric aerosol. They amount to almost a quarter either in concentration or in volume of the total aerosol particles. There is no decrease in the concentration of PBAP in the winter season, as expected, simply the composition of the PBAP changes over the year.

The biological composition of especially the giant particles shows seasonal variations within the individual size distributions due to the phenology. Within the different size classes there exist distinct peaks caused by the various flowering periods of the plants.

The dependence of biological particles on wind direction is connected with the locality of and the distance from their sources. The influence of rural regions leads to an increase in the large and giant biological particles like pollen and spores, whereas the influence of urban regions with industrial and municipal activities as sources for bacterial aerosol leads to a higher concentration of these smaller biological particles. The results shown are valid for an urban/rural influenced region. In marine air, for instance, the composition and the percentage of PBAB are different. Investigations on the Atlantic Ocean [53] show a percentage of PBAP to total aerosol of only 17% in number and 10% in volume.

More investigations are necessary for the different types of aerosol (desert, remote continental, etc.) where on account of the different source strengths of PBAP different size distributions and concentrations are expected. For the study of cloud processing, measurements of the concentration and the size distribution of PBAP below, in and above clouds are desirable.

ACKNOWLEDGEMENTS

Funding for this study was provided by the German Research Foundation through its "Sonderforschungsbereich 233: Chemistry and Dynamics of Hydrometeors". I wish to acknowledge Prof. R. Jaenicke for his support and discussion. I also wish to thank K. Peters for providing me with her data, and V. Dreiling for the mathematical calculations.

REFERENCES

1. Macher J. M. (1993). Biological Aerosol, 12. Annual Meeting of American Association for Aerosol Research, Tutorial Session 10, Oak Brook, Illinois Oct. 1993.
2. Rogge W. F., Hildemann L. M., Mazurek M. A., Cass G. R. (1993). Sources of Fine Organic Aerosol. 4. Particulate Abrasion Pruducts from Leaf Surfaces of Urban Plants, *Environ. Sci. Technol.* **27**, p. 2700–2711.
3. Strasburger, E. (1978). *Lehrbuch der Botanik*, Fischer Verlag, Stuttgart.
4. Schlegel H. (1984). *Allgemeine Mikrobiologie*, 6th edn., Thieme Verlag, Stuttgart, 571pp.

5. Heintzenberg J. (1989). Fine Particles in the Global Troposphere, A Review, *Tellus* **41B**, 149–160.
6. Dahl A. O. (1969). Wall Structure and Composition of Pollen and Spores, in: *Aspects of Palynology* edited by Tschudy R. H., Scott R. A., New York, pp. 35–48.
7. Stanley R. G., Linskens H. F. (1974). *Pollen—Biology, Biochemistry, Management*, Springer- Verlag, Heidelberg.
8. Knox R. B. (1979). *Pollen and Allergy*, Camelot Press Ltd, Southampton.
9. Block W., Worland R. (1993). Survival of Sub-zero Temperatures by Beetles in the Sub Antarctic. *Abstracts Sixth Intern. Conference on Biological Ice Nucleation*, Laramie, Wyo., USA, August 4–6.
10. Schnell R. C., Vali G. (1973). World-wide Source of Leaf-derived Freezing Nuclei, *Nature London*, **246**, 212–213.
11. Schnell R. C., Vali G. (1976). Biogenic Ice Nuclei: Part I, Terrestrial and Marine Sources, *J. Atmosph. Sciences*, **33**, 1554–1564.
12. Maki L. R., Galyan E. L., Chang-Chien M., Caldwell D. R. (1974). Ice Nucleation Induced by *Pseudomonas syringae*, *Applied Microbiology*, **28**, 456–459.
13. Sun F., Zhu H., He L., Zhang Y. (1993). Identification of Ice Nucleation Active Bacteria on Plants, *Abstracts Sixth Intern. Conference on Biological Ice Nucleation*, Laramie, Wyo, USA, August 4–6.
14. Maki L. R., Willoughby K. J. (1978). Bacteria as Biogenic Sources of Freezing Nuclei, *J. Applied Meterology*, **17**, 1049–1053.
15. Pouleur S., Richard C., Martin J.-G., Antoun H. (1992). Ice Nucleation Activity in *Fusarium acuminatum* and *Fusarium avenaceum*, *Applied and Environmental Microbiology*, **58**(9), 2960–2964.
16. Warren G., Wolber P. (1991). Micro Review—Molecular Aspects of Microbial Ice Nucleation, *Molecular Microbiology*, **5**(2), 239–243.
17. Durham O. C. (1941). The Volumetric Incidence of Atmospheric Allergenes, I. Specific Gravity of Pollen Grams, *J. Allergy*, **14**(6), 455–461.
18. Dingle A. N. (1966). Pollen as Condensation Nuclei, *J. Rech. Atmosph.*, **2**, 231–237.
19. Markgraf V. (1980). Pollen Dispersal in a Mountain Area, *Grana*, **19** 127–146.
20. Whitehead D. R. (1983). *Pollination Biology*, Academic Press, New York.
21. Spendlove J. C. (1974). Industrial, Agricultural and Municipal Microbial Aerosol Problems, *Developments in Industrial Microbiology*, **15**, 20–27.
22. Matthias-Maser S., Jaenicke R. (1995). Size Distribution of Primary Biological Aerosol Particles with Radii $> 0.2\mu m$ in an Urban /Rural Influenced Region, accepted for publication in *J. Atmospheric Research*.
23. Bourdillion R. P., Colebrooke L. (1946). Air Hygiene in Dressing Rooms for Burns or Major Wounds, *Lancet*, April 1946, 561–605.
24. Noble W. C. (1971). Dispersal and Acquisition of Microorganisms, in: *Proc. of Int. Conf. of Nosocomial Infection*, edited by Brachman, P. S. Atlanta Ga, American Hospital Assoc. Chicago., Aug.
25. Blanchard D. C., Syzdek L. D. (1972). Concentration of Bacteria in Jet-Drops from Bursting Bubbles, *J. of Geophysical Research*, **77**, (27), 5087–5099.
26. Lindemann J., Upper C. D. (1985). Aereal Dispersal of Epiphytic Bacteria over Bean Plants, *Applied and Environmental Microbiology*, **50**(5), 1229–1232.
27. Rempe H. (1937). Untersuchungen über die Verbreitung des Blütenstaubes durch Luftströmungen, *Planta*, **27**, 93–147.
28. Hamilton E. D. (1959). Studies on the Air Spora, *Acta Allergologica* **XIII**, 143–175.
29. Gregory P. H. (1945). Dispersion of Airborne Spores, *Transac. British Mycological Soc.*, **28**, 26–72.

30. Rüden H., Thofern E., Fischer P., Mihm U. (1978). Airborne Micro-organisms: Their Occurrences, Distribution and Dependence on Environmental Factors, Especially on Organic Compounds, *Pageoph.*, **116**, 335–350.
31. Andersen A. A. (1958). A New Sampler for Collection, Sizing and Enumeration of Viable Airborne Particles, *J. Bacteriology*, **76**, 471–484.
32. Ogden E. C., Raynor G. S., Hayes J. V., Lewis D. M., Haines J. H. (1974). *A Manual for Sampling Airborne Pollen*, Hafner Press, New York.
33. Matthias-Maser S., Jaenicke R. (1994). Examination of the Atmospheric Bioaerosol Particles with Radii $r > 0.2\,\mu m$, *J. Aerosol Science* **25**(8), 1605–1613.
34. Reiß J. (1983). Vereinfachte mikrobiologische Untersuchungsverfahren, III. Quantitative Bestimmung von Keimzahlen in der Luft, *Mikrokosmos*, 80–85.
35. May K. R. (1966). Multistage Liquid Impinger, *Bact. Review*, **30**(3), 559–570.
36. Maser R. (1988). Entwicklung und Eichung eines Sedimentations-Impaktionsmonitors zur Messung der Konzentration von groen Aerosolpartikeln, Diplomarbeit, Institut für Meteorologie, University of Mainz.
37. Nevalainen A., Willeke, K., Liebhaber F., Pastuszka J., Burge H. Henningson E. (1993). Bioaerosol Sampling, in *Aerosol Measurement, Principles, Techniques, and Application*, edited by Willeke, K., Barron P. A., Van Nostrand Reinhold, New York, pp. 471–492.
38. Griffiths W. D., Upton S. L., Mark D. (1993). An Investigation into the Collection Efficiency and Bioefficiencies of a Number of Aerosol Samplers, *J. Aerosol Science*, **24**, Suppl. 1, 541–542.
39. Alexander M. P. (1980). A versatile Stain for Pollen, Fungi, Yeast, and Bacteria, *Stain Technology*, **55**(1), 13–18.
40. Sheldon J. M., Lovell R. G., Mathews K. P. (1967). *A Manual for Clinical Allergy*, W. B. Saunders, Philadelphia.
41. Wittekind D. H. (1983). On the Nature of Romanowsky-Giemsa Staining and its Significance for Cytochemistry and Histochemistry: An Overall View, *Histochemical Journal* **15**, 1029–1047.
42. Matthias S. (1987). Ein Beitrag zur biogenen Komponente des atmosphärischen Aerosols in Mainz, Diplomarbeit, Institut für Meteorologie, University of Mainz.
43. Bovallius A., Roffey R., Henningson E. (1980). Long- range Transport of Bacteria, *Annals N. Y. Academy of Sciences*, **353**, 1980, 186–200.
44. Gregory P. H. (1973). *The Microbiology of the Atmosphere*, Leonard Hill Books, Aylesbury, 377pp.
45. Janzon L. -A. (1981). Airborne Pollen Grains under Winter Conditions, *Grana*, **20**, 183–185.
46. Linskens H. F., Jorde W. (1986). Pollentransport in groen Höhen—Beobachtungen während der Fahrt mit einem Gasballon, *Allergologie*, **9**(2), 55–58.
47. Fulton, J. D. (1966). Microorganisms of the Upper Atmosphere, III. Relationship between Altitude and Micropopulation, *Applied Microbiology*, **14**(2), 237–240.
48. Peters K. (1995). Die biogene Komponente des atmosphärischen Aerosols in Mainz unter Berücksichtigung von Jahreszeit und Witterung, Diplomarbeit, Institut für Physik der Atmosphäre, University of Mainz.
49. Willeke K., Whitby K. T. (1975). Atmospheric Aerosol: Size Distribution Interpretation, *J. Air Pollution Control Association*, **25**(5), 529–534.
50. Jaenicke R., Davies C. N. (1976). The Mathematical Expression of Size Distributions of Atmospheric Aerosol, *J. Aerosol Science*, **7**, 255–259.
51. Jaenicke R. (1988). Aerosol Physics and Chemistry, *Landolt-Börnstein Zahlenwerte und Funktionen aus Naturwissenschaft und Technik*, vol. 4 *Meteorologie*, edited by Fischer G., Springer-Verlag, Berlin, pp. 391–457.

52. Dreiling V. (1992). Bestimmung von Aerosolgröenverteilungen in Raum und Zeit aus integralen Parametern. Dissertation, Institut für Physik der Atmosphäre, University of Mainz.
53. Matthias-Maser S., Brinkmann J., Schneider W. (1998). A Contribution of Primary Biological Aerosol Particles to the Marine Atmosphere over the Atlantic Ocean, *Atmos. Environ.*, submitted.

11 Formation of Aerosol Particles from Biogenic Precursors

C. NICHOLAS HEWITT and BRIAN M. DAVISON

Lancaster University, UK

1 INTRODUCTION

The emission of gases from the biosphere into the atmosphere is one of the most important of all the processes that operate in the earth–atmosphere biogeo-chemical system. The flux rates of many biogenically produced gases equal or exceed their emission rates from anthropogenic sources and, in some cases, these compounds play a dominant role in controlling the composition of the atmosphere and the resultant chemical and physical processes that take place therein. For example, biogenic emissions of reactive hydrocarbons far exceed their anthropogenic emissions. These compounds are believed to be oxidized in the atmosphere, in part to carbon monoxide, which itself rapidly reacts with the hydroxyl radical. Thus biogenic hydrocarbon emissions influence the concentrations of hydroxyl radicals, which in turn directly influence the lifetimes of practically all pollutant gases in the troposphere.

Interactions between the biosphere and the chemistry and physics of the atmosphere are exceedingly complex and an understanding beyond the simplistic, as illustrated above, is only now beginning to emerge. One area where some slight progress has been made, but in which considerable uncertainties remain, is the production of aerosol particles within the atmosphere from gas-phase precursors emitted by the biosphere. At least three groups of gases are important in this respect: (i) reduced sulfur gases, especially dimethyl sulfide; (ii) reactive hydrocarbons, especially isoprene (C_5H_8) and the C_{10} family of

Atmospheric Particles, Edited by R. M. Harrison and R. Van Grieken.
© 1998 John Wiley & Sons Ltd.

monoterpenes; and (iii) basic gases, especially ammonia. In addition to secondary particle formation from gas-phase precursors, the biosphere also acts as a direct, primary source of particles to the atmosphere, but this process is considered elsewhere in the present volume.

Several distinct areas have to be addressed when considering the role of a biologically derived gas in the context of secondary aerosol particle formation. First, the biogenic gas source type, distribution and seasonal behaviour must be identified since these will directly influence the spatial and temporal distribution of emissions. Second, an understanding of the processes underlying the production and emission of the gas is required if flux estimates under past, present and future environmental conditions are to be made. Third, the chemistry of the gas and of its reaction products in the atmosphere must be understood in order that aerosol particle production rates can be predicted. Fourth, the physical and chemical characteristics of the resultant particles must be known in order to predict their behaviour, lifetimes, distributions and effects in the atmosphere.

For each of the three groups of gases mentioned above there are very considerable uncertainties in each of these four areas. Indeed, these uncertainties are so great that our present understanding of this topic is restricted to the barest minimum and it is not yet clear whether even this is fundamentally correct. Given these limitations, an attempt is made here to outline our present knowledge in these areas.

The formation or growth of particles from gases can occur by a number of processes, either as homogeneous processes involving vapour phase species only, or heterogeneous processes involving both vapour and solid phases.

For a single chemical species to condense to form new particles requires it to be supersaturated in the vapour phase so that addition of gas molecules to the molecular clusters formed exceeds their removal, until a critical size is achieved. The greater the supersaturation the smaller the cluster size required to start particle formation. An input of vapour is required to maintain the vapour pressure if nucleation is to be sustained. The thermodynamics of what is now termed classical nucleation theory was originally investigated by Volmer and Flood. [1] Homogeneous nucleation involving more than one species is more likely to occur as it does not require supersaturation of the individual components but only of the mixture as a whole.

Where particles are already present in the atmosphere there exists the possibility for scavenging of low vapour pressure gases on to the surface of the pre-existing aerosol, or heterogeneous condensation. The surface of an aerosol can also act as a site for gas-phase reactions or, in some cases, (e.g. SO_2 oxidation) the reaction may occur within the aerosol droplet itself. With a sufficiently high concentration of existing particles this process may exceed homogeneous nucleation so leading to particle growth rather than formation of new nuclei.

Modelling calculations [2, 3] suggest the rate of nucleation appears to be strongly dependent on temperature, vapour phase pressure of the components and relative humidity. Variations in these parameters and the particle surface area available for reactions control whether particle formation or particle growth will occur as condensible material becomes available in the atmosphere [4]. For example, Ferek et al. [5] observed new particle formation in the Arctic in the region of low-level stratus clouds which had scavenged the accumulation mode aerosol, so reducing the surface area for heterogeneous condensation.

Rapid increases in particle number densities from a few hundred to several thousand per cm^3 in the remote marine boundary layer over a time-scale of hours has been observed in a number of locations. [5–7] The failure of a binary homogeneous nucleation model [8] to reproduce such findings led Raes [9] to postulate that these observations were due to free troposphere nucleations and subsequent entrainment of the particles into the boundary layer rather than in situ formation. This illustrates the difficulties in verifying current theories of aerosol formation processes by experimental field evidence.

2 SECONDARY PARTICLE FORMATION FROM REDUCED SULFUR GASES

Several reduced sulfur compounds are emitted in appreciable quantities by the biosphere but two, dimethyl sulfide (DMS) and carbon disulfide (CS_2), are believed to act as precursors to aerosol particles in the atmosphere.

Dimethyl sulfide is one of the breakdown products of dimethyl sulfoniopropionate (DMSP), a compound involved in regulating cellular osmotic pressure in marine algae. Once released into sea-water, the gas transfers across the water–air interface, as a result of its appreciable concentration gradient. Indeed, relative to its DMS concentration in air, ocean surface waters are typically supersaturated by two orders of magnitude [10]. This transfer process has been described by Liss and Slater [11] in terms of the concentration gradient and a transfer velocity, which itself depends upon wind speed. This, and a variety of methods based upon ambient air concentration measurements, has allowed estimates of the global flux of DMS to be made. However, the strong temporal and spatial variations in DMS concentrations in both sea-water and air render such global flux estimates uncertain. Current predictions are in the range 8–51 Tg (S)y^{-1} with a best estimate of about 16 Tg (S) y^{-1}. [12] As the number of reliable DMS measurements increase, so will the accuracy of its flux estimates improve.

The spatial variability in the concentration of DMS in sea-water is illustrated by measurements made in the Atlantic Ocean when a range of 17–1700 ng (S) l^{-1} with a mean of 90 g (S) l^{-1} was observed [13]. The seasonality in concentrations is most enhanced in temperate latitudes. For example, average concentrations of 130 and 20 g (S) l^{-1} have been observed in summer and winter respectively in the waters of the north Pacific [14]. However, concentrations are generally higher in the more productive coastal waters and in the vicinity of river estuaries where extreme patchiness is commonly observed. [15]

The biochemical processes underlying dimethyl sulfoniopropionate production, and hence resulting in DMS in sea-water, are understood, as is the variability in production rates from different species of phytoplankton. [16] What is not clearly known is how dependent emissions are on air and sea-water temperature, solar flux and changing sea-water ecology. This severely limits the ability to predict changes in DMS production and flux rates in the future.

Concentrations of DMS in air are, of course, much lower than in sea-water and show a greater variability. For example, measurements made at Cape Grim, Tasmania, were in the range 30–380 ng (S) m^{-3} with a mean of 160 ng (S) m^{-3}. [17] Those made in Atlantic air at a coastal site in north-west Scotland over a 2-year period were in the range < 1–885 ng (S) m^{-3}, with a summer mean of 110 ng (S) m^{-3} and a mean during the spring phytoplankton bloom period of 490 ng (S) m^{-3}. [18] Winter values of < 1–10 ng (S) m^{-3} were seen. Similar seasonal trends have been observed elsewhere.

A diurnal cycle in the concentration of DMS has been observed in the Pacific Ocean, with mean daytime minima and night-time maxima of 120 and 200 ng (S) m^{-3} respectively. [19] A similar diurnal cycle was seen in the trade winds of the Caribbean, with mean daytime minima and night-time maxima of 56 and 83 ng (S) m^{-3} respectively. [20] This cycle is attributed to the rapid oxidation and removal of DMS by reaction with the photochemically produced hydroxyl radical (OH). In clean air OH concentrations follow a strong diurnal pattern, with essentially zero concentrations during the hours of darkness.

In continentally influenced air masses DMS concentrations are generally lower than in maritime air, due to a lack of sources and an increase in oxidant concentrations. The presence of the nitrate radical (NO_3) in polluted air during the night and its rapid reaction with DMS [21] may result in a reduction in amplitude of the diurnal cycle of DMS concentrations in such air masses.

Vertical profile measurements of DMS in the atmosphere suggest a steady decline in concentration with altitude within the boundary layer, followed by a rapid decrease in the free troposphere above. For example, a decrease in DMS concentrations under stable meteorological conditions in the vicinity of Barbados has been found, from 100 ng (S) m^{-3} at sea-level to 60 ng (S) m^{-3} at cloud top level at 1 km, with a rapid decline to a few ng (S) m^{-3} at 2 km, above the boundary layer. [22] Under more turbulent convective conditions the DMS concentrations in the free troposphere were an order of magnitude higher. Similar results have been observed elsewhere (e.g. [23, 24]). The short atmospheric lifetime of DMS of a few days [21] prevents its transfer to the free troposphere other than by rapid convective uplift.

In the atmosphere, the major loss mechanism for DMS is by reaction with OH, itself formed by a two-step photochemical process:

$$O_3 + hv \rightarrow O(^1D) + O_2$$

$$O(^1D) + H_2O \rightarrow 2OH$$

The reaction of DMS with OH has a temperature-dependent branching mechanism, with hydrogen abstraction being favoured over oxygen addition at higher temperatures: [25]
(i) hydrogen atom abstraction:

$$CH_3SCH_3 + OHO_2 \rightarrow CH_3SCH_2O_2 + H_2O$$

(ii) oxygen addition:

$$CH_3SCH_3 + OHO_2 \rightarrow CH_3S(O)CH_3 + HO_2$$

At 298 K approximately 70% of the overall reaction proceeds by abstraction. Measurements from the Antarctic confirm this, with the ratio of the major aerosol-phase products of each of the two channels, methane sulfonic acid and sulfate, increasing in colder air. [4, 26]

The subsequent reactions of $CH_3SCH_2O_2$ and $CH_3S(O)CH_3$ have been studied in laboratory smog chamber experiments and by modelling, [25, 27, 28] leading to the postulation of detailed reaction schemes for DMS in the atmosphere. Such a scheme [27] has been validated by field measurements. [29]

Both field measurements and laboratory experiments suggest that sulfate aerosol (SO_4^{2-}) is the major product of DMS oxidation, with smaller amounts of other products being formed. This sulfate may be present in the atmosphere as sulfuric acid (H_2SO_4) droplets or as fully or partially neutralized ammonium sulfate (($NH_4)_2SO_4$) or ammonium hydrogen sulfate (NH_4HSO_4). Other products include methane sulfonic acid (MSA), also largely present in the aerosol phase, and gas-phase sulfur dioxide (SO_2), dimethyl sulfoxide (DMSO: CH_3SCH_3) and dimethyl sulfone (DMSO$_2$: $CH_3SO_2CH_3$).

Particles produced in gas-to-particle conversion reactions at ambient temperatures are usually found in the smaller (Aitken nuclei) size range, with diameters $\leqslant 0.08 \, \mu m$. Particles in this size range efficiently act as nuclei for the condensation of low vapour pressure gaseous species or rapidly coagulate, so forming accumulation mode particles in the range 0.08–2 μm. Aitken nuclei consequently have very short lifetimes in the atmosphere, whereas accumulation mode particles have longer residence times and can be incorporated into cloud droplets followed by their removal by wash-out or rain-out. Because of their relatively large mass, accumulation mode particles typically contribute 50% or more of the total aerosol mass in ambient air, but only 5% or so of the total particle number density.

The formation of aerosol MSA and non-sea-salt SO_4^{2-} from DMS and their effects on the density of cloud condensation nuclei (CCN) has been the focus of much interest since it was postulated that they may play a key role in regulating the earth's radiation balance by their effects on cloudiness and on cloud albedo [30]. This hypothesis depends on the following assumed connections, some of which are more certain than others:

1. DMS is emitted from phytoplankton and reaches the atmosphere, where it is converted into particles which act as CCN.
2. The number density of CCN controls the number density of drops in stratiform clouds, and this in turn controls the amount of solar radiation reflected by stratiform clouds.
3. The radiation balance of the earth is sensitive to the reflectivity of stratiform clouds (and hence to DMS concentrations in marine air).
4. Positive or negative feedbacks between temperature and DMS emissions occur, so allowing phytoplankton to regulate the earth's climate system.

Several attempts have been made to investigate these links, for example between DMS emissions and non-sea-salt sulfate concentrations [31] and between DMS and CCN concentrations. [32, 33] In addition, satellite data of cloud cover, sea surface temperature and chlorophyll concentrations have been used to look (indirectly) at the direct link between DMS production and cloud albedo, without consideration of the intermediate physical and chemical steps [34]. While evidence is available to support some, at least, of these links, [35] other aspects of the DMS-climate hypothesis are far from proven. In fact a major test of the hypothesis was posed by suggesting that anthropogenic emissions of SO_2 in the northern hemisphere (NH) have not altered the cloud albedo of the NH relative to that of the southern hemisphere. [36]

A confounding factor may be that emissions of carbon disulfide (CS_2) produce, by oxidation, a concentration of carbonyl sulfide (COS) in the troposphere of about 700 ng (S) m^{-3}. [37] COS has an atmospheric lifetime measured in years [21, 38] and is transferred across the tropopause into the stratosphere. There it is photolized and oxidized to form a layer of sulfate aerosol, which itself has effects on the earth's radiation balance and hence on global climate [39]. Another confounding factor is the likely large nonlinearity of the dependence of new CCN particle production on the concentrations of both the gaseous precursors and the pre-existing ambient aerosol. It seems possible, for example, that small variations in DMS emission rate may cause variation in CCN concentrations when both are low, but in polluted air much higher concentrations of SO_2 may have little effect on CCN formation. [40] Thus, although the formation of SO_4^{2-} and MSA particles from the oxidation of DMS is easily demonstrated in the laboratory, and circumstantial evidence linking the two is available from field measurements, the role of DMS in the formation of condensation nuclei and cloud condensation nuclei in ambient air, and hence in influencing climate, is not yet proven.

3 SECONDARY PARTICLE FORMATION FROM BIOGENIC HYDROCARBONS

Probably the first suggestion that biogenic emissions of reactive gaseous hydrocarbons could contribute to the substantial production of aerosol particles was made by Went [41] who proposed that the photochemical products of organic emissions from trees were responsible for the blue haze often seen in rural areas on sunny days. Since then it has been shown that plants and trees do in fact emit a very wide variety of organic compounds of differing reactivities, and that some of these form aerosol-phase oxidation products. However, whether or not these make a significant contribution to the ambient aerosol is still a matter of discussion.

A large body of data is now available describing the emissions of volatile organic compounds (VOCs) into the atmosphere from the biosphere. Qualitative information on emissions is available for more than 600 plant species, of which nearly 300 are believed to emit isoprene and more than 50 to emit one or more of the monoterpene compounds. [42, 43] Most research in this area has to date focused on isoprene (2-methyl-1, 3-butadiene, C_5H_8) and the C_{10} monoterpenes because of the known gas-phase reactivities of these compounds and their relative ease of identification and quantification in air samples by the use of capillary gas chromatography with flame ionization detection. However, about 400 different VOCs are known to be emitted by plants [44] including aldehydes, ketones, organic acids, alcohols, alkanes and alkenes and the C_{10}–C_{40} terpenoids. For example, it has been found that 3-hexen-1-ol and 3-hexenylacetate were the dominant species emitted by a range of agricultural crops. [45] Several similar compounds were found from an investigation of emissions from 12 urban shade trees. [46] Over 50 different species including alcohols, ketones, aldehydes, esters and ethers have recently been identified by GC-MS from 11 different crop and grass species typically grown in Austria. [47]

A biosynthetic pathway for monoterpene formation has been proposed but their precise functions are unclear. Some may act as pollinator attractors, herbicides or pesticides or they may simply be waste or by-products with no specific physiological role. The pathway of isoprene biosynthesis is poorly understood, although an isoprene enzyme has been identified and a link with photosynthesis loosely established. The role of isoprene is not known, but it may be a means of losing excess energy under conditions of high temperature and/or light intensity.

Both temperature and light intensity strongly affect the emission rate of isoprene from plants, [48–50] while terpene emissions are exponentially influenced by temperature. [51] Evidence for a light dependence of monoterpene emissions is now emerging. These dependencies, established by both laboratory and field studies, have allowed the formulation of algorithms describing emission rates in terms of light intensity (PAR) and temperature.

Plant development and phenology are also known to influence isoprene emissions, at least for some species. [52] However, the effects of water deficit, air pollutant and other stresses, disease and insect attack, timber and crop harvesting, fire and other catastrophic events, death and decay and increased ultraviolet light flux on VOC emissions are all largely unknown. [43]

Various attempts have been made to estimate VOC emission rates from the biosphere on local, regional and global scales. For example, Hewitt and Street considered that emissions from the United Kingdom might lie in the range 50–100 kt yr^{-1}, with emissions of monoterpenes dominating those of isoprene. [43] On the global scale a model has been developed with a highly resolved spatial grid ($0.5° \times 0.5°$) which generates hourly average emissions. [53] Chemical species are grouped into four categories: isoprene, monoterpenes, other reactive VOCs (ORVOCs) and other, less reactive, VOCs (OVOCs). Reactive VOCs are defined as those with a chemical lifetime in the atmosphere of less than 1 day. Ecosystem-specific biomass data and emission factors, as well as the temperature and light dependence of emissions, were used to estimate emission rates from plants. Emissions from the oceans were estimated as a function of transfer velocities, determined from a general circulation model, and ocean colour satellite data used as a surrogate for phytoplankton concentrations in seawater. The annual global flux of VOCs estimated by the model is about 1100 Tg C, of which 38% is predicted to be isoprene, 12% monoterpenes, 25% ORVOCs and 25% OVOCs. Tropical, drought-deciduous and savanna woods contribute about half of all emissions, with other woodlands, croplands and shrublands each contributing 10–20% of the global total. The emissions of isoprene in temperate regions are much higher than previously estimated. For example, emissions for the UK predicted by the model are 84 kt (C) year^{-1} for isoprene, 120 kt (C) year^{-1} for the terpenes and 80 kt (C) year^{-1} for OVOCs. However, it should be noted that the uncertainties associated with these estimates are at least a factor of 3.

The ambient concentrations of natural VOCs have been measured at a variety of field sites, representing areas of different vegetation types. Concentrations of isoprene, α-pinene, β-pinene, δ-3-carene, limonene, camphene and myrcene up to 10 ppbv have been observed in forested regions, where they may account for a majority of the total VOC present in air. However, considerable variability in concentrations, both temporally and spatially, is found (see [42] for a summary of concentration data).

Once emitted into the atmosphere biogenic VOCs take part in an extremely complex sequence of reactions, initiated by the very reactive, but reasonably long-lived, oxidant ozone and the extremely reactive transient hydroxyl (OH) and nitrate (NO_3) radicals. A huge variety of products, of varying reactivities and hence lifetimes, have been identified. [42] Although many of the details of these reactions are not yet established, a general understanding of isoprene and monoterpene oxidation has emerged, particularly with respect to their role in the formation and/or removal of ozone. However, the contribution that bio-genic VOCs make to the organic aerosol observed in the atmosphere is not clear.

Several investigations have demonstrated that a significant fraction of α-and β-pinene reacting with ozone and/or OH can form aerosol-phase products, while isoprene produces rather less. [54–58] Indeed, aerosol-phase pinonalde-hyde, one of the main products of the α-pinene–ozone reaction, has been detected in forest air. [55]

Pandis *et al.* studied the time dependence of the number density and size distribution of particles formed from isoprene and β-pinene oxidation at rea-listic initial precursor concentrations in a photochemical smog chamber. [57] The data obtained were used to estimate the possible contribution made by biogenic VOC emissions to aerosol production in three different environments. In Los Angeles, an urban area with 33% natural vegetation, up to 50% of the secondary organic aerosol may arise from natural precursors. In Atlanta, Georgia, with 57% tree cover and where monoterpene emission rates of 300 t day^{-1} have been estimated, [59] a 10% yield of aerosol carbon following the oxidation of natural VOCs would produce $\sim$ 30 t day^{-1} of secondary aerosol, compared with only 2 t day^{-1} from anthropogenic precursor VOCs. The third case studied was a coniferous forest with monoterpene emissions totalling 2 mg m^{-2} hour^{-1} and a mixing height varying between 100 m at night and 1200 m during the early afternoon. It was predicted that, under static conditions, secondary aerosol concentrations resulting from biogenic VOC emissions could reach several tens of μg (C) m^{-3}.

The amount of ozone and secondary aerosol-phase products generated by individual organic components in a complex photochemical mixture has been estimated. [60] The greatest aerosol yields in this mixture were predicted to be those of the monoterpenes (> 700 μg m^{-3} ppmv^{-1} for α- and β-pinene).

Although it seems from laboratory experiments and model calculations that the monoterpenes have the potential to contribute significantly to particle loadings in both urban and rural environments, field measurements suggest that the major fraction of the ambient aerosol is composed of sulfate, with nitrate, elemental carbon, crustal materials and organic matter attributable to the combustion of fossil fuels and other anthropogenic sources. Reconciliation of laboratory experiments and field measurements is still, therefore, not poss-ible. Several possible causes of this exist. Most smog chamber type experiments have been carried out at unrealistically high precursor concentrations, in sys-tems where wall effects cannot be avoided, and the paucity of kinetic and mechanistic data for the reactions of most organic compounds hinders extrapolation from the laboratory. In addition the role of organic nitrates to the ambient aerosol has largely been overlooked (but see, for example, [61–63]). Recently the size distribution and elemental composition of aerosol particles in the Amazon basin have been obtained and factor analysis used to demonstrate that primary and secondary biogenic aerosol particles are responsible for most of the ambient aerosol loading in the region. [64]

4 AMMONIA AS A PRECURSOR TO SECONDARY AEROSOL PARTICLES

Ammonia is the most important and abundant alkaline compound found in the atmosphere and it plays a key role in aerosol chemistry. Gaseous ammonia (NH_3) is released at the earth's surface primarily through the decomposition of organic matter, both from agricultural practices on land and from ocean surface waters. Direct industrial sources, including fossil fuel combustion, are probably relatively insignificant on a global scale. For example, emissions from the UK are estimated to total 450 kt (N) yr^{-1}, of which 370 kt yr^{-1} is from livestock. Industrial sources, principally sewage works and sludge spreading, coal combustion, landfill sites, incineration, pets and the production of ammonia and nitrogen fertilizers, total 44 kt yr^{-1}. [65]

Globally, a flux of 40×10^{12} g year^{-1} has been estimated for emissions from land. [66, 67] Emissions of ammonia from the oceans have been estimated from the rate of neutralization of sulfate aerosol and this gives a flux of ~ 10 mmol m^{-2}day^{-1} in areas of high chlorophyll concentration, or (10–80) $\times 10^{12}$ g yr^{-1} globally. [68]

There are rather few reliable measurements of the concentrations of ammonia in the troposphere, but they are thought to range from < 0.1 ppbv above the boundary layer to > 10 ppbv in the vicinity of sources. [69]

Ammonia is extremely soluble in water and is very reactive with acidic compounds. It is therefore very efficiently removed from the atmosphere by interactions with aqueous and acid aerosols. In maritime air, sulfate is normally fully or partially neutralized by ammonium, presumably as a direct result of ammonia emissions from the ocean surface. Only in the Southern Ocean in air masses derived from Antarctica has unneutralized sulfuric acid aerosol been observed in the absence of ammonium sulfate or ammonium hydrogen sulfate. In polar air, accumulation mode sulfate, present in the form of H_2SO_4, has been found nearly completely free of neutralization by ammonium, with ammonium to sulfate ratios of less than 0.1. In comparison, maritime non-sea-salt sulfate aerosol has a molar ratio of 0.8–1.0. [4]

Ammonia may also play an important role in determining secondary aerosol composition in urban areas. For example, the predominant lead-containing compounds found in urban air are $PbSO_4(NH_4)_2SO_4$ and $PbBrCl(NH_4)_2BrCl$.

These are believed to arise from reaction of PbBrCl, emitted by vehicles using leaded fuel, with ammonium sulfate present in the ambient aerosol [70] (see also chapter 3).

5 CONCLUSIONS

The role of gaseous precursors of biogenic origins in the formation of secondary aerosol particles in the atmosphere is not well understood. Certainly dimethyl sulfide and ammonia emissions from the oceans play a significant part in controlling the composition of the maritime aerosol, but whether or not they influence the global climate is not yet clear. Biogenic hydrocarbons from plants have the potential to form secondary particles and may cause a localized loss of visibility, but again their regional and global scale effects are not fully understood.

REFERENCES

1. Volmer, M. and H. Flood (1934). "Formation of droplets in vapours" *Phys. Chem., Abt. A*. **170** 273–285.
2. Jaecker-Voirol, A. and P. Mirabel (1989). "Heteromolecular nucleation in the sulfuric acid–water system." *Atmos. Environ.* **23A** 2053–2057.
3. Raes, F., A. Saltelli and R. V. Dingenen (1992). "Modeling formation and growth of $H_2SO_4 - H_2O$ aerosols—uncertainty analysis and experimental evalution." *J. Aerosol Sci.* **23** 759–771.
4. Davison, B. M., C. N. Hewitt, C. O'Dowd, J. A. Lowe, M. H. Smith, M. Schwikowski, U. Baltensperger and R. M. Harrison (1996). "Dimethyl sulfide, methane sulfonic acid and physicochemical aerosol properties in Atlantic air from the United Kingdom to Halley Bay". *J. Geophys. Res.*, **101**, 22855–22874.
5. Ferek, R. J., P. V. Hobbs, L. F. Radke, J. A. Herring, W. T. Sturges and G. F. Cota (1995). "Dimethyl sulfide in the arctic atmosphere." *J. Geophys. Res.* **100** 26093–26104.
6. Covert, D. S., V. N. Kapustin, P. K. Quinn and T. S. Bates (1992). "New particle formation in the marine boundary layer." *J. Geophys. Res.* **97** 20581–20589.
7. Hoppel, W. A., G. M. Frick, J. Fitzgerald and R. E. Larson (1994). "Marine boundary layer measurements of new particle formation and the effect non-precipitating clouds have on aerosol-size distribution." *J. Geophys. Res.* **99** 14443–14459.
8. Raes, F. and R. V. Dingenen (1992). "Simulations of condensation and cloud condensation nuclei from biogenic sulfur dioxide in the remote marine boundary layer." *J. Geophys. Res.* **97** 12901–12912.
9. Raes, F. (1995). "Entrainment of free tropospheric aerosols as a regulating mechanism for cloud condensation nuclei in the remote boundary layer." *J. Geophys. Res.* **100** 2893–2903.
10. Andreae, M. O. (1986). "The ocean as a source of atmospheric sulphur compounds" in P. Buat-Menand, Ed., *The Role of Air–Sea Exchange in Geochemical Cycling*. D. Reidel, Dordrecht, 331–362.

11. Liss, P. S. and P. G. Slater (1974). "Flux of gases across the air–sea interface". *Nature* **247** 181–184.
12. Rodhe H. and J. Langner (1993). "Atmospheric concentration of DMS and its oxidation products estimated in a global 3-D model" in Restelli, G. and G. Angeletti (eds.) *Dimethylsulfide Oceans, Atmosphere and Climate* Kluwer Academic Publishers, Dordrecht, 333–343.
13. Barnard, W. R., M. O. Andreae, W. E. Watkins, H. Bingermer and H. W. Georgii (1982). "The flux of dimethyl sulphide from the oceans to the atmosphere". *J. Geophys. Res.* **87** 8787–8793.
14. Bates, T. S., J. D. Cline, R. H. Gammon and S. R. Kelly-Hansen (1987). "Regional and seasonal variations in the flux of oceanic dimethyl sulphide to the atmosphere". *J. Geophys. Res.* **92** 2930–2938.
15. Turner, S. M., G. Malin and P. S. Liss (1989). "Dimethyl sulphide and (dimethyl-sulphonio)propionate in European coastal and shelf waters". *Biogenic Sulphur in the Environment.* ACS symposium series 393.
16. Liss, P. S., G. Malin, S. M. Turner and P. M. Holligan (1994). "Dimethyl sulphide and Phaeocystis: a review". *J. Marine Systems* **51**, 41–53.
17. Andreae, M. O., R. J. Ferek, F. Bermond, K. P. Byrd, R. T. Engstrom, S. Hardin, P. D. Houmere, F. LeMarrec and H. Raemdonck (1985). "Dimethyl sulphide in the marine atmosphere". *J. Geophys. Res.* **90** 12891–12900.
18. Davison, B. M. and C. N. Hewitt (1992). "Natural sulphur species from the N. Atlantic and their contribution to the UK sulphur budget". *J. Geophys. Res.* **97** 2475–2488.
19. Andreae, M. O. and H. Raemdonck (1983). "Dimethyl sulphide in the surface ocean and the marine atmosphere: a global view". *Science* **211** 744–747.
20. Saltzman, E. S. and D. J. Cooper (1988). "Shipboard measurements of atmospheric dimethyl sulphide and hydrogen sulphide in the Caribbean and Gulf of Mexico". *J. Atmos. Chem.* **14** 191–209.
21. Hewitt, C. N., and B. M. Davison (1988). "Review: The lifetimes of organosulphur compounds in the troposphere". *App. Organomet. Chem.* **2** 407–415.
22. Ferek, R. J., R. B. Chatfield and M. O. Andreae (1986). "Vertical distribution of dimethyl sulphide in marine atmosphere". *Nature* **320** 514–516.
23. Luria, M., C. C. van Valin, D. L. Wellman and R. F. Pueschel (1986). "Contribution of the Gulf area natural sulphur to the North American sulphur budget". *Environ. Sci. Technol.* **20** 91–95.
24. Andreae, M. O., H. Berresheim, T. W. Andreae, M. A. Kritz, T. S. Bates and J. T. Merrill (1988). "Vertical distribution of dimethyl sulphide, sulphur dioxide, aerosol ions and radon over the Northeast Pacific Ocean". *Atmos. Chem.* **6** 149–173.
25. Hynes, A. J., P. H. Wine and D. H. Semmes (1986). "Kinetics and mechanism of OH reactions with organic sulphides". *J. Phys. Chem.* **90** 4148–4156.
26. Berresheim, H. (1987). "Biogenic sulphur emissions from the subantarctic and Antarctic oceans". *J. Geophys. Res.* **92** 13245–13262.
27. Yin, F., D. Grosjean and J. H. Seinfeld (1990). "Photooxidation of dimethyl sulphide and dimethyl disulphide. 1: mechanism development". *J. Atmos. Chem.* **11** 309–364.
28. Koga, S. and H. Tanaka (1993). "Numerical study of the oxidation process of dimethylsulfide in the marine atmosphere". *J. Atmos. Chem.* **17** 201–228.
29. Davison, B. M. and C. N. Hewitt (1994). "Elucidation of the tropospheric reactions of biogenic sulfur species from a field measurement campaign in NW Scotland". *Chemosphere* **28** 543–557.

30. Charlson, R. J., J. E. Lovelock, M. O. Andreae and S. G. Warren (1987). "Oceanic phytoplankton, atmospheric sulphur, cloud albedo and climate". *Nature* **326** 655–661.
31. Ayers, G. P. and J. L. Gras (1991). "Seasonal relationships between cloud condensation nuclei and aerosol methane sulphonate in marine air". *Nature* **353** 834–835.
32. Hegg, D., L. Radke and P. Hobbs (1991). "Measurements of Aitken nuclei and cloud condensation nuclei in the marine atmosphere and their relationship to the DMS–cloud–climate hypothesis". *J. Geophys. Res.* **96** 18727–18733.
33. O'Dowd, C., J. A. Lowe, M. H. Smith, B. M. Davison, C. N. Hewitt and R. M. Harrison (1997). "Sulfate CCN and biogenic sulfur emissions in and around Antarctica." *J. Geophys. Res.* **102**, 12839–12854.
34. Falowski, P., Y. Kim, Z. Kolber, C. Wilson, C. Wirick and R. Cess (1992). "Natural versus anthropogenic factors affecting low-level cloud albedo over the North Atlantic". *Science* **256** 1311–1313.
35. Hobbs, P. V., D. A. Hegg and R. J. Ferek (1993). "Recent field studies of sulfur gases: particles and clouds in clean marine air and their significance with respect to the DMS–cloud–climate hypothesis" in Restelli, G. and G. Angeletti (Eds.) *Dimethylsulphide: Oceans, Atmosphere, and Climate*, Kluwer, Dordrecht, 345–353.
36. Schwartz, S. E. (1988). "Are global cloud albedo and climate controlled by marine phytoplankton?" *Nature* **336** 441–445.
37. Crutzen, P. (1976). "The possible importance of CSO for the sulfate layer of the stratosphere". *Geophys. Res. Lett.* **3** 73–76.
38. Johnson, J. and H. Harrison (1986). "Carbonyl sulfide concentrations in the surface waters and above the Pacific Ocean". *J. Geophys. Res.* **91** 7883–7888.
39. Carroll, M. A. (1985). "Measurements of COS and CS_2 in the free troposphere". *J. Geophys. Res.* **90** 10483–10486.
40. Charlson, R. J. (1993). "Gas-to-particle conversion and CCN production" in Restelli, G. and G. Angeletti (Eds.) *Dimethylsulphide: Oceans, Atmosphere, and Climate* Kluwer, Dordrecht, pp. 275–286.
41. Went, F. W. (1960). "Organic matter in the atmosphere, and its possible relation to aerosol formation". *Proc. Natl. Acad. Sci. U.S.A.* **46** 212–221.
42. Fehsenfeld, F., J. Calvert, R. Fall, P. Goldan, A. B. Guenther, C. N. Hewitt, B. Lamb, S. Liu, M. Trainer, H. Westberg and P. Zimmerman (1992). "Emissions of volatile organic compounds from vegetation and the implications for atmospheric chemistry". *Global Biogeochemical Cycles* **6** 389–430.
43. Hewitt, C. N. and R. A. Street (1992). "A qualitative assessment of the emission of non-methane hydrocarbon compounds from the biosphere to the atmosphere in the U.K.: present knowledge and uncertainties". *Atmos. Environ.* **26A** 3069–3077.
44. Graedel, T. E. (1979). "Terpenoids in the atmosphere". *Rev. Geophys.* **17** 937–947.
45. Arey, J., A. M. Winer, R. Atkinson, S. M. Aschmann, W. D. Long and C. L. Morrison (1991). "The emission of (z)-3-hexen-1-ol, (z)-3-hexenylacetate and other oxygenated hydrocarbons from agricultural plant species". *Atmos. Environ.* **25A** 1063–1075.
46. Corchnoy, S. B., J. Arey and R. Atkinson (1991). "Hydrocarbon emissions from twelve urban shade trees of the Los Angeles, California, air basin". *Atmos. Environ.* **26B** 339–348.
47. Koenig, G. M. Brunda, H. Puxbaum, C. N. Hewitt, S. C. Duckham and J. Rudolph (1995). "Relative contribution of oxygenated hydrocarbons to the total biogenic VOC emissions of selected mid-European agricultural and natural plant species". *Atmos. Environ.* **29** 861–874.

48. Tingey, D. T., M. Manning, L. C. Grothaus and W. F. Burns (1979). "The influence of light temperature on isoprene emission rates from live oak". *Physiol. Plant.* **47** 112–118.

49. Isodorov, V. A., I. G. Zenkevich and B. V. Ioffe (1985). "Volatile organic compounds in the atmosphere of forests". *Atmos. Environ.* **19** 1–8.

50. Monson, R. K. and R. Fall (1989). "Isoprene emission from aspen leaves: Influence of environment and relation to photosynthesis and photorespiration". *Plant Physiol.* **90** 267–274.

51. Lamb, B., H. Westberg and G. Allwine (1985). "Biogenic hydrocarbon emissions from deciduous and coniferous trees in the United States", *J. Geophys. Res.* **90** 2380–2390.

52. Grinspoon, J., W. D. Bowman and R. Fall (1991). "Delayed onset of isoprene emissions in developing velvet bean leaves". *Plant Physiol.* **97** 170–174.

53. Guenther, A., C. N. Hewitt, D. Erickson, R. Fall, C. Geron, T. Graedel, P. Harley, L. Klinger, M. Lerdau, W. A. McKay, T. Pierce, B. Scholes, R. Steinbrecher, R. Tallamraju, J. Taylor and P. Zimmerman (1995). "A global model of natural volatile organic compound emissions". *J. Geophys. Res.* **100** 8873–8892.

54. Kamens, R., M. Gery, H. Jeffries, M. Jackson and E. Cole (1982). Ozone–isoprene reactions: product formation and aerosol potential". *Int. J. Chem. Kin.* **10** 955–975.

55. Yokouchi, Y. and Y. Ambe (1985). "Aerosols formed from the chemical reaction of monoterpenes and ozone". *Atmos. Environ.* **19** 1271–1276.

56. Hatakeyama S., K. Izumi, T. Fukuyama and H. Akimoto (1989). "Reactions of ozone with α-pinene and β-pinene in air: yields of gaseous and particulate products." *J. Geophys. Res.* **94** 13013–13024.

57. Pandis, S. N., S. E. Paulson, J. H. Seinfeld and R. C. Flagan (1991). "Aerosol formation in the photooxidation of isoprene and β-pinene". *Atmos. Environ.* **25A** 997–1008.

58. Zhang, S-H., M. Shaw, J. H. Seinfeld and R. C. Flagan (1992). "Photochemical aerosol formation from α-pinene and β-pinene". *J. Geophys. Res.* **97** 20717–20729.

59. Chameides, W. L., R. W. Lindsay, J. Richardson and C. S. Kiang (1988). "The role of biogenic hydrocarbons in urban photochemical smog: Atlanta as a case study". *Science* **241** 1473–1475.

60. Bowman, F. M., C. Pilinis and J. H. Seinfeld (1995). "Ozone and aerosol productivity of reactive organics". *Atmos. Environ.* **29** 579–590.

61. Palen, E. J., D. T. Allen, S. N. Pandis, S. Paulson, J. H. Seinfeld and R. C. Flagan (1993). "Fourier transform infrared analysis of aerosol formed in the photooxidation of 1-octene". *Atmos. Environ.* **27A** 1471–1477.

62. Mylonas, D. T., D. T. Allen, S. H. Ehrmann and S. E. Pratsinis (1991). "The sources and size distribution of organonitrates in Los Angeles aerosol". *Atmos. Environ.* **25A** 2855–2861.

63. Nielsen, T., A. H. Egelø, K. Granby and H. Skov (1994). "Atmospheric occurrence of particulate organic nitrates". Poster presented at the Eurotrac Symposium 1994.

64. Artaxo, P. and H.-C. Hansson (1995). "Size distribution of biogenic aerosol particles from the Amazon Basin". *Atmos. Environ.* **29** 393–402.

65. Sutton, M. A., C. J. Place, M. Eager, D. Fowler and R. I. Smith (1995). "Assessment of the magnitude of ammonia emissions in the United Kingdom". *Atmos. Environ.* **29** 1393–1411.

66. Wheeler, P. A., D. L. Kirchman, M. R. Landry and S. A. Kokkinakis (1989). "Diel periodicity in ammonium uptake in the oceanic subarctic Pacific: Implications for interactions in microbial food webs". *Limnol. Oceanog.* **34** 1025–1033.

67. Dawson, G. A. (1977). "Atmospheric ammonia from undisturbed land". *J. Geophys. Res.*, **21** 3125–3133.
68. Clarke, A. D. and J. N. Porter (1993). "Pacific marine aerosol: equatorial gradients in sulfate, ammonium and chlorophyll" in Restelli, G. and G. Angeletti, (Eds.) *Dimethylsulphide: Oceans, Atmosphere, and Climate*: Kluwer, Dordrecht, pp. 287–296.
69. Ziereis, H. and F. Arnold (1986). "Gaseous ammonia and ammonium ions in the free troposphere". *Nature* **321** 503–505.
70. Biggins, P. D. E. and R. M. Harrison (1979). "Identification of lead compounds in urban air". *Environ. Sci. Technol.* **13** 558–565.

12 Source Inventories for Atmospheric Trace Metals

JOZEF M. PACYNA

Norwegian Institute for Air Research, Kjeller, Norway

Atmospheric Particles, Edited by R. M. Harrison and R. Van Grieken.
© 1998 John Wiley & Sons Ltd.

1 INTRODUCTION

Many trace metals may create adverse effects on environmental and human health due to their toxicity and bioaccumulation in various environmental compartments. Whether these effects occur depends on the concentration levels of trace metals and their chemical forms in various compartments of the environment, as well as on the chemical, physical and biological conditions of a given environmental compartment.

A number of studies have been carried out to assess behaviour of trace metals in the environment. In general, these studies concluded that:

—many trace metals are ubiquitous in various raw materials, such as fossil fuels and metal ores, as well as in industrial products,

—some trace metals evaporate entirely or partially from raw materials during the high-temperature production of industrial goods, combustion of fuels, and incineration of municipal and industrial wastes, entering the ambient air with exhaust gases,

—releases to other environmental compartments (e.g. spills to water bodies, landfills, sewage lagoons, holding ponds) may also result in volatilization and entrainment of several trace metals,

—while emitted to the atmosphere, trace metals are subject to transport within air masses and migration through the ecosystem which cause perturbations of their geochemical cycles not only on a local scale but also on regional and even global scales,

—long-range transport of trace metals may also occur through interactions of these pollutants with other environmental compartments, e.g. through ocean currents,

—deposition of trace metals in areas surrounding the emission sources, as well as *en route* deposition during their long-range transport has reached values which in certain regions had exceeded the maximum permissible values,

—while entering the terrestrial and aquatic environments in the emission regions as well as far away from them, trace metals accumulate in soil and water reservoirs creating environmental problems known as "ecological bomb" areas, and

—uptake of trace metals by terrestrial and aquatic organisms and their metabolism leads to biomagnification of heavy metals in the environment.

Although numerous studies have been carried out on the assessment of the trace metal impact on the individual compartments of the environment, much less is known on fluxes of these pollutants. In general, it has been agreed that trace element fluxes can originate either from natural or anthropogenic sources. Natural sources are related primarily to the geological presence of trace metals in the crustal material and are brought to the air during various physical, chemical, biological and meteorological processes with no direct influence of humans. Release of trace metals during various industrial production processes and disposal of wastes is regarded as anthropogenic emissions. Obviously, it is often difficult to differentiate between the natural and anthropogenic origin of trace metals, particularly when those such as mercury have a potential of re-evaporation after being deposited into the aquatic or terrestrial surfaces. One measure of the anthropogenic origin of atmospheric trace metals in a given receptor is the estimation of enrichment factors (EFs) of trace metals. These factors present concentration increase of a given metal in the ambient aerosol, compared to its concentrations in certain reference material, such as crustal rocks or soils. This relation is standarized with the use of information on concentrations of certain reference metals, such as Al, Si, Ti or Sc. Thus, the enrichment factor of the element X can be estimated using the formula

$$\mathrm{EF}(x) = (X/\mathrm{Ref})\mathrm{aerosol}/(X/\mathrm{Ref})\mathrm{reference\ material}$$

where X/Ref is the concentration ratio of element X to a reference metal. Values of EF near unity suggest that crustal erosion is the primary source of element X and that its geochemical cycle has not been altered by emissions from anthropogenic sources. Values much higher than unity, e.g. higher than 100, imply the importance of sources other than crustal erosion. These sources include mostly various anthropogenic activities. However, some natural processes, such as volcanic eruptions, biogenic processes in sea-water, and forest fires can also contribute to this enrichment of trace metal concentrations. Besides, some trace

Table 1. Enrichment factors of trace metals on particles < 1.0 µm diameter in the Norwegian Arctic during winter episodes of long-range transport of air pollution

Element	Enrichment factor (Ti, earth crust)	Element	Enrichment factor (Ti, earth crust)
Ag	200–1000	Ni	30–80
As	700–3000	Pb	2500–4000
Cd	500–6000	Sb	1000–3200
Cr	30–100	Se	3700–36 000
Cu	30–100	V	35–50
Ga	8–20	W	20–130
In	40–200	Zn	300–900
Mo	60–150		

Table 2. World-wide atmospheric emissions of trace metals from natural sources (in 10^3 t year^{-1}) after Nriagu [2]

Source category	As	Cd	Co	Cr	Cu	Hg	Mn	Mo	Ni	Pb	Sb	Se	V	Zn
Wind-borne soil particles	0.3–5.0	0–0.4	0.6–7.5	3.6–50.0	0.9–15.0	0–0.1	42–400.0	0.1–2.5	1.8–20.0	0.3–7.5	0.1–1.5	0–0.4	1.2–30.0	3.0–35.0
Sea-salt spray	0.2–3.1	0–0.1	0–0.1	0–1.4	0.2–6.9	–	0–1.7	0–0.4	0–2.6	0–2.8	0–1.1	0–1.1	0.1–7.2	0–0.9
Volcanoes	0.2–7.5	0.1–1.5	0–1.9	0.8–29.0	0.9–18.0	0–2.0	4.2–80.0	0–0.8	0.9–28.0	0.5–6.0	0–1.4	0.1–1.8	0.2–11.0	0.3–19.0
Wild forest fires	0–0.4	0–0.2	0–0.6	0–0.2	0.1–7.5	0–0.1	1.2–45.0	0–1.1	0.1–4.5	0.1–3.8	0–0.5	0–0.5	0–3.6	0.3–15.0
Biogenic processes	0.4–7.5	0–1.7	0–1.3	0.1–2.2	0.1–6.4	0–2.7	4.1–55.5	0–1.0	0.1–1.7	0–3.4	0–1.3	0.6–14.3	0.1–2.4	0.4–16.0
Total	1.1–23.5	0.1–3.9	0.6–11.4	4.5–82.8	2.2–53.8	0–4.9	51.5–582.2	0.1–5.8	2.9–56.8	0.9–23.5	0.1–5.8	0.7–18.1	1.6–54.2	4.0–85.9

metals, such as Hg and Se, are at least partly emitted in a gas phase and the EF method is then difficult to use. Enrichment factors for a few trace metals, calculated with their concentrations measured at Ny Ålesund, Spitsbergen are presented in Table 1 for example. Very high values of EFs for Ag, As, Cd, Pb, Sb, Se and Zn indicate the anthropogenic origin of these pollutants, at least at remote locations.

The aim of this chapter is to present major source categories for atmospheric trace metals and to assess metal fluxes from these sources. Natural and anthropogenic sources are considered separately. Methods of emission estimates for trace metals are described and major regional and global emission inventories are reviewed.

2 NATURAL SOURCES

An assessment of trace metal emissions from natural sources is needed in order to discuss the extent of regional and global contamination of the environment by these pollutants. It is generally assumed that the principal natural sources of trace metals include wind-borne soil particles, volcanoes, sea-salt spray, and wild forest fires. A few emission inventories became available at the end of the 1970s attempting to summarize our knowledge on fluxes of trace metals from these sources (e.g. [1]). Then years later Nriagu [2] published a paper on global emissions of trace metals from natural sources on the basis of information presented in the 1970 reviews and new data which became available during the 1980s. A short summary of Nriagu's work, still considered as the state of the art, is presented in Table 2.

Soil-derived dust accounts for over 50% of the total Cr, Mn and V emissions, as well as for 20–30% of the Cu, Mo, Ni, Pb, Sb and Zn released annually to the atmosphere. Extensive studies on the formation and transport of wind-blown dust make this source relatively well known, compared to other natural sources. Several cases of long-range transport of the Saharan dust have been documented by direct measurements. [3] The description of generation mechanisms of soil dust and its entrainment in the airstream is well known. The Saharan desert seems to play an important role in the northern hemispheric dust cycle, providing about half of it. [4] The chemical composition of the Saharan dust was found to be quite constant (e.g. [5]) making the estimates of the trace metal amounts formed and transported with the Saharan dust quite reliable. Long-range transport of Asian dust was also studied (e.g. [5]). For example, the mass of Asian desert dust transported into the Arctic region was found as a major contributor to the total Arctic aerosol (e.g. [6]). The chemical composition of the Asian dust was also measured, making it possible to assess the flux of trace metals from this source. The above-mentioned flux estimates were used to present the emission data in Table 2.

Volcanic emanations can account for 40–50% of the total natural Cd and Hg and 20–40% of the total natural As, Cr, Cu, Ni, Pb and Sb emitted annually. Data on trace metal emissions during both volcanic eruption and venting in periods between eruptions are collected throughout the world; however, the metal concentrations in volcanic gases and particles are much more variable than the concentrations in the soil dust from various regions in the world. Therefore, assessments of emission fluxes of trace metals from volcanoes are much less accurate than the estimates of fluxes with soil dust.

Recent studies have shown that particulate organic matter is the dominant component of atmospheric aerosols in non-urban areas (e.g. [7, 8]) and that over 60% of the airborne trace metals in forested regions can be attributed to aerosols of biogenic origin (e.g. [9]). Indeed, Nriagu [2] estimated that biogenic sources contribute, on average, over 50% of Se, Hg and Mo, and from 30 to 50% of the As, Cd, Cu, Mn, Pb and Zn, to the total atmospheric emissions from all natural sources. Selenium in the marine aerosol, similarly to sulfur, can originate from its gaseous precursors as a result of gas-to-particle conversions. Methylation processes in the aquatic and terrestrial environments result in the re-emission of Hg on fine particles. The estimates of trace metal fluxes from biogenic activities are even less accurate than the estimates of the volcanic fluxes.

Sea-salt aerosols seem to account for < 10% of atmospheric trace metals from natural sources. The production mechanisms for sea-spray aerosols are quite well known (e.g. [10]). Two of these mechanisms are important for release of trace metals with sea-salt: "bubble bursting", important primarily for Cd, Cu, Ni, Pb and Zn, and gas exchange, important for As. The calculations for bubble bursting are often carried out using the trace metal concentrations in surface ocean waters and the enrichment of trace metals in the atmospheric sea-salt particles. The potential significance of As vapour emissions is usually assessed from ambient measurements of the metal in marine aerosols.

In certain parts of the world forest fires are the major emission sources of particles and particle-containing trace metals. In general, more than 10% of atmospheric Cu, Pb and Zn from natural sources can originate from this source. [2] However, the accuracy of emission fluxes of trace metals is rather low and can be compared to the accuracy of flux estimates for biogenic sources.

In summary, atmospheric fluxes of trace metals from natural sources can be quite significant even on a global scale. This significance can be even bigger in certain regions, e.g. in the area of volcanoes or forest fires. However, little has been done over the past two decades to improve the quality of flux estimates. Thus, these estimates are often inaccurate and difficult to compare with the estimates of anthropogenic fluxes.

3 ANTHROPOGENIC SOURCES

Energy generation, various industrial processes, the use of metals, disposal of wastes, and vehicular traffic have brought a serious increase of trace metal emissions to the atmosphere during the last few decades.

3.1 ENERGY GENERATION

Combustion of fossil fuels to produce electricity and heat is the main source of anthropogenic emissions of atmospheric Be, Co, Hg, Mo, Ni, Sb, Se, Sn and V [11] and an important source of As, Cr, Cu, Mn and Zn.

Electric power stations emit half of the total amounts of trace metals generated by the combustion of fossil fuels. Most of these emissions arrive from conventional thermal power plants. Many of them are so-called co-generation power plants, producing both the electricity and heat. In general, the amount of emissions from a conventional thermal power plant depends on: [12]

—the content of trace metals in the fuels,
—the physical and chemical properties of trace metals during combustion,
—technological conditions of a burner, and
—the type and efficiency of emission control equipment.

3.1.1 Affinity of Trace Metals for Pure Coal and Mineral Matter

More than 50 years ago it was suggested that various chemical compounds included in coal have either a high organic or inorganic affinity. In the meantime several studies have been carried out to define this affinity for individual trace metals to greater degree of detail as it was realized that this property has a major impact on the concentrations and chemical forms of the metals measured during coal combustion (e.g. [13]). The following conclusions have been drawn from these studies. Trace metals can be described as:

1. Associated with the organic fraction of coal, including organic sulfur, Br, Ge, Be, Sb and B.
2. Mainly associated with the inorganic fraction, including the sulfide-forming metals Zn, As, Cd, Fe, Zr, Hg, Pb, Hf, Mn and pyritic sulfur.
3. Metals that could be associated with either or both fractions, including Al, Si, Ti, V, Mo, K, P, Ga, Ca, Cr, Co, Ni, Cu, Mg and Se.

The association of this last group of metals is highly variable and intermediate. From among these metals, P, Ga, Ti and V tend to be associated with the organic fraction while Co, Ni, Cr, Se and Cu are mostly allied with the other metals having inorganic affinity.

Affinity of trace metals for pure coal and mineral matter has been an important factor when discussing their content in various types of coals. Although there are large differences between metal concentrations in the same type of coals from various coalfields reaching a few orders of magnitude, data from the literature (e.g. [13, 14]) suggest that lignite and subbituminous coals are less contaminated by the inorganic associated trace metals than medium-, low- and high-volatility bituminous coals. For example, coal from the Western basin in the United States is less contaminated than the coals from Eastern and Illinois basins. In Europe, German brown coal is on average cleaner than bituminous and subbituminous coals from the Czech Republic and Poland (e.g. [14, 15]).

3.1.2 Behaviour of Trace Metals during Combustion of Fossil Fuels

Several studies have been carried out to describe volatilization of trace metals during combustion of fossil fuels and following condensation on fine particles or on the surfaces of fly and bottom ashes (e.g. [15]). Combustion temperature in the boiler is one of the key parameters affecting the amounts of trace metals released. The higher the temperature in a boiler, the larger the discharges of volatile metals. For example, larger amounts of the inorganic associated trace metals are emitted into the air from conventionally fired boiler systems, burning fuel at temperatures higher than 1650 K, as compared with fluidized-bed systems with temperatures about 1100 K. The volatilization–condensation models for various trace metals are well described in the literature (e.g. [16]).

The load of the burner affects the emissions of trace metals in such a way that for low load and full load emissions are the largest, while for a 50% load the emission rate can be less than the maximum rate by a factor of 2.

3.1.3 Impact of Control Equipment on Physical and Chemical Forms of Trace Metals

Trace metals contained in exhaust gases are treated by various types of control equipment used to reduce the quantity of gaseous and particulate pollutants. Different types of control equipment result in alterations of chemical but primarily physical forms of trace elements as various control systems are quite selective in removal of fine particles. As a result of these alterations, trace metals differ with respect to penetration rate through the control equipment. Two control systems are the most frequently used to remove particles from the exhaust gases in coal-fired power plants: electrostatic precipitators (ESPs) and wet scrubbers.

The ESPs are very efficient at removing all types of particles < 0.01 μm diameter and can tolerate gases with temperature as high as 450 K. [17] Concerning the particle size distribution from the ESPs in coal-fired power plants, it has been concluded that the particle mass containing trace metals are concentrated mostly in two size ranges: (i) at *ca.* 0.15 μm diameter, and (ii) between 2 and 8 μm diameter (e.g. [18]).

The removal of particles (and trace metals with them) by wet scrubbers is practically independent of temperature, although high operating temperatures increase the water consumption and result in the formation of a steam plume. Generally, wet scrubbers are more efficient at removing trace metals from exhaust gases in coal-fired power plants than ESPs (e.g. [19]).

Fabric filters are sometimes used in coal-fired power plants. The collection efficiency is always very high, and even for particles of 0.01 μm diameter exceeds 99%. However, the lifetime of fabric filters is very dependent upon the working temperature and their resistance to chemical attack by corrosive elements in exhaust gases. The temperature of exhaust gases often exceeds the temperature tolerance for fabric filter material and therefore limits the fabric filter application. If used, a bimodal particle size distribution is observed at the outlet of the installation.

Flue gas desulfurization (FGD) installations also remove trace metals from exhaust gases. Most of the research in this field was directed towards the removal of mercury. The results suggest that between 20 and 60% of gaseous mercury in the exhaust gases from coal-fired power plants can be removed through the desulfurization installations. The efficiency of Hg removal depends mostly on the type of the installation (e.g. [20]). The decrease of trace metal concentrations in exhaust gases after passing through the FGD units is shown in Figure 1 based on the literature review of data from Japan, the United States, and the former Federal Republic of Germany. It can be concluded that a one order of magnitude decrease in concentrations after passing the emission control installation was measured for some metals, and particularly As, Cd, Pb and Zn.

3.2 INDUSTRIAL PROCESSES

The largest emissions of atmospheric As, Cd, Cu, In and Zn arrive from the pyrometallurgical processes employed in the production on non-ferrous metals, such as lead, copper and zinc. The type of technology employed in smelters, refineries and other operations, such as roasting, the content of trace metals in ores and scrap material and the type and efficiency of emission control equipment are the most important parameters affecting the quantity of these emissions.

3.2.1 Non-ferrous Metal Manufacturing

The production of copper is briefly described here to illustrate generation of trace metal emissions from non-ferrous manufacturing.

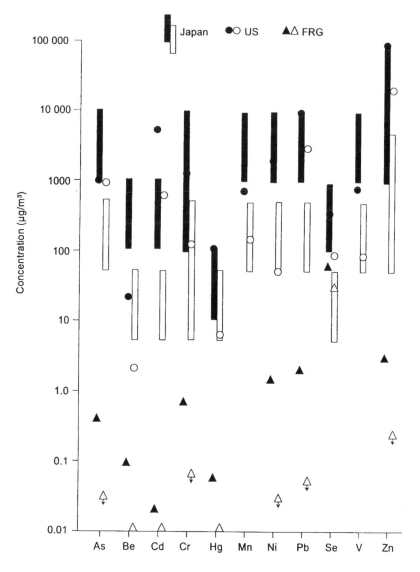

Figure 1. Results of investigations carried out in Japan, the United States and the former Federal Republic of Germany to assess the efficiency of trace element removal by wet FGD systems. The solid bars, circles and triangles represent concentrations before the FGD system.

In the traditional pyrometallurgical copper smelting process, the blister copper is refined in an anode furnace, cast into "anodes" and sent to an electrolytic refinery for further impurity elimination. The currently used copper smelters process ore concentrates by drying them in fluidized bed driers and then converting and refining the dried product in the same manner as the traditionally used process.

Concentrates usually contain 20–30% Cu. In roasting, charge material of copper mixed with a siliceous flux is heated in air to about 920 K, eliminating 20–50% of sulfur and portions of volatile trace metals. The roasted product, calcine, serves as a dried and heated charge for the smelting furnace.

In the smelting process, calcines are melted with siliceous flux in a flash smelting furnace to produce copper matte, a molten mixture of cuprous sulfide, ferrous sulfide and some trace metals. Matte contains usually between 35 and 65% of copper. Heat required in the smelting process comes from partial oxidation of the sulfide charge and from burning external fuel. Several smelting technologies are currently used in the copper industry, including reverberatory smelting, flash smelting (two processes are currently in commercial use: the INCO process and the OUTOKUMPU process), and the Noranda and electric processes.

In the reverberatory process heat is supplied by combustion of oil, gas or pulverized coal. The temperature in the furnace can reach 1800 K. Flash furnace smelting combines the operations of roasting and smelting to produce a high-grade copper matte from concentrates and flux. Most of flash furnaces use the heat generated from partial oxidation of their sulfide charge to provide much or all of the energy required for smelting. The temperature in the furnace reaches between 1500 and 1600 K. The Noranda process takes advantage of the heat energy available from the copper ore. The remaining thermal energy is supplied by oil burners, or by coal mixed with the ore concentrates. For the smelting in electric arc furnaces, heat is generated by the flow of an electric current in carbon electrodes lowered through the furnace roof and submerged in the slag layer of the molten bath (e.g. [21]). Concerning emissions of air pollutants from the smelting operations, all the operations described above emit trace metals.

The final step in the production of blister copper is converting. The rest of the iron and sulfur in the matte are removed in this process leaving molten blister copper. Blister copper usually contains between 98.5 and 99.5% pure copper with the rest consisting of trace metals, such as Au, Ag, Sb, As, Bi, Pb, Ni, Se, Te and Zn. There are various converting technologies applied in copper production. The temperature in the converting furnace reaches about 1400 K.

Electrostatic precipitators are the common particulate matter control devices employed at copper smelting facilities. The control efficiency of ESPs often reaches about 99%. However, most of trace metals are condensed on very fine particles, e.g. < 1.0 μm diameter, and the control efficiency for these particles is less, reaching only about 97% (e.g. [12]).

Processes generating emissions of trace metals during the manufacture of lead and zinc are in general similar to those described above for copper production. The lead and zinc are produced through processes such as roasting, sintering (lead), smelting and refining. The temperatures involved in these processes, as well as concentrations of various trace metals as impurities of ores, are similar to those mentioned above.

The trace metal emissions from secondary non-ferrous metal production are significantly lower than the emissions from primary works. Trace metals are introduced to the furnace of secondary works together with the scrap material. The chemical composition of input scrap and the type and efficiency of emission control installations are two major factors that affect emissions of trace metals from this source.

3.2.2 Other Industries

Other major anthropogenic sources of atmospheric trace metals include high-temperature processes in steel and iron manufacturing and cement production.

Trace metals are emitted to the atmosphere during three major operations in iron and steel mills: (i) coke production, (ii) pig iron manufacture in iron works (mainly sintering), and (iii) steel-making processes using basic oxygen (BO), electric arc (EA) and open hearth (OH) furnaces. Among the steel-making technologies, the EA process produces the largest amounts of trace metals and the trace metal emission factors for this process are about one order of magnitude higher than those for the BO and OH processes. [12] The EA furnaces are used primarily to produce special alloy steels or to melt large amounts of scrap for reuse. The scrap, which often contains trace metals, is processed in electric furnaces at very high temperatures. Most of the emitted particles containing trace metals have a size within the range from 0.1 to 5.0 µm diameter.

Much less scrap is used in the OH and BO furnaces, where mostly pig iron (molten blast furnace metal) is charged. This results in lower emissions of trace metals.

Major air pollution problems around cement plants are caused by emissions of fine particles containing various trace metals. These particles are formed during the high-temperature operations in kilns and driers. Two major technologies are employed in cement plants: either dry type or wet type of operations. The former process gives higher emissions of trace metals, mostly due to the operation of ESPs. The maximum resistivity of cement dust from dry kilns occurs at temperatures between 500 and 600 K, when the collecting efficiency of ESPs drops. Other types of the emission control installations are rarely used in the cement industry. The use of fabric filters in cement plants is limited by the temperature of exhaust gases. Large consumption of water and huge amounts of dust to be collected limit the possibilities for wet scrubber applications (e.g. [17]).

3.3 APPLICATION OF METALS

Trace metals are used as additives to produce or improve various products, particularly within the manufacturing, construction and chemical industries. For example, since 1970 the world demand for lead has increased by 25% to a record level of more than 5.6 Mt in 1990. OECD countries accounted for 65% of world demand in 1990, with the central and eastern European countries consuming 21%. Asia is now the third largest and fastest-growing region. During the last two decades, Asian demand for lead increased by a factor of 6 and now accounts for 9% of world demand. [22] Other trace metals often used as additives include Cd, As, Zn, Sb, Bi, Te, V and Hg.

Application of trace metals used to produce or improve various industrial goods often results in atmospheric emissions. Again, an example of Pb applications is presented here to illustrate this environmental problem.

The following lead additives are often in use:

- Lead oxide
- Lead sulfate
- Lead phosphate
- Lead phthalate
- Lead stearate
- Lead chromate
- Red lead
- White lead
- Calcium plumbate
- Lead naphthenate

These additives are used to manufacture the following products:

- Gasoline
- Ceramics
- Paint
- Glass
- Plastics
- Selected alloys, including:
  —brass
  —solder
  —lead shot (used in ammunition and production of alloys)
  —lead weights (used by the fishing industry)

The use of lead as a gasoline additive is considered later in this chapter.

Lead monoxide, also known as litharge, plumbous oxide and yellow lead oxide has been used in storage batteries, ceramics, pigments and paints. Its application has been less than the use of other lead compounds, including lead carbonate, sulfate and chromate. Lead monoxide has been manufactured on the basis of metallic lead (e.g. partially oxidized powered lead, pig lead, molten lead), or lead compounds such as lead carbonate and lead nitrate.

Lead tetroxide, also known as lead oxide red, red lead and lead orthoplumbate, has been used in the manufacture of colourless glass and in ship paint. This compound is also used in medicine and cement for special applications. Lead tetroxide has been produced on the basis of lead monoxide at temperatures reaching 770 K.

Lead sulfate is used in storage batteries and as a paint pigment. Many countries have implemented legislation concerning lead paint that prevents the use of lead sulfate in paint. Under these rules, lead sulfate may only be used in paint for certain historic buildings and art preservation.

Lead carbonate application is similar to that of lead sulfate. The chemical is used as a paint pigment. Similar restrictions are imposed on the use of lead carbonate as on the application of lead sulfate. Lead carbonate can be used in paint only for some restoration purposes. Basic lead carbonate of variable composition is known as white lead. White lead is the oldest and most important lead pigment, also used in putty and ceramics. This compound is also used for corrosion control and for protection of outdoor work.

Lead chromate pigments are manufactured from lead chloride. Lead chromate is used as a paint pigment. Orthorhombic lead chromate pigments with high colour strength are used in plastics and in coatings. These pigments have a ratio of lead chromate to lead sulfate of about 100 : 1.

Lead phosphate is used as a stabilizer in plastics. Lead stearate is used as a lacquer and varnish drier, and in high- pressure lubricants. Lead naphthenate is a resinous material used as a paint and varnish drier, wood preservative, catalyst, insecticide and lubricating oil additive. Calcium plumbate is used as an oxidizer in the manufacture of glass and matches.

There are also other compounds of lead which are used as pigments, stabilizers, varnish and paint driers, in manufacturing of glass and ceramics, as well as plastics. These compounds include: lead acetate, lead antimonite, lead borate, leaded zinc oxide, lead molybdate, lead nitrate, lead oleate, lead thiosulfate, lead oxychloride, lead silicate, lead vanadate, lead titanate and lead tungstate.

The amount of Pb emissions to the atmosphere during the manufacture of lead additives and the product manufacture is often described as the Pb emission quantity in relation to the production unit (so-called emission factors). As an example, Table 3 presents Pb emission factors for the production and application of lead additives. [23]

Table 3. Lead emission factors for the production and application of lead additives

No.	Source category	Unit	Emission factor
1.	Manufacture of lead additives		
	lead monoxide	g Pb t^{-1} × 15 lead oxide produced	65.0
	lead tetroxide (red lead)	g Pb lead tetroxide produced	50.0
	lead carbonate (white lead)	g Pb lead carbonate produced	28.8
	lead chromate	g Pb lead chromate produced	6.5
2.	Product manufacture glass:		
	—low leaded (< 18% PbO) glass	g Pb glass produced	10.0–30.0
	—crystal glass	g Pb crystal glass	60.0
	—ceramics	g Pb ceramic products	310.0
	plastic	g Pb stabilizer produced	150.0
	brass and bronze production		
	—blast furnace	g Pb ingots	16.0
	—crucible furnace	g Pb ingots	10.0
	—cupola furnace	g Pb ingots	65.0
	—reverberatory furnace	g Pb ingots	60.0
	—rotary furnace	g Pb ingots	60.0
	—ferroalloy	g Pb Pb-containing ferroalloy	53.0
	solders	g Pb Pb consumed	130.0
	other metal fabricating industries (e.g. other alloys, cables)	g Pb Pb consumed	2000.0

3.4 WASTE DISPOSAL

At present, the most widely adopted methods of waste disposal include:

—landfill (with or without control),
—incineration,
—disposal at sea,
—recovery/recycling,
—composting,
—export.

Incineration of municipal and industrial hazardous wastes, the next most common method of waste disposal after landfilling, is an important source of trace metal emissions to the atmosphere. In some countries, for example Sweden, Japan and Switzerland, incineration has even overtaken landfill as the predominant disposal option for municipal wastes, accounting for up to 75% of this waste stream.

3.4.1 Municipal Waste Incineration

Municipal solid waste is the unwanted material collected from households and commercial organizations. It consists of a mix of materials; paper, plastics, food scraps, glass and defunct household appliances. The composition and quantity of wastes produced per person vary with the effectiveness of the material recovery scheme in place and with the affluence of the neighbourhood from which it is collected.

Municipal waste is incinerated to reduce its volume to save landfill costs and, in some instances, to recover energy from its combustion either for heating or electricity generation. There are many different furnace designs for the incineration of municipal wastes. A range of grate designs and fluidized beds are in use. [24] However, the principal influences on the level of trace metal emissions from incinerators are the waste- burning capacity of the incinerator, the way in which it is operated, the degree of abatement fitted to the plant, and the content of trace metals in the input wastes.

3.4.2 Incineration of Sludges from Waste Water Treatment

Sewage sludge is incinerated to reduce its volume (reduce disposal cost) and, in some instances, to recover energy from its combustion either for heating or electricity generation. Sewage sludge is the organic residue after the biological treatment of municipal wastes. As produced it is mainly water and must be dewatered to a moisture content below 30% before it will burn. Several dewatering processes are available, centrifuges, belt or plate processes. There are three main designs of furnace used for sludge incineration: rotary kiln, fluidized bed and multiple hearth. [24] However, the principal influences on the amount of trace metal emissions to the atmosphere are the degree of pollution abatement equipment fitted to the plant and the content of trace metals in the input material to the incinerator.

It can be expected that incineration will continue to gain popularity as a disposal option in the future because the alternative options, landfill and sea disposal, are coming under increasing public opposition and regulatory pressure. New incinerators for special/hazardous wastes are being built in many countries. [25]

3.5 VEHICULAR TRAFFIC

The largest sources of trace metals entering the environment from burning petroleum products are tetraethyl lead and other gasoline additives, diesel fuel combustion, metal compound additives for lubricants, worn metals that accumulate in spent lubricants, and automobile tyres. By far the most important is combustion of leaded gasoline.

Lead constitutes over 20% of the total mass of fine particles emitted from cars burning leaded gasoline. Approximately 75% Pb contained in leaded gasoline is emitted directly to the atmosphere. Leaded gasoline is usually regarded as containing $> 0.4\,\mathrm{g}\,\mathrm{Pb}\,\mathrm{l}^{-1}$ of gasoline, while low-leaded gasoline would contain about $0.15\,\mathrm{g}$ Pb l^{-1} gasoline. Unleaded gasoline does not contain lead additives; however, it may contain between 10 and 15 mg Pb l^{-1} gasoline as a result of the Pb content in crude oil. The percentage contribution of unleaded gasoline to the total use of gasoline in Europe by the end of 1993 is presented in Table 4 [26]. It is a policy of many countries to phase out both leaded and low-leaded gasoline from the market. Therefore, significant reductions of the Pb emissions from gasoline combustion can be expected in the near future.

Concerning the emission of other trace metals from internal combustion engines, Table 5 presents the chemical composition of fine particle emissions from highway vehicles, [27] indicating quite a number of trace metals which can be of concern. These values should be regarded as an example obtained from one study. The chemical composition may vary from one study to another; however, major metals and their content in the aerosol around highways should be comparable to that given in Table 5.

4 PREPARATION OF EMISSION INVENTORIES

Fluxes of air pollutants are presented on the basis of emission inventories. An atmospheric emissions inventory is a data base of information on:

—emissions measurements,
—emission factors,
—individual emission sources,
—activity statistics,
—emissions estimates derived from the above data [28].

In addition, an emission data base includes information on accuracy of emission estimates and/or measurements, reporting procedures, transparency and verification procedures as well as cost estimates.

For consideration of global and regional issues inventories and emission estimates are structured hierarchically. For example, regions are aggregated into countries and countries into groups of countries within international inventories. Individual emission sources are aggregated into larger groupings, such as activities, subsectors (or subcategories) and sectors (or categories).

In the majority of cases emission measurements, carried out in various ways, form the basis of emissions estimation procedures. An exception is a calculation procedure involving chemical analysis of input and output materials. Measured data can be used for emission estimates directly or indirectly. Direct use is made

Table 4. Consumption of gasoline in Europe in 1993[26]

	Gasoline consumption (Mt)	Unleaded gasoline (% of gasoline)
Austria	2.8	100
Belgium	2.9	40
Denmark	1.8	90
Finland	2.0	89
France	16.0	40
Germany	32.4	85
Greece	2.6	23
Italy	16.3	25
Netherlands	3.6	80
Norway	1.6	80
Portugal	1.8	17
Spain	9.1	19
Sweden	4.2	76
Switzerland	4.2	75
United Kingdom	23.5	51

when stack tests are performed by means of continuous or spot monitoring at individual sites. Indirect use is made of emission measurements whenever this information is transformed to an emission factor or included in a special

Table 5. Chemical composition of fine aerosol emissions from highway vehicles (%)[27]

Trace metal	Gasoline autos and trucks (leaded fuel)	Automobiles (unleaded fuel)	Diesel engine	Tyre tread	Brake lining
Al	0.1	0.1	0.3		
Br	8.2		v.s.		
Ca		0.2	0.8		5.5
Cl	5.4		1.7		
Cu	v.s.	v.s.	0.7		
Fe	0.3	0.1	1.3		
Pb	21.1		0.1		
Mg					8.3
Mn		v.s.	v.s.		
Ni		v.s.			
K		0.1			
Si	0.1	0.5	0.2		15.4
Na			0.4		
V			v.s.		
Zn	v.s.	0.1	0.2	1.0	

v.s. = very small.

calculation procedure, e.g. a chemical mass balance. The emission factor is often defined as the amount of a given material (gases, particles, vapour, chemical compounds, etc.) generated during the consumption of a unit of raw materials (fossil fuels, ores, etc.) or the production of a unit of industrial goods (electricity, non-ferrous and ferrous metals, etc.). For emissions from transportation, emission factors are also defined as the amount of a given material generated per distance driven. [29]

4.1 MAJOR STEPS IN EMISSION DETERMINATION

Initial steps in emission determination include the selection of:

—substances to be inventoried,
—source categories,
—determination procedures,
—source resolution,
—source activity data.

Concerning trace metals, the most commonly inventoried ones are As, Cd, Hg, Pb and Zn. This is due to their effects on environmental and human health, as well as their ubiquitous appearance in the environment. Recently, the UN Economic Commission for Europe (ECE) Task Force on Heavy Metals has agreed to consider Hg, Cd and Pb as priority metals for the international protocol on emission reductions, which is now in the final stage of preparation.

There are various source category splits within different emission inventories carried out at present. Two approaches are often considered: engineering and economic. The former approach of source category split is based on the emission generation processes, while the latter relates to major socio-economic activities. An example of the engineering approach used jointly in the UN ECE European Monitoring and Evaluation Programme (EMEP) and the Commission of the European Communities (CEC) CORINAIR Programme is presented in Table 6.

Depending on the circumstances, sources can be treated in emission inventories individually or collectively. The individual approach, often called a source-by-source approach, relates to point sources such as power plants, refineries, waste incinerators and airports for which site-specific activity and emission data are available. The trend is for more sources to be provided as point sources as legislative requirements extend to more source types and pollutants. The collective approach predominantly relates to sources comprising large numbers of small emitters, for which the emission conditions are relatively similar. These sources are often called area sources. The collective approach is also applied to estimate emissions from line sources. In some inventories, vehicle emissions from road transport, railways, inland navigation,

Table 6. CORINAIR/EMEP common source sector split*

Source sector	Cd	Hg	As	Cr	Cu	Ni	Pb	Zn	Se
Public power, co-generation and district heating	X	XX	XX	X	XX	XX	X	X	XX
Commercial, institutional and residential combustion	X	XX	X	X	X	X	X	X	X
Industrial combustion	X	XX	X	X	X	X	X	X	X
Production processes	XX	XX	XX	XX	XX	X	XX	XX	X
Extraction and distribution of fossil fuels									
Solvent use					X	X	X	X	
Road transport	X				X	X	XX	X	
Other mobile sources and machinery	X				X	X	XX	X	
Waste treatment and disposal	XX	XX	X	X	X	X	X	XX	X
Agriculture			X					X	
Nature		X	X		X		X	X	X

*XX = Major source category.

shipping or aviation are provided for sections along the line of the road, railway track, sea-lane, etc.

Selection of source activity data, that is, how frequently do sources generate emissions, is an important factor for establishing an accurate emission inventory. In general, activity data should be linked to the emission generation process as closely as possible. For example, emission characteristics for combustion processes depend on the type of fuel. Thus, activity data must be reported separately for different fuels instead of using total energy consumption or production. Whenever relevant activity data are not available or not freely accessible, special measurement campaigns might need to be carried out.

In the final step, emission factors and source activity data are used to calculate emissions. Emission factors for trace metals are now becoming available from emission factor guidebooks. The Emission Inventory Guidebook is now being prepared by the UN ECE Task Force on Emission Inventories in co-operation with the CEC with the aim to provide an up-to-date comprehensive summary of emission inventory methodology for each of the pollutants and sources to be quantified [30]. The guidebook is structured in chapters each presenting information on emissions of various pollutants from a given source category. This information is given within a common format which is a crucial feature of the guidebook, designed to ensure that users can readily locate and understand the essential aspects of the area covered. There are also supplementary documents including various computer programs to estimate emissions from mobile sources, as well as from sources with temperature-driven emissions, such as Hg re-emission.

4.2 VERIFICATION OF EMISSION DATA

Interest in programs and approaches to verify or validate emission data has increased in response to the rigorous demands being placed upon emission inventories. Concepts for emissions inventory verification are now being reviewed and modified particularly within the UN ECE Task Force on Emission Inventories. [31] Since most emissions data for trace metals are estimates, it is often impossible to derive statistically meaningful quantitative error bounds for inventory data. Frequently it is possible, however, to provide ranges that bound the likely minimum and maximum for an emission estimate or to develop a quantitative data quality parameter to assess the relative confidence that can be associated with various estimates.

Efforts to improve emission estimation reliability concentrate on three major topics:

—accuracy of activity data,
—level of detail of procedures applied,
—quality of emission factors.

One of the most common methods of emission validation for trace metals is the application of the emission inventories in air quality modelling (e.g. [32]). Regional emission inventories (e.g. European-wide) seem to have an accuracy of 25% for Pb, 50% for Hg and Cd and about 100% for other metals.

4.3 SPATIAL RESOLUTION OF EMISSION DATA

Spatial resolution of emission data is needed when the emission data are input to air quality models. If source data are not available at the required level of spatial distribution, e.g. geographical location and emission quantity for a given point source, some method of allocating total emission fluxes is needed. Examples of allocation parameters used in such cases include population density, number of industrial employees, industrial energy use, length of traffic lanes, and proportion of arable land to total land area.

Allocation data available from statistics always relate to administrative regions. However, modellers need allocation into rectangular grid systems. This latter requirements can be satisfied through transformation of emissions data into the grid cells using information on the geographical location of point sources and allocation parameters that assume that the emissions determined collectively (emissions from area and line sources) can be regarded as homogeneously distributed within the individual administrative units. The decision on the appropriate allocation or transformation method largely depends on the share of emissions already determined individually and allocated correctly.

4.4 SPECIES RESOLUTION

Trace metal studies may require the specification of individual chemical forms of Hg and other metals because of the species differences in bioaccumulation, biomagnification and toxicity. Therefore, preliminary approaches of separate emission inventories for various chemical species have been made for Hg (e.g. [29]). Concerning other metals, some information has been published for As, Cd and Pb (e.g. [33]).

4.5 OTHER ASPECTS OF EMISSION INVENTORIES

Heavy metals can be emitted to the atmosphere on particles of various sizes. When efficient emission control equipment is in place, e.g. ESPs, wet scrubbers, or a combination of both, heavy metals enter the atmosphere on very fine particles, which are then subject to various agglomeration/accumulation processes in the ambient air. Obviously, less efficient control equipment would also allow coarser particles to be emitted (e.g. [17]). The proportion of metals between fine and coarse particles is different for various metals. For example, lead is generally associated with submicrometre particles, while cadmium size distribution may contain submicrometre, as well as supermicrometre aerosol.

As the size of heavy metal-containing particles determines the toxicity of atmospheric emissions, the migration of pollutants through individual environmental media, and the long- range transport of air pollutants, it should be of great importance to be able to assess quantitatively the emissions of heavy metals generated on various sizes of particles from at least major source categories. The apportionment of these emissions to major source categories can be quite a complicated task as various emission sources, such as power plants, waste incinerators and smelters can employ the same type of control equipment, allowing the same size particles to be emitted into the atmosphere. In such cases, the variability of heavy metal amounts in a certain size of particles emitted from various source categories can be assigned to the variability of heavy metal content of raw (input) materials, e.g. fossil fuels, wastes and ores. The latter information is often very difficult to obtain. Therefore, accurate information on size-differentiated emissions of various heavy metals is largely lacking in the literature and needs urgent improvement.

5 REGIONAL EMISSION INVENTORIES

Emission inventories for trace metals have been prepared on a local, regional and global scale. Emission inventory for a single point source, a group of point sources or area sources within a geographical region of a given country is often regarded as an inventory on a local scale. The European-wide emission inventory is considered as a regional emission inventory. The northern hemispheric emission inventory is regarded as a global emission inventory. Major approaches to present regional and global emission inventories are discussed in this section.

5.1 EUROPEAN EMISSION INVENTORIES FOR TRACE METALS

The first attempt to estimate atmospheric emissions of trace metals from anthropogenic sources in the whole of Europe was completed at the beginning of the 1980s [34]). This survey included information on atmospheric emissions of As, Be, Cd, Co, Cr, Cu, Mn, Mo, Ni, Pb, Sb, Se, V, Zn and Zr. Earlier emission estimates have dealt with either a single metal (often either Cd or Pb) or a given source category (e.g. combustion of fossil fuels to produce electricity and heat). The 1980 estimates were made on the basis of emission factors calculated for major source categories, separately, and the statistical information for the 1979/1980 reference years.

A few years later the 1980 European emission inventory was updated, completed and emission gridded. The 1982 version of this emission inventory was prepared by Axenfeld et al. [35] for As, Cd, Hg, Pb and Zn. An update of emission factors has been carried out and the results are shown in Figure 2. Later on, a further update has been prepared for Pb emissions (1985 reference year) and Hg (1990 reference year). Spatial resolution of the 1990 Hg emission data within the EMEP grid system of 150 by 150 km is presented in Figure 3 [36] as an example of spatial aggregation of emission data from sources in Europe.

The most recent version of the European emission inventory for As, Cd, Pb and Zn has been estimated for the reference years 1991/1992. A summary of these estimates is presented in Table 7. [37]

A historical review of emissions of Cd, Pb and Zn in Europe during the period from 1955 to 1987 has been attempted at the International Institute for

Table 7. Emission estimate for As, Cd, Pb and Zn in Europe (in t year^{-1}) at the beginning of the 1990s[37]

Source category	As	Cd	Pb	Zn
Stationary fuel combustion	860	280	3 560	6 400
Non-ferrous metal manufacturing	1 300	390	8 250	13 800
Road transport	–	60	41 900	110
Iron and steel production	160	85	3 100	8 500
Waste disposal	5	30	260	450
Other sources	250	50	1 020	3 730
Total	2 575	895	58 090	32 990

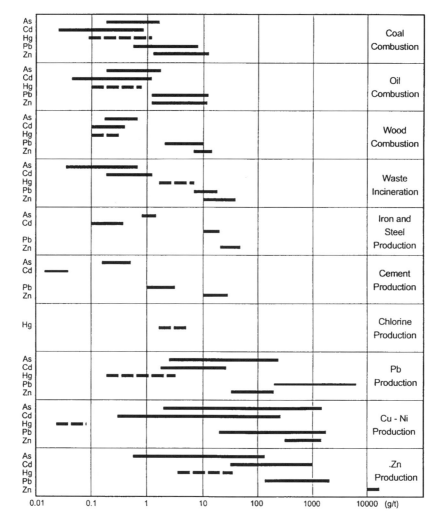

Figure 2. Emission factors for As, Cd, Hg, Pb and Zn used to estimate European emissions [35]

Applied System Analysis (IIASA). [38] A summary of these data is shown in Figure 4 together with the emissions at the beginning of the 1990s. [37] The largest emissions of Cd and Zn, emitted mostly from industrial sources, are estimated in Europe for the late 1960s. At the beginning of the 1970s the first efficient electrostatic precipitators (ESPs) were installed at major electric power plants, smelters and cement plants. As a result, emissions of particles and thus heavy metals decreased significantly from these sources. Concerning lead emis-

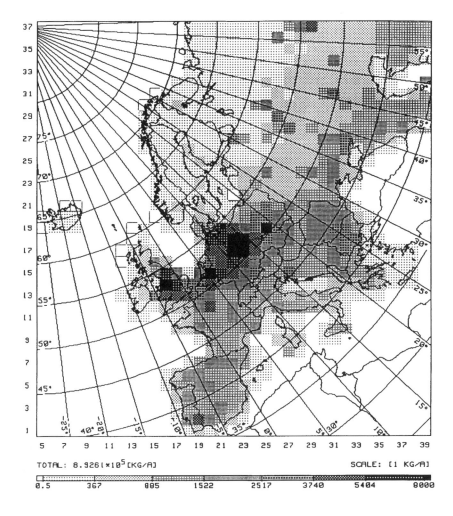

Figure 3. Spatial resolution of the 1990 Hg emissions in Europe within the EMEP grid system of 150 km by 150 km [36]

sions, obviously the part of lead emissions from industrial sources followed the decreasing trend described for Cd and Zn. However, the major source of Pb, combustion of leaded gasoline, was still increasing its emissions at the beginning of the 1970s, as the low-leaded gasoline was only introduced in Europe at the end of the 1970s (mostly in Germany). Unleaded gasoline appeared in the European market in small amounts in the first half of the 1980s.

Further completion and improvement of the above-mentioned IIASA study are now in progress. However, it can now be added that a significant decrease

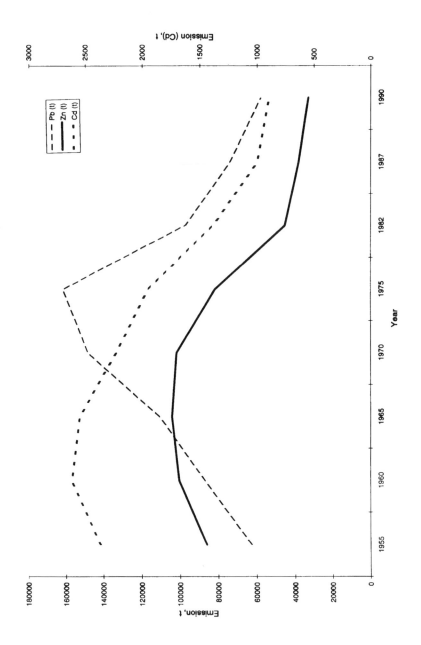

Figure 4. Historical trend of atmospheric emissions of Cd, Pb and Zn in Europe from 1955 to the beginning of the 1990s [38]

in the heavy metal emissions in Eastern Europe, perhaps up to 25%, has been noticed during the period from 1990 to 1994. Installation of efficient emission control equipment, as well as a general decrease of economic activity in this part of Europe as a result of political changes in the region and the transfer from centrally planned to market-oriented economies can be seen as the major reasons for the decrease in pollutant emissions to the environment, including atmospheric heavy metals. Changes in emissions are also reflected in decreasing air concentrations and deposition of heavy metals in Eastern Europe and other regions in Europe, affected by emissions in the former, e.g. southern Scandinavia. This particularly applies to mercury. It should be added, however, that the changes in atmospheric deposition are lower than the emission changes, as the source–receptor processes are not linear.

The following conclusions have been drawn from the experience on emission inventorying for trace metals in Europe:

1. Pyrometallurgical processes in the primary non-ferrous metal industries are the major sources of atmospheric As, Cd and Zn in Europe. Combustion of coal in utility and industrial boilers is the major source of anthropogenic Hg and combustion of leaded gasoline is still the major source of Pb. Little information is available on emission of trace metals from various diffuse sources.
2. The emission estimates were based on emission factors. It was recommended that emissions from major point sources, such as electric power plants, heat-producing plants and primary smelters should be measured and not calculated on the basis of emission factors. This recommendation is valid for the above-mentioned point sources of emissions world-wide.
3. More national data on emission quantities and emission factors is necessary in order to improve the quality of emission inventories in Europe. This is specially important for sources within various applications of metals and incineration of wastes.

5.2 TRACE METAL EMISSION INVENTORIES FOR THE FORMER SOVIET UNION

Estimates of atmospheric emissions of As, Cd, Cu, Cr, Mn, Ni, Pb, Sb, Se, V and Zn for 12 major source regions in the former Soviet Union were prepared by Pacyna on the basis of emission factors and statistical data for the reference year 1979/1980. [39] A summary of these data for major source regions is presented in Table 8. Emissions from non-ferrous and ferrous metal production, fossil fuel combustion and gasoline combustion were estimated to dominate the total emissions in these regions.

Recently Sivertsen et al. [40] have assessed the level of air pollution in the border areas of Norway and Russia, reporting the Cu and Ni emissions from

Table 8. Anthropogenic emissions of air pollutants in the former Soviet Union in 1979 (i t)

Region	As	Cd	Cu	Cr	Mn	Ni	Pb	Sb	Se	V	Zn
Kola Peninsula	165	29	235	122	106	645	745	23	26	122	180
Pechora basin	11.5	3.5	49	81	74	73	198	12	22	66	56
Norilsk area	246	26	402	31	28	935	742	25	33	130	262
Urals	551	145	1 435	1 390	1 160	1 620	9 530	94	180	3 000	3 920
Yakutsk area	6.5	2	20	46	42	30	130	4	9	20	26
Moscow area	16	6	81	66	56	300	3 620	6	22	861	74
Donetsk area	63	40	273	1 050	1 100	506	3 340	57	112	995	2 520
Kuznetsk area	429	262	154	456	375	138	3 940	24	60	158	8 830
Fergana area	980	274	1 830	21	40	210	4 540	71	93	766	4 550
Caucasus	253	54	641	6	30	75	2 100	25	32	215	266
Leningrad area	4	2	13	17	15	80	1 090	2	3	233	20
Baikal area	55	14	157	41	39	50	715	11	17	33	88
Total	2 780	860	5 290	3 320	3 070	4 660	30 700	350	610	6 730	20 800

sources on the Kola Peninsula in 1989 to be *ca.* 310 and 510 t, respectively. These data are in good agreement with emissions presented in Table 8. However, very recent information (e.g. [41,42]) suggests that these emissions can be about one order of magnitude higher. Indeed, there is a need to verify information on trace metal emissions on the Kola Peninsula and other source regions in the former Soviet Union as these emissions contribute substantially to the contamination of the environment on a global scale.

5.3 TRACE METAL EMISSION INVENTORIES IN THE UNITED STATES AND CANADA

Emission data for toxic air pollutants in the United States are much less comprehensive than those for the criteria air pollutants, such as sulfur dioxide, nitrogen oxides, carbon monoxide, VOCs (excluding certain non-reactive organic compounds), particulate matter less than $10\,\mu m$, lead, and total suspended particulate matter. The major reason for that is the fact that an extensive and long-term monitoring and emissions tracking programme similar to that for criteria pollutants has not been established for air toxics (except for Pb). The US Environmental Protection Agency (EPA) is, however, beginning development of a county-wide toxics database, bringing together elements from different toxics programmes [43].

There are two different programmes in the United States collecting information on emissions of toxics:

—Toxic Release Inventory (TRI) estimates (currently for over 300 chemicals in 20 chemical categories) submitted annually since 1987 to EPA by certain manufacturing facilities,
—national inventories for specific pollutants prepared by EPA to support special studies called for by the 1990 Clean Air Act Amendment (CAAA).

The EPA inventories are more comprehensive than the TRI in that they attempt to identify and quantify all source categories and air emissions of toxics, whether from manufacturing facilities, commercial facilities, mobile sources, or residential and consumer sources. These inventories also include emissions from facilities with fewer than 10 employees and emissions from sources with very low concentrations of toxics, which are exempted from the TRI reporting requirements.

Mercury and cadmium are the two trace metals for which national emissions have been prepared, in addition to Pb which is a criteria pollutant. Mercury is one of the seven pollutants specified in the section of the Clean Air Act (CAA), which requires the EPA to identify source categories and subcategories of hazardous air pollutants (HAPs) in urban areas that pose a threat to human health. Emission estimates for both Hg and Cd are prepared within the

Locating and Estimating (L&E) document series([44] and [45] for Hg and Cd, respectively). The L&E documents provide a compilation of available information, including emission factors, on the sources and emissions of specific toxic air pollutants.

The Hg and Cd emissions from the L&E documents are presented in Table 9 together with emissions of Pb. A historical trend of Pb emissions in the United States is shown in Figure 5. A major decrease in lead emissions appeared in the 1970s when both the ESPs and low- and unleaded gasoline was introduced.

Environment Canada has initiated several projects on emission inventory development for trace metals; Pb, Hg, Cd and As have been given a priority in these projects. The contribution of atmospheric emissions of these compounds from various regions to the total emissions in Canada is shown in Figure 6. In Table 10 emissions of some trace metals from major source categories are presented on the basis of data from Jacques [46] and Voldner and Smith. [47] Emission estimates for Hg and Pb have been revised to account for major changes in consumer patterns over the years. New emission data for 1993 will be available in 1996.

Table 9. Emission estimates for atmospheric Cd, Hg and Pb in the United States for various reference years (t year^{-1})

Source	Cd 1990[45]	Hg 1990[44]	Pb 1994[43]
Fossil fuel combustion	242	125	444
Industrial processes	45	35	1735
Metal application	3	19	84
Waste disposal	17	118	762
Transportation	–	5	1436
Total	307	302	4461

Table 10. Anthropogenic emissions of selected trace metals from major source categories in Canada in 1982 (t)[46]

Source category	As	Cd	Cr	Cu	Hg	Ni
Fuel combustion/ stationary sources	19	53	39	10	8	232
Industrial process	451	260	20	1672	16	559
Transportation	–	4	40	4	–	13
Solid waste incineration	1	5	–	2	2	1
Miscellaneous sources	–	–	–	–	5	–
Total	471	322	69	1688	31	845

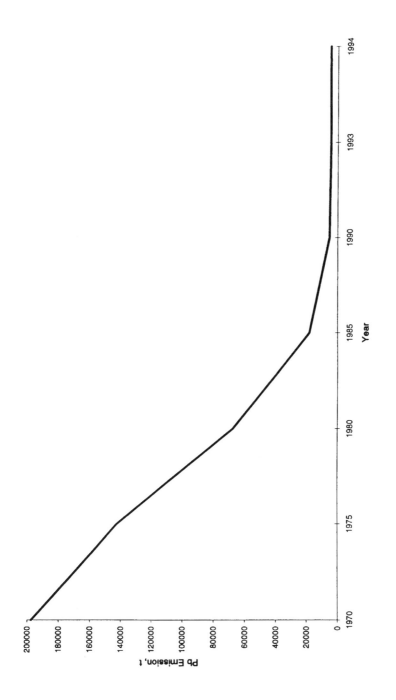

Figure 5. Historical trend of Pb emissions to the atmosphere in the United States [43]

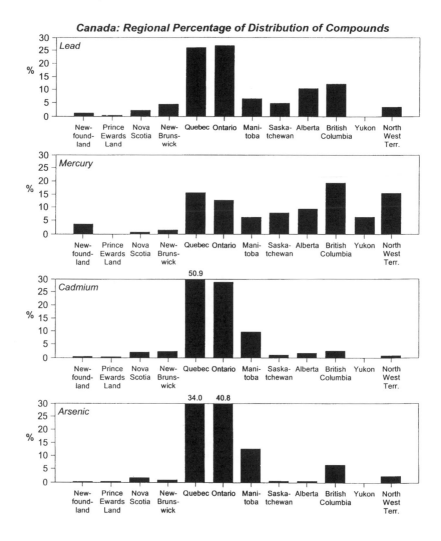

Figure 6. Contributions of trace metal emissions from various provinces in Canada to the total national emissions [46]

6 GLOBAL EMISSION INVENTORIES

Regional emission inventories can be used to prepare a global emission inventory for trace metals. However, the first quantitative world-wide estimate of the annual industrial input of 16 trace metals into the air, soil and water has been made using information on emission factors for major source categories and statistical data on the global production of industrial goods and the global consumption of raw materials. [11] This inventory has been made only for major source categories with no spatial resolution whatsoever. As such, the survey has been very useful in assessing fluxes of trace metals to the global environment and in evaluating the perturbations of geochemical cycles of these pollutants in the environment due to human activities.

However, the first global inventory of trace metals by Nriagu and Pacyna has been of less value for modellers studying transport and migration of trace metals through the environment. Gridded emission inventories are needed for such purposes. It was not until the beginning of the 1990s when the Global Emission Inventories Activities (GEIA) programme was established within the IGBP International Global Atmospheric Chemistry (IGAC) programme. The ultimate goal of GEIA is to develop and/or certify emission inventories for all species of interest to air quality policy-makers and modellers on a 1° by 1° global grid system. [48] One of the GEIA projects is on global emission inventories for trace metals. So far, the major focus has been placed on global atmospheric emissions of lead. Information on global emissions of mercury is also completed.

The GEIA emission inventory for Pb has been prepared for the reference year 1989 [49]. In Europe, information on the Pb emissions in 1989, estimated by national authorities, was available from the Scandinavian countries, the United Kingdom, the former Federal Republic of Germany, the Netherlands, Austria and Poland. These estimates were accepted for the GEIA inventory. For the rest of the European countries the 1982 emission estimates, mentioned earlier in this chapter, were updated first for the year 1985 and then for the year 1989. Information on lead emissions in the United States and Canada in 1985 was obtained through the programmes mentioned earlier in this chapter. Several emission inventory programmes have been carried out in Asia. The Pb emission data presented at the International Symposium on Emission Inventory and Prevention Technology for Atmospheric Environment in Tsukuba, Japan [50] were used in the global Pb inventory. The following annual quantities of Pb were emitted from gasoline combustion in the mid-1980s: 4100 t in Japan, 3000 t in Thailand, 8400 t in China and 1300 t in India.

After completion of the global Pb emission inventory it was concluded that between 150 000 (minimum scenario) and 210 000 t of the element (maximum scenario) are emitted annually to the atmosphere. The largest emissions were estimated for gasoline combustion, about 62 and 70% of the total emissions in the maximum and minimum scenarios, respectively. The estimates of 1989 Pb emissions by sectors (the maximum scenario) are presented in Figure 7. Emissions from the European and Asian sources each contribute about one-third of the global emissions followed by emissions from sources in North America. The

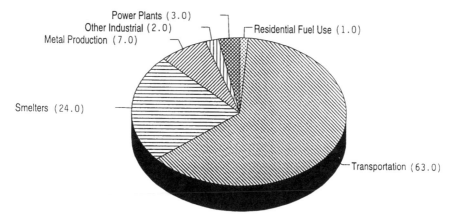

Power Plants (3.0)
Other Industrial (2.0)
Metal Production (7.0)
Residential Fuel Use (1.0)
Smelters (24.0)
Transportation (63.0)

Figure 7.   Global emissions of atmospheric Pb in 1989: the maximum scenario

spatial distribution of total lead emissions, maximum scenario, within the GEIA grid system of 1 ° by 1 °, is presented in Figure 8. [49]

The quantitative assessment of global Hg emissions in 1985 is now completed within GEIA on the basis of earlier estimates by Nriagu and Pacyna. [11] Information from various research programmes and national authorities and international organizations in Europe and North America is collected, e.g. the UN ECE Task Force on Heavy Metals Emissions. Emission factors and statistical data on industrial goods production and raw material consumption were used to estimate emissions in other parts of the world.

Preliminary results indicate that combustion of fossil fuels, mostly coal to produce electricity and heat, is the major source of atmospheric emissions of Hg on a global scale, followed by refuse incineration. [49]

7 CONCLUDING REMARKS

Significant progress has been made during the past two decades in the assessment of emission sources and source fluxes for trace metals world-wide. This progress accounts for both the development of assessment tools and presentation of quantitative evaluations.

Results of emission estimates for trace metals, particularly the estimates of the European and North American emissions, can be discussed taking into account their completeness, transparency and comparability.

Completeness of emission estimates is related to a list of compounds inventoried and a list of source categories considered. The following can be concluded:

Figure 8. Spatial distribution of total lead emissions: maximum scenario within the GEIA grid system of 1° by 1° [49]

1. The most important trace metals, with a view to their toxic and environmental effects, are included in present emission inventories. More information is needed on emissions of various chemical and physical forms of some of the metals, particularly mercury.
2. Major sources of emissions of heavy metals are considered in present inventories, prepared on a regional and global scale. The list of these sources may not be satisfactory when preparing emission inventories on a country or local administrative scale.
3. As some metals cycle through the environment, it is not sufficient to address atmospheric emissions alone. A multimedia approach, accounting for releases to all environmental compartments, must be adopted.
4. Information on emission beyond the UN ECE region is limited.

Transparency of emission data relates mainly to the estimation methodologies and reporting procedures. The following can be concluded:

1. Most of the emission estimates have been made on the basis of emission factors or emission rates. There is a fairly good understanding of parameters affecting emission factors for trace metals. Preliminary efforts have been made to present emission factors in a form of emission estimation guidebook. Further work is needed to improve the quality of information presented in current guidebooks and to develop other methods for emission estimates, e.g. models relating the amount of emissions to the meteorological, technical, physical and chemical conditions during the emission-generating processes.
2. More measurements are needed to assess the concentrations of heavy metals in exhaust gases and then to elaborate emission rates in a transparent way.
3. Harmonization of emission reporting is needed between major research programmes with respect to definition of sources and compounds.
4. Easily accessible emission data bases should be established.

Comparability is regarded as one of the means to verify/validate emission data. The following conclusions can be drawn in this field:

1. Limited information in Europe and North America indicates an agreement within a factor of 2 between emission factors for major trace metals from major source categories, such as combustion of fuels and non-ferrous metal production, assuming that the same types of emission-generating processes are considered and basic (with no emission control) emission factors are compared.
2. More work is needed in order to improve sampling techniques and analytical methods for trace metals, particularly Hg.

REFERENCES

1. J. M. Pacyna, in *Toxic Metals in the Atmosphere*, Edited by J. O. Nriagu and C. I. Davidson, pp 33–52, Wiley, New York (1986).
2. J. O. Nriagu, *Nature (London)*, **338**, 47–49 (1989).
3. SCOPE, *Saharan Dust. Scientific Committee on Problems of the Environment*, Wiley, New York (1979).
4. C. Junge, in *Saharan Dust*, SCOPE 14, Wiley, New York (1979).
5. K. A. Rahn, R. D. Borys, G. E. Shaw, L. Schutz and R. Jaenicke, in *Saharan Dust*, SCOPE 14, Wiley, New York (1979).
6. J. M. Pacyna and B. Ottar, *Atmos. Environ.* **23**, 809–815 (1989).
7. R. W. Talbot, M. O. Andreae, T. W. Andreae and R. C. Harriss, *J. Geophys. Res.* **93**, 1499–1508 (1988).
8. P. Artaxo, H. Storms, F. Bruynseels and van R. E. Van Grieken, *J. Geophys. Res.* **93**, 1605–1615 (1988).
9. W. H. Zoller, in: *Changing Metals Cycles and Human Health*. Edited by J. O. Nriagu, pp 27–41, Springer-Verlag, Berlin (1983).
10. D. C. Blanchard, *Progress Oceanog.* **1**, 71–202 (1963).
11. J. O. Nriagu and J. M. Pacyna, *Nature (London)* **333**, 134–139 (1988).
12. J. M. Pacyna, in *Control and Fate of Atmospheric Trace Metals*. Edited by J. M. Pacyna and B. Ottar, pp 15–32, Kluwer Academic Publishers, Dordrecht (1989).
13. H. J. Gluskoter, in *Trace Elements in Fuel*. Edited by Babu, pp 1–22, Advances in Chemistry Series, Washington (1975).
14. H. J. Gluskoter, R. R. Ruch, W. G. Miller, R. A. Cahill, G. B. Dreher and J. K. Kuhn, *Trace Elements in Coal: Occurrence and Distribution*. Illinois State Geological Survey Circular 499, Urbana (1977).
15. J. M. Pacyna, *Coal-fired Power Plants as a Source of Environment Contamination by Trace Metals and Radionuclides*, Technical University of Wroclaw Publ., Wroclaw (1980).
16. J. W. Kaakinen, Trace Element Study on a Pulverized Coal-fired Power Plant, Ph.D. Dissertation, University of Colorado, Boulder (1979).
17. J. M. Pacyna, in *Atmospheric Pollution*, pp. 93–104, Hemisphere Publ. Corp., Washington (1987).
18. D. S. Ensor, S. Cowen, R. Hooper and G. Markowski, *Evaluation of the George Neal No. 3 Electrostatic Precipitator*, EPRI Rept. FP-1145, Electric Power Research Institute, Palo Alto (1979).
19. D. S. Ensor, S. Cowen, A. Shendrikar, G. Markowski and G. Waffinden, *Kramer Station Fabric Filter Evaluation*, EPRI Rept. CS-1669, Electric Power Research Institute, Palo Alto (1981).
20. J. M. Pacyna and G. J. Keeler, *Water, Air & Soil Pollut.* **80**, 621–632 (1995).
21. UN ECE, *Heavy Metals Emissions*, Task Force on Heavy Metals Emissions State-of-the-Art Report, Prague (1994).
22. OECD, *Risk Reduction Monograph No. 1: Lead*, Organization for Economic Co-operation and Development, Paris (1993).
23. J. M. Pacyna, *Emissions Factors for Lead from Sources using Lead Additives in the United Kingdom*, A Report for the Environmental Resources Management (ERM) in the UK, Hagan (1995).
24. P. J. Coleman, in *Atmospheric Emission Inventory Guidebook*, 2nd Draft, UN Economic Commission for Europe, Geneva (1995).
25. H. Yakowitz, *Resources Conservation and Recycling* **8**, 131–178 (1993).
26. EFOA, *The European Fuel Oxygenates Association Newsletter* **12**, 3 (1994).

27. J. M. Pacyna, in *Toxic Metals in the Atmosphere*, Edited by J. O. Nriagu and C. I. Davidson, pp 1–32, Wiley, New York (1986).
28. G. McInnes, J. M. Pacyna and H. Dovland eds. *Proc. the First Meet. of the Task Force on Emission Inventories*, EMEP/CCC Rept. 4/92, Norwegian Institute for Air Research, Lillestrøm (1992).
29. J. M. Pacyna and T. E. Graedel, *Annu. Rev. Energy Environ.* **20**, 265–300 (1995).
30. G. McInnes, ed. *The Atmospheric Emission Inventory Guidebook*. European Environment Agency Task Force, Copenhagen (1995).
31. J. D. Mobley and M. Saeger, *Proc. the Second Meet. of the Task Force on Emission Inventories*, EMEP/CCC Rept. 8/93, pp. 137–153. Norwegian Institute for Air Research, Lillestrom (1993).
32. J. M. Pacyna, A. Bartonova, P. Cornille and W. Maenhaut, *Atmos. Environ.* **23**, 107–114 (1989).
33. J. M. Pacyna, in *Lead, Mercury, Cadmium and Arsenic in the Environment*. Edited by T. C. Hutchinson and K. M. Meema, pp. 69–88, Wiley, Chichester (1987).
34. J. M. Pacyna, *Atmos. Environ.* **18**, 41–50 (1984).
35. F. Axenfeld, J. Munch, J. M. Pacyna, J. A. Duiser and C. Veldt, *Test-Emissionsdatenbasis der Spurenelements As, Cd, Hg, Pb, Zn und der speziellen Organischen Verbindungen y-HCH (Lindan), HCB, PCB und PAK für Modellrechnungen in Europa*. Umweltforschungsplan des Bundesministers für Umwelt Naturschutz und Reaktorsicherheit, Luftreinhaltung, Forschungsbericht 104 02 588, Dornier GmbH Report, Friedrichshafen (1992).
36. F. Axenfeld, J. Munch and J. M. Pacyna, *Europaische Test-Emissionsdatenbasis von Quicksilber-Komponenten für Modellrechnungen*. Umweltforschungsplan des Bundesministers für Umwelt Naturschutz und Reaktorsicherheit, Luftreinhaltung, Forschungsbericht 104 02 726, Dornier GmbH Report, Friedrichshafen (1991).
37. J. M. Pacyna, data in preparation for the International Institute for Applied Systems Analysis, Laxenburg (1996).
38. K. Olendrzynski, S. Anderberg, J. Bartnicki, J. M. Pacyna and W. Stigliani, *Atmospheric Emissions and Depositions of Cadmium, Lead and Zinc in Europe during the Period 1955–1987*, IIASA Rept. WP-95-35, International Institute for Applied Systems Analysis, Laxenburg (1995).
39. NILU, *Emission Sources in the Soviet Union*. NILU Rept. TR-4/84, the Norwegian Institute for Air Research, Lillestrom (1984).
40. B. Sivertsen, T. Makarova, L. O. Hagen and A. Baklanov, *Air Pollution in the Border Areas of Norway and Russia*. NILU Rept. OR-8/92, the Norwegian Institute for Air Research, Lillestrom (1992).
41. V. Ya Pozniakov in *Aerial Pollution in Kola Peninsula*. Edited by M. V. Kozlov, E. Haukioja and V. T. Yarmishko, pp 16–20. Proceedings of the International Workshop, St. Petersburg, 14–16 April (1992).
42. I. V. Lyangusova, Accumulation of Chemical Elements in Ecosystems of Pine Forests in Kola Peninsula. Ph.D. thesis, V. L. Komarov Bot. Inst., Moscow (1990).
43. US EPA, *National Air Pollutant Emission Trends*, EPA Rept. 454/R-95-011, The US Environmental Protection Agency, Research Triangle Park, NC (1995).
44. US EPA, *Locating and Estimating Air Emissions from Sources of Mercury and Mercury Compounds*, EPA Rept. 454/R-93-023, US Environmental Protection Agency, Research Triangle Park, NC (1993).
45. US EPA, *Locating and Estimating Air Emissions from Sources of Cadmium and Cadmium Compounds*, EPA Rept. 454/R-93-040, US Environmental Protection Agency, Research Triangle Park, NC (1993).

46. A. P. Jacques, *Summary of Emissions of Antimony, Arsenic, Cadmium, Copper, Lead, Manganese, Mercury, and Nickel in Canada.* Environmental Analysis Branch, Conservation and Protection, Environment Canada, Ottawa (1987).
47. E. Voldner and L. Smith, *Production, Usage and Atmospheric Emissions of 14 Priority Toxic Chemicals.* Report of the Joint Water Quality Board/Science Advisory Board/International Air Quality Advisory Board of the International Joint Commission. Presented at the Workshop on Great Lakes Atmospheric Deposition, 29–31 October (1986).
48. T. E. Graedel, T. S. Bates, A. F. Bouwman, D. Cunnold, J. Dignon, I. Fung, D. J. Jacob, B. K. Lamb, J. A. Logan, G. Marland, J. M. Pacyna, M. Placet and C. Veldt, *Global Biogeochemical Cycles*, **7**, 1–26 (1993).
49. J. M. Pacyna, M. T. Scholtz and Y.-F. Li, *Environ. Rev.*, **3**, 145–159 (1995).
50. ITIT, in *Proceedings of the ITIT International Symposium on Emission Inventory and Prevention Technology for Atmospheric Environment*, Tsukuba, Japan (1992).

13 Dry Deposition of Particles

MARIA J. ZUFALL AND CLIFF I. DAVIDSON
Carnegie Mellon University, USA

Atmospheric Particles, Edited by R. M. Harrison and R. Van Grieken.
© 1998 John Wiley & Sons Ltd.

1 INTRODUCTION

Atmospheric inputs of particulate pollutants to a region may cause environmental damage. For example, acidic particles reaching sensitive ecosystems can harm vegetation and degrade water quality. Particles depositing in urban areas can damage structures, such as buildings and monuments.

Airborne particles may be delivered to surfaces by wet or dry deposition. Wet deposition is that from precipitation, while dry deposition is that derived from methods other than precipitation. Despite its importance, much uncertainty remains in our understanding of the dry deposition process. Efforts have been made to quantify the amount of dry deposition by modelling and experimental methods. Models incorporate mathematical representations of physical processes and empirically derived formulations to estimate dry deposition. In general, these models are limited in their predictive ability due to our incomplete understanding of the many factors influencing deposition.

Methods of measurement are complicated by variations in terrain that affect deposition. Measurements at one location may not accurately reflect those at another nearby location. Experimental dry deposition estimates may be obtained by inferring deposition from airborne data or by collecting particles that have deposited on to a surface.

This chapter examines recent work in particle dry deposition including modeling efforts and experimental methods.

1.1 OVERVIEW OF THE DEPOSITION PROCESS

The deposition process can be idealized as taking place in three separate steps: aerodynamic transport, boundary layer transport and surface interactions. Aerodynamic transport refers to particles carried from any height in the free atmosphere to the viscous sublayer. This transport is characterized mainly by turbulent diffusion and sedimentation. The viscous sublayer is a thin (several mm or smaller) layer between a surface and the free atmosphere which contains a steep velocity gradient. This velocity gradient exists because the velocity at a solid stationary surface must be zero. The sublayer consists of mainly laminar flow with intermittent bursts of turbulence. Transport across this layer, the second step, is governed by Brownian diffusion for small particles, and by interception, inertial forces, and sedimentation for larger particles. Electrical migration, thermophoresis and diffusiophoresis may also influence small particles under certain conditions. The third and final step in the process involves surface interactions. Once at the surface, particles may adhere or bounce off, depending on the characteristics of the contaminant and the surface. Any of these three steps may be rate-limiting, depending on the specific conditions of the atmosphere, surface and contaminant. [1]

1.2 MECHANISMS OF TRANSPORT

The previous section mentioned several mechanisms that are part of the dry deposition process: turbulent diffusion, sedimentation, Brownian diffusion, interception, inertial forces, electrical migration, thermophoresis and diffusio-phoresis. These mechanisms and their importance to the deposition process are discussed further here.

Turbulent diffusion utilizes turbulent energy to transport particles along a concentration gradient. As long as particles are continually deposited there will be a concentration gradient between the atmosphere and a perfect sink surface. This mechanism of transport is most important in the free atmosphere where turbulence is greatest.

Transport by sedimentation is due to the force of gravity. A drag force, due to the viscosity of air, opposes the acceleration of gravity and the two forces on the particle reach equilibrium relatively quickly ($<<$ 1 s for a 1 μm spherical particle). At equilibrium the particle will fall at a constant velocity. Sedimentation transport is greatest for large ($>$ 1 μm) particles since the settling velocity strongly depends on particle size. [2] The equation for sedimentation also includes a buoyant force which acts in the same direction as the drag force, but this is usually very small for airborne particles.

Brownian diffusion is similar to turbulent diffusion, as it also depends on the concentration gradient. With Brownian diffusion, the driving force is the random thermal energy of air and particles. Brownian diffusion is much weaker than turbulent diffusion and only becomes significant when turbulent diffusion is weak, such as in the viscous sublayer. This mechanism mainly applies to gases and submicron particles, as transport by Brownian diffusion decreases with increasing particle size. [2]

Interception and inertial forces may transport large ($>$ 1 μm) particles across the viscous sublayer. A particle with a diameter greater than the sublayer height moving with the mean motion of the air will be deposited by interception when it collides with an obstacle. This occurs when the particle is travelling on an air streamline that passes within one particle radius of the obstacle. Inertial forces may lead to impaction and turbulent inertial deposition. Impaction occurs when the particle leaves the streamline since it cannot follow rapid changes in the airflow, and collides with an obstacle. Turbulent inertial deposition occurs when inertial energy derived from the component of airflow perpendicular to the surface (turbulent eddies) carries a particle close to the surface.

Electrical migration is the movement of particles due to electrostatic attraction or repulsion. For most atmospheric conditions, Chamberlain [3] showed that this transport mechanism is minor in comparison to other mechanisms.

Thermophoresis is the movement of small particles in a temperature gradient driven from higher to lower temperature regions. [2] For temperature gradients found over most natural surfaces, thermophoresis effects are minor and can be neglected. [3]

Diffusiophoresis refers to movement of particles in a mixture of two (or more) gases along a concentration gradient in the direction of the diffusive flux of the heavier gas. [4] This may become important over water surfaces, where a humidity gradient exists due to evaporation of water from the surface. The movement of the water molecules drives aerosol particles away from the surface. [5] The importance of diffusiophoresis as a transport mechanism away from water surfaces is not well understood.

Once a particle has traversed the viscous sublayer, it will interact with the surface. Depending on the characteristics of the contaminant and surface, a particle may stick or bounce off. There may also be subsequent chemical reactions.

Particle deposition by any combination of these transport mechanisms largely depends on atmospheric, surface and particle characteristics. Examples of atmospheric parameters include wind speed, humidity, stability and temperature. In addition, micrometeorological parameters related to the surface such as friction velocity, roughness height and zero-plane displacement will affect the deposition. Surface properties which influence deposition include chemical and biological reactivity, geometry of roughness elements, terrain characteristics and wetness. Characteristics of the depositing particles also affect deposition; the most important of these include size, shape, density, reactivity, hygroscopicity and solubility. [6]

Dry deposition models attempt to account for these transport mechanisms and physical/chemical properties of a system. Unfortunately, none of the available models has been verified over the wide range of conditions likely to be encountered, and thus it is difficult to assess model uncertainties. In contrast to modeling, many researchers have conducted laboratory and field experiments to estimate dry deposition empirically. These measurements have the advantage of providing data for specific lab and field conditions, but there are questions about the extent to which these results can be generalized.

The rest of this chapter is divided into six sections. First, dry deposition of particles to vegetation is discussed, looking at both mathematical models and experimental results reported in the literature. Then we discuss dry deposition to bodies of water, again presenting mathematical models and experimental results. Dry deposition to urban surfaces is then briefly explored. Fourth, we address special issues associated with deposition of coarse particles. Resuspension of particles from surfaces is then considered, followed by a short section on critical loads.

2 DRY DEPOSITION TO VEGETATION

This section begins with a brief summary of models developed for particle deposition to vegetation. Some of this information has been taken from the literature review of Davidson and Wu. [1] Experimental data published since the 1990 review paper are also presented.

2.1 MODELING DRY DEPOSITION TO VEGETATIVE CANOPIES

Dry deposition is usually quantified in terms of a flux F ($g\,cm^{-2}\,s^{-1}$):

$$F = v_d(z)C(z) \tag{1}$$

where v_d is the deposition velocity ($cm\,s^{-1}$), and C the airborne concentration of particles ($g\,cm^{-3}$). v_d and C are functions of height z, although F is assumed to be invariant with z.

2.1.1 Aerodynamic Transport

Transport of particles in the free atmosphere occurs by turbulent diffusion and gravitational sedimentation. The minor influence of Brownian diffusion may also be included. The flux due to either diffusion process is modelled as a coefficient times the concentration gradient. The turbulent diffusion coefficient K ($cm^2\,s^{-1}$) of a contaminant above the vegetation canopy is assumed to be the same as the kinematic eddy viscosity of the air, whose expression is derived by analogy to the transport of air momentum:

$$K = ku_*(z - d) \tag{2}$$

where k is von Karman's constant and is equal to 0.4, u_* the friction velocity ($cm\,s^{-1}$), and d the zero plane displacement (cm). Within the canopy, the turbulent transfer coefficient is taken to be

$$K = khu_*\exp\left[-n\left(1 - \frac{z}{h}\right)\right] \tag{3}$$

where h is the height of the canopy, and n an empirical coefficient dependent on characteristics of the canopy.

In a non-adiabatic atmosphere, the expression on the right-hand side of each equation is divided by a correction factor ϕ_c to account for instability. Correction factors are determined experimentally, and are a function of $\zeta = (z - d)/L$, where L is the Monin–Obukhov length. Expressions for stability factors have been determined by Businger *et al.*, [7] Wesely and Hicks [8], and Hicks *et al.* [9]

The Brownian diffusivity D of a contaminant ($\text{cm}^2\,\text{s}^{-1}$) is defined as

$$D = \frac{kTc}{3\pi\mu d_p} \tag{4}$$

where k is the Boltzmann constant ($\text{g cm}^2\,s^{-2}\text{K}^{-1}$), T the temperature (K), c is the Cunningham correction factor, μ is the dynamic viscosity of air ($\text{g cm}^{-1}\text{s}^{-1}$), and d_p the physical diameter of the particle (cm).

For particles of sizes typically encountered in the atmosphere, the transport due to sedimentation by gravity is usually taken as the Stokes settling velocity:

$$v_g = \frac{\rho_p g c d_p^2}{18\mu} \tag{5}$$

where ρ_p is the particle density (g cm^{-3}). This formulation assumes that the density of air is negligible compared to the particle density so that the buoyant force can be neglected. The flux due to settling is calculated as the sedimentation velocity times the concentration.

Since model calculations assume spherical, unit density particles, the aerodynamic diameter is introduced to apply these models to real particles. The aerodynamic diameter is defined as the diameter of a spherical, unit density particle with the same motion characteristics as the actual particle. [10]

The total flux to the surface can be represented as (where downward flux is positive)

$$F = (D + K)\frac{dC}{dz} + v_g C(z) \tag{6}$$

Use of equation (6) requires several empirical parameters, since it is difficult to determine the boundary condition for $C(z)$ and there are complexities in estimating values of some parameters in the equation. For example, determining $C(z)$ requires estimates of particle removal by leaves, stems and other parts of the canopy. Collection efficiencies of canopy elements are discussed in section 2.1.2.

To estimate dry deposition, it is also necessary to determine the velocity profile. The velocity profile above the canopy can be calculated by analyzing momentum transport:

$$u(z) = \frac{u_*}{k} \ln \left(\frac{z - d}{z_0} \right) \quad \text{for } z > h \tag{7}$$

where z_0 is the roughness height (cm). Within a canopy, the wind speed profile is

$$u(z) = u(h)\exp\left[-n\left((1 - \frac{z}{h})\right)\right] \quad \text{for } z < h \tag{8}$$

where $u(h)$ is the wind speed at the height of the canopy.

There are distinct differences between the shapes of the wind profile $u(z)$ and the contaminant concentration profile $C(z)$ above the surface, as shown in Figure 1. The shape of the wind and concentration profiles can be used to identify the region where there is the greatest resistance to air momentum flux and contaminant flux, respectively. Under most conditions, there is a continual flux of momentum downward towards the surface by turbulent eddies; the wind speed decreases starting well above the surface, as momentum is rapidly destroyed at the surface by friction. However, the concentration profile remains relatively constant until very near the surface because most particles are only slowly transported through the viscous sublayer. Most of the gradient in momentum therefore occurs well away from the surface where resistance to momentum transport is greatest, while most of the gradient in particle concentration occurs in the viscous sublayer where resistance to particle mass transport is greatest.

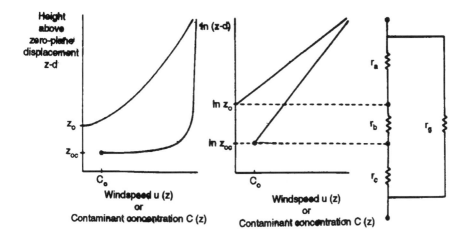

Figure 1. Wind speed and contaminant concentration as functions of height above the zero-plane displacement of a rough surface. Hypothetical sinks for air momentum and contaminant concentration are denoted by $(z - d) = z_0$ and $(z - d) = z_{oc}$, respectively. (From Davidson, C. I. and Wu, Y.-L., In Lindberg, S. E., Page, A. L., and Norton, S. A., eds., *Acidic Precipitation*, Springer-Verlag, 1990. With permission.)

2.1.2 Boundary Layer Transport: Collection Efficiency by Canopy Elements

To fully represent the particle dry deposition flux, equation (6) is combined with a formulation representing the flux of contaminants removed at the surface. This is determined from the rate at which particles are transported parallel to the ground integrated over the height of interest, combined with the efficiency of collection. The particle transport parallel to the ground is determined from the wind speed profile multiplied by the concentration profile.

The efficiency of collection depends on the area available for collection and the efficiency of the mechanisms which deposit the particle on the receptor. A "receptor" is a portion of the surface that will collect a contaminant, either a horizontal area or a vertical protrusion from a horizontal plane. Due to the complexity of any natural surface, this efficiency is extremely difficult to model. Mathematical models have been created to represent the area available for collection. One such model describes the surface as a field made up of blades of grass, where each blade is considered as an individual receptor. [11] Another model, in this case for a pine forest, uses a Gaussian distribution of foliage to determine the area available for collection. [12–14] Slinn [15] created a more general model in which a leaf area index was used to describe the surface area of vegetation per unit volume at a specific height. Finally, Wiman and Agren [16] created a model for transport and deposition through an open field to a forest where the particles were deposited. The ranges of deposition velocities resulting from these models are presented in Figure 2.

The collection efficiency depends on the surface and particle characteristics as well as the specific mechanisms. Brownian diffusion, interception and impaction are the main mechanisms by which particles are transported across the viscous sublayer. For a vegetative canopy, each leaf, stem or other canopy element may have a viscous sublayer. The collection efficiency of particles transported by Brownian diffusion can be represented by analogy to heat flow and is dependent on the diffusion coefficient D. The efficiency of interception depends on the ratio of particle to receptor size. Impaction efficiency is a function of the Stokes number and depends on the wind speed and specific characteristics of the particle and surface. All of the individual efficiencies can be combined to give an overall efficiency. [1]

2.1.3 Surface Interactions: Particle Rebound

Some contaminants may come into contact with a surface but do not stick to it. This bounce-off is usually characterized by a fractional value representative of the surface and type of contaminant.

Bounce-off occurs when the kinetic energy of the particle after impact is large enough to overcome surface attraction forces. Wu *et al.* [17] found that bounce-off is dominated by energetic particles driven by turbulent fluctuations, and

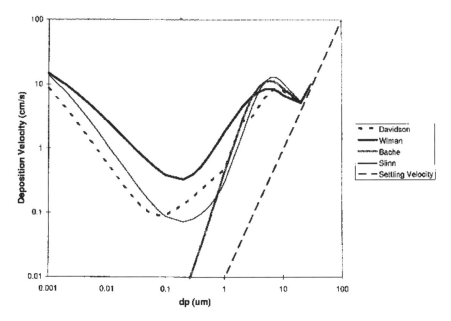

Figure 2. Deposition velocities from the models of Davidson *et al.*, [11] Slinn, [15] Bache [13] and Wiman and Agren. [16] (Redrawn from Ruijgrok, W., Davidson, C. I., and Nicholson, K. W., *Tellus*, **47B**, 587, 1995)

that the rebound fraction increases as either wind speed or particle size becomes greater. This analysis assumed that the attractive forces between the particle and surface are due only to van der Waals' forces. The fraction of rebound is determined from the probability that the particle velocity v_i just before reaching the surface is greater than the critical velocity v_i^*. Based on conservation of energy and the energy of van der Waals' attraction, v_i^* is calculated to be

$$v_i^* = \frac{u_*[A(1 - e^2)/18\pi\mu\nu h_0]}{(\tau_p^+)^{0.5}} \tag{9}$$

where A is the Hamaker constant ($\mathrm{g\,cm^2\,s^{-2}}$), e the coefficient of restitution, h_0 the smallest distance between the particle and the surface (cm), and τ_p^+ the non-dimensionalized particle relaxation time:

$$\tau_p^+ = \frac{\tau u_*^2}{\nu} \tag{10}$$

$$\tau = \frac{\rho_p c d_p^2}{18\mu} \tag{11}$$

where v is the kinematic viscosity of air (cm^2 s^{-1}). Predictions of the rebound fraction for a Monte Carlo simulation (50 runs) of the probability of $v_i > v_{I^*}$ are shown in Figure 3.

Wind tunnel experiments were also conducted to determine bounce-off fractions experimentally. Uranine particles were generated and discharged into the wind tunnel. Rebound fractions were calculated from the ratio of deposition rates to greased and ungreased Teflon plates. The results of the experimental fraction of rebound are presented with calculated results in Figure 3.

Paw U [18] suggests that the critical rebound velocity should be calculated from the vertical component of the particle velocity, not from the particle speed. Through wind tunnel studies with *Lycopodium* spores depositing on leaves, it was found that rebound began to occur at wind speeds of several m s^{-1} for flow parallel to the surface. However, for flow perpendicular to the surface, particle bounce-off occurred at speeds of 0.7 m s^{-1}. These results are consistent with earlier theories which used the kinetic energy based on the vertical component of the particle velocity. [19–21]

2.1.4 Turbulent Bursts

Although many models assume that the viscous sublayer consists of entirely laminar flow, this is not the case. There are bursts of turbulent air which move

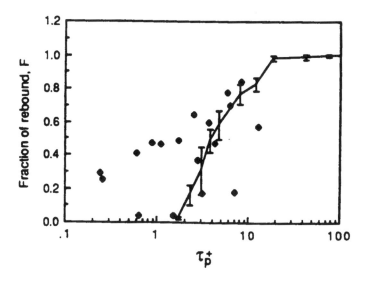

Figure 3. Model predictions for rebound fraction (solid line with one standard deviation error bars) compared with results measured in a wind tunnel. (From Wu, Y.-L., Davidson, C. I., and Russell, A. G., *Aerosol Sci. Technol.*, **17**, 231, 1992. With permission.)

swiftly through the viscous sublayer, carrying particles with them. Turbulent bursts may be eddies from the free atmosphere that extend down to the surface [17] or eddies created due to surface roughness. [22] Particles transported by these bursts are not affected by the viscous sublayer and move at approximately the same rate as in the aerodynamic transport region. Recently, developments have been made in quantifying this phenomenon.

Wu et al. [17] developed a model for particle dry deposition which includes the effects of turbulent bursts. In this model, particles near the centre of a turbulent burst deposit, while those far from the centerline are transported back into the main flow regime. The average lateral distance between bursts λ determines the probability of a particle depositing by this method. The limiting distance between the particle and center of the burst within which deposition occurs, y_{lim}, is determined by the trajectory of a particle which deposits $\lambda/4$ from the centreline. The fraction of particles in the viscous sublayer that deposits is $2y_{\text{lim}}/\lambda$, and the particle deposition velocity is $2w_0 y_{\text{lim}}/\lambda$, where w_0 is the instantaneous vertical velocity at the viscous sublayer edge. Figure 4 shows the results of this model when compared to measured results for a vertical, sticky surface.

Cleaver and Yates [22] developed estimates of turbulent burst size, spacing and rate of occurrence. On average, turbulent bursts have a diameter of $20(v/u_*)$, are laterally spaced $135(v/u_*)$ apart, and occur every $75(v/u_*^2)$.

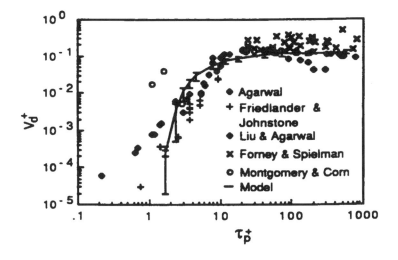

Figure 4. Modelled deposition velocities (solid line with one standard deviation error bars) compared with measured results reported in the literature. [91–95] (From Wu, Y.-L., Davidson, C. I., and Russell, A. G., *Aerosol Sci. Technol.*, **17**, 231, 1992. With permission.)

The effect of turbulent bursts on deposition rates depends on particle size. Particles in the viscous sublayer greater than $10\,\mu m$ are deposited swiftly due to gravity sedimentation and are not affected by turbulent bursts. Smaller particles carried by turbulent bursts will move at a speed comparable to that in the aerodynamic transport region. These small particles will have a deposition velocity up to 20 times greater than if they were only influenced by Brownian diffusion in the viscous sublayer.

2.1.5 Alternative Models using Resistance to Transport

Resistance models for dry deposition are used to calculate the dry deposition velocity based on the resistance to transport $(s\,cm^{-1})$ for each step in the process:

$$v_d = \frac{1}{r_a + r_b + r_c} + v_g \tag{12}$$

where r_a is the resistance to aerodynamic transport, r_b is the resistance to transport through the boundary layer and r_c is resistance to collection by the surface. Note that r_a does not include the effects of gravity.

The expression for aerodynamic resistance is based on equation (2), combined with the equation for flux in the absence of gravity $(F = K\,dC/dz)$:

$$r_a = \frac{\ln\left(\frac{z-d}{z_0}\right) - \Psi_c}{ku_*} \tag{13}$$

where Ψ_c is the stability correction factor for contaminants. The correction factor is introduced after integration of the equation for flux. Note that for an adiabatic atmosphere $(\Psi_c = 0)$, combining equation (13) with equation (7) yields $r_a = u(z)/u_*^2$.

The boundary layer resistance incorporates many mechanisms including eddy and Brownian diffusion, interception, inertial motion and sedimentation. The total effect of all of these mechanisms, when combined with aerodynamic transport, can be represented by a contaminant sink z_{0c}. The contaminant sink is the height at which the concentration is effectively zero for cases where the particles stick to the surface without bounce-off:

$$r_b = \frac{\ln(z_0/z_{0c})}{ku_*} \tag{14}$$

z_{0c} is usually very difficult to determine, since it depends on the various collection efficiencies of the different mechanisms. Instead of equation (14), a simple empirical equation has been developed: [9]

$$r_b = \frac{2(Sc/Pr)}{ku_*} \tag{15}$$

where the Prandtl number Pr accounts for the difference between heat and momentum transport:

$$Pr = v/\kappa \qquad (16)$$

where κ is the thermal diffusivity cm^2 s^{-1}).

The Schmidt number Sc accounts for the difference between momentum and mass transport: [9]

$$Sc = v/D \qquad (17)$$

The canopy resistance depends on the type of vegetation at the surface. Resistances for individual parts of the canopy may be summed according to the "big leaf" model: [9]

$$\frac{1}{r_c} = \frac{1}{r_s(Sc/Pr) + r_m} + \frac{1}{r_{cut}} + \frac{1}{r_{soil}} \qquad (18)$$

where r_s is the stomata resistance, r_m the mesophyll resistance, r_{cut} the cuticle resistance, and r_{soil} the resistance of the soil. The model of Kramm *et al.* [23] provides a similar formulation but also includes the resistance of nearly dry soil, and resistances of water due to leaf and soil wetness.

2.2 EXPERIMENTAL RESULTS

Dry deposition flux measurements may be made using either surface analysis methods or atmospheric flux methods. Surface analysis methods are those which analyze material accumulated over time on a surface. Examples of this type of measurement include foliar extraction, throughfall and stemflow methods, use of surrogate surfaces, and snow sampling. Atmospheric flux methods are those in which the flux is estimated from measurements of contaminants in the atmosphere. These include eddy correlation, the profile method, the tracer method, aerometric balance studies, eddy accumulation and variance methods. In this section, we describe these categories and provide recently published results. Additional descriptions and data for earlier measurements are provided by Davidson and Wu. [1]

2.2.1 Surface Analysis Methods

These methods provide a direct measurement of the amount of contaminant delivered to a surface. However, it is difficult to extrapolate the results at one location to an entire region.

2.2.1.1 Foliar extraction This involves "washing" of a plant surface to obtain material that has accumulated by dry deposition over an exposure period. The measurements are calibrated against leaves where the amount of contaminant deposited prior to the exposure period is known. One potential problem with the method is that some of the chemical species extracted using this method may be due to plant leaching, not dry deposition. In addition, it may be difficult to differentiate whether a chemical species existed in particulate or gaseous form before depositing. It is also difficult to extrapolate the flux to an entire canopy since leaves in different positions in a canopy will collect different amounts of atmospheric material. Therefore, large amounts of data are needed for reliable results. The results of two recent foliar washing studies are presented below.

Bytnerowicz et $al.$ [24] performed foliar extraction of NO_3^-, SO_4^{2-}, PO_4^{3-}, Cl^-, F^-, NH_4^+, Ca^{2+}, Mg^{2+}, Na^+, K^+, Zn^{2+}, Fe^{3+}, Mn^{2+}, Pb^{2+} and H^+ to pine branches in a subalpine zone at Eastern Brook Lake Watershed, Sierra Nevada, California. On each tree picked for analysis, two branches 5–10 cm long at 1–1.5 m above the ground were used. One of these branches was coated with Teflon to isolate the surface from the interior of the foliage. Overall, the amount of material collected by the Teflon-covered branches was less than that collected by the exposed branches, due to ions leached from the foliage. The total deposition flux to the entire forest was based on forest inventory data and leaf area indices. Flux to the ground was also estimated.

Shanley [25] made measurements of dry deposition of Ca^{2+}, Mg^{2+}, Mn^{2+}, Pb^{2+} and SO_4^{2-} to spruce foliage in the Black Forest, Germany. This experiment used branches of high-elevation Norway spruce near the top of the canopy. The branches were rinsed with deionized water and the resulting water samples were analized. Corrections for leaching were made in two ways. One method used correction factors from other experiments, while the second assumed that loadings during a clean air period were all from leaching. These two methods provided similar results. The results were compared to measurements made in nearby petri dishes. The comparisons showed wide variations and significant differences over the course of the experiment. Deposition velocities determined from foliar washing ranged from 0.034 to 0.78 $cm\,s^{-1}$, with a median of 0.33 $cm\,s^{-1}$. Figure 5 shows a comparison between deposition velocities to foliage and petri dishes.

2.2.1.2 Throughfall Throughfall is the water which falls to the ground beneath plant canopies during precipitation. [26] Chemical species concentrations in precipitation are subtracted from concentrations in throughfall to determine the dry deposition. The method assumes that material accumulated on surfaces by dry deposition since the previous storm is completely washed off by the precipitation. Throughfall can thus provide an estimate of the flux to an entire canopy. The measurements depend on the amount and strength of the rain, and the length of the preceding dry period. [27] Because the method relies

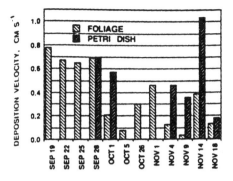

Figure 5. Deposition velocities of SO_4^{2-} to foliage and petri dishes in the Black Forest, Germany. (From Shanley, J. B., *Atmos. Environ.*, **23**, 403, 1989. With permission.)

on precipitation to remove contaminants, those particles irreversibly absorbed by the plant surface are not measured. In addition, leaching from foliage may contribute contaminants to the collected rain-water. As with foliar washing, this method cannot differentiate between particulate and gaseous forms of a contaminant.

In a study in the Hubbard Brook Experimental Forest, throughfall was used to estimate the dry deposition of sulfur. [28] Twenty-four collectors, each consisting of two 21 cm diameter funnels, were mounted 1 m above the ground. Throughfall results were compared with inferential estimates made from average measured airborne concentrations and deposition velocities calculated from the Hicks *et al.* [9] resistance model. The throughfall study estimated a flux of 3.3 kg S ha^{-1}year^{-1}, while the inferential estimate yielded 2 kg S/ha^{-1}year^{-1}.

Although there is much uncertainty associated with this method, researchers continue to use it since it provides an estimate of dry deposition flux over complex canopies. [26, 27, 29, 30] It is also often used in conjunction with other methods for comparison and corroboration. [31–33]

2.2.1.3 Surrogate surfaces   Surrogate surface measurements involve the use of artificial collectors. These surfaces may be aerodynamically designed so that flow characteristics can be analyzed. Other surfaces with less well- known airflow characteristics such as petri-dishes, buckets and Teflon plates may also be used. Surrogate surfaces offer a convenient method of measurement, but their results are difficult to extrapolate to natural canopies.

A relatively large number of surrogate surface deposition studies have been reported in the literature. Studies up to the late 1980s have been summarized by

Davidson and Wu. [1] Here we present results from a limited number of recent studies.

Santachiara *et al.* [34] used surrogate surfaces to measure deposition of aerosol particles in a cornfield near Bologna, Italy. Nuclepore filters (0.2 μm, 47 mm in diameter) were placed on the bottom and top of square plates and upward-facing polysterene petri dishes (53 mm inner diameter, 10 mm rim). The deposition fluxes and airborne concentrations were measured at heights of 160 and 60 cm (the canopy height). The collectors on the top and bottom of the plates were used to measure the downward and upward fluxes of particles, respectively. At the 160 cm level, the deposition velocities on the upward-facing Nuclepore filter were generally greater than those on the downward-facing filter with the exception of K^+, Na^+ and NH_4^+. At 60 cm the deposition velocities were approximately equal for Cl^- and NH_4^+, but were greater on the upward-facing filter for NO_3^-, Na^+, Ca^{2+} and Mg^{2+}. Surprisingly, the reverse was true for K^+ and SO_4^{2-}. Deposition velocities for the petri dishes were generally greater than for the flat plate at the same height; this is consistent with an earlier finding that deposition velocity increases as rim height increases, attributed to enhanced turbulence and decreased bounce-off due to the rim. [35] Results of this study are presented in Figure 6.

In a study of deposition of polychlorinated dibenzo-*p*-dioxins and dibenzo-furans (PCDD/F), Koester and Hites [36] used inverted frisbees and square flat plates with glass collection surfaces to measure dry deposition at two sites in Indiana. PCDD/F are toxic compounds which enter the atmosphere as by-products of certain combustion sources. The aerodynamically designed inverted frisbee samplers created minimal wind flow disturbances. The depth of the sampler and a coating of mineral oil minimize bounce-off. This study found that the dry deposition flux was inversely correlated with the temperature. At cooler temperatures more PCDD/F are adsorbed to particles which dominate their deposition flux. Deposition velocities for individual species at the two sites are presented in Figure 7. The average dry deposition velocity was found to be 0.2 cm s^{-1}.

2.2.1.4 Snow sampling This method uses increases in contaminant concentration in surface snow over time to estimate accumulation by dry deposition. Sampling times are at least 1–2 days to allow for significant accumulation. The method attempts direct measurement of dry deposition to a natural surface, although processes other than deposition can affect the concentrations. Sublimation, migration within the snow pack, and chemical changes may also affect the data. Furthermore, the results may be difficult to interpret due to heterogeneity in concentrations and the inability to distinguish between gaseous and particulate forms of chemical compounds.

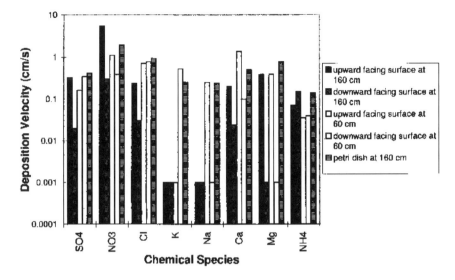

Figure 6.   Deposition velocities measured in a cornfield by Santachiara *et al.* [34]

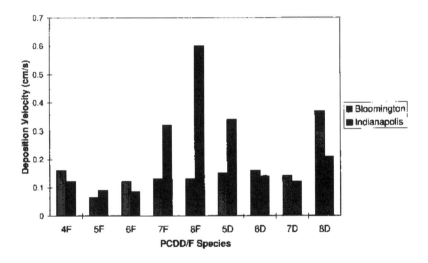

Figure 7.   Dry deposition velocities for PCDD/F species from Koester and Hites [36]

Snow samples were collected at Summit, Greenland after a 3-month period of accumulation in July, 1993 and 1994. [37] The samples were collected at 2 cm vertical intervals and then analyzed by ion chromatography for major anions and cations. The contaminant concentrations measured in the snow were compared with those predicted using a model for snow, fog, and dry deposition. The ratios of modeled to measured concentrations are 0.85 and 0.95 for SO_4^{2-} for the 1993 and 1994 field seasons respectively, and are 0.45 and 0.90 for Ca^{2+} for the 1993 and 1994 field seasons respectively.

2.2.2 Atmospheric Flux Methods

Atmospheric flux methods can be used to determine dry deposition from measurements of airborne concentrations and micrometeorological data. The short sampling times allow for better time resolution than other methods. Since these methods estimate flux from turbulent transfer only, they work best when applied to small particles which are less affected by sedimentation. [38] The methods require adequate fetch and uniform canopy characteristics for accurate results.

2.2.2.1 Eddy correlation Eddy correlation measurements may be taken from a tower or an aircraft. This technique involves measurement of the vertical wind velocity w and the contaminant concentration C. Each of these two properties is divided into a time-averaged component ($\bar{w}$, $\bar{C}$) and a deviation from the time-averaged component (w', C'). The deviations are caused by random turbulent fluctuations. The dry deposition flux is calculated as the time averaged value of the product, $\overline{w'C'}$. The vertical wind fluctuations are measured with a fast-response wind sensor, and the particle concentration fluctuations are measured with a sensor for the contaminant. Although the method provides a wealth of high-resolution flux data, sensors are expensive, are subject to noise and detection limit problems, and are not available for all contaminant species.

Eddy correlation measurements were made for small particles at high wind speeds at the Boulder Atmospheric Conservatory. [39] This study used an active-scattering aerosol spectrometer probe to measure particle concentration, and a sonic anemometer to measure vertical wind speed from a tower. For particles with a mean diameter of 0.2 μm, the deposition velocity was found to be $0.6 - 1.0$ cm s^{-1} for an unstable period with a mean horizontal wind speed of 8.7 m s^{-1}, and 0.1–0.4 cm s^{-1} for a stable period with mean wind speed of 6.4 ms^{-1}. These deposition velocities were compared to those estimated from the model of Slinn. [15] The measured results were nearly a factor of 2 greater than the modeled results.

2.2.2.2 Concentration profile The dry deposition flux of a contaminant is determined in this method from $F = K(dC/dz)$, where K is the estimated eddy diffusivity of the contaminant, and dC/dz is the measured concentration gradient. The eddy diffusivity can be estimated by analogy to sensible heat. The concentration gradient is obtained by measuring the concentration at more than one height. Correction factors are included for non-adiabatic atmospheric conditions. This method does not require fast response sensors as with eddy correlation. However, the results are very sensitive to the precision of the concentration data. In addition, number concentration and size-specific mass concentration measurements may not accurately represent deposition flux because deposition of small particles on to large particles will be measured as deposition to the surface. Furthermore, it is desirable to keep sampling times limited to periods of unchanging meteorology (generally a few hours at most).

The profile technique was used to measure dry deposition of small particles to a short grassy surface. [40] Wind speed and temperature profiles were obtained from measurements taken at four heights between 0.15 and 2 m above the surface. Aerosol concentrations were measured at six heights over this same range using a Nuclepore polycarbonate prefilter and a PTFE membrane at each height. The prefilter was used to exclude particles greater than 2 μm. Sampling occurred discontinuously over 50 separate intervals with a mean duration of 162 minutes, during which atmospheric conditions were neutral 29 times, unstable 17 times and stable 4 times. The gradients were determined from the least squares regression method, and the deposition velocity was calculated to be 0.08 cm s^{-1} for neutral and unstable conditions and 0.25 cm s^{-1} for stable conditions.

2.2.2.3 Tracers When a non-reacting, non-depositing tracer is released simultaneously with a contaminant of interest, the amount of dry deposition of the contaminant may be obtained. The method is useful over a wide range of distances (100 m–100 km), and sampling times may range from 1 hour to 1 day. It is assumed that dispersion of the non-depositing species is identical to that of the depositing species. This technique is primarily used for intensive field studies rather than routine dry deposition monitoring.

Lead deposition over Los Angeles was estimated by the dual tracer method using CO as the conservative species. [41] Vehicle emissions in a tunnel were used as the base Pb/CO ratio. The fraction of aerosol remaining airborne was calculated, as were deposition velocities over different size ranges. Results are shown in Figure 8.

2.2.2.4 Aerometric balance studies This method consists of taking measurements of contaminant concentrations by aircraft around the perimeter of a region. The data are used to estimate inflow and outflow from the region. When combined with estimates of contaminant loss by chemical

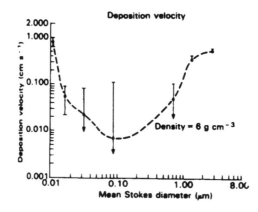

Figure 8. Deposition velocity vs particle size measured using the dual tracer method. (From Main, H. H., and Friedlander, S. K., *Atmos. Environ.*, **24A**, 103, 1990. With permission.)

reactions, the amount lost to dry deposition can be determined. This method works best over large regions and for contaminants with slow reaction rates. However, extensive aircraft sampling is necessary to construct the mass balance.

2.2.2.5 Eddy accumulation This technique measures the difference in contaminant concentration in updrafts and downdrafts to determine the flux from dry deposition. Two samplers are used: one operates only when the vertical wind velocity is downward, while the other sampler operates only when the wind is directed upward. The method obtains direct flux data over long averaging times for species without fast-response sensors. However, accurate flow controllers and precise chemical analyses are required since differences in concentrations in updrafts and downdrafts are likely to be small. Total averaging times of several days or longer may be needed.

2.2.2.6 Variance Since turbulent fluctuations in temperature, humidity and contaminant mass concentration are highly correlated, relationships can be developed to determine contaminant flux based on measurements of the other variables. When the sensible heat flux F_s is known, the contaminant flux is estimated as $F = F_s \sigma_c / \sigma_T$, where σ_c is the standard deviation of the contaminant concentration, and σ_T is the standard deviation of the temperature. The particle dry deposition flux may also be determined from the water vapour flux F_w, where $F = F_w \sigma_c / \sigma_w$. Here σ_w is the standard deviation of the water vapour concentration.

Aerometric balance studies, eddy accumulation and variance methods are not yet widely used for measuring atmospheric particle dry deposition, so no recent results are presented for these methods.

A summary of measured deposition velocities from recent experiments is presented in Figure 9.

3 DEPOSITION TO WATER SURFACES

Much of this section has been reproduced from Zufall and Davidson. [42]

Quantifying particle dry deposition to water presents difficulties not found in modelling dry deposition to vegetation. Water is an essentially smooth surface with intermittent wave action. These waves create bursting bubbles, spray formation and irregular geometries which are difficult to model due to their transient nature. In addition, the high humidity near the surface, the temperature difference between air and water, and movement of the water surface create complex situations which do not lend themselves easily to mathematical formulations.

The high humidity near the water surface promotes hygroscopic growth of certain types of particles. [43] These effects are most pronounced over fresh water, where the relative humidity may approach 100%, while over salt water, the relative humidity is limited to approximately 98.3% due to Raoult's law. Hygroscopic growth occurs rapidly and reduces the dependence of deposition velocity on particle size. [44]

The temperature difference between the air and water also affects dry deposition. In the spring and summer, the water is usually cooler than the air leading to stable conditions at the surface. During autumn and winter, the water temperature is warmer than the air which creates unstable conditions near the water surface. This increases turbulent mixing which enhances dry deposition.

The presence of waves, spray formation and bursting bubbles provides a formidable challenge for deposition modelling. The waves break up the viscous sublayer at the surface. The presence of waves and spray enables capture of particles above the horizontal water surface and provides more surface area to collect particles. The formation and influence of waves are strongly influenced by wind speed, and can be highly variable over short periods of time.

Another parameter specific to water is the slippage of the water surface. Unlike solid surfaces where a no-slip condition applies and the horizontal velocity equals zero, the surface of the water moves at approximately 3–5% of the mean wind speed at 10 m. [45, 46] This motion effects transport through the region just above the water surface.

Wind speeds tend to be higher over water than over nearby land. Although dry deposition rates are enhanced if winds become sufficiently strong to pro-

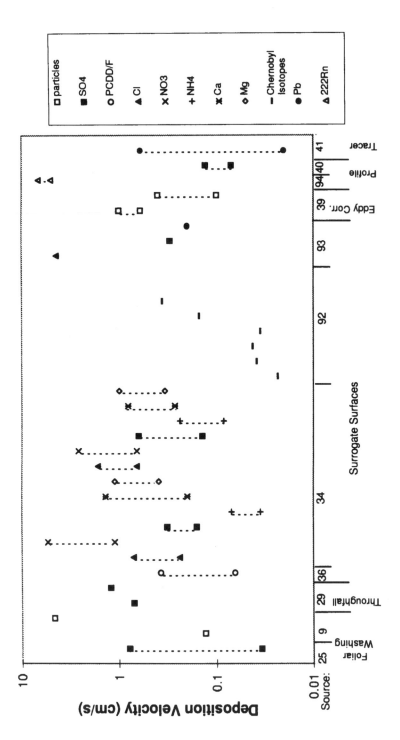

Figure 9. Deposition velocities over vegetative surfaces from different experimental methods. Data from the following investigators are included: Shanley, [25] Nicholson and Watterson, [96] Lovblad, [29] Koester and Hites, [36] Santachiara *et al.*, [34] Aoyama and Hirose, [97] Foltescu *et al.*, [98] Sievering, [39] Schery and Whittlestone, [99] Allen *et al.* [40] and Main and Friedlander [41]

duce breaking waves, increases in wind speed shorten transit time over a body of water, which may reduce total deposition. [47]

Several researchers have tried to incorporate these effects to develop a comprehensive model of particle deposition to natural water surfaces. Here we present in detail the models of Slinn and Slinn [45, 46] and Williams [44] and briefly describe several other models. We also describe several experimental studies.

3.1 DRY DEPOSITION MODELING TO WATER

3.1.1 Model of Slinn and Slinn [45, 46]

Slinn and Slinn developed a simplified model for natural water surfaces which takes into account air and slip velocity fluctuations, as well as hygroscopic growth. The model considers two layers, the "constant flux layer" far from the surface, and the "deposition layer" or viscous sublayer adjacent to the surface. For simplicity, the flux is also assumed to be constant within the viscous sublayer.

Dry deposition in the constant flux layer is dominated by turbulent transfer and sedimentation by gravity. Sedimentation can be modelled from Stokes law, as presented in equation (5). The particle deposition velocity from turbulent transfer may be obtained from analogy with a deposition velocity for momentum transfer:

$$v_{dm}(z) = C_d u(z) \tag{19}$$

where C_d is the drag coefficient:

$$C_d = \frac{u_*^2}{u(z)^2} \tag{20}$$

For solid surfaces, v_d is generally smaller than v_{dm} due to added resistance to account for particle transport across the viscous sublayer. However, because the surface of the water can slip, the surface may not be a perfect sink for momentum so that v_{dm} is reduced slightly; thus v_d for particles can be greater than v_{dm}. The deposition velocity for momentum may provide a reasonable estimate of that portion of v_d due to turbulent transfer. However, it must be recognized that particle deposition is often rate limited by transfer in the viscous sublayer, not by turbulent transfer mechanisms.

The flux at the top of the constant flux layer (arbitrary reference height) can be modeled as

$$F_h = k_c'(C_h - C_\delta) + v_{gd} C_h \tag{21}$$

where C_h is the concentration (g cm^{-3}) at the reference height h, C_δ the concentration at the height of the viscous sublayer δ, v_{gd} is the settling velocity for dry particles, and k'_c is the turbulent transfer coefficient for the constant flux layer excluding the effects of gravity (cm s^{-1}), developed from analogy to momentum transfer:

$$k'_c = \frac{C_d u_h}{1 - k} \tag{22}$$

Particle transfer in the viscous sublayer near the interface is much more complex. If the interface were solid, turbulent transfer must vanish at the interface because the vertical and horizontal components of the air velocity become zero (no-slip condition). However, at an air–liquid interface, the water may slip, so the horizontal component may be non-zero. This may suppress transfer of momentum at the surface to the point where viscosity dominates momentum transfer. Because the water surface may be moving and thus reduce the velocity gradient, the viscous sublayer thickness δ may be less than that for a solid surface, although the latter may be used as an estimate.

Atmospheric turbulence is assumed to have negligible direct influence on particle transport in this layer. Particles are assumed to possess their "wet radius", which is the equilibrium size from hygroscopic growth in the region of high humidity near the surface.

Transport across the viscous sublayer is approximated by Joukowski's "caterpillar-tread" model of deposition to smooth, sticky surfaces. This model defines deposition velocity as

$$v_d = v_g - \beta m' + u*^2 \, (Sc^{0.6} + 10^{-3/St})/(ku) \tag{23}$$

where u is the mean value of the horizontal wind (m s^{-1}) and $\beta m'$ is a mean slip velocity caused by diffusiophoresis. β is a constant equal to 1000 cm^3 g^{-1} and m' is the rate of water evaporation from the surface (g cm^{-2}s^{-1}). St is the number or impaction parameter defined by

$$St = \frac{\rho_p u c d_p^2}{18\mu L} \tag{24}$$

where L is a characteristic length of the receptor. [2] Note the similarity between equations (5), (11), and (24).

Equation (23) is modified in two ways to better represent the process: (i) the dependence of flux on D is changed to $D^{1/2}$ and (ii) the wet particle radius is used instead of dry radius. This leads to

$$k'_D = -\beta m' + C_d u_h (Sc^{-0.5} + 10^{-3/St})/k \tag{25}$$

where k'_D is the transfer coefficient for the viscous sublayer, excluding the effects of gravity (cm s^{-1}). The flux across the viscous sublayer can be approximated by

$$F_\delta = k'_D(C_\delta - C_0) + v_{gw}C_\delta \tag{26}$$

where v_{gw} is the Stokes settling velocity using the wet particle radius. It is assumed that the concentration at the interface C_0 is zero, and that there is no resuspension. The flux is assumed to be constant across both layers, so that $F_h = F_\delta$. With these assumptions, the deposition velocity can be solved from

$$\frac{1}{v_d} = \frac{1}{k_c} + \frac{1}{k_D} - \frac{v_{gd}}{k_c k_D} \tag{27}$$

In this equation k_c is the overall transfer coefficient for the constant flux layer including the effects of gravity (cm s^{-1} = $k'_c + v_{gd}$ and k_D is the overall transfer coefficient for the viscous sublayer including the effects of gravity (cm s^{-1}) = $k'_D + v_{gw}$. The resulting deposition velocities from this equation are presented as a function of particle size in Figure 10.

This model has been used by many researchers to predict the amount of dry deposition to a water surface (e.g. Baeyens et al. [48], Dulac et al. [49, 50], Rojas et al. [51, 52]).

3.1.2 Model of Williams [44]

The Williams model of dry deposition of particles to natural water surfaces includes effects of spray formation in high winds, effects of particle growth due

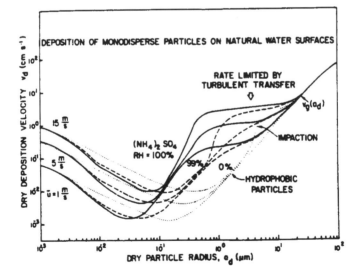

Figure 10. Plots of modelled deposition velocity for three mean wind speeds u and three relative humidities. (From Slinn, S. A., and Slinn, W. G. N., *Atmos. Environ.*, **14**, 1013, 1980. With permission.)

to high relative humidities, the variation of turbulent transfer with wind speed, air/water temperature differences, and surface roughness. It uses a two-layer model similar to Slinn and Slinn. [45,46]

Particle transfer in the constant flux layer is governed by turbulent transfer and gravitational settling. In the viscous sublayer, transfer can follow two parallel paths in addition to gravitational settling. One path is controlled by the boundary layer resistance r_b across the viscous sublayer from air to the smooth water surface. The other path is transfer to that fraction of the water surface which is broken by the formation of spray and bubbles. In addition, gravitational settling in the viscous sublayer is adjusted for particle growth due to high relative humidity.

The main difference between this model and that of Slinn and Slinn [45,46] is the introduction of broken surface effects. The fraction of area that has a broken surface is represented by α. This value is strongly dependent on wind speed and can be calculated as [53]

$$\alpha = 1.7 \times 10^{-6} u(10)^{3.75} \tag{28}$$

where $u(10)$ is the wind speed at a height of 10 m. The flux in the constant flux layer is represented by

$$F(z) = (1 - \alpha)k_{as}(C(z) - C_{\delta s}) + \alpha k_{ab}(C(z) - C_{\delta b}) + v_{gd}C(z) \tag{29}$$

where k_{as} is the transfer coefficient in the constant flux layer for a smooth surface and k_{ab} the analogous term for a broken surface. Note that the flux is written as a function of height, although in steady state this is assumed constant throughout the layer and is equivalent to F_h defined earlier. The turbulent transfer coefficient is given by [54]

$$k_{as} = \frac{\kappa u_*}{[\ln(z/Z_{0cs}) - \Psi_h(z/L)]} \tag{30}$$

A similar expression defines k_{ab} where the roughness length for a smooth surface z_{0cs} is replaced with z_{0cb}, that for a broken surface. These parameters are interrelated and depend on the wind speed and air/water temperature difference. The flux through the deposition layer is represented by

$$F_\delta = (1 - \alpha)k_{ss}(C_{\delta s} - C_0) + \alpha k_{bs}(C_{\delta b} - C_0) + (1 - \alpha)v_{gw}C_{\delta s} + \alpha v_{gw}C_{\delta b} \tag{31}$$

where k_{ss} is the smooth surface transfer coefficient due to diffusion and impaction and is taken from Slinn and Slinn [45] and Slinn. [55] k_{bs} is the broken surface transfer coefficient which depends on scavenging by impaction and coagulation with spray droplets, but its formulation is entirely unknown.

This model assumes that the surface is a perfect sink for particles so that C_0, the concentration at the interface, equals zero, and that the system is at steady state so that $F_h = F_\delta$. Figure 11 gives results of the deposition velocity from this model.

Model results suggest that stability effects are important only at low wind speeds, and even then only stable temperature gradients need be considered. In addition, broken surface transfer has a more pronounced effect than particle growth in the deposition layer.

Figure 12 shows a comparison of the Slinn and Slinn [45,46] model and the Williams [44] model at low and high wind speeds. At low wind speeds, the models are similar. However, Slinn and Slinn's approximation of the deposition velocity of intermediate size particles by the turbulent transfer coefficient produces higher deposition velocities than the Williams [44] model. At high wind speeds, where the effects of bursting bubbles are most pronounced, the model of Williams [44] results in much higher deposition velocities for small particles.

3.1.3 Other Water Deposition Models

The model of Fairall and Larsen [56] represents the dry deposition process to an open ocean through three layers: viscous sublayer, turbulent surface layer and mixed layer. The mixed layer extends above the surface layer to approximately 1000 m. The model formulation is similar to that of Slinn and Slinn,

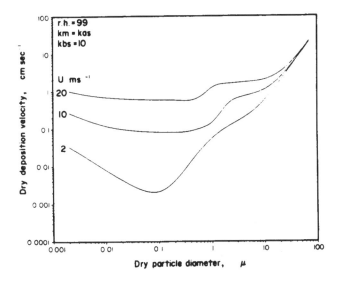

Figure 11. Plots of modelled deposition velocity for three different wind speeds, relative humidity of 99%, the lateral transfer coefficient k_m equal to k_{as}, and no difference between the air and water temperatures. (From Williams, R. M., *Atmos. Environ.*, **16**, 1933, 1982. With permission.)

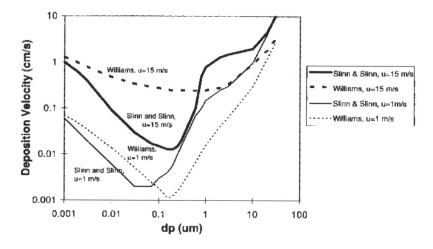

Figure 12. Comparison of Slinn and Slinn [45,46] and Williams [44] models at high and low wind speeds

[45,46] but takes into account the possibility of non-zero surface concentration. The surface source provides a replenishment of particles from the water to the atmosphere.

Gatz [57] developed a model for particles resuspended from urban surfaces transferred horizontally to a lake. This model and experimental results found that a significant portion (up to 25%) of atmospheric pollutant loading to Lake Michigan is due to resuspended particles.

Other studies have used modified versions of the models presented here or new formulations (e.g. Hummelshoj *et al.*, [58] Rojas *et al.*, [52] Quinn *et al.* [59]).

3.2 MEASUREMENTS OF DEPOSITION TO WATER SURFACES

Experimental deposition studies provide estimated amounts of particle deposition to specific bodies of water and also help to validate models. Measurements have been conducted using surrogate surfaces placed over the water and using wind tunnels with trays of water. No techniques have yet been employed to measure directly deposition to a natural water surface. The following are the results of recent experimental measurements of dry deposition to water.

3.2.1 Surrogate Surfaces

Dulac *et al.* [50] conducted experiments on dry deposition of aluminium to the north-western Mediterranean Sea. The flux was measured using an open hemispheric plexiglass collection device, $0.1 \, m^2$ in surface area with a 10 cm high neck, covered by 1 mm nylon mesh. Airborne size distributions of particles were measured with a high-volume cascade impactor. Observed deposition velocities ranged from 1.2 to $7 \, cm \, s^{-1}$. The observed values were compared to calculations using the Slinn and Slinn [45,46] model. This model was applied using several different methods to determine the effect of the large size fraction. When the deposition velocity was calculated using a single mass-median diameter, it underestimated the observed values by at least an order of magnitude. To account for the influence of large particles, v_d was also calculated by: (i) using a correction factor from the upper tail of the mass particle size distribution; (ii) calculating a mean v_d for every impactor stage i assuming an average diameter d_i and weighting the overall v_d : $v_d = \sum C_i v_{di} / C$ where there are six stages; and (iii) dividing a fitted lognormal particle size distribution into 100 successive size intervals i, each corresponding to 1% of the total mass and weighting the v_d over these intervals. The third method provided the best agreement to observed values. This suggests that the dry deposition fluxes are controlled by the gravitational settling of the largest particles ($> 10 \, \mu m$). This also suggests that the current aerosol sampling techniques underestimate aerosol mass concentrations because they do not correctly sample particles $> 10 \, \mu m$. This can result in a significant underestimation of deposition fluxes calculated with size distribution data.

3.2.2 Wind Tunnel Studies

Wind tunnel studies by Larsen *et al.* [60] determined deposition velocities to water under controlled conditions. MgO particles were generated and allowed to circulate in a wind tunnel. Measurements were made for wind speeds of 3 and $7 \, m \, s^{-1}$, with and without the presence of bursting bubbles, for particle diameters of 0.5 and 1 μm. Results of this study are presented in Figure 13. The deposition velocity increased by a factor of 10 for the higher wind speed, increased slightly when the bubblers were activated, and barely changed between the two particle sizes.

Figure 14 presents a summary of deposition velocities to water surfaces found by recent researchers through both models and experiments. The elements are ordered by increasing mass-median diameter according to Milford and Davidson. [61,62]

4 DRY DEPOSITION IN URBAN AREAS

The concentrations and fluxes of airborne pollutants in urban areas are generally higher compared to less populated regions due to the proximity of pollutant sources. For example, Noll *et al.* [63] found fluxes of Ca, Cd, Zn,

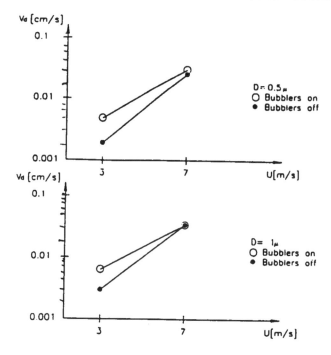

Figure 13. Deposition velocities measured in a wind tunnel, with and without bursting bubbles. (From Larsen, S. E., Edson, J. B., Mestayer, P. G., Fairall, C. W., and De Leeuw, G., *In* P. Borrell, P. M. Borrell, and W. Seiler, eds. *Transport and Transformation of Pollutants in the Troposphere*, SPB Academic Publishing, 1991. With permission.)

Pb and Cu to be 10–300 times greater near Chicago than at a nearby rural site. These high levels of contaminants harm vegetation and contribute to the discolouring and decay of man-made structures. In addition to the economic consequences of building degradation, the irreplaceable loss of buildings and structures of historical or artistic importance is a concern (Saiz-Jimenez, [64] Gould *et al.*, [65] Sherwood *et al.* [66]).

The heterogeneous structure of a city complicates the quantification of dry deposition. Each structure disturbs wind flow patterns, altering the direction and speed of the wind, and in some cases, introducing vortices near the building surface. When buildings are close to each other, their individual flow fields interact, and when very close together, the main flow skims over the tops of structures. [67] The particular type of surface also affects the amount of deposition. Some building materials are more reactive with the airborne contaminants and therefore enhance deposition.

In addition to deposition variations from building to building, deposition on a single structure varies widely at different locations depending on wind pat-

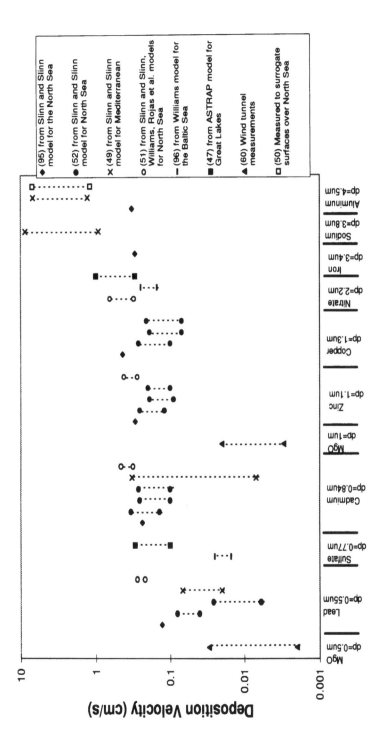

Figure 14. Deposition velocities found for water surfaces from both measured and modelled results. Data from the following investigators are included: Ottley and Harrison, [100] Rojas *et al.*, [51,52] Dulac *et al.*, [49,50] Lindfors *et al.*, [101] Shannon and Voldner, [47] Larsen *et al.*, [60] Values of d_p are mass-median aerodynamic diameters taken from literature reviews of Milford and Davidson, [61,62] with exception of $d_p = 0.5\ \mu m$ and $d_p = 1\ \mu m$ for MgO taken from Larsen *et al.* [60]

terns, exposure, humidity and temperature. Edges, corners, projections and carved ornaments usually increase turbulence and exposure to pollutants, but may also protect portions of the surface, depending on the airflow patterns. These also increase the extremes of temperature and moisture ranges, due to shading from solar radiation and sheltering from precipitation. [66] In general, as the complexity of the form increases, dry deposition to the material increases.

Despite the complexity and uncertainty associated with quantifying dry deposition to urban areas, relatively accurate models of deposition to both entire urban areas and single buildings have been developed. These models are based on the same principles as those to vegetative surfaces. Estimates of dry deposition to urban areas have also been made through experimental methods.

4.1 MODELING DRY DEPOSITION TO URBAN AREAS

To represent deposition to an entire urban area, models may use simple area averaged meteorological characteristics. Urban deposition models have been developed in terms of the resistance analogy model. Aerodynamic resistance can be calculated from

$$r_a(z) = \frac{u(z)}{u_*^2} \Psi(z/L) \qquad (32)$$

where Ψ is a correction factor for atmospheric stability slightly different from the correction factors introduced in section 2.1.5.

Boundary layer resistance in an urban area depends on the complex flow patterns around a building and on the structure's roughness characteristics. Based on theoretical considerations and wind tunnel studies, the following formulation has been developed:

$$r_b = \frac{\ln(z_0/z_{0c})}{ku_*} = \frac{c(Re_*)^m(Sc)^n}{ku_*} \qquad (33)$$

where m, n and c are constants estimated to be 0.45, 0.8 and 0.9, respectively by Owen and Thompson [68] or 0.25, 0.5 and 2.9, respectively by Brutsaert. [69,70]

The surface resistance component depends on aerodynamics of the building surfaces, the microclimate, and the chemistry of the surface and the contaminant. Because it depends on the specific nature of the contaminant and surface, there is no general formulation to describe the surface resistance. Values of r_c have been determined experimentally for certain surfaces and substances.

4.2 EXPERIMENTAL MEASUREMENTS OF DRY DEPOSITION TO URBAN SURFACES

Dry deposition to urban areas is generally measured directly from the surface to be studied or from surrogate surfaces. Sherwood *et al.* [66] present a literature survey on the topic. A few additional recent studies are summarized here. Wu *et al.* [71] measured deposition to various parts of an equestrian bronze statue in Gettysburg National Military Park, PA. Deposition fluxes were measured on the statue's right flank, left flank, back of horse, and belly and on closely located surrogate surfaces. The deposition velocity for SO_4^{2-} was greatest for the surrogate surface ($0.3\,\mathrm{cm\,s^{-1}}$), followed by the back of the horse ($0.19\,\mathrm{cm\,s^{-1}}$), the left flank ($0.18\,\mathrm{cm\,s^{-1}}$), the belly ($0.09\,\mathrm{cm\,s^{-1}}$) and the right flank ($0.038\,\mathrm{cm\,s^{-1}}$).

A study conducted at the Cathedral of Learning on the University of Pittsburgh campus measured deposition to both vertical and horizontal surfaces (Gould *et al.*, [65] Lutz *et al.*, [72] Etyemezian *et al.* [73]). Airborne concentrations were measured using Teflon Zeflour filters. Greased and ungreased Teflon surfaces were placed on aerodynamically designed airfoils to measure the flux of contaminants to horizontal surfaces and on the sides of the building to determine deposition to vertical facing surfaces. Deposition velocities of SO_4^{2-} to horizontal surfaces ranged from 0.1 to $0.6\,\mathrm{cm\,s^{-1}}$, while the ratio of horizontal to vertical v_d varied from 1 to 6. Airflow patterns around the building were found to significantly influence deposition, as fluxes measured close to one another differed by a factor of 6.

Another study compared deposition rates between urban and rural areas in south-east Michigan. [74] The urban site was located in Warren, Michigan, 7 km north of Detroit, while the rural site was located 66 km north of Warren in Lapeer, Michigan. Deposition fluxes were measured in polyethylene buckets and airborne concentrations were measured with Teflon filters. Deposition velocities in the urban region were found to be 0.63 $\mathrm{cm\,s^{-1}}$ for SO_4^{2-}, $0.69\,\mathrm{cm\,s^{-1}}$ for NO_3^-, $3.1\,\mathrm{cm\,s^{-1}}$ for Cl^-, $2.3\,\mathrm{cm\,s^{-1}}$ for Ca^{2+}, $1.9\,\mathrm{cm\,s^{-1}}$ for Mg^{2+}, $2.3\,\mathrm{cm\,s^{-1}}$ for Na^+ and $1.9\,\mathrm{cm\,s^{-1}}$ for K^+. Deposition fluxes in the urban region were consistently higher than those in the rural area by a factor of 2.

Experimental measurements of contaminant fluxes to an entire city are more difficult to obtain. As evidenced by measurements made on the equestrian statue and on the Cathedral of Learning, deposition rates may vary greatly on a single structure, in addition to variations from building to building. Also, the pollutant airborne concentrations may vary at different locations due to numerous sources in the area. Although urban dry deposition is an important problem, much uncertainty remains in its analysis.

5 COARSE PARTICLES

Investigators have provided evidence of the importance of large particles in affecting overall mass dry deposition of a chemical species. For example, Davidson and Friedlander [75] and Davidson et al.[35] estimated that the bulk of the mass deposition of certain trace metals and SO_4^{2-} was due to a small fraction of airborne particles. Coe and Lindberg [76] analyzed particle dry deposition using a scanning electron microscope. Although 40% of the deposited number of particles were less than 2 µm in diameter, the deposited particle mass was dominated by particles greater than 10 µm.

Recently several deposition studies have focused on larger ($d_p > 5$ µm) particles. These studies found that despite the small number of these large airborne particles, their deposition accounts for a substantial portion of total dry deposition (Holsen et al., [77] Holsen and Noll, [78] Baeyens et al., [48] Dulac et al., [50] Zufall and Davidson[79]). Dry deposition modeling formulations which use total particle mass concentrations in all size ranges and a single deposition velocity tend to underestimate the contribution of coarse particles and thus the dry deposition. [78] The importance of the large size fraction has prompted the development of new sampling techniques for large particles [80] and new coarse particle models. [81]

Noll and Fang [81] developed a model to predict dry deposition of atmospheric coarse particles over a smooth horizontal surface. This model considers large (> 5 µm) particles to be deposited by sedimentation and inertia. The deposition velocity is defined as

$$v_{di} = v_{gi} + v_{Ii} \tag{34}$$

where v_{gi} and v_{Ii} are the settling velocity and inertial deposition velocity, respectively, for size range i. The inertial deposition velocity is calculated from

$$v_{Ii} = \varepsilon u_* \tag{35}$$

where ε is the effective inertial coefficient determined through experiment to be

$$\varepsilon = 1.12e^{-30.36/d_p} \tag{36}$$

This expression was developed by comparison of deposition to the top and bottom of a flat plate. The deposition velocity to the underside of the plate was considered to be equal to the inertial deposition velocity minus the sedimentation velocity. The variable ε was plotted with data from the top and bottom of the deposition plate as a function of particle diameter, which led to equation (36). This analysis assumes that atmospheric turbulence is sufficient to provide a uniform particle concentration at the top of the viscous sublayer near the plate surface. Results of this model, applied only to $d_p > 5$ µm, are shown in Figure 15.

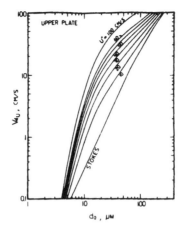

Figure 15. Deposition velocity as a function of size for coarse particles. (From Noll, K. E. and Fang, K. Y.-P., *Atmos. Environ.*, **23**, 585, 1989. With permission.)

A study by Holsen and Noll [78] compared different models with experimental results (fluxes measured to a smooth surrogate surface) to assess the importance of the large size fraction. Results showed that both the Slinn and Slinn [45] model and the Noll and Fang [81] model had results closer to experimental measurements when the particle size distributions were taken into account. The modeled dry deposition flux was 0.033–4% of the measured flux when average particle concentrations and average deposition velocities were used in the Slinn and Slinn model. When particle size distributions were taken into account, the Slinn and Slinn [45] model predicted values 10–33% of the measured flux, while the Noll and Fang [81] model for coarse particles overpredicted the measured flux by two to five times except for the largest particles in the distribution. A comparison of the Slinn and Slinn [45] model, Noll and Fang [81] model and measured flux distributions is given in Figure 16.

6 RESUSPENSION

Once particles are deposited, they may be resuspended into the atmosphere if their lift force is greater than the forces holding the particle to the surface. Significant resuspension of particles decreases the total amount of dry deposition. Note that the initial suspension of particles from a surface occurs by the same process as described here for resuspension.

Wu *et al.* [82] determined that particle resuspension occurs in two steps. The particle is first moved by the drag force, causing a decrease in the surface

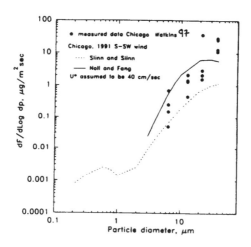

Figure 16. Flux distributions as modelled by Slinn and Slinn, [45] Noll and Fang [81] and measured to a flat plate, Watkins. [102] (From Holsen, T. M. and Noll, K. E., *Environ. Sci. Technol.*, **26**, 1807, 1992. With permission.)

adhesion force due to surface roughness. Next, the lift force moves the particle away from the surface, carrying it into the free atmosphere.

The amount of resuspension depends on free stream velocity, relative humidity, particle characteristics and surface characteristics. In addition, the amount of previously deposited material can be an important factor.

The free stream air velocity has a major effect on resuspension according to wind tunnel experiments which used uranine particles, polymer microspheres, *Lycopodium* spores, paper mulberry pollen and Johnson grass pollen ranging in size from 5 to 42 μm. [82] Experiments showed that there was little if any resuspension at wind speeds of 1.2 and 1.8 m s⁻¹. At 4 m s⁻¹ there was only occasional resuspension. The fraction of particles resuspended has been found to increase with wind speed to the power n, where n may range from 1 to 6 (e.g. Sehmel [6]).

Increases in relative humidity were found to decrease resuspension, due to an increase in the adhesion force. This effect is most pronounced for lower wind speeds. At a 62% humidity the fraction of particles resuspended at wind speeds of 5, 6, 7 and 8 m s⁻¹ was found to be 0.26, 0.49, 0.73 and 0.84, respectively. At 78% humidity, the fraction of particles resuspended at those wind speeds was found to be 0.02, 0.07, 0.37 and 0.81. [82]

Particle characteristics greatly influence the amount of resuspension from a surface. Larger particles are found to resuspend more easily because the drag force increases more quickly than the adhesive force and because of greater protrusion into the turbulent airstream. The increase in resuspension for larger particle sizes is most notable at high wind speeds. [83] This suggests that large particle resuspension may be dominated by brief wind gusts. Particle shape and composition may also be important. Resuspension was less for 42 μm Johnson grass pollen particles than for 30 μm *Lycopodium* spores due to the irregular shape of grass pollen, as well as differences in particle composition. [82]

The type of surface also affects the amount of particle resuspension. Resuspension from glass, plexiglass, and the upward-facing and downward-facing surfaces of an oak leaf were compared. [82] The largest amount of resuspension occurred from the glass surface followed by the top of the oak leaf, plexiglass, and the down-facing surface of the oak leaf. The differences were most likely due to increased roughness and electrostatic effects. However, Nicholson [83] found that resuspension rates from concrete and grass surfaces were relatively similar, suggesting that the surface effects on resuspension are complex and not well understood.

Finally, the amount of previously deposited material affects resuspension. In experiments with initially particle-laden surfaces, the number of particles remaining on a surface was found to decrease exponentially [82] or according to t^{-n}. [84] Initially ($t < 1$ min) all of the easily resuspended particles are removed, followed by the slower resuspension of the remaining particles.

These factors can be combined into an empirical resuspension rate Λ, equal to the fraction of surface deposit removed per unit time. Studies in wind tunnels and at natural sites have been conducted to determine values for Λ. Results of two wind tunnel studies are shown in Figure 17. Values several orders of magnitude lower than those in the figure have been reported for resuspension of radionuclides in the contaminated areas of Russia and Belarus. Gavrilov *et al.*[85] reported values for Λ of approximately $10^{-9.6}$ s^{-1}.

A study of resuspension in deciduous and pine forests found resuspension rates to be on the order of $10^{-6} - 10^{-5}$ s^{-1}. [86] These rates were determined from surrogate surface methods and throughfall. The surrogate surfaces were placed at different heights within and above the canopy to determine the vertical variation of resuspension. The resuspension rates were only slightly higher above the canopy than within the crown. The decrease in resuspension rates with depth into the canopy was probably due to the lower wind speeds within the canopy. However, the resuspension rates did not decrease proportionally to the wind speed, suggesting that large, infrequent turbulent eddies in the canopy were responsible for a portion of the resuspension.

Once the particles are resuspended their trajectories are determined by the flow characteristics of the turbulent airflow, not the velocity of their

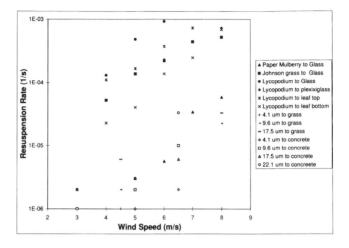

Figure 17. Resuspension rates from wind tunnel studies of Wu *et al.* [82] and Nicholson [83]

release. [82] These resuspended particles may then be transported and redeposited.

7 CRITICAL LOADS

An important application of dry deposition knowledge involves quantifying the flux of pollutants to determine if they will lead to an ecosystem reaching its critical load. Critical load is defined as an exposure to one or more pollutants above which there are significant harmful effects on specified sensitive elements of the environment. [87] By determining critical loads, goals can be set for future pollutant inputs to an ecosystem based on available technology and minimal

Table 1. Critical loads for total acid deposition (mol ha^{-1} year^{-1})

Source	Forests	Surface waters
De Vries,[89] Netherlands	1100–1400	400
Warfvinge and Sverdrup,[90] Sweden	1640	1225
Nilsson and Grennfelt,[87] Sweden (empirical data)	700–1100*	
Nilsson and Grennfelt,[87] Sweden (mass balance method)	200–1400*	

*These values are for nitrogen only.

harm to the ecosystem. This concept was originally developed for acidification and eutrophication, but may be used for other pollutant problems. [88]

The critical acid load for forest soils CL_{acid} (moles ha^{-1} year^{-1}) can be formulated as

$$CL_{acid} = BC_w + ALK - BC_t - AC_N \qquad (37)$$

where BC_w is the weathering of the base cations, ALK is the lowest acceptable output of alkalinity from the system, BC_t is the permanent uptake of base cations, and AC_N is the acidity input due to conversion of nitrogen within the soil. [88] The lowest acceptable output of alkalinity is the main criteria in determining critical loads to forest soils. This value defines a chemical status in the soil which must be maintained and may be determined from the concentration of Al, the Al/Ca ratio, or the NH_4/K ratio in the soil water. [89]

The critical acidifying load may also be determined from the critical load for each acidifying species.

$$CL_{acid} = CL(SO_x) + CL(NO_x) + CL(NH_4) - CL(BC) \qquad (38)$$

where $CL(x)$ is the critical deposition load for the species x. [88] The critical load of each species is determined from mass balance methods and ecosystem changes or from empirical methods.

Critical loads to surface waters are calculated to maintain a pH of at least 5.3 and to allow maximum concentrations of Al and NO_3 from European Community drinking water standards of 0.2 and 50 mg L^{-1} respectively. [88] Results of several critical load analyses are presented in Table 1.

Although these values are developed from a steady-state analysis and are presented as a yearly average, short-term effects can be very important. Seasonal, diurnal and precipitation variations can greatly alter the state of an ecosystem. In addition, the critical load depends largely on the actual deposition and individual ecosystem.

8 SUMMARY

While many advances have been made in the study of dry deposition, uncertainties still remain. This chapter has considered the current state of research on particle dry deposition to vegetation, water and urban surfaces.

Particles are transported through the free atmosphere mainly by turbulent transfer and gravitational settling. The particles are then transferred by interception, impaction or Brownian diffusion through a relatively quiescent viscous sublayer to a surface. Once at the surface the particles may stick or bounce off, depending on the atmosphere, particle and surface characteristics. At a later date, the particle may also resuspend into the atmosphere.

Dry deposition may be quantified by modeling or experiments. The latter may include field tests or laboratory experiments such as wind tunnel studies. The deposition velocity, defined as the contaminant mass flux divided by its airborne concentration, depends largely on particle size, wind speed and a variety of other characteristics of the particle, surface and atmosphere.

The dry deposition flux can be measured using surface accumulation methods or by atmospheric flux techniques. The former refers to experiments which measure accumulation of contaminants at a particular location and extrapolate this flux to an entire area. Atmospheric flux methods use airborne concentrations and atmospheric conditions to infer deposition. While certain models and experimental methods provide reasonable estimates of dry deposition for specific circumstances, none have been proven to provide consistently accurate results over a wide range of conditions.

Atmospheric inputs can cause substantial harm to an ecosystem or structure, and dry deposition may account for a substantial portion of these inputs. The critical load is the amount of atmospheric pollutant load an ecosystem can handle without substantial harm. A greater understanding of the dry deposition process assists in determining this critical load and averting damage.

ACKNOWLEDGEMENTS

The authors would like to thank IUPAC for their support and the book editors and other authors for their helpful comments. This work was funded in part by EPA contract CR 82-2054-01-1 through a subcontract with the University of Michigan.

SYMBOLS

A Hamaker constant ($g\,cm^2\,s^{-2}$)

ACN   Acidity input due to conversion of nitrogen within the soil ($moles\,ha^{-1}year^{-1}$ or $kg\,km^{-2}year^{-1}$)

ALK   Lowest acceptable output of alkalinity from system ($moles\,ha^{-1}year^{-1}$ or $kg\,km^{-2}year^{-1}$)

BC_t Permanent uptake of base cations from system ($moles\,ha^{-1}year^{-1}$ or $kg\,km^{-2}year^{-1}$)

BC_w Weathering of base cations added to system ($moles\,ha^{-1}year^{-1}$ or $kg\,km^{-2}year^{-1}$)

C Airborne concentration of a contaminant ($g\,cm^{-3}$)

c Cunningham correction factor

C' Turbulent fluctuating component of airborne contaminant concentration ($g\,cm^{-3}$)

$\bar{C}$ Time-averaged airborne contaminant concentration ($g\,cm^{-3}$)

C_d Drag coefficient

C_δ Airborne contaminant concentration at height of viscous sublayer ($g\,cm^{-3}$)

C_h Airborne contaminant concentration at reference height ($g\,cm^{-3}$)

C_i Airborne contaminant concentration of particles in size interval i ($g\,cm^{-3}$)

CL_{acid} Critical acid load (moles $ha^{-1}year^{-1}$ or $kg\,km^{-2}year^{-1}$)

C_0 concentration at interface between air and a water surface ($g\,cm^{-3}$)

D Brownian diffusivity of the contaminant ($cm^{-2}\,s^{-1}$)

d Zero-plane displacement (cm)

d_p Particle diameter (cm or μm)

e Coefficient of restitution

F Flux or dry deposition rate of a contaminant ($g\,cm^{-2}s^{-1}$)

F_δ Flux through deposition layer ($g\,cm^{-2}s^{-1}$)

F_h Flux at top of constant flux layer in Slinn and Slinn[45, 46] model ($g\,cm^{-2}s^{-1}$)

F_s Sensible heat flux ($calories\,cm^{-2}s^{-1}$)

F_w Water vapour flux ($g\,cm^{-2}s^{-1}$)

g acceleration due to gravity ($cm\,s^{-2}$)

h Height of the canopy or reference height (cm)

h_0 Smallest distance between the particle and the surface (cm)

K Turbulent diffusion coefficient (cm^2s^{-1})

k Von Karman's constant = 0.4

k Boltzmann's constant = 1.38 e-23 J/K . . .

k'_c Constant flux layer transfer coefficient excluding gravity effects in the Slinn and Slinn model [45,46] ($cm\,s^{-1}$)

k'_D Viscous sublayer transfer coefficient excluding gravity effects in the Slinn and Slinn model [45, 46] ($cm\,s^{-1}$)

k_{ab} Constant flux layer transfer coefficient for broken surface in model of Williams [44] ($cm\,s^{-1}$)

k_{as} Constant flux layer transfer coefficient for a smooth surface in model of Williams [44] ($cm\,s^{-1}$)

k_{bs} Broken surface transfer coefficient in model of Williams [44] ($cm\,s^{-1}$)

k_c Overall transfer coefficient for constant flux layer including gravity, in Slinn and Slinn model [45, 46] ($cm\,s^{-1}$)

k_D Overall transfer coefficient for the viscous sublayer, including gravity, in Slinn and Slinn model [45, 46] ($cm\,s^{-1}$)

k_{ss} Smooth surface transfer coefficient in model of Williams [44] ($cm\,s^{-1}$)

L Monin–Obukhov length (cm)

m' Rate of water evaporation from the surface ($g\,cm^2s^{-1}$)

n	Empirical parameter in equation for u, for $z < h$
Pr	Prandtl number $= \nu/\kappa$
r_a	Resistance to aerodynamic transport (s cm^{-1})
r_b	Resistance to boundary layer transport (s cm^{-1})
r_c	Resistance to collection by the surface (s cm^{-1})
r_{cut}	Cuticle resistance (s cm^{-1})
r_m	Mesophyll resistance (s cm^{-1})
r_p	Particle density (g cm^{-3})
r_s	Stomata resistance (s cm^{-1})
r_{soil}	Resistance of the soil (s/cm)
Sc	Schmidt number $= \nu/D$
St	Stokes number or impaction parameter
T	Temperature (°C or K)
t	Time (s)
u	Mean value of horizontal wind speed (cm s^{-1})
$u(h)$	Wind speed at height of canopy (cm s^{-1})
$u*$	Friction velocity (cm s^{-1})
u_s	Initial velocity of a particle due to momentum transfer (cm s^{-1})
v_d	Dry deposition velocity (cm s^{-1})
v_{di}	Deposition velocity for size interval i (cm s^{-1})
v_{dm}	Deposition velocity for momentum transfer (cm s^{-1})
v_g	Settling velocity due to gravity (cm s^{-1})
v_{gd}	Settling velocity for dry particles (cm s^{-1})
v_{gi}	Settling velocity for size interval i (cm s^{-1})
v_{gw}	Settling velocity using wet particle radius (cm s^{-1})
v_i	Particle velocity just before reaching surface (cm s^{-1})
v_i^*	Critical velocity for rebound to occur (cm s^{-1})
v_{Ii}	Inertial deposition velocity for each particle size interval i (cm s^{-1})
w	Vertical wind velocity (cm s^{-1})
w'	Turbulent fluctuating component of vertical wind velocity (cm s^{-1})
w_0	Instantaneous vertical velocity at the edge of the viscous sublayer (cm s^{-1})
$\bar{w}$	Time-averaged vertical wind velocity (cm s^{-1})
y_{lim}	Limiting distance between the particle and centre of the turbulent burst within which deposition occurs (cm)
z	Height above surface (cm)
z_0	Roughness height (cm)
z_{oc}	Contaminant sink (cm) as described in resistance model
z_{ocb}	Contaminant roughness length for a broken surface (cm) from Williams [44]
z_{ocs}	Contaminant roughness length for a smooth surface (cm) from Williams [44]
α	Fraction of area of a water surface that has a broken surface

β	Mean slip velocity constant for diffusiophoresis $= 1000\,\mathrm{cm^3\,g^{-1}}$
δ	Height of viscous sublayer (cm)
ε	Effective inertial coefficient in model of Noll and Fang [81]
ϕ_c	Stability-dependent correction factor applied to eddy transport of contaminant mass
κ	Thermal diffusivity $(\mathrm{cm^2\,s^{-1}})$
λ	Average lateral distance between turbulent bursts (cm)
μ	Dynamic viscosity of air $(\mathrm{g\,cm^{-1}s^{-1}})$
ν	Kinematic viscosity $(\mathrm{cm^2\,s^{-1}})$
σ_c	Standard deviation of airborne contaminant concentration $(\mathrm{g\,cm^{-3}})$
σ_T	Standard deviation of temperature ($^\circ$C or K)
σ_w	Standard deviation of water vapour concentration $(\mathrm{g\,cm^{-3}})$
τ	Particle relaxation time (s)
τ_p^+	Non-dimensionalized particle relaxation time
Ψ	Stability correction factor
Ψ_c	Stability-dependent correction factor in the equation for r_a for contaminant mass
ζ	Dimensionless height $= \frac{(z-d)}{L}$

REFERENCES

1. Davidson, C. I. and Wu, Y.-L. (1990). Dry deposition of particles and vapors, *In* S. E. Lindberg, A. L. Page, and S. A. Norton, eds. *Acidic Precipitation*, Springer-Verlag, New York.
2. Friedlander, S. K. (1977). *Smoke, Dust and Haze*, John Wiley and Sons, New York.
3. Chamberlain, A. C. (1960). Aspects of deposition of radioactive and other gases and particles, *In* E. G. Richardson, ed. *Aerodynamic Capture of Particles*, Pergamon, New York, 63.
4. Waldmann, L. and Shmitt, K. H. (1966). Thermophoresis and diffusiophoresis of aerosols, *In* C. N. Davies, ed. *Aerosol Science*, Academic Press, London and New York, 137.
5. Marwil, E. S. and Lemmon, E. C. (1985). Aerosol simulation including chemical and nuclear reactions, *In Proceedings of the 1985 Summer Computer Simulation Conference*. 22–24 July, North Holland, Amsterdam, The Netherlands, 446.
6. Sehmel, G. A. (1980). Particle and gas dry deposition: a review, *Atmos. Environ.*, **14**, 983.
7. Businger, J. A., Wyngaard, J. C., Izumi, Y., and Bradley, E. F. (1971). Flux–profile relationships in the atmospheric surface layer, *J. Atmos. Sci.*, **28**, 181.
8. Wesely, M. L. and Hicks, B. B. (1977). Some factors that affect the deposition rate of sulfur dioxide and similar gases on vegetation, *J. APCA*, **27**, 1110.
9. Hicks, B. B., Baldocchi, D. D., Hosker, R. P. Jr., Hutchison, B. A., Matt, D. R., McMillen, R. T., and Satterfield, L. C. (1985). *On the Use of Monitored Air Concentrations to Infer Dry Deposition*, NOAA Technical Memorandum ERL ARL-141, Silver Springs, MD.
10. Flagan, R. C. and Seinfeld, J. H. (1988). *Fundamentals of Air Pollution Engineering*, Prentice-Hall, Inc. Englewood Cliffs, NJ, 307.

11. Davidson, C. I., Miller, J. M., and Pleskow, M. A. (1982). The influence of surface structure on predicted particle dry deposition to natural grass canopies, *Water Air Soil Pollut.*, **18**, 25.
12. Bache, D. H. (1979). Particle transport within plant canopies I. A framework for analysis, *Atmos. Environ.*, **13**, 1257.
13. Bache, D. H. (1979). Particulate transport within plant canopies II. Prediction of deposition velocities, *Atmos. Environ.*, **13**, 1681.
14. Peters, K. and Eiden, R. (1992). Modeling the dry deposition velocity of aerosol particles to a spruce forest, *Atmos. Environ.*, **26A**, 2555.
15. Slinn, W. G. N. (1982). Predictions for particle deposition to vegetative canopies, *Atmos. Environ.*, **16**, 1785.
16. Wiman, B. L. B. and Agren, G. I. (1985). Aerosol depletion and deposition in forests—a model analysis, *Atmos. Environ.*, **19**, 335.
17. Wu, Y.-L., Davidson, C. I., and Russell, A. G. (1992). A stochastic model for particle deposition and bounceoff, *Aerosol Sci. Technol.*, **17**, 231.
18. Paw U, K. T. (1992). Rebound and reentrainment of large particles, *In* S. E. Schwartz and W. G. N. Slinn, eds. *Precipitation Scavenging and Atmosphere Exchange Processes*, Proceedings, 5th International Conference on Precipitation Scavenging and Atmosphere–Surface Exchange Processes, Richland, WA, 15–19 July 1153.
19. Dahneke, B. (1971). The capture of aerosol particles by surfaces, *J. Coll. Interface Sci.*, **37**, 342.
20. Dahneke, B. (1975). Further measurements of the bouncing of small latex spheres, *J. Coll. Interface Sci.*, **51**, 58.
21. Paw U, K. T. (1983). The rebound of particles from natural surfaces, *J. Coll. Interface Sci.*, **93**, 442.
22. Cleaver, J. W. and Yates, B. (1973). Mechanism of detachment of colloidal particles from a flat substrate in a turbulent flow, *J. Colloid Interface Sci.*, **44**, 464.
23. Kramm, G., Dlugi, R., Dollard, G. J., Foken, T., Molders, N., Muller, H., Seiler, W., and Sievering, H. (1995). On the dry deposition of ozone and reactive nitrogen species, *Atmos. Environ.*, **29**, 3209.
24. Bytnerowicz, A., Dawson, P. J., Morrison, C. L., and Poe, M. P. (1992). Atmospheric dry deposition on pines in the Eastern Brook Lake watershed, Sierra Nevada, California, *Atmos. Environ.*, **26A**, 3195.
25. Shanley, J. B. (1989). Field measurements of dry deposition to spruce foliage and petri dishes in the Black Forest, F. R. G., *Atmos. Environ.*, **23**, 403.
26. Lindberg, S. E., Cape, J. N., Garten, C. T. Jr., and Ivens, W. (1992). Can sulfate fluxes in forest canopy throughfall be used to estimate atmospheric sulfur deposition? A summary of recent results, *In* S. E. Schwartz and W. G. N. Slinn, eds. *Precipitation Scavenging and Atmosphere Exchange Processes*, Proceedings, 5th International Conference on Precipitation Scavenging and Atmosphere–Surface Exchange Processes, Richland, WA, 15–19 July, 1367.
27. Erisman, J. W. (1993). Acid deposition onto nature areas in the Netherlands: part II. throughfall measurements compared to deposition estimates, *Water Air Soil Pollut.*, **71**, 81.
28. Lovett, G. M., Likens, G. E., Nolan, S. S. (1992). Dry deposition of sulfur to the Hubbard Brook Experimental Forest: A preliminary comparison of methods, *In* S. E. Schwartz and W. G. N. Slinn, eds. *Precipitation scavenging and atmosphere exchange processes*, Proceedings, 5th International Conference on Precipitation Scavenging and Atmosphere–Surface Exchange Processes, Richland, WA, 15–19 July, 1391.

29. Lovblad, G., (1992). Mapping of deposition over Sweden, *In* S. E. Schwartz and W. G. N. Slinn, eds. *Precipitation Scavenging and Atmosphere Exchange Processes*, Proceedings, 5th International Conference on Precipitation Scavenging and Atmosphere–Surface Exchange Processes, Richland, WA, 15–19 July, 1533.

30. Beier, C., Gundersen, P., and Rasmussen, L. (1992). A new method for estimation of dry deposition of particles based on throughfall measurements in a forest edge, *Atmos. Environ.*, **26A**, 1553.

31. Friedland, A. J., Miller, E. K., Battles, J. J., and Thorne, J. F. (1991). Nitrogen deposition, distribution and cycling in a subalpine spruce–fir forest in the Adirondacks, New York, USA, *Biogeochemistry*, **14**, 31.

32. Miller, E. K., Panek, J. A., Friedland, A. J., Kadlecek, J., and Mohnen, V. A. (1993). Atmospheric deposition to a high- elevation forest at Whiteface Mountain, New York, USA, *Tellus*, **45B**, 209.

33. Butler, T. J. and Likens, G. E. (1995). A direct comparison of throughfall plus stemflow to estimates of dry and total deposition for sulfur and nitrogen, *Atmos. Environ.*, **29**, 1253.

34. Santachiara, G., Prodi, F., and Vivarelli, F. (1991). Dry deposition measurements of aerosol particles in a corn field, *J. Aerosol Sci.*, **22**, S577.

35. Davidson, C. I., Lindberg, J., Schmidt, A., Cartwright, L. G., and Landis, L. R. (1985). Dry deposition of sulfate onto surrogate surfaces, *J. Geophys. Res.*, **90**, 2123.

36. Koester, C. J., and Hites, R. A. (1992). Wet and dry deposition of chlorinated dioxins and furans, *Environ. Sci. Technol.*, **26**. 1375.

37. Bergin, M. H., Davidson, C. I., Dibb, J. E., Jaffrezo, J. L., Kuhns, H. D., and Pandis, S. N. (1995). A simple model to estimate atmospheric concentrations of irreversibly deposited aerosol chemical species based on snow core chemistry at Summit, Greenland, *Geophys. Res. Letters*, **22**, 3517.

38. Nicholson, K. W. (1988). The dry deposition of small particles: a review of experimental measurements, *Atmos. Environ.*, **22**, 2653.

39. Sievering, H. (1987). Small-particle dry deposition under high wind speed conditions: eddy flux measurements at the Boulder Atmospheric Observatory, *Atmos. Environ.*, **21**, 2179.

40. Allen, A. G., Harrison, R. M., and Nicholson, K. W. (1991). Dry deposition of fine aerosol to a short grass surface, *Atmos. Environ.*, **25A**, 2671.

41. Main, H. H., and Friedlander, S. K. (1990). Dry deposition of atmospheric aerosols by dual tracer method I. area source, *Atmos. Environ.*, **24A**, 103.

42. Zufall, M. J. and Davidson, C. I. (1997). Dry deposition of particles to water surfaces, *In* J. Baker, ed. *Atmospheric Deposition of Contaminants to the Great Lakes and Coastal Waters*, SETAC Press.

43. Fitzgerald, J. W. (1975). Approximation formulas for the equilibrium size of an aerosol particle as a function of its dry size and composition and the ambient relative humidity, *J. Appl. Met.* **14**, 1044.

44. Williams, R. M. (1982). A model for the dry deposition of particle to natural water surfaces, *Atmos. Environ.*, **16**, 1933.

45. Slinn, S. A. and Slinn, W. G. N. (1980). Predictions of particle deposition on natural waters, *Atmos. Environ.*, **14**, 1013.

46. Slinn, S. A. and Slinn, W. G. N. (1981). Modeling of atmospheric particulate deposition to natural waters, *In* S. J. Eisenreich, ed. *Atmospheric Pollutants in Natural Waters*, Ann Arbor Science Publishers, Ann Arbor, MI, 23.

47. Shannon, J. D. and Voldner, E. C. (1992). Deposition of S and NO_x nitrogen to the Great Lakes estimated with a regional deposition model, *Environ. Sci. Technol.* **26**, 970.

48. Baeyens, W., Dehairs, F., and Dedeurwaerder, H. (1990). Wet and dry deposition fluxes above the North Sea, *Atmos. Environ.*, **24A**, 1693.
49. Dulac, F., Buat-Menard, P., Ezat, U., Melki, S., and Bergametti, G. (1989). Atmospheric input of trace metals to the western Mediterranean: uncertainties in modeling dry deposition from cascade impactor data, *Tellus*, **41B**, 362.
50. Dulac, F., Bergametti, G., Lasno, R., Remoudaki, E, Gomes, L., Ezat, U. and Buat-Menard, P. (1992). Dry deposition of mineral aerosol particles in the marine atmosphere: significance of the large size fraction, *In* S. E. Schwartz and W. G. N. Slinn, eds. *Precipitation Scavenging and Atmosphere Exchange Processes*, Proceedings, 5th International Conference on Precipitation Scavenging and Atmosphere–Surface Exchange Processes, Richland, WA, 15–19 July, 841.
51. Rojas, C. M., Injuk, J., Van Grieken, R. E., and Laane, R. W. (1993). Dry and wet deposition fluxes of Cd, Cu, Pb and Zn into the southern bight of the North Sea, *Atmos. Environ.*, **27A**, 251.
52. Rojas, C. M., Van Grieken, R. W., and Laane, R. W. (1993). Comparison of three dry deposition models applied to field measurements in the southern bight of the North Sea, *Atmos. Environ.*, **27A**, 363.
53. Wu, J. (1979). Oceanic whitecaps and sea state, *J. Physical Oceanography*, **9**, 1064.
54. Hess, G. D. and Hicks, B. B. (1975). The influence of surface effects on pollutant deposition rates over the Great Lakes, In *Proc. Second Fed. Conf. on Great Lakes*, 239.
55. Slinn, W. G. N. (1976). Dry deposition and resuspension of aerosol particles—A new look at some old problems. *In Surface exchange of particulate and gaseous pollutants*, ERDA Symp. Ser. 38, 1.
56. Fairall, C. W. and Larsen, S. E. (1984). Dry deposition, surface production and dynamics of aerosols in the marine boundary layer, *Atmos. Environ.*, **18**, 69.
57. Gatz, D. F. (1992). Large-particle resuspension from lake shore industrial areas: potential for dry deposition loading of near-shore waters of Lake Michigan, *In* S. E. Schwartz and W. G. N. Slinn, eds. *Precipitation Scavenging and Atmosphere Exchange Processes*, Proceedings, 5th International Conference on Precipitation Scavenging and Atmosphere–Surface Exchange Processes, Richland, WA, 15–19 July, 1195.
58. Hummelshoj, P., Jensen, N. O., and Larsen, S. E. (1992). Particle dry deposition to a sea surface. *In* S. E. Schwartz and W. G. N. Slinn, eds. *Precipitation Scavenging and Atmosphere Exchange Processes*, Proceedings, 5th International Conference on Precipitation Scavenging and Atmosphere–Surface Exchange Processes, Richland, WA, 15–19 July, 829.
59. Quinn, T. L., Ondov, J. M. and Holland, J. Z. (1992). Dependence of deposition velocity on the frequency of meteorological observations for the Chesapeake Bay, *J. Aerosol Sci.*, **23**, S973.
60. Larsen, S. E., Edson, J. B., Mestayer, P. G., Fairall, C. W., and de Leeuw, G. (1991). Sea spray and particle deposition: an air/water tunnel experiment and its relation to over-ocean conditions, *In* P. Borrell, P. M. Borrell, and W. Seiler, eds. *Transport and Transformation of Pollutants in the Troposphere*, Proceedings of EUROTRAC Symposium '90, SPB Academic Publishing, The Netherlands.
61. Milford, J. B. and Davidson, C. I. (1985). The sizes of particulate trace elements in the atmosphere—a review, *J. APCA*, **35**, 1249.
62. Milford, J. B. and Davidson, C. I. (1987). The sizes of particulate sulfate and nitrate in the atmosphere—a review, *J. APCA*, **37**, 125.
63. Noll, K. E., Yuen, P.-F., and Fang, K. Y.-P (1990). Atmospheric coarse particulate concentrations and dry deposition fluxes for ten metals in two urban environments, *Atmos. Environ.*, **24A**, 903.

64. Saiz-Jimenez, C. (1993). Deposition of airborne organic pollutants on historic buildings, *Atmos. Environ.*, **27B**, 77.

65. Gould, T. R., Lutz, M. R., Davidson, C. I., Finger, S., Small, M. J. (1993). *Influence of Atmospheric Pollutant Concentrations and Deposition Rates on Soiling of a Limestone Building Surface*, Progress report for the Preservation Assistance Division, National Park Service, U. S. Department of the Interior.

66. Sherwood, S. I., Gatz, D. F., and Hosker, R. P. Jr. (1990). *Acidic Deposition:* SOS/T-Report 20: *Processes of Deposition to Structures*, NAPAP. GPO. Washington, D. C.

67. Hussain, M. and Lee, B. E. (1980). *An Investigation of Wind Forces on Three-dimensional Roughness Elements in a Simulated Atmospheric Boundary Layer Flow. Part II: Flow over Large Arrays of Identical Roughness Elements and the Effect of Frontal and Side Aspect Ratio Variations*, Univ. of Sheffield report BS 56.

68. Owen, P. R., and Thompson, W. R. (1963). Heat transfer across rough surfaces, *J. Fluid Mech.* **15**, 321.

69. Brutsaert, W. (1975). The roughness length for water vapor, sensible heat, and other scalars, *J. Atmos. Sci.*, **32**, 2028.

70. Brutsaert, W. (1982). *Evaporation into the Atmosphere*, D. Reidel Publishing Co., Dordrecht.

71. Wu, Y.-L., Davidson, C. I., Dolske, D. A., and Sherwood, S. I. (1992). Dry deposition of atmospheric contaminants: The relative importance of aerodynamic, boundary layer, and surface resistances, *Aerosol Sci. Technol.*, **16**, 65.

72. Lutz, M. R., Gould, T. R., Etyemezian, V., Davidson, C. I., Finger, S., Small, M. J. (1994). *Influence of Atmospheric Pollutant Concentrations and Deposition Rates on Soiling of a Limestone Building Surface 1991–1994*. Progress report for the Preservation Assistance Division, National Park Service, U. S. Department of the Interior.

73. Etyemezian, V., Davidson, C. I. Finger, S. (1995). *Influence of Atmospheric Pollutants on Soiling of a Limestone Building Surface*. Progress report for the National Park Service, U. S. Department of the Interior.

74. Dasch, J. M. and Cadle, S. H. (1985). Wet and dry deposition monitoring in southeastern Michigan, *Atmos. Environ.*, **19**, 789.

75. Davidson, C. I. and Friedlander, S. K. (1978). A filtration model for aerosol dry deposition: application to trace metal deposition from the atmosphere, *J. Geophys. Res.*, **83**, 2343.

76. Coe, J. M. and Lindberg, S. E. (1987). The morphology and size distribution of atmospheric particles deposited on foliage and inert surfaces, *J. APCA*, **37**, 237.

77. Holsen, T. M., Noll, K., Fang, G., Lee, W., Lin, J., and Keeler, G. J. (1993). Dry deposition and particle size distributions measured during the Lake Michigan urban air toxics study, *Environ. Sci. Technol.*, **27**, 1327.

78. Holsen, T. M. and Noll, K. E. (1992). Dry deposition of atmospheric particles: Application of current models to ambient data, *Environ Sci. Technol.* **26**, 1807.

79. Zufall, M. J. and Davidson, C. I. (1995). Dry deposition of atmospheric contaminants to Lake Michigan, *In* R. D. Vidic and F. G. Pohland, eds., *Innovative Technologies for Site Remediation and Hazardous Waste Management*. Proceedings of the National Conference of the American Society of Civil Engineers. Pittsburgh, Pa, 628.

80. Noll, K. E. and Fang, K. Y.-P. (1986). A rotary impactor of size selective sampling of atmospheric coarse particles, Presented at the 79th annual meeting of the air pollution control association, Minneapolis, MN, 22–27 June.

81. Noll, K. E. and Fang, K. Y.-P. (1989). Development of a dry deposition model for atmospheric coarse particles, *Atmos. Environ.*, **23**, 585.

82. Wu, Y.-L., Davidson, C. I., and Russell, A. G. (1992). Controlled wind tunnel experiments for particle bounceoff and resuspension, *Aerosol Sci. Technol.* **17**, 245.

83. Nicholson, K. W. (1993). Wind tunnel experiments on the resuspension of particulate material, *Atmos. Environ.*, **27A**, 181.

84. Wen, H. Y., Kasper, G., and Udischas, R. (1989). Short and long term particle release from surfaces under the influence of gas flow, *J. Aerosol Sci.* **20**, 923.

85. Gavrilov, V. P., Klepikova, N. V., Troyanova, N. I., and Rodean, H. C. (1995). Stationary model for resuspension of radionuclides and assessments of ^{137}Cs concentration in the near-surface layer for the contaminated areas in the Bryansk Region of Russia and Belarus, *Atmos. Environ.*, **29**, 2633.

86. Wu, Y.-L., Davidson, C. I., Lindberg, S. E., and Russell, A. G. (1992). Resuspension of particulate chemical species at forested sites, *Environ Sci. Technol.* **26**, 2428.

87. Nilsson J. and Grennfelt, P. (1988). Critical loads for sulfur and nitrogen, *Nord 1988*:15, Nordic Council of Ministers, Copenhagen.

88. Grennfelt, P. A. (1992). Critical loads and deposition monitoring, *In* S. E. Schwartz and W. G. N. Slinn, eds. *Precipitation Scavenging and Atmosphere Exchange Processes*, Proceedings, 5th International Conference on Precipitation Scavenging and Atmosphere–Surface Exchange Processes, Richland, WA, 15–19 July, 1543.

89. De Vries, W. (1993). Average critical loads for nitrogen and sulfur and its use in acidification abatement policy in the Netherlands, *Water Air Soil. Pollut.*, **68**, 399.

90. Warfvinge, P. and Sverdrup, H. (1992). Calculating critical loads of acid deposition with profile—A steady-state soil chemistry model, *Water Air Soil Pollut.*, **63**, 119.

91. Agarwal, J. K. (1975). Aerosol sampling and transport, Ph.D. Dissertation, University of Minnesota, Minneapolis, MN.

92. Friedlander, S. K. and Johnstone, H. F. (1957). Deposition of suspended particles from turbulent gas streams, *Ind. Eng. Chem.*, **49**, 1151.

93. Liu, B. Y. H. and Agarwal, J. K. (1974). Experimental observation of aerosol deposition in turbulent flow, *J. Aerosol Sci.*, **5**, 145.

94. Forney, L. J. and Spielman, L. A. (1975). Aerosol rebound from viscous coatings in turbulent flow, *J. Colloid Interface Sci.*, **52**, 468.

95. Montgomery, T. L. and Corn, M. (1970). Aerosol deposition in a pipe with turbulent air flow, *J. Aerosol Sci.*, **1**, 185.

96. Nicholson, K. W. and Watterson, J. D. (1992). Dry deposition of particulate material onto wheat, *In* S. E. Schwartz and W. G. N. Slinn, eds. *Precipitation Scavenging and Atmosphere Exchange Processes*, Proceedings, 5th International Conference on Precipitation Scavenging and Atmosphere–Surface Exchange Processes, Richland, WA, 15–19 July, 1003.

97. Aoyama, M. and Hirose, K. (1992). Particle size dependent dry deposition velocity of the Chernobyl radioactivity, *In* S. E. Schwartz and W. G. N. Slinn, eds. *Precipitation Scavenging and Atmosphere Exchange Processes*, Proceedings, 5th International Conference on Precipitation Scavenging and Atmosphere–Surface Exchange Processes, Richland, WA, 15–19 July, 1581.

98. Foltescu, V. L., Isakson, J., Selin, E., and Stikans, M. (1994). Measured fluxes of sulfur, chlorine and some anthropogenic metals to the Swedish west coast, *Atmos. Environ.*, **28**, 2639.

99. Schery, S. D. and Whittlestone, S., (1995). Evidence of high deposition of ultrafine particles at Mauna Loa observatory, *Atmos. Environ.*, **29**, 3319.

100. Ottley, C. J. and Harrison, R. M., (1993). Atmospheric dry deposition flux of metallic species to the North Sea, *Atmos. Environ.*, **27A**, 685.

101. Lindfors, V., Sylvain, M. J., and Danski, J. (1992). Dry and wet deposition of sulfur and nitrogen compounds over the Baltic Sea, *In* S. E. Schwartz and W. G. N. Slinn, eds. *Precipitation Scavenging and Atmosphere Exchange Processes*, Proceedings, 5th International Conference on Precipitation Scavenging and Atmosphere–Surface Exchange Processes, Richland, WA, 15–19 July, 855.
102. Watkins, L. A. (1986). Master's Thesis, Illinois Institute of Technology.

14 Wet Processes Affecting Atmospheric Aerosols

S. G. JENNINGS

University College Galway, Ireland

Atmospheric Particles, Edited by R. M. Harrison and R. Van Grieken.
© 1998 John Wiley & Sons Ltd.

1 INTRODUCTION

The origin of atmospheric particles can be either anthropogenic (man-made) or natural. Primary particles are those which are emitted into the atmosphere as particles such as black carbon and organic particles in smoke plumes, soil dust particles, sea-spray and volcanic particles, etc. Secondary particles are formed from gas-to-particle conversion processes in the atmosphere—such as sulphates (from SO_2), nitrates (from NO_x) and secondary organics (from gaseous hydrocarbons).

One key property of the atmospheric aerosol apart from size is chemical composition. In nearly all cases, the dry fine particle mass is dominated by five main types of chemical species: sulfates—normally with ammonium or hydrogen cations, organics, nitrates—normally with ammonium cations, soil dust and elemental carbon.

Another key property of the ambient aerosol is its hygroscopicity. The atmosphere contains a mixture of (water) soluble and (water) insoluble aerosol particles. Under normal ambient conditions, water constitutes a crucially important component of the aerosol particle—and of its mass. The hygroscopic or deliquescent properties of some of the main aerosol constituents (sulfates, nitrates and some organics) result in aerosol particles accumulating a significant fraction of water. [1–4] A reduction in the atmospheric concentration of hygroscopic aerosol species leads to a reduction in the water content of the atmospheric aerosol.

Although removal of aerosol particulate material from the atmosphere by precipitation is episodic by its nature, nevertheless the importance of wet deposition as a powerful aerosol removal process is evident from the following considerations:

(i)   The close relation between deposition levels of radioactive material following the Chernobyl nuclear reactor accident with the amount of precipitation falling during the passage of the main plume. [5]
(ii) Wet (and dry) acid deposition of sulfur and nitrogen species is causing damage to aquatic and terrestrial ecosystems. [6]
(iii) The enhancement of visibility, due to the cleansing of the atmosphere of aerosol, is observed frequently.

Sulfate aerosol constitutes the principal aerosol chemical species in the atmosphere and so is a good surrogate for the atmospheric aerosol in general. In-cloud scavenging together with wet deposition by precipitation is the major removal process of sulfate aerosol particles from the atmosphere [7] though the efficiency of the process can only be estimated crudely. It is estimated that wet deposition accounts for $\approx 60\%$ of the total deposition of sulfur. [8] Similar figures were predicted for the USA and Europe. [9] Wet deposition rates of sulfur in China are generally larger than those in the USA. Emissions of oxides of sulfur together with those of nitrogen oxides contribute 70 and 30% respectively to the acidity of precipitation. [10]

This chapter contains the following outline of sections describing a variety of wet processes affecting atmospheric aerosols. Section 2 deals briefly with the effect of water vapour on aerosol particles in regards to the activation and formation of aerosol particles and their growth (or evaporation). The condensation of water vapour upon cloud condensation nuclei (CCN) resulting in the so-called nucleation scavenging of aerosol material is also described in section 2. The transformation of aerosol particles below cloud through scavenging by liquid and solid precipitation is presented in sections 3 and 4. Theoretical modelling and experimental testing (in the laboratory and in the field) of a number of aerosol scavenging processes are also described in these sections. The bulk scavenging of aerosol particles by precipitation using the so-called scavenging ratio and scavenging rate is discussed in section 5. The effect of clouds on the processing of aerosol is considered in section 6. Modelling studies and some field validation studies of the scavenging of aerosol in water and ice clouds are presented in section 7. The final section concludes with some requirements and recommendations for further research on the study of wet processes affecting atmospheric aerosols, with particular emphasis on the wet deposition of aerosol particles.

2   THE EFFECT OF WATER VAPOUR ON AEROSOL PARTICLES

2.1   NUCLEATION AND GROWTH OF AEROSOL PARTICLES

Atmospheric particles, once formed, may either grow or evaporate depending on the water vapour content of the environment and on the volatility of the particles. Some atmospheric constituents are considerably more volatile (for example ammonium chloride) than others such as sodium chloride. A wide range of relative humidity is encountered in the atmosphere ranging from close to 0% in hot arid regions to saturation under foggy or cloudy conditions. A relative humidity value between 30 and 40% is frequently used to delineate between a dry aerosol and a wet aerosol particle.

A hygroscopic aerosol particle will generally consist of soluble material (solute) contained in the solution and its concentration will determine its equilibrium relative humidity. If the relative humidity of the environment is higher than that of the aerosol particle, then the particle will grow and become more liquefied. An evaporating solution particle leads to a more saturated solution and then to a dry particle. Phase transition occurs from solution to solid or from solid to solution (deliquescence point) which tends to be more abrupt. Values of relative humidity at which transition of phase occurs for a number of atmospheric inorganic salts [11,12] are given in Table 1.

Table 1. Relative humidity (%) at which phase transition occurs at 298 K for a number of atmospheric inorganic salts[11,12]

Constituent	Deliquescence relative humidity (%)
$(NH_4)_2SO_4$	79.9 ± 0.5
NH_4Cl	77
NaCl	76.3 ± 0.1
Na_2SO_4	84.2 ± 0.4
NH_4NO_3	61.8
NH_4HSO_4	39
$NaNO_3$	74.3 ± 0.4

When the relative humidity decreases after being above the deliquescence point, the droplet size decreases at approximately the same rate that it grows until after the relative humidity falls again below the deliquescence point. The size continues to decrease but at a lesser rate than it initially grew, thus exhibiting hysteresis. The decrease in size continues until the particle reaches the "dry" size at about 30% relative humidity.

The homogeneous nucleation of a liquid droplet from the vapour alone, in the absence of an aerosol nucleus, requires a supersaturation of the order of 340% which is not realizable under normal atmospheric conditions. Instrumentation such as Aitken, Nolan–Pollak and TSI condensation nucleus counters operate under such a range of high degree of supersaturation to detect the total number concentration of particles greater than a minimum diameter of order 3–10 nm. However, only a relatively small fraction of atmospheric particles act as condensation nuclei in the natural cloud (where the prevailing supersaturation is of an order of 1% or less) or fog condensation process and are known as cloud condensation nuclei (CCN). Supersaturation is a measure of the excess water vapour about saturation (100% relative humidity) so that, for example, 100.5% relative humidity is equivalent to 0.5% supersaturation.

The increasing importance of CCN, from the standpoint of global climate, is an added reason to examine the effect of water vapour on aerosol particles. Water vapour has its greatest effect on soluble aerosol material. Nucleation on soluble aerosol particles is well described in standard textbooks [13–15] in terms of the Köhler curves. The maximum or critical supersaturation, S_c, of the particle is the important aerosol parameter since its value determines whether or not an aerosol nucleus is activated. Only particles with $S_c <$ maximum supersaturation can partake in the condensation growth of the particles to cloud droplets. The remainder are inactivated and are often referred to as haze particles. Wet haze particles are described in more detail in Chapter 15.

It is shown that S_c is proportional to the square root of the molecular weight of the solute particle and inversely proportional to the square root of the van't

Hoff factor. [13,14] Thus, for example, sodium chloride particles possess S_c values about 20% less than ammonium sulfate particles of the same size and consequently will be activated more readily. The activation of particles is dependent also on the water-soluble fraction of the aerosol. At least an order of magnitude greater critical supersaturation is required for activating an insoluble aerosol particle as compared to a soluble aerosol particle of the same size. [16]

The growth (or evaporation) of an aerosol particle is governed by the difference between the vapour pressure of the particle and the ambient vapour pressure and is described in detail in Chapter 6.

The increase in the mass of water, acquired by an aerosol particle at different relative humidities, has been evaluated [16] for NaCl and $(NH_4)_2SO_4$ constituents for a range of salt masses (from 10^{-16} to 10^{-12} g equivalent to particle radii from 0.022 to about 0.5 μm). The results are summarized in Table 2.

The NaCl particles acquire about a sixfold increase in mass (increase of 1.8 in size) due to water uptake for the increase in relative humidity to 90%. The corresponding increase for $(NH_4)_2SO_4$ particles is about a 2.4-fold increase in water mass uptake (an increase of 1.33 in size). This is about a factor 2.5 less than that for NaCl which is mainly due to the much larger molecular weight of $(NH_4)_2SO_4$.

The relative increase in aerosol particle size caused by the condensation of water vapour on particles of initial radius $a_0 > 0.1$ μm as a function of relative humidity is shown in Figure 1 for a range of aerosol chemical constituents. [18] The compiled figure [18] is based on measurements of aerosol particle growth [2,19] and of aerosol scattering coefficient [20] with relative humidity and of calculated aerosol growth for external mixtures of 60% pure salt constituents and 40% insoluble aerosol material. [2] It is seen that particle growth is fairly modest (< factor of 2) for relative humidity < 90% as also inferred from Table 2. Plots of mean aerosol particle size as a function of relative humidity have also been published by other workers. [4,16,21] Particle radii increase by a factor of as much as 2.5 as relative humidity approaches 100%. In the case of mixed particles the growth (and evaporative) rates are more complicated because the particle may go through several stages of multi-phase equilibria before forming a droplet. [4]

It is shown [22] that H_2SO_4 particles can hold more than their weight in water at 50% relative humidity—thus even moderate values of relative humidity can influence the mass (and size) of the ambient aerosol. Therefore it has become increasingly important to adopt uniform aerosol sample equilibration standards and procedures in the sampling of the atmospheric aerosol. The World Meteorological Organization (WMO) Global Atmosphere Watch (GAW), 1991, recommend aerosol sampling to take place with the aerosol at a relative humidity of < 40%. [23]

Relative humidity as well as having an effect on aerosol particle size also affects aerosol haze content. Summer-time maxima in aerosol extinction coeffi-

Table 2. Increase of water uptake by NaCl and $(NH_4)_2SO_4$ particles for relative humidity of 80 and 90%.[16,17] Taken from Pruppacher and Klett[14]

Relative humidity (%)		NaCl			$(NH_4)_2SO_4$		
Dry aerosol particle mass (g)	–	1.0×10^{-16}	1.0×10^{-14}	1.0×10^{-12}	1.0×10^{-16}	1.0×10^{-14}	1.0×10^{-12}
Dry aerosol radius (µm)	–	0.022	0.103	0.480	0.022	0.096	0.449
Mass of water acquired by particle (g)	80	3.1×10^{-16}	3.2×10^{-14}	3.3×10^{-12}	1.05×10^{-16}	1.1×10^{-14}	1.1×10^{-12}
	90	5.3×10^{-16}	5.8×10^{-14}	6.1×10^{-12}	2.1×10^{-16}	2.4×10^{-14}	2.4×10^{-12}

cient (minima in visibility) over the region south of the Great Lakes and east of the Mississippi in the USA is attributed [24] to the greater aerosol sulfate concentrations in the eastern USA which interact with the higher (relative) humidity of the eastern USA to produce dense sulfate water hazes. The National Park Service aerosol data [24] show that sulfates are nearly twice as

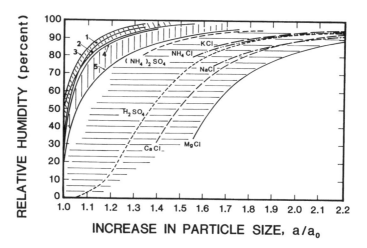

Figure 1. Relative increase in particle size caused by water condensation on particles of initial radius $a_0 > 0.1\,\mu m$ as a function of relative humidity, for different chemical compositions of the particles. [18] (1), diagonal hatching, average growth curves for individual particles: curve 1, [19] curves 2 and 3 [20] for $p = 2$ and 3, respectively, where the scattering coefficient $b \sim a^p$ in which a is particle radius. (2), vertical hatching, average growth curves of samples of atmospheric aerosol particles; curves 4 and 5 for particle density $\rho_p = 1.0$ and $2.5\,g\,cm^{-3}$, respectively. (3), horizontal hatching, calculated curves representing external mixtures of 60% pure salt and 40% insoluble matter. [2] (Reproduced by permission of the author, W. George N. Slinn.)

high in the summer as during the remainder of the year. It is concluded that trends in seasonal sulfur emissions provide a plausible explanation for the observed seasonal trends of atmospheric extinction coefficient over the eastern United States. [25] However, such qualitative comparisons do not necessarily provide conclusive evidence of a cause–effect relationship.

2.2   THE NUCLEATION SCAVENGING PROCESS

Aerosol particles are incorporated into a cloud through vertical motion into its base and by lateral entrainment at its edges. A fraction of the aerosol population contained in the rising air parcel entering the cloud at its base will contain CCN upon which condensation of water vapour will occur. This process results in the conversion of aerosol material to cloud droplets and is known as nucleation scavenging.

As soon as condensation occurs, depletion of water vapour occurs and continues until the relative humidity falls back towards the saturated value of 100% relative humidity. Many particles do not become activated and remain as interstitial unactivated particles in the cloud.

One needs to know the distribution of critical supersaturation, S_c, or the so-called supersaturation spectrum among the aerosol particle population in order to predict the outcome of the cloud condensation process. The number concentration of CCN increases with supersaturation and the CCN spectrum follows quite well the relationship

$$N \propto cS^k \qquad (1)$$

at least in the vicinity of the maximum supersaturation obtained during the condensation stages, where c is the "concentration" and k is the "slope parameter".

Assuming a CCN activity spectrum of the form (1), it has been shown [26] that the concentration of CCN with critical supersaturation $S \leq Sc$ follows approximately the relationship

$$N \propto c^{2/k+2} U^{3k/2k+4} \qquad (2)$$

where U is the updraft speed. This relationship predicts that for large values of the slope parameter, k, the concentration of activated cloud droplets is largely controlled by the rate of cooling, U, and is quite insensitive to the concentration, c, of the nucleating particles. For small k, the aerosol nuclei concentration, c, determines N predominantly, becoming proportional to c as $k \to 0$.

In the atmosphere, measured nuclei spectra [27] are found to give k values typically ≈ 0.5 or less (for maritime air) in the range from about 0.05–2% supersaturation of relevance to the particle nucleation process. CCN number concentration can vary over several orders of magnitude and is dependent on proximity to sources and meteorological factors such as air mass type. [26] Over land, only a small fraction (usually less than about 20%) of the total aerosol number concentration serve as CCN at supersaturation of 1% or less. However in maritime air, particularly in the more remote regions, the fraction of CCN to total particle number can range from approximately 0.2 to around 0.35. [28,29]

CCN concentration over continents typically range from 100 to 1000 cm^{-3}, while over oceans they range from a few tens to a few hundreds cm^{-3}. The only long-term measurement of CCN is that from 1980 to the present in the clean Southern Ocean air at Cape Grim, Tasmania. [28]

It is expected that aerosol particles which consist of water-soluble, hygroscopic substances are more likely to act as CCN. The relatively little work done on CCN composition suggests that most CCN are likely composed of $(NH_4)_2SO_4$ or possibly NH_4Cl. [30,31] Despite the general consensus in recent years that the primary source of CCN in marine air is sulfate aerosol, recent measurements [32] have provided evidence that sea-salt CCN can be the dominant source of activated CCN in the marine stratocumulus environment, particularly in high wind conditions.

Some substances such as elemental carbon, EC, are hydrophobic and chemically inert at source. However, it is pointed out [33] that EC is likely to become coated with hygroscopic substances as the EC containing air parcel ages with distance from the source. Thus EC becomes more amenable for removal via nucleation scavenging. Such a "hygroscopic coating" process is also likely to occur with age for other initially non-hygroscopic substances in the atmosphere and thus aid in their removal (via nucleation).

2.3   ICE-FORMING NUCLEI (IN)

Heterogeneous nucleation of ice in supercooled water or in a supersaturated environment is facilitated by the presence of atmospheric ice nuclei (IN), with nucleation occurring more readily on surfaces having a lattice structure geometrically similar to that of ice. It appears that the atmospheric aerosol is quite uniform with respect to its ability to initiate the ice phase, as evidenced by the relatively close agreement in number concentration with geographic location. [14] Median concentration of IN vary from about $1\ l^{-1}$ (at $-10°C$) to around $1000\ l^{-1}$ (at $-20°C$).

Clay mineral particles appear to exhibit a good degree of ice-nucleating ability, as evidenced from the residue of ice and snow crystals. [34] Quite frequently, just one solid particle was found in the central portion of a snow crystal. These central particles had diameters between 0.1 and 15 μm with a mode between 0.4 and 1 μm and were deemed to be instrumental in the nucleation of the crystals. These findings suggest that larger-sized, including coarse mode aerosol particles—often associated with desert and arid regions—are a major source of IN. Biogenic (organic) aerosols also contribute to the IN population [35] but both urban regions and oceans appear to be relatively poor sources of IN.

3   SCAVENGING BY LIQUID PRECIPITATION

3.1   BELOW-CLOUD SCAVENGING PROCESSES

The concentration level of aerosol particles would increase rapidly in the atmosphere if there were no removal mechanisms. Removal of atmospheric aerosol particles by precipitation is one of the most important natural processes in the transformation of aerosols. It is often referred to as precipitation or below-cloud scavenging. Both liquid (fog, cloud droplets and raindrops) and solid (ice crystals, snow, graupel and hail elements) precipitation can be involved in this removal process. Scavenging by liquid precipitation is treated in this section.

Greenfield [36] first combined scavenging processes in the atmosphere: Brownian diffusion, turbulent shear diffusion and inertial impaction and found that the overall scavenging coefficient had a broad and distinctive minimum for aerosol particles between about 0.05 and $2.0 \mu m$ in radius. This minimum is referred to generally as the "Greenfield gap" and is due to Brownian diffusion dominating particle capture for $r < 0.05 \mu m$ and due to inertial impaction dominating particle capture for $r > 2 \mu m$. Subsequent investigators [37, 38] include the phoretic processes of thermophoresis and diffusiophoresis in addition to the other scavenging mechanisms referred to above and obtain similar results. The resultant gap is shown in Figure 2, where the variation of the scavenging rate with particle radius is shown. [37] Collision efficiency values for inertial impaction [39] are used to determine the contribution of inertial impaction to the scavenging rate. More recent studies [40, 41] include electrical forces due to electrical charges on the drop and the aerosol particles, as well as the presence of electric fields. Thus, in summary, aerosol particles can be removed from the atmosphere through the following processes:

(i)   Brownian diffusion through coagulation of aerosol particles with liquid or solid precipitation.
(ii) Collision of aerosol particles with liquid or solid precipitation caused by phoretic forces: thermophoresis and diffusiophoresis.
(iii) Inertial impaction of aerosol particles on precipitation elements, caused by the hydrodynamic interaction of the aerosol particles with either liquid or solid precipitation, with both falling free under gravity.
(iv) Electrical forces which can promote precipitation—these electrical forces exist if the colliding entities are electrically charged or if an external electric field is present.

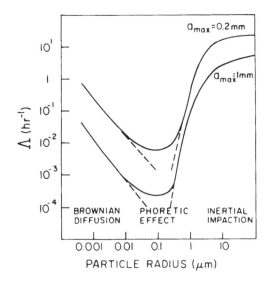

PARTICLE RADIUS (μm)

Figure 2. Scavenging rate (hour^{-1}) as a function of aerosol particle size for Brownian diffusion, inertial capture, and thermo-and diffusiophoresis; [37] $\Delta T = T_\infty - T = 3\,°C$, precipitation rate $R = 10\,mm\,hour^{-1}$; raindrop size distribution $n(a)da = (10^{-4}R/6\pi a_{max}^7)a^2 \exp(-2a/a_{max})da$, with R in cm s^{-1}; drop terminal velocity $V_\infty = 8000a(s^{-1})$ with a in cm. Collision efficiency values for inertial impaction based on earlier work [39] are used to calculate the contribution of inertial impaction to the scavenging rate. (Reproduced by permission of the authors, W. George N. Slinn and J. M. Hales, American Meteorological Society and D. Reidel Publishing Company.)

Small aerosol particles will attach themselves to drops by Brownian diffusion. The scavenging rate Λ_B due to Brownian diffusion of aerosol particles of radius $r \ll a$ (cloud droplet radius) is given [14] by the expression

$$\Lambda_B = \frac{1.35w_L D}{\bar{a}^2} \qquad (3)$$

for a Khrgian and Mazin cloud drop size distribution, where w_L is the cloud liquid water content, $\bar{a}$ the average cloud droplet radius and D the diffusion coefficient of the aerosol particles. The half-life of the aerosol particles is given therefore by

$$t_{1/2} = \frac{(\ln 2)\bar{a}^2}{1.35w_L D} \qquad (4)$$

So, for $w_L = 1\,g\,m^{-3}, \bar{a} = 10\,\mu m, r = 0.01\,\mu m, D = 1.56 \times 10^{-8}\,m^2\,s^{-1}$ (at $T = 15°C, p = 1000\,mb$), the half-life $t_{1/2}$ becomes equal to about 40 minutes. Therefore, Aitken or CCN nucleation particles will be scavenged efficiently by water droplets (of cloud size) within the span of about an hour.

The mechanism of thermophoresis refers to motion of aerosol particles described due to a thermal gradient between the scavenging drop and the particle. [42] The direction of motion of the particles is towards the lower temperature region. The process of diffusiophoresis [43] refers to aerosol particle motion due to concentration gradients of constituents (such as water vapour). The amount of water vapour is an important factor in governing the relative strength of thermophoretic and diffusiophoretic forces.

The effect of Brownian motion, thermophoresis and diffusiophoresis on the scavenging of aerosol particles for an evaporating water droplet (radius $10 \,\mu m$) at 98% relative humidity and for a water drop growing at a supersaturation of 0.3% has been studied. [44] Phoretic processes have little dependence on particle size (in the range $0.01-1 \,\mu m$) and dominate Brownian motion for particle radius $> 0.05 \,\mu m$ in the case of evaporating drops. Values are also provided of the collection kernel ($cm^3 \, s^{-1}$) for a water drop in air collecting aerosol particles by Brownian diffusion and phoretic forces for particle radii in the range $0.01-1 \,\mu m$ and for drops in the radius range $5-1000 \,\mu m$. [44]

3.2 THEORETICAL WORK ON SCAVENGING BY LIQUID PRECIPITATION

A full numerical treatment of the capture of the aerosol particle due to combined action of inertial forces, phoretic and electrical forces by means of a particle trajectory method but neglecting Brownian diffusion, has been developed. [40] This spans an aerosol particle radius from 0.5 to $10 \,\mu m$. The calculations were performed for a rather limited number of liquid drops: of radius 42, 72, 106, 173, 309 and $438 \,\mu m$ falling at terminal velocity in air. The calculations were carried out for ambient conditions of $10 \,°C$, 900 mb and 100, 95, 75 and 50% relative humidity. The water drops and aerosol particles carried electric charges of $0.2a^2, 2r^2$ where a is the drop radius and r is the aerosol particle radius. The external electric field was assumed to range between 0 and $3 \times 10^5 \, V \, m^{-1}$ (of thundercloud dimension).

Extended calculations have been made [45] to compute collision efficiencies over a large number of drop sizes using a finer integration mesh size and leading to more accurate flow fields around the collector drop than hitherto obtained. In addition, separate temperature and vapour density fields were determined, accounting for the difference between the Prandtl and Schmidt numbers rather than using the same dimensionless field for both quantities, as treated earlier. [40]

For aerosol particles with radius $r > 0.5 \,\mu m$ the efficiency $E(a,r)$ with which a drop of radius a collides with an aerosol particle of radius r is determined by integrating the equation of motion of an aerosol particle under the simultaneous influence of gravity, air drag, electrical forces and thermophoretic and diffusiophoretic forces:

$$m\frac{dv}{dt} = mg^* - \frac{6\pi\eta r(v-u)}{1+\propto N_{Kn}} + F_{th} + F_{df} + F_e \qquad (5)$$

where m, v and r are the mass, velocity and radius of the aerosol particle respectively. The efficiency is computed from an analysis of the trajectory of the aerosol particles moving past the water drop assuming that the flow around the aerosol particle does not affect the drop motion (justified for ratio of particle mass to collector drop mass $< 10^{-3}$). In the above equation $g^* = g(\rho_p - \rho_a)/\rho_p$, where g is the acceleration due to gravity, ρ_p is the density of the aerosol particle, ρ_a is the density of air, u is the velocity of air around the collector water drop, $N_{Kn} = \lambda_a/r$ is the Knudsen number, λ_a is the mean free path of air molecules ($\lambda_a = 0.065\,\mu m$ at standard pressure and temperature), F_{th} is the thermophoretic force, F_{df} is the diffusiophoretic force, F_e is the electrical force and the slip-flow Cunningham correction factor $\propto$ is given by

$$\propto = 1 + 1.26N_{kn} + 0.44\exp(-1.07/N_{Kn}) \qquad (6)$$

The particle trajectory is obtained by numerically integrating Equation (5) in the reference frame of the drop. From a knowledge of the particle trajectory around the drop, the collision efficiency given by Equation (7) below is calculated,

$$E(a,r) = \frac{\pi y_c^2}{\pi(a+r)^2} \qquad (7)$$

where y_c is the largest initial horizontal offset the particle can have from the drop axis and still collide with the drop. Further details of the calculational procedures are given elsewhere. [40,45]

A second model applicable to smaller aerosol particles ($r < 0.5\,\mu m$) considers aerosol particle capture due to combined action of Brownian motion, phoretic forces and electric charge and was developed [41] through solving the convective diffusion equation. Inertial effects were neglected and the collision efficiencies were found from a determination of the flux of particles to the scavenging drop falling at terminal velocity in air of specific atmospheric conditions. This so-called flux model was extended [46] to include the effect of external electric fields.

The combined particle trajectory method [40] and the particle flux method [41] yield values of the efficiency with which water drops of a range of sizes collide with aerosol particles with radii varying from 0.001 to 10 μm and the composite collision efficiency values for a 310 μm radius water drop are shown in Figure 3. [14] The "Greenfield gap" and the sensitivity of the depth, width and position of the gap to phoretic and electrical forces can be seen. The electric force due to the oppositely charged aerosol particle and drop has the most pronounced effect on the filling in of the gap. However, the "Greenfield gap" remains unfilled despite the imposition of quite large electrical forces. There appears to be no adequate scavenging mechanism for this gap regime. How-

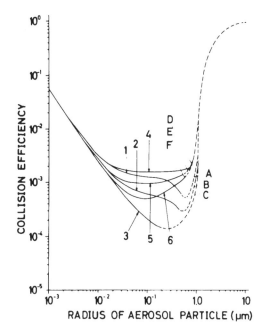

Figure 3. Effect of electric charges on the efficiency with which a 310 μm radius water drop collides with aerosol particles of $\rho_p = 2\,g\,cm^{-3}$ in air of various relative humidities Φ_v; and of 10 °C and 900 mb. Composite results combining the particle trajectory method [40] (dashed lines) with the particle flux model [41] are shown. Curves 1, 2, 3, i.e. A, B, C are for $\tilde{Q}_a = 0$, $\tilde{Q}_r = 0$ and $\Phi_v = 50$, 75 and 95% respectively. Curves 4, 5, 6, i.e. D, E, F are for $\tilde{Q}_a = \pm 2.0$, $\tilde{Q}_r = \pm 2.0$ and $\Phi_v = 50$, 75 and 95% respectively; where $\tilde{Q}_a = Q_a/a^2$ and $\tilde{Q}_r = Q_r/r^2$. From Pruppacher and Klett. [14] (Reproduced by permission of the authors, Hans P. Pruppacher and James D. Klett, and the publisher D. Reidel Publishing Company.)

ever, the gap may be bridged at least partially through nucleation scavenging of the CCN and IN aerosol population.

3.3   EXPERIMENTAL VERIFICATION OF SCAVENGING THEORIES

Most scavenging work has been performed in the laboratory and most experimental work has suffered from deficiencies. For example, some experiments have been carried out with drops not at their terminal velocity or held stationary by means of a support. In general the electrical charge of both drop and particle has not been monitored and often the ambient relative humidity and temperature have not been controlled. Ideally, monodisperse aerosol particles should be used but this has been rare. Earlier work [47] probably represents one of the better experimental investigations to determine the efficiency with which both uncharged and charged aerosol particles are collected by drops. Good agreement was obtained between experiment and prediction as shown in Figure 4. However, the theory was only tested for one particle size (0.25 μm) and for an uncommonly low relative humidity value (23%). An experimental study of the collection efficiency for three drop radii of 0.62, 0.82 and 0.98 mm for AgCl aerosol particles of radius 0.15, 0.20, 0.25, 0.36 and 0.45 μm has been carried out. [48] These experiments represent a wider variation in particle size than hitherto examined.

Laboratory measurements of the efficiency, E, with which water drops of radii between 400 and 500 μm carrying electrical charge, collect monodisperse aerosol particles of radii between 0.35 and 0.88 μm have been made. [49] The measured values of E were within the range 1–6% and agree favourably with those made by other workers [48,50] in the overlap region of the measurements, although the values of the above experiments overestimate the predicted values [41] by a factor of about 50.

Calculated and measured values of collection efficiency, E, with which water drops (in the radius range 100–1000 μm) collect submicrometre aerosol

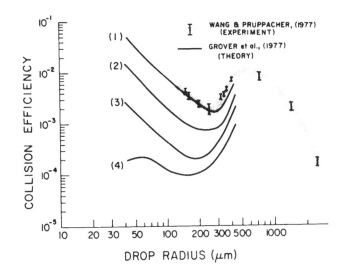

Figure 4. Variation of the efficiency with which water drops collide with aerosol particles of $r = (0.25 \pm 0.03)$ μm, $\rho_p = 1.5\,\mathrm{g\,cm^{-3}}$ in air of 1000 mb, as a function of drop size and relative humidity Φ_v of the air. Experimental results [47] are due to inertial forces, phoretic forces and Brownian diffusion, in air of $\Phi_v = 23\%$, $T = 22\,°C$, and $p = 1000$ mb. Theory considers inertial and phoretic effects only (for such a particle size the collection efficiency due to Brownian diffusion is about one order of magnitude lower than that due to phoretic effects); (1) $\Phi_v = 20\%$, (2) $\Phi_v = 75\%$, (3) $\Phi_v = 95\%$, (4) $\Phi_v = 100\%$. [47] From Pruppacher and Klett. [14] (Reproduced by permission of the authors, Hans P. Pruppacher and James D. Klett, and the publisher D. Reidel Publishing Company.)

particles are generally in the range 10^{-3}–10^{-4} in the absence of electrical forces. [40,47,51–53] When the aerosol particles are charged or are located in an electric field of magnitude comparable with values found in thunderstorms, the values of E are increased to around 10^{-2} or greater. [40, 47–50, 54, 55] In addition, experimentally determined values generally exceed the theoretical ones, typically by an order of magnitude or so. Frequently, the experimental set-up does not allow for fine control of the electrical charge on the drop and aerosol particles. For example, the measurement of the individual charge of submicrometre aerosol particles is not easy to achieve and consequently is frequently omitted in scavenging experiments. Neither is it easy to simulate the likely low charges carried by the natural aerosol. Thus, there remain considerable uncertainties and omissions in most laboratory scavenging experiments to date. In addition, inherent difficulties associated with simulating natural conditions in the laboratory are characteristic of laboratory investigations.

However, field experiments can also possess some major deficiencies and difficulties: vertical distributions of aerosol particles are desirable and are normally not known; the occurrence of spatial and temporal inhomogeneities in precipitation and air mass are added complications, so that sampling must be done rapidly.

There is generally a lack of reliable field studies of the scavenging of aerosol particles by precipitation. Airborne measurements are described [56] in coal-powered plant plumes before and after rain showers for which drop size distributions were obtained. Large discrepancies prevail between measured (over an order of magnitude greater than theory) scavenging efficiencies and theory for submicrometre aerosol particles, although the measurements agree well with theory for aerosol particles with radius $> 0.5\,\mu m$. A large field campaign has been conducted and scavenging efficiencies of aerosol particles by hydrometeors have been measured [57] during frontal precipitation passages in winter time. Observations were made at three ground stations along the slope of Mount Rigi, central Switzerland. It was also found [57] that particles $< 1\,\mu m$ radius were 10–100 times more efficiently removed than predicted by theory and that supermicrometre-sized particles are less efficiently scavenged than predicted. It is suggested [57] that phoretic forces acting on submicrometre particles may be underestimated and that electrical charges on both aerosol particles and drops (for which there are no measurements in the above field experiments) may also be underestimated. Schumann (pers. comm., 1989) suggests that melting raindrops some degrees colder than the surrounding air, may create local supersaturation in the flow field of the drop. Under such conditions thermophoretic and diffusiophoretic forces will reinforce each other by providing the flux of vapour to the drop and thereby causing an increase in the collision efficiency, thus possibly reducing the discrepancy between his measurements and prediction.

4   SCAVENGING BY SOLID PRECIPITATION

4.1   THEORETICAL COMPUTATIONS

The scavenging of aerosol particles by solid precipitation has not been studied as much, as compared to liquid precipitation scavenging. Theoretical work is more complicated than for liquid drops because of the irregular shape of ice elements. The efficiency with which aerosol particles are collected by ice crystal plates for particle radii from 0.001 to 10 μm has been studied. [58, 59] A flux model [41] has been used to calculate the collection efficiency of aerosol particles with radii < 0.5 μm by columnar ice crystals. [60] A flux model [41] and a trajectory model have been combined to yield collection efficiency of aerosol particles for radii 0.001–10.0 μm by falling columnar ice crystals. [61]

4.2   EXPERIMENTAL WORK ON THE SCAVENGING OF AEROSOL PARTICLES BY ICE ELEMENTS

4.2.1   Laboratory Studies Involving the Ice Phase

Numerous laboratory studies have been carried out to determine the scavenging efficiency of aerosol particles by ice crystals—which include work [62] using simulated snowflakes; work [63] using natural snowflakes and laboratory aerosol, but possible important phoretic effects were not examined; and work [64] using simulated discs and stellar, dendritic ice crystal shapes. Other laboratory work [65, 66] includes the use of natural snow and artificial aerosol particles— here snow was all of stellar shape; and the use [67] of a wide range of natural snow crystal shape colliding with aerosol particles of mean radius 0.75 μm. Laboratory experiments were also carried out [68] to measure the scavenging efficiencies of small ice plates grown in a cloud chamber which were then allowed to fall through a dense aerosol. A laboratory study has been conducted [69] into the efficiency with which aerosol particles of radii between 0.1 and 0.25 μm are scavenged by snowflakes and it was found that the collection efficiency remained independent of flake size and agreed quite well with a study [70] in the overlapping region.

A series of laboratory experiments [63, 65–67, 69] have been shown [69] to yield scavenging efficiencies which are within an order of magnitude of each other. They are also in reasonable agreement with a number of theoretical studies. [58–61]

4.2.2   Field Studies Involving the Ice Phase

Field measurements of the scavenging of aerosol particles by snow crystals have been carried out by a host of workers including work [71] which found that the measured collection efficiency by snow crystals falling through a polluted layer in Sapporo, Japan was one to two orders of magnitude larger than could be accounted for by hydrodynamic effects alone—they postulate that the differences may be due to turbulence effects or Brownian diffusion. [71]

Other fieldwork includes the study of rimed snow [72–76] and that of snow scavenging of water-soluble aerosols. [77] All showed an enhancement of scavenging efficiencies of submicrometre aerosol particles by one to three orders of magnitude above that predicted by scavenging theories. The collection efficiency, E, of charged and uncharged natural snow crystals in the removal of laboratory aerosol has also been measured. [78] It was found [78] that E was increased by about an order of magnitude due to the snow crystal charge.

5   PRECIPITATION BULK SCAVENGING OF AEROSOL PARTICLES

In the study of precipitation bulk sampling, it has been the practice for some 25 years to use the so-called scavenging ratio (SR) which is defined as follows:

$$\text{SR} = \frac{\text{Concentration of the aerosol substance in precipitation}(C)}{\text{Concentration of the aerosol substance in air}(K)\text{at the ground}} = \frac{C}{K} \quad (8)$$

The basic premise underlying the use of the SR is that the ground-level aerosol concentration is a measure of the aerosol concentration in air aloft from which the aerosol is removed by precipitation. The concept of SR presumes that there is a simple linear relation between C and K, although this assumption is not necessarily valid, as shown [79] from examination of multi-year sulfate data sets from remote Canadian locations. The SR gives little insight into the factors that affect wet deposition, not least since atmospheric aerosol concentrations measured at or near ground level may bear no relation to the concentrations entering a cloud system or contained below cloud.

If most of the particulate sulfate particles entering a cloud is incorporated into cloud water there will be a one-to-one relationship between the amount of sulfate in the cloud water and that contained in the air before entering the cloud and the use of an SR would be justified under these conditions. However if, for example, cloud water sulfate is also produced by aqueous phase oxidation of gaseous SO_2, then there will be nonlinearity between C and K and the usefulness of SR is weakened.

Nevertheless, SRs are often used [80] to characterize and parameterize the removal flux. When calculated from data averaged over long time periods (e.g. a year or more), SRs are reasonably consistent (within a factor of 2) over extensive marine regions. For instance, it is found [80] that SRs calculated from one-year moving average nss-SO_4^{2-} concentrations at Bermuda range from 160 to 240 (for 4 years of data) and those at Barbados from 230 to 270 (for 3 years of data). This consistency suggests that, once SRs have been characterized, average annual concentrations of aerosols or precipitation may be reasonably parameterized over large areas from those in other areas. Scavenging processes are, however, complex and vary greatly over time and space.

The scavenging rate for both in-cloud and below cloud scavenging can be written [18] as

$$\Psi(a,r) = \int_0^\infty \varepsilon E(a,r)A(a)[V(a) - v(r)]N(a)\mathrm{d}a \tag{9}$$

where $v(r)$, usually negligible, is the gravitational fall speed of the aerosol particle of radius r, a the size of the precipitation hydrometeor (whether water drop or snowflake), $N(a)$, $V(a)$ and $A(a)$ are hydrometeor number concentration, terminal velocity and effective cross-sectional area. $E(a,r)$ and ε are the aerosol particle/hydrometeor collision and retention efficiencies, and εE is known as the collection efficiency (ε is normally taken to be unity). Several physical processes contribute to particle/hydrometeor collisions and values of $E(a,r)$ are fully determinable as described in sections 3.2 and 4.1 for both liquid drops and ice elements. The scavenging rate is a weighted average of the collision (collection) efficiency over all hydrometeor sizes.

Using the assumption that the hydrometeor size distribution does not change with time, the number concentration of aerosol particles is related to the initial concentration by the precipitation scavenging rate $\Psi(a,r)$ through

$$N(r) = N_0(r)\exp(-\Psi(a,r)t) \tag{10}$$

where $N(r)$ is the concentration (per unit volume) of aerosol particles with radii between r and $r + \mathrm{d}r$ and $N_0(r)$ is the initial concentration. One can calculate precipitation scavenging rates using equation (9) for various aerosol particle sizes, for particular hydrometeor size distributions and under specified meteorological conditions.

The percentage of aerosol particles scavenged (equal to $1 - N(r)/N_0(r)$) can also be calculated. It is found [45] that the percentage scavenged by a Marshall–Palmer distribution for the precipitation drop spectrum is less than 1% under all conditions for particle radius $< 1\ \mu\mathrm{m}$ and is generally of the order of 0.1% or less for the submicrometre aerosol fraction. However, these predictions appear to be underestimated as compared to measurements as discussed in section 3.3.

Knowledge of the hydrometeor size distribution will permit calculation of the aerosol scavenging rate through use of Equation (9). Because of the inherent uncertainties in the hydrometeor and microphysical characteristics, a bulk parameterization approach to wet removal of aerosol has frequently been made. While it has the virtue of simplicity, it has the disadvantage of obscuring the full details of wet removal processes. Such a bulk parameterization approach has been adopted [18, 81] through the use of the following proposed approximation to the precipitation scavenging rate:

$$\psi = \frac{cp\bar{E}(r, R_{\mathrm{m}})}{R_m} \tag{11}$$

where c is a numerical factor of order unity, p the precipitation rate (mm hour^{-1}), E the aerosol particle/precipitation element collision efficiency using a mean precipitation element size, R_m.

Measured rain scavenging rates, Ψ, for aerosol particles generally increase in an approximate linear manner with precipitation rate in accordance with Equation (11). [18] Values of Ψ range between about 10^{-4} and 10^{-3}s^{-1} over the precipitation rate range, p, from 1 to 10 mm hour^{-1}. A similar range of values for Ψ are reported [82] for aerosol particles and radioactive iodine plumes. Field studies have been carried out [56] on the scavenging of aerosol particles by precipitation near Centralia, Washington, USA in May 1976. The aerosol size distribution was measured just prior to and just following scavenging by a rain shower. The measurements yielded scavenging rates varying from 3.2×10^{-4}s^{-1} for aerosol particles of radius 0.1 μm to 1.2×10^{-3}s^{-1} for particles of radius 0.01 μm.

6 PROCESSING OF AEROSOL BY NON-PRECIPITATING CLOUDS

Non-precipitating cloud cycles are likely to experience several cycles (of the order of 10–25) of evaporation and recondensation before a precipitating cloud system develops. [83, 84] Therefore the aerosol is likely to be cycled through more than between about 5–12 non-precipitating clouds on average before the particles are probably removed by precipitating scavenging. Thus clouds act as sources of aerosol material as well as being sinks. A residence time of the order of 5 days for aerosol material with respect to precipitation is estimated. [84] The residence time of aerosol particles with respect to cloud condensation is much shorter—of the order of 6–12 hours—before being incorporated into cloud droplets through the cloud condensation process. [84]

It is estimated that atmospheric aerosols are cycled three times, on a global average, through the cloud system. [84] This cycling of aerosols through non-precipitating clouds which are frequently present towards the top of the boundary layer is a major factor in shaping the submicrometre aerosol size distribution.[85] This cycling of clouds results in the repeated evaporation of cloud droplets and their probable subsequent re-formation as described below.

It is predicted[86] that aerosol particles which comprise both interstitial and evaporated cloud droplets are removed primarily by condensational growth and interstitial cloud scavenging with time constants of 1–10 hours and about 1 day for particle sizes < 0.01μm and 0.02–0.03 μm respectively. The smaller unactivated interstitial aerosol particles diffuse to the cloud droplets, transferring their mass to the cloud droplets which possess much larger surface areas.

When the cloud droplets later evaporate, the resulting residues are larger than the original cloud nuclei on which the cloud droplets had formed and these can participate further in the nucleation of subsequent clouds. These processes are seemingly a major factor in shaping the aerosol size distribution.

Evidence for the appearance of reprocessed cloud aerosol has been found [86] from the measurements of aerosol sulfate and aerosol scattering coefficient in cloud-free regions.

The partitioning of aerosol particles between cloud droplets and interstitial air was measured [87] for clouds on Mt Kleiner Feldberg (825 m above sealevel) in the Taunus mountains, north-west of Frankfurt am Main, Germany. On average 87, 42 and 64% of the aerosol accumulation mode volume were found to be incorporated into cloud droplets for three separate cloud events. The sulfate found in the cloud water was concluded to have originated from the aerosol particles and not from in-cloud chemical reactions. [87] Nucleation scavenging and entrainment were concluded to be the most plausible processes influencing the partitioning of the aerosol particles in the presence of cloud. [87]

7   MODELLING STUDIES OF WET PROCESSES AFFECTING ATMOSPHERIC AEROSOLS

7.1   SCAVENGING OF AEROSOL IN WATER CLOUDS

Further insight into the wet deposition of atmospheric aerosol pollutants is provided by the performance of microphysical modelling studies. Here removal of aerosol particles by in-cloud microphysical processes is considered and is confined in the first instance to warm clouds. The removal of aerosol particles by the process of condensation of water vapour, known as nucleation scavenging, received little attention prior to the work of Flossman and co-workers. [88–90] Field measurements [91] in cumulus and strato-cumulus clouds in western Washington state, USA indicated, however, that most of the aerosol number (between 80 and 92%) is incorporated into cloud water just above the cloud base by nucleation scavenging. Preliminary field measurements [92] suggested that nucleation scavenging efficiencies (ratio of aerosol material in cloud water compared to air) for sulfate in cumulus, strato-cumulus and stratus clouds was $63 \pm 53\%$.

The subsequent fate of aerosol particles once they are scavenged by cloud droplets through the condensation process is not yet determined experimentally, principally because of the small quantity of aerosol involved—typically 10^{-9} per volume of a typical cloud droplet. This at least is remedied partially through modelling—now described in some detail in the following subsections.

Theoretical models have been formulated [88,90] which incorporate processes that govern the wet deposition of atmospheric aerosol particles. The model, [88] which is concerned exclusively with warm clouds, treated the removal of aerosol particles by condensation of water vapour (inhomogeneous nucleation of CCN from available water vapour). In addition, inactivated aerosols were allowed to be collected by impaction scavenging through the mechanisms of Brownian diffusion, inertial, hydrodynamic forces and through phoretic and electrical forces, as discussed earlier (section 3.1) when considering below-cloud scavenging. The following conclusions [88] were drawn:

(A) The aerosol particle mass removed by nucleation scavenging is larger by several orders of magnitude than that of the remaining interstitial aerosol removed by impaction.
(B) The mass mixing ratio of aerosol, scavenged through nucleation by cloud droplets, strongly increases with decreasing cloud droplet size so that greater contamination occurs for the smallest sized cloud droplets.
(C) If adequate time is available for the cloud to grow via the collision–coalescence process, the aerosol mass scavenged (via nucleation) within the cloud water is redistributed such that it is contained primarily in the precipitation size range of the drops (100–1000 μm) so that the aerosol will be removed via precipitation from the cloud to ground.

Despite some shortcomings (admitted by the authors) in the model [88] such as an unrealistic aerosol size distribution, an unrealistically assumed uniform aerosol (water-soluble) composition and the use of a simple parcel model, the authors [88] feel that the main conclusions of the work hold, as reiterated [93] in their reply to subsequent comments on their work. [94]

The detailed microphysical model [88] linked by a simple entraining air parcel model was linked more realistically to a two-dimensional convective warm cloud dynamic model in later work. [90] This treatment also permitted a study to be made of the contribution of below-cloud scavenging to total scavenging. The contribution amounted to about 40%, with nucleation scavenging contributing about 60% to the aerosol mass in the rain coming to ground. Field observations of below-cloud scavenging for sulfate particles show considerable variability ranging between about 10% [95] and 70%. [96]

Detailed knowledge or prediction of the impaction efficiency is important in order to compare aerosol concentration in individually measured [97] precipitation drops for example. Indeed it was shown [89] that the in-cloud nucleation scavenging process followed by the collision and coalescence of the droplets led to a uniform size-independent distribution of aerosol pollutant concentration. Thus below-cloud impaction scavenging is an important process which has to be considered in order to reproduce, for example, realistic aerosol pollutant concentrations in the individual raindrops. However, as pointed out,[98] detailed knowledge of the collection efficiencies between interacting drops and aerosol particles is not required to model the scavenging of an average cloud—rather a parameterized approach is adequate.

7.2 SCAVENGING OF AEROSOL IN CLOUDS WHICH INCLUDE THE ICE PHASE

The effect of the ice phase on in-cloud (convective) removal of atmospheric aerosol particles by nucleation and impaction scavenging has been treated. [99] The transfer of aerosol mass from the liquid phase into the ice phase by drop freezing and scavenging of aerosol mass by riming of graupel is concluded [99] to dominate the scavenging of aerosol mass—due to the riming of snow crystals, as well as due to the impaction of aerosol particles on to ice particles. They [99] find that the efficient uptake of aerosol mass by riming of graupel is due to the fact that the graupel acquire drops which previously themselves have very efficiently scavenged aerosol particles by nucleation scavenging. It is found [99] that, even allowing for the effects of the ice phase, scavenging of aerosol particles by drop nucleation dominates impaction scavenging above all other scavenging mechanisms. Impaction scavenging of aerosol particles by snow crystals is found to be the least efficient scavenging mechanism. Also, it is found [99] that aerosol particle mass becomes redistributed inside mixed ice–water clouds, in such a manner that the main aerosol mass is associated with the main condensed water mass, i.e. it is associated with the graupel mass if riming is the dominant process of precipitation formation and it is associated with the liquid water mass if collision and coalescence of water drops are the main process for the formation of precipitation. However, it is also concluded that great uncertainties still exist for scavenging of aerosol particles. [99]

7.3 VALIDATION BY FIELD MEASUREMENTS

The model predictions cannot be validated directly with field measurements (of sulfate aerosol) in the cloud and precipitation for the Hawaiian convective cloud case study, [90] since none are available at the prescribed model time. However, field study CLEOPATRA, carried out west of Munich, Germany in July 1992, permitted air, cloud and precipitation measurements to be compared [100] with the model. [90] It was found that virtually all aerosol particles with radii greater than $0.2\,\mu m$ became activated to form cloud droplets. [100] The cut-off in interstitial aerosol number concentration was quite sharp, in quite good agreement with predictions. [88,101] This limiting size was found, using an air parcel model, to be independent of the chemical composition of the particles and strongly dependent on the prevailing super-saturation. [101] However, a similar cut-off was not observed in the interstitial aerosol measurements—rather a more gradual fall-off in concentration occurred. [87]

In addition to the nucleation scavenging mechanism, aqueous phase transformation of gaseous species (for example SO_2) through scavenging can also occur followed by its oxidation to aerosol particulates (for example SO_4^{2-}). Evidence has been found, through measurement, for the scavenging of both sulfate and nitrate in cumulus, stratus and strato-cumulus clouds. [102] Removal efficiencies (by mass) of between 80 and 98% of sulfur in stratus cloud have been measured for which the dominant removal mechanism is attributed to nucleation scavenging. [103] Observations also support high (as great as 75%) uptake of sulfate aerosol and the light-scattering aerosol fraction by cloud droplets. [86]

It was found that scavenging of SO_4^{2-} accounted for all of the snow particulate matter in 67% of the events. [104] They [104] also found that the scavenging of gaseous HNO_3 by snow is much more efficient than particulate NO_3 and accounts for 80–90% of snow nitrate particulate. Of the sulfate processed by a modelled two-dimensional warm marine cloud, 87% was scavenged by nucleation of sulfate aerosol. [105] Significant differences between the scavenging behaviour of nitrate and sulfate have been found in modelling studies [106] and reflect some basic chemical and microphysical differences. For example, HNO_3 has a much greater solubility than SO_2 which causes it to be readily dissolved in cloud (and precipitation) drops as soon as they are nucleated, whereas SO_2 oxidizes to SO_4^{2-} only in the presence of oxidizing agents such as H_2O_2 and O_3 (see also Chapter 3).

About a two orders of magnitude variation in the snow-out scavenging rate (from 10^{-6} to 10^{-4} s^{-1}) for an aerosol particle radius range from 0.1 to 1 μm is obtained. [107] Little is yet known of relations between the scavenging rates and precipitation intensity from field measurements. The effects of ice particle melting and aggregation on below-cloud scavenging has not yet been investigated.

One might expect that ice crystals should be more efficient scavengers of aerosol particles than water drops for equivalent precipitation contents, because of their larger surface to volume ratio. The collection efficiencies of aerosol particles captured by water drops, columnar ice crystals and planar ice crystals are compared in order to determine which collector is more efficient in scavenging. [108] Use is made of the volume swept out by the scavenger per unit time as the basis for comparison. The comparison shows that ice plates are the most efficient scavengers of aerosol particles between 0.01 and 1 μm, while water drops and ice columns are more efficient scavengers of particles > 1 μm.

8 SOME CONCLUSIONS AND SOME UNRESOLVED PROBLEMS

Considerable progress has been made on the study of the removal of atmospheric aerosols through wet deposition via the condensation or ice nucleation mode (nucleation scavenging) and via impaction, i.e. through attachment of aerosol particles to liquid or solid hydrometeors by mechanisms such as Brownian diffusion, inertia, hydrodynamic forces, phoretic and electrical forces. However, considerable gaps in our knowledge of precipitation scavenging remain and some of these provide the basis for future research into the removal of aerosol particles through wet deposition.

Some of the outstanding (unresolved) problems and gaps in the study of wet deposition of aerosol particles are as follows:

1. There is a severe lack of validation of current scavenging theories. There has been a paucity of field case scavenging studies and the few that have been carried out have shown large disparity with theory, as described in sections 3, 4 and 7. This may partially be due to the incomplete parameterization of the ambient conditions and of the interacting particles and precipitation elements. The lack of quantitative and reliable field data is due also to the difficulty of monitoring the continually changing ambient conditions. Thus, there is considerable scope to mount some carefully planned field scavenging experiments by means of detailed case studies. One possibility is to utilize fixed stations at different altitudes, for example at Mount Rigi, Switzerland or at the field observatory of Kleiner Feldberg in Germany. Simultaneous aerosol particle size instrumentation and precipitation spectrometers should ideally be operated at each site in the region. The use of a radar to infer precipitation size clearly would be advantageous. Another approach is to choose a well-characterized chimney stack emission plume for example and to determine scavenging coefficients for selected precipitation events. The extent to which wet removal processes influence the size distribution and composition of aerosols also needs to be determined.

2. Despite some few good laboratory investigations of aerosol particle scavenging, the theories have not been adequately tested for a wide range of aerosol particle or precipitation (liquid) size, nor for a wide range of humidity and temperature conditions. The theories have not been checked for typical atmospheric aerosol constituents. No experiments have been carried out in the presence of an external electric field for example, while only a few measurements have been made with a limited range of known particle and precipitation electrical charge. Thus there is ample scope for some definitive, well-controlled laboratory work in this area.

3. Much work needs to be done in the area of ice scavenging of aerosol particles. Inconclusive experimental results pertain to measurements of collection efficiencies of aerosol particles by different crystal habits. Studies of electrically charged ice elements have received scant attention. The theoretical predictions for a variety of electrically charged ice shapes interacting with charged particles have not been checked. More definitive experimental studies are needed for verifying the various theories.

4. It is difficult to assess the correctness or accuracy of available scavenging theories by either liquid or solid precipitation in the absence of definitive measurements. Scavenging models have been successfully coupled with two-dimensional models for convective clouds for the study of in-cloud scavenging: applications to below-cloud scavenging remain to be done. Up to now, aerosol particles have been modelled as spherical particles—there is scope for theoretical (and experimental) work on the scavenging of non-spherical particles. Further theoretical work on scavenging by regularly shaped ice elements needs to be done. There is also scope to extend the theoretical scavenging work by liquid precipitation to a much wider range of atmospheric pressures and temperatures than hitherto examined (up to now all calculations have been performed at 900 mb, 10 °C). Work also needs to be done on the scavenging of aerosol particles by non-spherical larger-sized liquid drops.

5. Clouds appear to have a profound effect on aerosol characteristics, but much more work needs to be done in the mutual interactions between aerosols and clouds. The mechanisms by which clouds of various types modify, via cloud processing, the chemical and physical properties of tropospheric aerosols need to be more fully determined and better understood. This will require knowledge of the relationships between cloud properties (both physical and chemical) and the properties of the aerosol particles formed or modified by the clouds. Some of these deficiencies will be addressed within the International Global Atmospheric Chemistry (IGAC's) new focus on atmospheric aerosols, particularly within the new activity 1: Aerosol Characterization and Process Studies. [109]

GLOSSARY OF PRINCIPAL TERM DEFINITIONS

Term	Definition
Brownian diffusion	Random motion of a particle, in which the particle displacement in time t is proportional to $\sqrt{t}$
Cloud condensation nuclei	Aerosol particle which will act as a centre (nucleus) of a (cloud) droplet from the vapour under natural condensation conditions
Collection efficiency	Product of retention (coalescence) efficiency and collision efficiency
Collision efficiency	Ratio of the collision cross-section to the geometric cross-section
Condensation nuclei	Aerosol particles of order 3 nm or greater in size which nucleate for supersaturation values much greater than found in natural cloud

Critical supersaturation	The aerosol particle(s) will grow providing the ambient supersaturation exceeds the critical supersaturation
Deliquescence	Phase transition from solution to solid or from solid to solution
Deliquescence point	Relative humidity at which phase transition occurs
Diffusiophoresis	Aerosol particle movement due to concentration gradients of parameters (water vapour for example) in a gaseous mixture
Graupel	Rimed product of an ice particle
Haze	Suspension of aerosol particles giving rise to visibilities between 1 and 10 km
Homogeneous nucleation	Nucleation of an aerosol particle from the vapour alone and in the absence of aerosol particles
Hydrometeor	Precipitation element
Ice nucleus	Aerosol particle which will act as the centre of growth of an ice element under natural condensation conditions
Inertial impaction	Impaction due to gravitational acceleration
Mass mixing ratio	Ratio of the mass of substance to that of air ($g\,kg^{-1}$)
Nucleation	Formation of an embryo of a condensed (liquid or solid) phase from the vapour
Nucleation scavenging efficiency	Ratio of aerosol material in cloud water compared to air
Precipitation element (hydrometeor)	Water drop or ice element above about 150 μm in radius such that it reaches the ground before evaporation occurs
Saturation	100% relative humidity
Scavenging ratio	Ratio of concentration of aerosol substance in precipitation to that in air at the ground
Solute	Solid aerosol material contained in an aerosol solution
Supersaturation	a measure of the excess water vapour above 100% relative humidity
Thermophoresis	Motion of aerosol particles due to a temperature gradient between a liquid drop (ice element) and the particle
van't Hoff factor	A measure of the degree of ionic dissociation in solution
WMO	World Meteorological Organization

LIST OF PRINCIPAL SYMBOLS

a	Cloud droplet radius or water drop radius or hydrometeor
A	Effective cross-sectional area of hydrometeor
CN	Condensation nuclei
C	Concentration of aerosol material in precipitation
CCN	Cloud condensation nuclei
D	Diffusion coefficient of an aerosol particle
E	Collision efficiency
EC	Elemental carbon
F_{th}	Thermophoretic force on aerosol particle
F_{df}	Diffusiophoretic force on aerosol particle
F_e	electrical force on aerosol particle
GAW	Global Atmosphere Watch
i	van't Hoff factor
IN	Ice nuclei
K	Concentration of aerosol material in air
m	Mass of aerosol particle
N	Number concentration (number per unit volume)
N_0	Initial number concentration
N_{Kn}	Knudsen number $(= \lambda_a/r)$
p	Pressure, precipitation rate (mm hour^{-1})
r	Radius of aerosol particle
R	Precipitation rate (mm hour^{-1})
R_m	Mean precipitation element size
S	Supersaturation (%)
S_c	Critical supersaturation (%)
SR	Scavenging ratio
$t_{1/2}$	Half-life of aerosol particle
T	Temperature
U	Updraft speed
u	Airflow velocity around a collector water drop
w_L	Cloud water content (mass per unit volume)
v	Gravitational fall speed (terminal velocity) of aerosol particle
V	Gravitational fall speed (terminal velocity) of hydrometeor
α	Slip-flow Cunningham correction factor
ϵ	Retention (coalescence) efficiency of an aerosol particle colliding with a hydrometeor
ϵE	Collection efficiency
λ_a	Mean free path of air molecules
Λ_B	Scavenging rate (s^{-1}) due to Brownian diffusion of aerosol particles
ρ_a	(Mass) density of air

ρ_p Bulk density of aerosol particle
Φ_v Relative humidity
Ψ Scavenging rate (s^{-1}) for both in-cloud and below-cloud scavenging

REFERENCES

1. Winkler, P., and Junge, C. E. (1972). The growth of atmospheric aerosol particles as a function of relative humidity, Part 1. Method and measurement at different locations, *J. Rech. Atmos.*, **6**, 617.
2. Winkler, P. (1973). The growth of atmospheric particles as a function of the relative humidity, II. An improved concept of mixed nuclei, *J. Aerosol Sci.*, **4**, 373.
3. Stelson, A. W., and Seinfeld, J. H. (1981). Chemical mass accounting of urban aerosol, *Environ. Sci. Technol.*, **15**, 671.
4. Tang, I. N., Wong, W. T., and Munkelwitz, H. R. (1981). The relative importance of atmospheric sulphates and nitrates in visibility reduction, *Atmos. Environ.*, **15**, 2463.
5. Clark, M. J., and Smith, F. B. (1988). Wet deposition of Chernobyl releases, *Nature*, **332**, 245.
6. Rodhe, H., Grennfelt, P., Wisniewski, J., Agren, C., Bengtsson, G., Hultberg, H., Johansson, K., Kauppi, P., Kucera, V., Oskarsson, H., Pihl Karlsson, G., Rasmussen, L., Rosseland, B., Schotte, L., Selldén, G., and Thörnelöf, E. (1995). *Acid Reign '95?—Conference Summary Statement*. Summary statement from the 5th International Conference on Acidic Deposition Science and Policy, Göteborg, Sweden, 26–30 June 1995, 1.
7. Penner, J. E., R. J. Charlson, R. J., Hales, J. M., Laulainen, N., Leifer, R., Novakov, T., Ogren, J. A., Radke, L. F., Schwartz, S. E., and Travis, L. (1993). *Quantifying and Minimising Uncertainty of Climate Forcing by Anthropogenic Aerosols*, Report No. DOE/NBB-0092T, US Dept. of Energy, Washington, DC, USA.
8. Langner, J., and Rodhe, H. (1991). A global cloud three-dimensional model of the tropospheric sulfur cycle, *J. Atmos. Chem.*, **13**, 225.
9. Badr, O., and Probert, S. D. (1994). Atmospheric sulphur: trends, sources, sinks and environmental impacts, *Applied Energy*, **47**, 1.
10. Dunmore, J. (1987). Acid rain in Europe, In *European Environmental Yearbook*, Cutrera, A., (Ed.), Docter, Milan, 665 pp.
11. Tang, I. N., and Munkelwitz, H. R. (1993). Composition and temperature dependence of the deliquescence properties of hygroscopic aerosols, *Atmos. Environ.*, **27A**, 467.
12. Warneck, P. (1988). *Chemistry of the Natural Atmosphere*, Academic Press, San Diego, USA, 757 pp.
13. Twomey, S. A. (1977). *Atmospheric Aerosols*, Elsevier, Amsterdam–Oxford–New York, 302 pp.
14. Pruppacher, H. R., and Klett, J. D. (1978). *Microphysics of Clouds and Precipitation*, D. Reidel Publishing Company, Dordrecht: Holland, Boston, 714 pp.
15. Fletcher, N. H. (1962). *The Physics of Rainclouds*, Cambridge University Press.
16. Hänel, G. (1976). Radiative effects of atmospheric aerosol particles as functions of the relative humidity at thermodynamic equilibrium with the surrounding moist air, *Advances in Geophys.*, **19**, 73.
17. Low, R. D. H. (1969). *A Comprehensive Report on Nucleation Condensation Nuclei*, AD 691700, ECOM-5249, Atmospheric Sciences Laboratory, White Sands Missile Range, NM 88005, USA.

18. Slinn, W. G. N. (1984). Precipitation scavenging, *Atmospheric Science and Power Production*, Randerson, D., (Ed.), National Technical Information Service, U.S. Dept. of Commerce, Springfield, VA, USA, 466.

19. Junge, C. E. (1952). Die Konstitution des atmosphärischen Aerosols, *Ann. Meteorol.*, **5**, 1.

20. Charlson, R. J., Pueschel, R. F., and Ahlquist, N. C. (1969). The integrating nephelometer as an instrument for studying atmosphere aerosols. In *Proceedings of the Seventh International Conference on Condensation and Ice Nuclei*, Prague and Vienna, 292.

21. Sloane, C. S. (1984). Optical properties of aerosols of mixed composition, *Atmos. Environ.*, **18**, 871.

22. Pierson, W. R., Brachaczek, W. W., Gorse, R. A., Japar, S. M., Norbeck, J. M., and Keller, G. J. (1989). Atmospheric acidity measurements on Allegheny mountains and the origins of ambient acidity in the northeastern United States, *Atmos. Environ.*, **23**, 431.

23. WMO, GAW report (1991). Report of the WMO meeting of experts to consider the aerosol component of GAW, coordinated by Jennings, S. G., *WMO/TD-No. 485, Report No. 79*, 1.

24. Trijonis, J. C. (1990). Existing conditions for visibility/aerosols in *Visibility: Science and Technology*. Report 24, National Acid Precipitation Assessment Program, Government Printing Office, Washington D. C., 20402-9325, USA, 24-57–24-68.

25. Husar, R. (1990). Historical visibility trends, In *Visibility: Existing and Historical Conditions—Causes and Effects*. Acidic Deposition State of Science and Technology. Report 24, National Acid Precipitation Assessment Program, Government Printing Office, Washington D. C., 20402–9325, USA, 24-68–24-76.

26. Twomey, S. A. (1959). The nuclei of natural cloud formation: the precipitation in natural clouds and the variation of cloud droplet concentration, *Geofisica. Pura et Appl.*, **43**, 243.

27. Twomey, S. A. (1993). Radiative properties of clouds, in *Aerosol Effects on Climate*, Jennings, S. G. (Ed.), The University of Arizona Press, Tucson and London, 275.

28. Gras, J. L. (1995). CN, CCN and particle size in Southern Ocean air at Cape Grim, *Atmos. Res.*, **35**, 233.

29. Jennings, S. G., Geever, M., McGovern, F. M., Francis, J., Spain, T. G., and Donaghy, T. (1997). Microphysical and physio-chemical characterisation of atmospheric marine and continental aerosol at Mace Head, *Atmos. Environ.* **31**, 2795.

30. Twomey, S. A. (1968). On the composition of cloud nuclei in the Northern United States. *J. Rech. Atmos.*, **3**, 281.

31. Dinger, J., Howell, H. B., and Wojciechowski, T. H. (1970). On the source and composition of cloud condensation nuclei in the subsident air mass over the North Atlantic, *J. Atmos. Sci.*, **27**, 1791.

32. O'Dowd, C. D., Smith, M. H., Consterdine, I. E., and Lowe, J. A. (1997). Marine aerosol, sea-salt, and the marine sulphur cycle: a short review, *Atmos. Environ.* **31**, 73.

33. Ogren, J. A., Groblicki, P. J., and Charlson, R. J. (1984). Measurement of the removal rate of elemental carbon from the atmosphere, *Science of the Total Environ.*, **36**, 329.

34. Kumai, M. (1976). Identification of nuclei and concentrations of chemical species in snow crystals sampled at the South Pole, *J. Atmos. Sci.*, **33**, 833.

35. Schnell, R. C., and Vali, G. (1976). Biogenic ice nuclei. Part 1: Terrestrial and marine sources, *J. Atmos. Sci.*, **33**, 1554.

36. Greenfield, S. M. (1957). Rain scavenging of radioactive particulate matter from the atmosphere, *J. Meteorol.*, **14**, 115.
37. Slinn, W. G. N., and Hales, J. M. (1971). A revaluation of the role of thermophoresis as a mechanism of in and below cloud scavenging, *J. Atmos. Sci.*, **28**, 1465.
38. Pilat, M. J., and Prem, A. (1976). Calculated particle collection efficiencies of single droplets including inertial impaction, Brownian diffusion, diffusiophoresis and thermophoresis, *Atmos. Environ.*, **10**, 13.
39. Zimin, A. G. (1964). Problems in nuclear meteorology, Karol, I. L., and Malahov, S. G. (Eds.), *USAEC Report AEC-tr- 6129*, State Publ. House for Literature, 139.
40. Grover, S. N., Pruppacher, H. R., and Hamielec, A. E. (1977). A numerical determination of the efficiency with which spherical aerosol particles collide with spherical water drops due to inertial impaction and phoretic and electrical forces, *J. Atmos. Sci.*, **29**, 1655.
41. Wang, P. K., Grover, S. N., and Pruppacher, H. R. (1978). On the effect of electric charges on the scavenging of aerosol particles by clouds and small raindrops, *J. Atmos. Sci.*, **35**, 1735.
42. Brock, J. R. (1962). On the theory of thermal forces acting on aerosol particles, *J. Colloid. Sci.*, **17**, 768.
43. Hidy, G. M., and Brock, J. R. (1970). *The Dynamics of Aerocolloidal Systems*, Pergamon Press, London.
44. Young, K. C. (1974). The role of contact nucleation in ice initiation in clouds, *J. Atmos. Sci.*, **31**, 768.
45. McGann, B. T., and Jennings, S. G. (1991). The efficiency with which drizzle and precipitation sized drops collide with aerosol particles, *Atmos. Environ.*, **25A**, 791.
46. Wang, P. K., and Pruppacher, H. R. (1980). The effect of an external electric field on the scavenging of aerosol particles by cloud drops and small raindrops, *J. Colloid and Inter. Sci.*, **75**, 286.
47. Wang, P. K., and Pruppacher, H. R. (1977). An experimental determination of the efficiency with which aerosol particles are collected by water drops in saturated air, *J. Atmos. Sci.*, **35**, 1735.
48. Lai, K. Y., Dayan, N., and Kerker, M. (1978). Scavenging of aerosol particles by a falling water drop, *J. Atmos. Sci.*, **35**, 674.
49. Byrne, M. A., and Jennings, S. G. (1993). Scavenging of sub-micrometre aerosol particles by water drops, *Atmos. Environ.*, **27A**, 2099.
50. Barlow, A. K., and Latham, J. (1983). A laboratory study of the scavenging of submicron aerosol by charged raindrops, *Quart. J. Roy. Meteorol. Soc.*, **109**, 763.
51. Hampl, V., Kerker, M., Cooke, D. D., and Matijevic, E. (1971). Scavenging of aerosol particles by a falling water droplet, *J. Atmos. Sci.*, **28**, 1211.
52. Kerker, M., and Hampl, V. (1974). Scavenging of aerosol particles by a falling water droplet and calculation of washout coefficient, *J. Atmos. Sci.*, **31**, 1368.
53. Beard, K. V. (1974). Experimental and numerical collision efficiencies for submicron particles scavenged by small raindrops, *J. Atmos. Sci.*, **33**, 1595.
54. Adam, J. R., and Semonin, R. G. (1970). Collection efficiencies of raindrops for submicron particulates, *Precipitation Scavenging*, AEC Symposium series, **22**, 1313.
55. Grover, S. N., and Beard, K. V. (1975). A numerical determination of the efficiency with which electrically charged cloud drops and small raindrops collide with electrically charged spherical particles of various densities, *J. Atmos. Sci.*, **32**, 2156.
56. Radke, L. F., Hobbs, P. V., and Eltgroth, M. W. (1980). Scavenging of aerosol particles by precipitation, *J. Appl. Meteorol.*, **19**, 715.
57. Schumann, T. (1989). Large discrepancies between theoretical and field-determined scavenging coefficients, *J. Aerosol Sci.*, **20**, 1159.

58. Martin, J. J., Wang, P. K., and Pruppacher, H. R. (1980). A theoretical study of the effect of electric charges on the efficiency with which aerosol particles are collected by ice crystal plates, *J. Colloid. Interface Sci.*, **78**, 44.

59. Martin, J. J., Wang, P. K., and Pruppacher, H. R. (1980). A theoretical determination of the efficiency with which aerosol particles are collected by simple ice crystal plates, *J. Atmos. Sci.*, **37**, 1628.

60. Wang, P. K., and Pruppacher, H. R. (1980). The efficiency with which aerosol particles of radius less than 1 µm are collected by columnar ice crystals, *Pure and Appl. Geophys.*, **118**, 1090.

61. Miller, N. L., and Wang, P. K. (1989). Theoretical determination of the efficiency of aerosol particle collection by falling columnar ice crystals, *J. Atmos. Sci.*, **46**, 1656.

62. Starr, J. R., and Mason, B. J. (1966). The capture of airborne particles by water drops and simulated snow crystals, *Quart. J. Roy. Meteor. Soc.*, **92**, 490.

63. Knutsen, E. O., Sood, S. K., and Stockman, J. D. (1976). Aerosol collection by snow and ice crystals, *Atmos. Environ.*, **10**, 395.

64. Prodi, F., Caporaloni, M., Santachiara, G., and Tampieri, F. (1981). Inertial capture of particles by obstacles in form of disk and stellar crystals, *Quart. J. Roy. Meteorol. Soc.*, **107**, 699.

65. Murakami, M., Kikuchi, K., and Magono, C. (1985a). Experiments on aerosol scavenging by uncharged snow crystals. Part I: collection efficiency of uncharged snow crystals for micron and submicron particles, *J. Meteor. Soc., Japan*, **63**, 119.

66. Murakami, M., Kikuchi, K., and Magono, C. (1985). Experiments on aerosol scavenging by natural snow crystals. Part II: Attachment rate of 0.1 µm dia. particles to stationary snow crystals, *J. Meteor. Soc., Japan*, **63**, 130.

67. Sauter, D. P., and Wang, P. K. (1989). An experimental study of the scavenging of aerosol particles by natural snow crystals, *J. Atmos. Sci.*, **46**, 1650.

68. Bell, D. A., and Saunders, C. P. R. (1990). The scavenging of high altitude aerosol by small ice crystals, *Atmos. Environ.*, **25A**, 801.

69. Mitra, S. K., Barth, U., and Pruppacher, H. R. (1990). A laboratory study of the efficiency with which aerosol particles are scavenged by snow flakes, *Atmos. Environ.*, **24A**, 1247.

70. Sood, S. K., and Jackson, M. R. (1970). Precipitation scavenging, U.S. Dept. of Commerce, NTIS, *CONF-700601*, Washington, DC, USA, 121.

71. Magono, C., Endoh, T., Harimaya, T., and Kubota, S. (1974). A measurement of the scavenging effect of falling snow crystals on the aerosol concentration, *J. Meteor. Soc. Japan*, **52**, 407.

72. Graedel, T. E., and Franey, J. P. (1975). Field measurements of submicron aerosol washout by snow, *Geophys. Res. Lett.*, **2**, 325.

73. Magono, C., Veno, F., and Kubota, S. (1975a). Observations of aerosol particles attached to falling snow crystals, Part I. *J. Fac. Sci., Hokkaido University* Series V11, **4**, 93.

74. Magono, C., Endoh, T., and Itasaka, M. (1975b). Observations of aerosol particles attached to falling snow crystals, Part II. *J. Fac. Sci., Hokkaido University* Series V11, **4**, 103.

75. Magono, C., Endoh, T., Veno, F., Kubota, S., and Itasaha, M. (1979). Direct observations of aerosol particles attached to falling snow crystals, *Tellus*, **31**, 102.

76. Murakami, M., Hiramatsu, C., and Magono, C. (1981). Observation of aerosol scavenging by falling snow crystals at two sites of different heights, *J. Meteor. Soc. Japan*, **59**, 763.

77. Murakami, M., Kimura, T., Magono, C., and Kikuchi, K. (1983). Observation of precipitation scavenging for water soluble particles, *J. Meteor. Soc. Japan*, **61**, 346.

78. Murakami, M., and Magono, C. (1983). An experimental study of electrostatic effect on aerosol scavenging by snow crystals. Precipitation scavenging, dry deposition and resuspension, **1**, *Precipitation Scavenging*, coordinated by Pruppacher, H. R., Semonin, R. G., and Slinn, W. G. N., 541.

79. Barrie, L. A. (1992). Scavenging ratios: Black magic or a useful scientific tool?, *Precipitation Scavenging Processes*, coordinated by Schwartz, S. E., and Slinn, W. G. N., Hemisphere Publishing Corporation, Washington, Philadelphia, London, 403.

80. Galloway, J. N., Savoie, D. L., Keene W. C., and Prospero, J. M. (1993). The temporal and spatial variability of scavenging ratios for nss sulfate, nitrate, methanesulfonate and sodium in the atmosphere over the North Atlantic Ocean, *Atmos. Environ.*, **27A**, 235.

81. Slinn, W. G. N. (1977). Precipitation scavenging: some problems, approximate solutions and suggestions for future research, In *Precipitation Scavenging (1974)*, Proc. of Symp. at Champaign, Illinois, 14–18 October 1974, pp. 1–60, CONF-741003, National Technical Information Service, U.S. Dept. of Commerce, Springfield, VA, USA.

82. Brenk, H. D., and Vogt, K. J. (1981). The calculation of wet deposition from radioactive plumes, *Nuclear Safety*, **22**, 362.

83. Pruppacher, H. R. (1986). The role of cloud physics in atmospheric multiphase systems: ten basic statements, in *Chemistry of Multiphase Atmospheric Systems*, NATO ASI Ser., Vol. 6, Jaeschke, W. (Ed.), Springer-Verlag, New York, 133.

84. Pruppacher, H. R., and Jaenicke, R. (1995). The processing of water vapour and aerosols by atmospheric clouds, a global estimate, *Atmos. Res.*, **38**, 283.

85. Hoppel, W. A., Fitzgerald, J. W., Frick, G. M., Carson, R. E., and Mack, E. J. (1990). Aerosol size distributions and optical properties found in the marine boundary layer over the Atlantic Ocean, *J. Geophys. Res.*, **95**, 3659.

86. ten Brink, H. M., Schwartz, S. E., and Daum, P. H. (1987). Efficient scavenging of aerosol by liquid-water clouds, *Atmos. Environ.*, **21**, 2035.

87. Hallberg, A., Noone, K. J., Ogren, J. A., Svenningsson, I. B., Flossmann, A. I., Wiedensohler, A., Hansson, H. C., Heintzenberg, J., Anderson, T., Arends, B. G., and Maser, R. (1994). Phase partitioning of aerosol particles in clouds at Kleiner Feldberg, *J. Atmos. Chem.*, **19**, 107.

88. Flossmann, A. I., Hall, W. D., and Pruppacher, H. R. (1985). A theoretical study of the wet removal of atmospheric pollutants. Part 1: The re-distribution of aerosol particles captured through nucleation and impaction scavenging by growing cloud drops, *J. Atmos. Sci.*, **42**, 582.

89. Flossmann, A. I., Pruppacher H. R., and Topalian, J. H. (1987). A theoretical study of the wet removal of atmospheric pollutants: Part II: The uptake and re-distribution of $(NH_4)_2SO_4$ particles and SO_2 gas simultaneously scavenged by growing cloud drops, *J. Atmos. Sci.*, **44**, 2912.

90. Flossmann, A. I., and Pruppacher, H. R. (1988). A theoretical study of the wet removal of atmospheric pollutants. Part III: the uptake, redistribution, and deposition of $(NH_4)_2SO_4$ particles by a convective cloud using a two-dimensional cloud dynamics model, *J. Atmos. Sci.*, **45**, 1857.

91. Radke, L. (1983). Preliminary measurements of the size distribution of cloud interstitial aerosol. *Precipitation Scavenging, Dry Deposition and Resuspension*, Vol. 1, Elsevier, Amsterdam, 71.

92. Hegg, D. A. and Hobbs, P. V. (1983). Preliminary measurements on the scavenging of sulfate and nitrate by clouds. *Precipitation Scavenging, Dry Deposition and Resuspension*, Vol. 1, Elsevier, Amsterdam, 78.

93. Flossmann, A. I., and Pruppacher, H. R. (1989). Reply, *J. Atmos. Sci.*, **46**, 1870.

94. Ogren, J. A., and Charlson, R. J. (1989). Comments on "A theoretical study of the wet removal of atmospheric pollutants. Part 1: The re-distribution of aerosol particles captured through nucleation and impaction scavenging by growing cloud drops and Part II: The uptake and re-distribution of $(NH_4)_2SO_4$ particles and SO_2 gas simultaneously scavenged by growing cloud drops", *J. Atmos. Sci.*, **46**, 1867.

95. Scott, B. C. (1978). Parameterisation of sulfate removal by precipitation, *J. Appl. Meteorol.*, **17**, 1375.

96. Petrenchuk, O. P., and Selenzera, E. S. (1970). Chemical composition of precipitation in regards of the Soviet Union, *J. Geophys. Res.*, **75**, 3627.

97. Bächmann, K., Haag, I., and Röder, A. (1993). A field study to determine the chemical content of individual raindrops as a function of their size, *Atmos. Environ.*, **27A**, 1951.

98. Flossmann, A. I. (1993). The effect of the impaction scavenging efficiency, on the wet deposition by a convective warm cloud, *Tellus*, **45B**, 34.

99. Alheit, R. R., Flossmann, A. I., and Pruppacher, H. R. (1990). A theoretical study of the wet removal of atmospheric pollutants. Part IV: The uptake and redistribution of aerosol particles through nucleation and impaction scavenging by growing cloud drops and ice particles, *J. Atmos. Sci.*, **47**, 870.

100. Wurzler, S., Responder, P., Flossmann, A. I., and Pruppacher, H. R. (1994). Simulation of the dynamics, microstructure and cloud chemistry of a precipitating and a non-precipitating cloud by means of a detailed 2-D cloud model, *Beitr. Phys. Atmos.*, **67**, 313.

101. Ahr, M., Flossmann, A. I., and Pruppacher, H. R. (1989). On the effect of the chemical composition of atmospheric aerosol particles on the nucleation scavenging and the formation of a cloud interstitial aerosol, *J. Atmos. Chem.*, **9**, 465.

102. Hegg, D. A., Hobbs, P. V., and Radke, L. F. (1984). Measurements of the scavenging of sulphate and nitrate in clouds, *Atmos. Environ.*, **18**, 1939.

103. Sievering, H., Van Valin, C. C., Barrett, E. W., and Pueschel, R. F. (1984). Cloud scavenging of aerosol sulphur: Two case studies, *Atmos. Environ.*, **18**, 2685.

104. Cadle, S. H., VandeKopple, R., Mulana, P. A., and Muhlbaier Dasch, J. (1990). Ambient concentrations, scavenging ratios and source regions of acid related compounds and trace metals during winter in northern Michigan, *Atmos. Environ.*, **24**, 2981.

105. Flossmann, A. I. (1994). A 2-D spectral model simulation of the scavenging of gaseous and particulate sulfate by a warm marine cloud, *Atmos. Res.*, **32**, 233.

106. Wurzler, S., Flossmann, A. I., Pruppacher, H. R., and Schwartz, S. E. (1995). The scavenging of nitrate by clouds and precipitation, *J. Atmos. Chem.*, **20**, 259.

107. Sparmacher, H., Fülber, K., and Bonka, H. (1993). Below-cloud scavenging of aerosol particles: particle-bound radionuclides—experimental, *Atmos. Environ.*, **27A**, 605.

108. Wang, P. K. (1992). Comparison between the collection efficiency of aerosol particles by water drops and ice crystals. Precipitation scavenging and atmosphere–surface exchange, Vol. 1 *The Georgii Volume: Precipitation Scavenging Processes*, coordinated by Schwartz, S. E., and Slinn, W. G., Hemisphere Publishing Corporation, Washington, Philadelphia, London, 87.

109. *IGAC Activities Newsletter* (1995). Activities of the International Global Atmospheric Chemistry Project, Newsletter Issue No. 2, September 1995. IGAC Core Project Office, MIT Bldg. 24–409, Cambridge, MA 02139–4307, USA.

15 Condensed Water Aerosols

JOST HEINTZENBERG

Institute for Tropospheric Research, Leipzig, Germany

1 INTRODUCTION

A definition of the *condensed water aerosols* comprises hazes, fogs and clouds. They consist of solid and/or liquid particles suspended by the carrier gas air.

Atmospheric Particles, Edited by R. M. Harrison and R. Van Grieken.

Because of their short residence time in the atmosphere, one may argue whether precipitation elements, i.e. raindrops, snow, hail and graupel, should be included in this definition as well. The present text touches upon precipitation elements only marginally. Any atmospheric particle with water as the dominating constituent is traditionally called a *hydrometeor*. In the present context, certain hazes (defined below) fall under this category. Thus, we shall use the term "condensed water aerosols" that comprise both hydrometeors and wet haze.

The meteorologist divides hydrometeors into *cloud elements* below 100 μm drop radius and *precipitation elements* which are larger and thus have a good chance of reaching the ground before they evaporate. Within three level categories of altitude meteorology differentiates four main categories of clouds:

1. *Cumulus*: clouds with significant vertical development.
2. *Stratus*: layer clouds.
3. *Nimbus*: rain clouds, and
4. *Cirrus*: fibrous ice clouds.

All categories can be combined to form subgroups, e.g. strato-cumulus or nimbo-stratus, except for category 4 (rain does not fall out of ice clouds). These subgroups describe the macroscopic cloud features in greater detail or specify certain stages of cloud development. The typical features of the different cloud types have been illustrated. [158] Their macroscopic and microscopic features have been introduced [147] and discussed in detail. [40, 113]

Horizontal visibility is used in meteorology to differentiate between haze and fog. In *hazes* the visibility is between 1 and 10 km. Surface air with visibilities below 1 km is called *fog*. Thus, macroscopically, fog is nothing but a layer cloud with sufficiently long surface contact to dampen vertical motions. Consequently, fog-covered regions look very similar to strato-cumulus areas in satellite images in terms of their structural and optical appearance. [179] We shall see below, though, that this surface contact causes microscopic and chemical properties that differentiate fogs from other clouds. Because of the occurrence of dissimilar optical phenomena, condensed water aerosols with a horizontal visibility between 0.5 and 1 km have been termed *mist*. [36]

There are special atmospheric conditions where the visibility definition of haze and fog breaks down. In desert areas visibilities can be well below 1 km without fog being present. Here, dry, coarse wind-blown dust is the visibility-limiting aerosol. Other exceptions can be found in heavily polluted urban regions like Los Angeles. Here, equally low visibilities have been observed in dense dry aerosols of mainly organic composition termed "photochemical smog" (smog = smoke + fog). However, in the present context we are not concerned with these dry aerosols.

2   WET HAZE

At relative humidities (RH) well below 100% there are micrometre and sub-micrometre-sized atmospheric particles containing more than 50% water. On average over central Europe, this is the case for particles with 0.1 μm < particle radius< 1 μm at RH> 90%. [185] For coarse particles (> 1 μm particle radius), the corresponding value of RH is between 85 and 90%. Containing that much condensed water in a strongly curved body requires highly hygroscopic aerosol material. [85] This material can be

1. Mineral acids like sulfuric acid or nitric acid,
2. Soluble inorganic salts such as magnesium chloride, sodium chloride, sulfates, nitrates, or
3. Polar organic material.

The first of the three groups is the dominating wet aerosol material in the stratosphere, forming the so-called Junge aerosol layer between 15 and 25 km altitude, [87] or (at lower temperatures), polar stratospheric clouds. [62, 171, 172] Soluble organic and inorganic material (groups two and three), on the other hand, are major components of tropospheric aerosol particles. [184] When their *deliquescence point* is reached by the ambient RH they dissolve and form haze droplets. Deliquescence points of common pure compounds in the aerosol can be found in Chapter 14. Mixtures of salts may decrease the deliquescence points. [168] In the total suburban aerosol, on average 50–60% water-soluble material were found without systematic seasonal variability. [184] The total variability, on the other hand, was considerable, spanning a range of 30 to over 80% water-soluble material.

Besides hygroscopic aerosol material, physical particle properties such as capillaries and contact surfaces in particle aggregates can lead to local reductions in saturation vapour pressure over haze particles. Through such structural effects considerable water uptake is expected to take place in wettable aerosol particles and in mixed particles below the lowest deliquescence point of a mixture of soluble salts in a particle. [185]

Metastable states of supersaturated solutions in atmospheric particles, [29, 148] and chemically "bound" water [142] can also contribute to significant mass fractions of water in aerosol particles at RH values below the actual deliquescence points.

Wet haze particles have features which are of interest to the atmospheric chemist. Due to the water uptake their surface increases considerably, (by more than 60% according to our definition of condensed water aerosols). This increase in (wet) surface facilitates the uptake of reactive gases such as SO_2. Since more than half of the particle mass consists of water and dissolved aerosol material, liquid-phase chemical reactions of the dissolved aerosol material and deposited gases are likely to occur. Some of these reactions will form compounds such as sulfates which remain stable in the dry particle at lower RH. There is experimental evidence [72] for such aerosol formation processes to occur in the accumulation size range (0.05 μm particle radius < 0.5 μm). However, there is no theory available for the chemical processes in the highly concentrated multi-phase systems of wet haze particles.

3   WARM CLOUDS AND FOGS

Low and medium high (< 5 km) clouds at lower latitudes and lower clouds and fogs at all latitudes in summer consist mostly of liquid water with some dissolved and undissolved impurities. Their main microphysical formation processes are condensation of water vapour with subsequent kinematic coagulation and coalescence of the cloud drops. Under favourable conditions (the vertical extent of the cloud often being the controlling factor), the drops will grow until precipitation is initiated. The physical processes in these clouds have been discussed in detail. [144]

3.1   MICROPHYSICAL CHARACTERISTICS OF WARM CLOUDS

Because of the similarities of warm and mixed-phase clouds this section includes some of the parameters and characteristics of mixed-phase clouds. The total liquid water content (LWC), or total condensed water content (TWC) per volume of cloudy air when including mixed-phase clouds, is the basic microphysical parameter of clouds. Total number concentrations and average drop radius are other basic characteristics for which typical values are listed in Table 1.

Table 1   Basic physical characteristics of different types of clouds. TWC = Total condensed water content, N = Number concentration of hydrometeors, R = most important radius range.

Type	TWC ($g\,m^{-3}$)	N (cm^{-3})	R (μm)
Fog	0.05–0.5	1–10×10^3	0.5–30
Stratus and stratocumulus†	0.05–0.5	20–500	0.5–30
Fair-weather cumulus			
(a) continental†	0.1–0.5	1–2000	2–30
(b) marine†	0.05–0.5	10–400	5–50
Thunderstorm	2	75	2–4000
Cirrus clouds	10^{-3}–2	10^{-4}–5×10^3	$\leq$ 2–4000*

† The range of N was taken from a compilation in 159.
* Crystal length.

A recent review of Russian data [114] presented TWC measurements for over 10^5 km of flight path in continental mid-latitude clouds. For the temperature range of $+10\,°C$ to $-50\,°C$ average TWC values varied between 0.19 and $0.012\,g\,m^{-3}$, respectively. The author provided a temperature parametrization of the TWC.

The shape of the size distribution is one of the major unresolved problems in cloud microphysics. Condensational growth predicts narrow distributions while measurements show broad and often multimodal shapes. One line of reasoning seeks the explanation in the incomplete modelling of small-scale processes such as inter-drop competition, turbulent variations of atmospheric state parameters and variability in the composition of cloud drop nuclei. Entrainment of dry air (cf. section 5) is forwarded as another argument for the observed shape of the drop population. Measurements from 1980 of drop size distributions in warm clouds from 56 publications up to 1975 were tabulated in [155] who discussed their representativeness in [157]. From a total length of 2502 flight kilometres in stratiform clouds it was determined that about 14% of the cloud space was occupied by bimodal distributions. [97] Since these features were closely connected with turbulent motions that author suggested the use of the data on drop population in studies of turbulence in stratiform clouds. In convective clouds the fraction of space occupied by bimodal distributions varies between several per cent to 50%. [93, 177]

3.2   CHEMICAL CHARACTERISTICS OF WARM CLOUDS

Clouds form on aerosol particles. During their lifetime soluble gases are scavenged by the drops and irreversible reactions of the scavenged material take place in the liquid phase. As a result of the variability of the input aerosol and of the trace gases in the atmosphere hundreds of chemical compounds and related chemical reactions have been identified in condensed water aerosols (cf. the review [54]). One of the most important chemical features of clouds is the fact that major oxidation pathways in the atmosphere go through the condensed phase. As an example, global modelling studies determined that about 80% of the SO_2 oxidation to sulfate takes place in cloud water. [105]

The major constituents of fog and cloud water have been listed for different locations. [43] The author also presents an extensive discussion of minor constituents and cloud processes. Table 2 enlarges the range of settings in his Table III which is representative of a large data set with the cleanest cloud sample taken during flow from the Arctic in the study reported. [126]

Besides these major components, practically the whole range of trace constituents of the atmospheric aerosol has been analyzed in cloud water.

The bulk data in Table 2 give an incomplete picture of the chemical variability of clouds. In a recent review, [127] it was pointed out that the evolution of a cloud should lead to systematic drop-size dependencies in composition.

Table 2. Major trace constituents in condensed water aerosols of warm and mixed-phase clouds in different settings. Concentrations are given in $\mu eq\,l^{-1}$

Species	Highly polluted fog[45]	Ground-based cloud data		Airborne cloud data		
		Central Europe[26]	Sub-Arctic Scandinavia[126]	SW USA[66]	NW USA[66]	Mixed phase[75]
Cl^-	130	53	0.4	0.01–760	n.d.	–
NO_3^-	1700	490	0.4	48–220	5–40	110
SO_4^{2-}	1600	300	1.0	96–400	15–150	240
H^+	4	23	3.8	0.3–25	3–79	–
Ca^{2+}	130	46	0.2	–	–	–
K^+	31	8	n.d.	1–69	–	–
Mg^{2+}	14	12	0.2	–	–	–
Na^+	40	34	0.4	1.7–410	–	–
NH_4^+	2900	660	0.6	56–230	–	75

n.d. = below detection limit; – = not determined.

However, in many cases, it may be impossible to observe the full range of chemical inhomogeneities because of experimental limitations. With the counterflow virtual impactor (CVI), size-dependent composition studies have been made. During the stage of stratiform cloud evolution where condensational growth dominates, an increase in solute concentration with droplet size was observed, [130] and larger drops nucleated on larger droplet nuclei. [68] Larger drops had different solute composition. [123] Single particle analyses of cloud drops have shown similar size dependencies in composition. [119, 120] Stratus cloud samples taken with impactors having two different lower cut radii (≈ 2 and $8\,\mu m$) indicated higher concentrations of sea-salt-related species in larger drops. [116] On the other hand, because of the low supersaturations coupled with very high aerosol concentrations in highly polluted fogs, equilibrium growth may dominate in this system. [127] Consequently, a decrease in solute concentration with drop size can be expected which was confirmed experimentally. [131] Details about the partitioning of different species in cloudy air, [60] and about the origin of cloud droplet nuclei, [52] have been derived from single particle analyses of CVI samples.

4   MIXED-PHASE AND ICE CLOUDS

Introducing a third phase complicates the atmospheric system considerably, in particular since the occurrence of the ice phase is not simply related to temperature but appears to be related to many factors which are not completely understood. Atmospheric liquid water easily supercools. Liquid clouds at

−20 °C are not uncommon. At −40 °C supercooled drops have been measured, [151] and there are reports of icing on aircraft down to −55 °C. There are two main reasons for this apparent discrepancy between laboratory and atmospheric behaviour of water. One, in condensed water aerosols the liquid phase is highly dispersed. Thus, every small drop must go through an ice nucleation process before a cloud is fully iced. Two, cloud water is not clean. It contains trace substances which reduce its freezing temperature.

4.1 MIXED-PHASE CLOUDS

At mid-latitudes this type of cloud is very frequent. If cloud top temperatures reach below −5 °C, ice particles are observed in these clouds. Besides this temperature, cloud age and cloud type influence the occurrence of the ice phase. The phase diagram of water reveals that supercooled drops and ice crystals cannot coexist in a stable cloud. Air which is saturated with respect to liquid water is always supersaturated with respect to ice. The absolute difference in saturation vapour pressure has a maximum between −10 and −15 °C (cf. Figure 1). This difference is the basis for the so-called Bergeron–Findeisen mechanism [14] for the initialization of precipitation in mixed-phase clouds: even a few ice particles can grow at the expense of the drops. After sufficient gain in fall speed they will grow additionally by collecting drops (riming) on their way down. The big and as yet unresolved question concerns the cause of those few ice crystals that start the process. It appears that certain material in condensed water aerosols helps the freezing of the liquid phase.

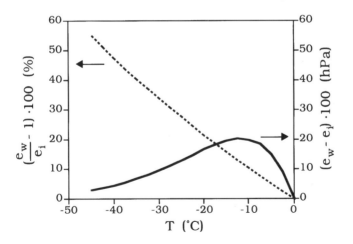

Figure 1. Absolute and relative differences in saturation vapour pressure between liquid (e_w) and solid (e_i) water as a function of temperature

However, we do not know which particle properties further the freezing process. Consequently, attempts to quantify the concentration of atmospheric ice nuclei have not been very reproducible. The latest instrumental development in this area has been reported in ref. 146.

4.2 CIRRUS CLOUDS

At cirrus temperatures water vapour saturation ratios with respect to ice are so high (cf. Figure 1) that no special particle properties are necessary to initiate ice nucleation. Thus, present models of cirrus formation assume homogeneous ice nucleation on the materials which are thought to be dominating in the upper troposphere, namely ammonium sulfate and sulfuric acid. [79]

Due to their optical properties and their widespread occurrence, cirrus clouds have a strong influence on the radiation balance of the earth and are even suspected of stabilizing tropical sea surface temperatures. [145] One major reason for their climatic importance lies in their persistence which was noted in an early review of the ice phase in the atmosphere. [178] More recent *in situ* data confirmed the early observations of several hours of lifetimes for cirrus crystals. [165] These findings and intensified studies of polar stratospheric clouds in relation to atmospheric ozone have initiated the search for physical and chemical surface processes on the surfaces of these condensed water aerosols which would explain their long lifetime in subsaturated parts of the atmosphere. [141]

4.3 MICROPHYSICAL CHARACTERISTICS

The microphysical characteristics of mixed-phase clouds are closer to those of warm clouds than those of cirrus clouds and thus are listed in section 2.1. Table 1 includes the range of basic microphysical parameters for cirrus clouds. A recent review of microphysical and optical cirrus properties has been presented. [67] The authors stress the importance of particle sizes below the threshold of traditional crystal measurements ($\approx 20\,\mu$m). In particular in cold cirrus, number concentrations tend to increase with decreasing particle size down to the threshold radius of present instrumentation ($\approx 2\,\mu$m). The highest observed crystal concentrations ($> 2.5\,\mu$m aerodynamic radius) are of the order of $10^4\,\text{cm}^{-3}$, [32, 165] which is still lower than the highest observed dry aerosol number concentrations in the cirrus region. [166] Thus, even higher crystal numbers are conceivable with lower size limits in crystal sensors.

4.4 CHEMICAL CHARACTERISTICS

Our knowledge of the chemical characteristics of condensed water aerosols with the ice phase present is even more limited than on the physical side. In mixed-phase clouds, all samplers have to struggle with icing problems while in cirrus clouds very small amounts of condensates are available for sampling. In Table 2 an example is given for the composition of mixed-phase clouds.

By number, submicrometre particles dominate in upper tropospheric aerosol samples. [56] They consist mainly of sulfuric acid. [18, 74, 161, 187] Residues of cirrus crystals were investigated, [56] which were collected by means of an impactor in autumn 1990 over Colorado and Wyoming. Their X-ray analyses yielded predominantly lighter (Na–S) elements with sulfur being the most prominent. The heavier elements (Cl–Ni) were found as well but in lower abundances.

On one occasion in January 1992, the CVI has been used to sample cirrus elements over the Alps for subsequent analyses. By means of single particle (cf. section 8.1) analyses, high abundances of minerals in micrometre-sized cirrus crystal residues were found. [69] As compared to interstitial and out-of-cloud particles the crystal residues were depleted in lighter elements (Na–S) and enriched in the heavier elements Ti–Cd. The high abundances of crustal elements were explained by the maximum of the aerosol cloud from the Pinatubo eruption being present in the sampling area at the time of the experiment.

5   MIXING PROCESSES IN THE CLOUD ENVIRONMENT

Assuming a saturated adiabatic ascent of the air above cloud base one can calculate a so-called adiabatic cloud water content LWC_{ad} as a function of time or altitude. Generally, measured values of LWC are considerably smaller than LWC_{ad}. This discrepancy is explained by the entrainment of air, mostly from the cloud top. The entrainment consists of a number of concurrent processes:

- part of the cloud water evaporates
- the cloudy air cools down
- new, unactivated aerosol particles and gases enter into the cloud
- in the extreme case, the cooling air mixture sinks below the cloud base

For the quantification of the degree of mixing the Paluch diagram has been developed. [53, 136, 137] Through conditional sampling of ozone in clouds, direct quantification of entrainment events has been accomplished. [91]

Different hypotheses have been formulated for the description of the mixing processes in clouds. In *homogeneous mixing* the time-scale of the turbulent diffusion of an entrained air parcel and for the molecular diffusion at the boundary between a parcel and its environment are small compared to the time-scale for the evaporation of cloud drops in the mixed, subsaturated air. As a result, cloud inhomogeneities due to the entrainment are evened out before the drops have evaporated. Consequently, the size distribution of the hydro-meteors does not change much.

In *inhomogeneous mixing* the time-scale of evaporation is fast compared to the time-scales of mixing. In that case, the drop size distribution and thus the whole development of the cloud is strongly affected by the entrainment. The frequently measured bimodality of drop populations (cf. section 2.1) would indicate a dominance of this type of mixing.

Entrainment of air with gaseous and particulate pollutants seriously affects the microphysical and chemical characteristics of clouds and fogs. The effect on drop size distributions has been investigated in modelling and experimental studies. [7–9, 80, 81, 138, 143, 169, 170, 174] As a result of the changes in drop population, initialization as well as rates of precipitation will be affected. [19, 28] Entrained soluble and reactive gases affect the liquid-phase chemical processes in clouds and thus the composition of precipitation and of the processed aerosol. [22, 41, 48] There is some evidence for the suspicion that mixing processes in clouds or at cloud bound-aries cause local supersaturations of condensable vapours other than water vapour. The resulting nucleation of particles would make the cloud environ-ment a source of new particles in the atmosphere. [65, 180] First attempts at incorporating local fluctuations into particle nucleation models have been reported for the case of sulfuric acid and simplistic cloud processing mechan-isms. [34, 90]

6   MICROPHYSICAL METHODOLOGY FOR CONDENSED WATER AEROSOLS

In this section selected techniques for the physical characterization of con-densed water aerosols are discussed. Instead of attempting a comprehensive survey [96, 156] as collected up to 1980, we focus on methods which have either become standard techniques or which have a potential use not yet fully exploited in the study of condensed water aerosols.

6.1   OPTICAL TECHNIQUES

Optical techniques are the backbone of today's microphysical cloud and fog studies because they allow *in situ* analysis of the volatile condensed water aerosols. The base was founded by a series of single particle scattering and optical shadow sensors constructed by Knollenberg since 1970. [11, 13, 94–96] With a combination of these sensors a radius range from about 1 µm to millimetres can be realized. From about 30 µm upwards, one- and two-dimen-sional particle images and concurrent grey-scale information on the particle shadow can be recorded. Thus, drops can be differentiated from solid particles by the shape of their shadow. Initially, these optical sensors were mainly used for measurements of drop size distributions. Later interpretations of the number-related results in terms of LWC revealed limitations of this technique. [2, 10, 12, 13, 23] By external fans and electronic adaptations the airborne scattering sensors have been modified for ground-based use.

Statistical uncertainties in measuring low number concentrations of large particles is a major limitation of the single particle techniques. More recent sensor developments reduce this problem by measuring optical effects of particle populations. With a laser diffraction technique a particle volume measuring instrument was constructed [49, 50] that is being deployed in ground-based versions and on airborne platforms [51] with sampling rates up to 5 kHz. The detection limit of the most sensitive version is on the order of milligrams of water per cubic metre which is sufficient for most fog and cloud studies in the lower atmosphere but insufficient for wet hazes. By employing two different optical weightings of the diffraction data, surface and volume proportional signals can be derived yielding as combined result an optically effective drop radius.

The inherent limitation of single particle sensors for drop size distributions is given by the small sample volume which is necessary to minimize the probabilities of coincidence. This limitation has been overcome in a recently developed cloud drop sensor which measures forward scattered light from a population of drops in a sample volume of the order of $10 \, cm^3$. [104] The instrument utilizes 240 scattering angles in the range 0.1–8.4 °.

With holographic imaging systems the lower size limit for imaging particle sensors has been reduced to about 5 μm, and the first holographic experiments reported. [162] Most recent applications in ground- based [20, 21] and airborne [104] sensors have been published. By incorporating a fibre optical sensor array into the holographic system, concurrent measurements of the scattering function in 10° steps over an unspecified range have been achieved. [103] The value of the holographic techniques under varying field conditions has yet to be established more firmly.

Instead of the complex holographic techniques, a simple video camera with a 10× objective lens has been used to record images of condensed water aerosols on balloon-borne sondes. [117, 118, 167] With this expendable system cloud particles from 3.5 to 1000 μm radius have been recorded up to 12 km altitude through a telemetry link to the ground.

For ground-based use there is an interesting optical technique that has not been widely used in studies of condensed water aerosols. The probe GBPP-100 (Particle Measuring Systems Inc., Boulder, CO) senses the shadowing effects of particles with radii between 100 and 6000 μm in an expanded laser beam in 62 size classes. The primary results are particle numbers per unit of time. With the local wind velocity these numbers can be converted to volumetric units. The value of the GBPP-100 lies in its capability of quantifying drizzle processes. As shown in

Figure 2, even in ground fog drizzle production is occurring and is connected to the microphysical properties of the fog. An error analysis of this instrument has been reported. [188]

6.2 ELECTRICAL TECHNIQUES

The electrical techniques for the characterization of condensed water aerosols can be divided into two groups. In the first class the particles *as they are* change the electrical parameters of a hot wire sensor upon their impact. The electrical change results from the cooling effect of the evaporating water of the impacted particles. An electrical signal proportional to the integral cooling effect of a hydrometeor population should be a measure of the liquid water content of the cloud. Among other cloud sensors, this type of instrument has been reviewed. [96] The most common representative of this group of instruments is the so-called J-W probe which initially was manufactured by the Johnson-Williams Company. Since this type of instrument is sensitive to changes in air speed, temperature and density, different compensatory measures have been taken to reduce this sensitivity. In the original sensor a compensating wire is oriented parallel to the airflow. The Australian Commonwealth Scientific and Industrial Research Organization redesigned the probe to reduce power consumption and to increase stability. [92] The resulting so-called King probe is widely used on aircraft. However, the stability problems have not been fully eliminated. A

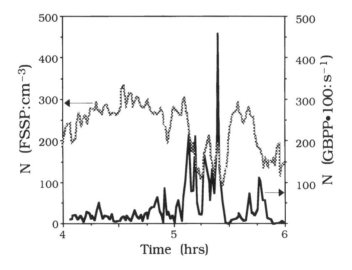

Figure 2.   Time series of drizzle drops (GBPP-100) measured with a PMS GBPP-100 at the surface and fog drops (FSSP- 100) measured with PMS FSSP-100 during a fog event at San Pietro Capofiume, Italy, 15 November 1994

comparison of this type of sensor with optical devices has been evaluated. [10, 139] The authors find that the J-W probe has a rather low frequency cut-off at ≈ 0.4 Hz while underestimating the water content by a factor which depends on the size distribution of hydrometeors.

Instead of sensing the integral cooling effect of impacting particles of a population, the individual cooling pulses of drops larger than 2 μm can be registered with a commercial hot wire flow meter to determine the mass size distribution of condensed water aerosols. [133] By impinging condensed water aerosols on a heated nichrome plate in an impactor, mass size distributions in the radius range 3 to ≈ 10 μm could be derived from the cooling signal of individual drops. [190] No larger set of field data has yet been reported with this technique.

In the second group of electrical techniques *the particles are charged* before processing and sensing them. In a first prototype [88] cloud drops are charged proportional to their surface upon entering the probe. They discharge when hitting an electrode. The resulting electrical pulse is registered as a measure of the droplet size. Applications of this type of sensor yielded particle size distributions in the radius range 2–30 μm in cumuli and fogs. A comparison of this technique with a cloud drop impactor gave reasonable agreement. [33] A much more sophisticated way of processing and analysing charged haze and cloud drops has been developed. [111] In this droplet aerosol analyzer (DAA) the drops are charged in a unipolar charger. After a diffusion drier the non-volatile residual particles remain. These particles are analyzed with various *in situ* methods such as tandem differential mobility analyzers [115] and optical particle counters. The method allows us to connect drop size with the subsequent results on the non-volatile material in the drops. The DAA has been validated in the field, [24] with good agreement between conventional instruments and the DAA. Its first field deployment gave unique results on drop size-dependent solute concentrations in orographic clouds. [112] While the DAA is much more complicated than any previous microphysical technique its basic principle rests on well-characterized electrical aerosol instrumentation. Because of the high sensitivity in terms of mass concentrations its high potential value for studies of wet haze justifies the high degree of sophistication.

6.3 REPLICATOR TECHNIQUES

The fabrication of permanent replicas is an old technique of preserving the shape of volatile objects. Shortly before exposing it to a condensed water aerosol, a collection substrate is coated with a solution of Formvar (a polyvinyl acetal resin) in ethylene dichloride or chloroform. Particles impacting on the surface will leave an imprint which remains permanently after the solution has dried provided that the solution evaporates before the particles. This technique had its first application to atmospheric samples. [152] More recent designs have developed the replicator principle to continuously operating laboratory [149] and airborne samplers [3, 61] which have lower radius limits of a few micrometers. Even in its present automated form replicators involve complicated sample processing steps which limit reliability and reproducibility of this technique.

7 SAMPLING TECHNIQUES FOR HYDROMETEORS

The sampling of hydrometeors is facilitated by the fact that the condensed water increases particulate mass concentrations by about three orders of magnitude as compared to dry aerosol particles and wet hazes. Inertial separation techniques can readily be applied to segregate the supermicrometre-sized condensed phase from the carrier gas. In most studies, on the other hand, the sampling is not done for the condensed water but for trace substances which were part of the dry nuclei or which became associated with the hydrometeor during its lifetime. The sampling of these highly diluted trace substances poses several problems. One, the risk of contamination increases with every surface which the sampling comes into contact with. Two, most sampling techniques involve significant changes in the thermodynamic environment of the hydrometeors which result in changed phase distributions of volatile species including water itself. In the present context precipitation elements such as rain or snow are not treated extensively. A survey of existing precipitation samplers and a short review of analytical techniques for precipitation analyses can be found. [163]

7.1 SAMPLING OF WARM CLOUDS

Initially, passive samplers were used which made use of the horizontal wind velocity to collect drops on natural [29] or artificial filaments [55] by means of inertial impaction. The efficiency of samplers of this type strongly depends on wind velocity both in terms of amount of collected material per time and concerning the smallest collected drop size. Thus, intepretation and compatibility of results are limited by the variable size and mass fractions of the hydrometeor population collected with this type of sampler.

Despite a rather negative evaluation, [73] the most frequently used airborne collector is the so-called Mohnen Slotted-Rod Cloud Water Collector. [186] It was developed initially for ground-based use and has been adapted rather uncritically for many airborne platforms.

The principle of a cyclone separator builds on accelerating the sample air in a vortex from which hydrometeors deviate because of their inertia. [27, 189] This concept has been used in a separator for airborne sampling of hydrometeors.

[175] The ram pressure developed by the aircraft is converted into a swirling motion by a fixed annular set of turning vanes. The centrifugal motion forces the drops to walls from where they are extracted into sampling vessels. Downstream of the sampler, the interstitial air can be measured or sampled. The lower cut radius of this device was estimated to be about 2.5 μm.

7.2 SAMPLING OF FOGS

While there are no major differences between clouds and fogs (cf. section 1) the latter often exhibit low LWCs which necessitate high-volume sampling devices. During such conditions, the small amount of condensed water (down to $\approx 50\,\mu g\,m^{-3}$) is dispersed over large numbers of small drops leading to low visibilities. Fog formation processes require rather low wind speeds. Thus, passive samplers do not collect many (mostly the largest) fog drops. Consequently, active high-volume impaction devices have been developed to allow the collection of millilitres of the most frequent fog drops per hour. Highly efficient multi-jet impactors in various designs have been utilized. [15] This design approach was compared to other samplers, [153] and optical fog water measurements. [16, 38] Sixteen units of this design were combined in a fog sampler to generate rapidly (≈ 15 minutes) sufficient amounts of fog samples for multi-component chemical analysis of drop radii larger than 2.5 μm. [37]

The development of a rotating arm collector employing the impaction principle has been reported [100] for the collection of fog and cloud droplets above 5 μm radius. Similar devices have been developed by other groups and were compared with each other. [25, 71] While the pH analyses of the compared samples showed good agreement, larger discrepancies were found for major ions. They were explained by varying evaporative losses. In this comparison, only three of the five samplers showed any agreement for LWC.

By using a combination of an optical fog detector and a rotating string impactor an automatic station for fog water collection was designed which is being used in the Po Valley, Italy. [44]

7.3 SAMPLING OF MIXED AND ICE CLOUDS

In all conventional cloud samplers, the condensed water is impinging on to surfaces on which it accumulates or from which it flows to some collection vessel. During this process it is unavoidable that the sample is affected by the contact surfaces, by the thermodynamic changes in the sampler and by particles of different composition mixing in the accumulating sample. While accepting one drastic thermodynamic change of the sample many of these sampling problems can be avoided. Based on this experimental compromise, the CVI was developed in the early 1980s. [128] With this device supermicrometre particles are separated inertially from the cloudy air and processed in an internal carrier gas of known composition and thermodynamic properties. Thus, the CVI can be used in liquid clouds as well as in ice clouds. The internal carrier gas is warm and dry. Consequently, the condensed water and volatile impurities are evaporating and can be sensed by appropriate gas sensors. The non-volatile residues can be sensed *in situ* by conventional aerosol instrumentation and can be collected by aerosol samplers for subsequent analyses. A small amount of internal clean gas is bled out of the CVI inlet to avoid contamination of the sampler. By changing the amount of counterflow, the smallest sampled particle size can be varied and thus size-dependent particle properties can be studied. [68, 122, 130]

Depending on the ratio of arrival velocity of the particles to the internal velocity of the CVI, concentration enrichments between 10 and 100 have been realized with this sampler. This enrichment greatly facilitates the detection of small concentrations of condensed water aerosols (such as in cirrus clouds) and of impurities in the condensed water in terms of the residue particles.

The CVI has been deployed in warm clouds, fogs and cirrus clouds (cf. the corresponding sections). Initial reports stressed its capability of measuring in real time the composition of condensed water aerosol. [124, 128, 129] Later, the CVI was combined with interstitial cloud measurements to quantify the partitioning of gases, [30, 125] and non-volatile particulate matter between interstitial air and the hydrometeors. [58, 59] The fact that a CVI requires the sampling, purification and control of several flows has prevented many researchers from applying this technique in their studies of condensed water aerosols. However, the mechanical design is simple, and sophisticated flow controls can rather easily be realised with commercial mass flow controllers and PC-based commercial software.

8   METHODOLOGY FOR THE ANALYSES OF CONDENSED WATER AEROSOLS

8.1   ANALYSES OF SOLUBLE MATERIAL

Trace substances in precipitation have been studied for more than 150 years, mostly soluble material like Cl and SO_4. Initially, colorimetric techniques were used such as the Thorin method introduced by Persson. [140] Today, ion chromatography (IC) is a standard commercial technique for bulk analyses down to sample volumes on the order of millilitres. With capillary electrophoresis (CE), [78] and micro-HPLC (high-pressure liquid chromatography), samples volumes could be reduced to microlitres. Detection limits of the ionic techniques for frequently analized components in atmospheric samples are listed in Table 3. With IC and micro-HPLC, major ions have been analyzed in individual raindrops. [4, 6]

Table 3. Detection limits of common components in condensed water aerosols in ionic analyses

Anion	Analysis	Detection limit (μeq l^{-1})	Cation	Analysis	Detection limit (μeq l^{-1})
Br^-	IC	1.25	Ca^{2+}	IC	0.5*
$C_2O_4^{2-}$	IC	0.68	Fe^{2+}	IC	1.8
$CH_3SO_3^-$	IC	0.2	Fe^{3+}	IC	2.7
HCO_2^-	IC	0.44	K^+	IC	0.25*
Cl^-	CE	0.28*	Mg^{2+}	IC	0.83*
F^-	CE	1.0	Na^+	IC	0.43*
$C_2H_5CO_2^-$	IC	0.27	NH_4^+	IC	0.56*
$CH_3CO_2^-$	IC	0.34			
NO_3^-	IC	0.16*			
SO_4^{2-}	IC	0.21*			

* Manufacturer's data.

For the sum of dissolved and undissolved metals in condensed water aerosols and precipitation samples atomic absorption spectroscopy (AAS) is the standard technique. [102] Recently, transition metals in size-classified rain samples have been studied with AAS. [5] Instrumental neutron activation analysis (INAA) [192] and particle induced X-ray emission (PIXE) [84] are other analytical techniques which have been applied to condensed water aerosols. They have been compared to AAS. [82]

Because of the high instrumental requirements dissolved organic components have been studied to a much smaller extent in condensed water aerosols. With IC, HPLC and fluorimetric techniques, fog water composition and phase partitioning of organic acids like formic, acetic and peruvic acids have been investigated. [39, 182, 183]

8.2 ANALYSES OF INSOLUBLE MATERIAL

In Junge's comprehensive monograph [86] on air chemistry, insoluble materials are listed under "rarely determined compounds". Even 25 years later, a review of inorganic composition of cloud and rain-water [176] does not list the undissolved matter in hydrometeors. However, amounts on the same order of magnitude as soluble matter had been reported quite early. In urban samples in the Berlin area, 23–84 mg l^{-1} were found on an annual basis. [106] One to 6 mg l^{-1} larger than 0.2 μm particle radius were found in individual rainfall events in Tsukuba, Japan. [1] These first results were based on simple filtration and gravimetric measurements on rain samples. The size distribution of insoluble particles in hydrometeors was unknown. No particular attention was paid to the submicrometre efficiency of filtration even though a rough estimate of the expected hydrosol number and mass size distributions from the input aerosol data would have suggested significant number-and mass fractions in the fine particle size range.

With the introduction of the Coulter counter to atmospheric research, size distributions down to about $0.3\,\mu m$ particle radius could be determined in water samples. The instrument is based on an electrochemical detection of particles suspended in an electrolyte. As the particles pass through a small aperture ($\approx 30\,\mu m$ orifice) in the measuring cell, they change the electrical resistance of the orifice which is sensed as a pulse. Size distributions in rain and fog samples have been given with this counter. [35, 98, 99]

The development of ultra-clean manufacturing technologies required monitoring techniques for high purity liquids. For that purpose optical particle counters have been developed that are able to size hydrosol particles down to about $0.1\,\mu m$ particle radius by means of their light scattering off a laser beam. One such instrument, the so-called IMOLV (Particle Measuring Systems Inc., Boulder, CO) has been characterized for its application in atmospheric research. [110] Since it consumes only 1 ml of liquid per minute, very small samples of hydrometeors can be analyzed with this sensor. In small rain and snow samples total number concentrations up to $10^7\,ml^{-1}$ have been measured. As an example of its potential use, Figure 3 gives grand average number and volume size distributions determined with the IMOLV in Stockholm, Sweden

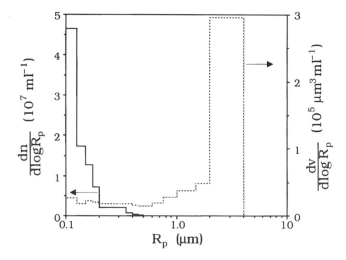

Figure 3.   Grand average number (solid line, left scale) and volume (dashed line, right scale) size distributions of insoluble particles in event samples of precipitation taken at the Department of Meteorology, Stockholm University, over the period of 1 year in 1990–1991

over a period of one year in 1990–1991. With its sensitivity, particle size distributions in individual raindrops or snow crystals can be measured after sufficient additions of particle-free water. It should be noted though that this technique requires freshly prepared particle-free water of the highest quality, sample lines and vessels which do not release particles. Particular consideration of bubbles in the sample is necessary because the optical principle cannot distinguish between particles and bubbles in the carrier water.

The measurements with the IMOLV showed increasing particle number concentrations down to its size limit which is given by the background light scattering of carrier liquid and optics. In order to extend the size range below the optically accessible limit, other techniques have been developed. One approach is based on spraying up a hydrosol sample into a particle-free carrier gas with subsequent analysis by means of aerosol sensors. Through appropriate sampling of the aerosol, the impurities in samples of condensed water aerosol can be enriched for trace analyses. [63] With tandem differential mobility analysers the resulting aerosol was measured. [181] With this system insoluble particles in fog/cloud water were determined. [99] Number size distributions down to 0.025 μm were measured with this arrangement. Care has to be taken though, to distinguish residues of dissolved material from insoluble particles in the sample. [121]

Besides aerosol generation, other techniques have been used for the pre-concentration of samples of condensed water aerosols. A chelating agent was explored [83] and a low-temperature ashing method was applied. [135] Freeze-drying for pre-concentration was used [70, 164] to extend the range of PIXE analysis to the sub-ppb level.

Two pathways have been explored for the chemical characterization of insoluble material in condensed water aerosols. The first path begins with an acidification of the sample with, for example, ultra-pure nitric acid down to a pH of about 1. [107] In this report particle size-dependent solubilities of four metals are also discussed. A literature survey is given [160] showing the solubilities of 10 trace elements measured frequently in atmospheric condensed water aerosols. After the acidification all particulate matter is assumed to have dissolved or additional steps of extraction and digestion are taken. [107] On the mobilized samples classical techniques like atomic absorption or emission spectrometry are used.

In the other approach, the filtered particulate matter is analyzed by PIXE. Sampling techniques and sample substrates can be optimized to yield a large number of trace components by means of the PIXE technique. [64] Besides PIXE, INAA has been used for multi-element characterizations of insoluble material in samples of condensed water aerosols. [101] Crystallographic information about the composition of insoluble material in condensed water aerosols has been derived by means of X-ray diffraction analysis. [76, 77]

Individual insoluble particles in samples of condensed water aerosols have been analyzed by energy-dispersive X-ray spectroscopy analysis (EDAX) in the electron microscope, [60, 132, 191] yielding relative mass fractions of elements with atomic numbers larger than 11. Single particle EDAX analizes have been combined with other physical measurements in a complex scheme of water examination [109] for the characterization of colloid particles in rain-water.

Further discussion of the above methods as applied to aerosol particles appears in Chapters 3–5.

9 DEPOSITION OF CONDENSED WATER AEROSOLS

A sink process removes particles from a well-defined atmospheric reservoir such as a cloud. On the way to the ground, gases and particles can be incorporated into the falling hydrometeors. This process is called *sub-cloud scavenging*. The opposite process takes place below high clouds in *virgae* (fall streaks) when falling hydrometeors evaporate before reaching the ground. [42, 150] Even though deposition is inhibited in this case, the precipitating cloud will cause a redistribution of atmospheric constituents. For the *wet deposition* of condensed water aerosols gravitational settling (sedimentation) and impaction on surface obstacles are the dominating sink processes. These processes of precipitation scavenging are discussed in detail in Chapter 3. In the case of surface impaction the term *cloud deposition* is used even though the meteorologist would use the

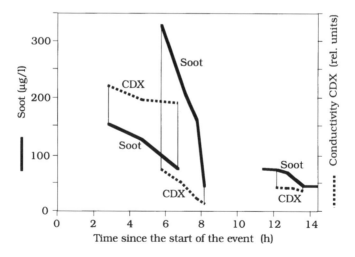

Figure 4. Temporal development of the insoluble component soot and soluble material, represented by the electrical conductivity (CDX) during the course of three frontal precipitation events over Hamburg, FRG, in 1990

term *fog deposition* because the cloud is in contact with the ground. A qualitative overview of the different pathways and mechanisms governing precipitation scavenging can be found. [57] In some studies the concept of cloud deposition is widened to include the condensation of dew, rime or frost on surface structures. The process is then termed *occult deposition*. [89] In the context of condensed water aerosols, this concept is avoided because it includes the codeposition of water and trace substances which are not part of the atmospheric condensed phase.

9.1 CLOUD DEPOSITION

If falling hydrometeors or cloud elements are being carried around surface obstacles larger particles cannot follow the curved streamlines around the structures and will be deposited by impaction. This process is particularly effective in coastal and mountain-top areas with high wind velocities where the wetting of the vegetation can deposit substantial amounts of cloud water[89] and related substances. [154] For mountains which reach above the cloud base, cloud deposition often exceeds the actual rainfall. For the widespread case of cloud water deposition a resistance model was developed [108] which accounts for impaction, sedimentation, evaporation and condensation on surfaces of forest canopies. This model was used [134] to compare the deposition of atmospheric acidity by cloud water to that by rain-water in polluted clouds over Germany. While cloud water contributed only about 15% of the total amount of water, the authors calculated a three to six times higher deposition of trace substances through cloud-water interception than via rain. The difference was due to the much higher concentrations of material in cloud drops than in rain. For soot smaller deposition fluxes were found in fogs than for other (soluble) particles. [31]

From gradient measurements with fast drop sensors like the FSSP-100 (Particle Measuring Systems, Inc., Boulder, CO) cloud deposition velocities have been derived. [47] Together with fast three-dimensional wind sensors, *in situ* measurements of cloud deposition fluxes have been made with fast optical cloud sensors in complex terrain. [17, 46, 173] Substantial downward liquid water fluxes of the order of 1 mm per 24 hours were found during cloud events. [173] Besides the expected downward fluxes due to impaction, the observed size dependence of the fluxes suggested upward fluxes of drops below 4 μm radius which were explained by phase change processes.

GLOSSARY

Abundance	Mass fraction of one analyzed component in relation to the sum of all analyzed components in a sample or particle
Atomic absorption spectroscopy (AAS)	Standard spectroscopic analysis for metals in solutions
Bergeron–Findeisen mechanism	Initialization of precipitation in mixed-phase clouds where ice particles can grow at the expense of the drops
Bimodality	Distribution with two relative maxima
Capillary electrophoresis (CE)	Wet-chemical analytical technique in which ions are separated by means of their migration speed through an electrolyte to which an electrical field is applied
Cirrus	Fibrous ice clouds
Cloud deposition	See wet deposition
Cloud element	Condensed water particle below 100 μm of water equivalent drop radius
Condensed water aerosols	Hazes, fogs and clouds. They consist of solid and/or liquid particles suspended by the carrier gas air
Condensed water content	Mass of condensed water (liquid and/or solid) per volume of moist air
Counterflow virtual impactor (CVI)	Inertial sampler of condensed water aerosols irrespective of their phase
Cumulus	Clouds with significant vertical development
Cyclone	Separator for condensed water aerosols in which the sample air is accelerated in a vortex from which hydrometeors deviate because of their inertia
Deliquescence	Phase transition from solid material to saturated solution
Deliquescence point	Relative humidity of deliquescence
Droplet aerosol analyzer (DAA)	Electrical sensor for the size-dependent liquid water content and trace composition of condensed water aerosols
Entrainment	Mixing of surrounding air into a cloud, mostly from the top

Fog	Surface air with visibilities below 1 km
Fog deposition	See wet deposition
Haze	Aerosol with a horizontal visibility between 1 and 10 km
High performance (pressure) liquid chromatography (HPLC)	Wet-chemical chromatographic technique with the stationary phase being near-monodisperse granulate and the mobile phase being processed at high pressures
Homogeneous mixing	Cloud mixing process in which the time-scale of the turbulent diffusion of an entrained air parcel and for the molecular diffusion at the boundary between a parcel and its environment are small compared to the time-scale for the evaporation of cloud drops in the mixed, subsaturated air
Hydrometeor	Any atmospheric particle with water as the dominating constituent
Hydrosol	Two-phase system with a solid phase suspended in a liquid phase
Impaction	Process in which particles are forced to move on curved streamlines from which they deviate on to some surface because of their inertia
Inhomogeneous mixing	Cloud-mixing process in which the time-scale of evaporation is fast compared to the time-scales of mixing
Instrumental neutron activation analysis (INAA)	Multi-element nuclear analytical technique in which γ-ray spectrometry is applied to samples which are irradiated with thermal neutrons
Ion chromatography (IC)	Wet-chemical analytical technique in which ions are separated by means of their migration speed through an electrolyte
J-W probe	Hot-wire sensor for the liquid water content of clouds
Liquid water content (LWC)	Mass of liquid water per volume of moist air
Mist	Condensed water aerosols with a horizontal visibility between 0.5 and 1 km
Nimbus	Rain clouds
Occult deposition	See wet deposition
Particle induced X-ray emission (PIXE)	Analytical technique for condensed elements with mass number $x > 12$ in which X-ray emissions are analized which are stimulated by exposing the sample(particle) to a high-energy particle beam

Precipitation elements	Condensed water particle above 100 μm of water equivalent drop radius, thus having a good chance of reaching the ground before it evaporates
Riming	Collection of cloud drops by falling ice crystals
Scattering	Redirection of electromagnetic radiation due to the interaction with molecules or particles
Stratus	Layer clouds
Sub-cloud scavenging	Incorporation of gases or particles into hydrometeors on the way between cloud base and the ground
Wet deposition	Transfer of dissolved and undissolved material to the earth's surface by precipitation. Transfer through impaction of cloud or fog droplets on to vegetation and other types of surfaces is often referred to as cloud or fog deposition, respectively. In the term "occult deposition" the concept is widened to include the condensation of dew, rime or frost on surface structures

REFERENCES

1. Ambe, Y., and Nishikawa, M. 1987. Variations in different sized water insoluble particulate matter in rain water. *Atmos. Environ.* **21**, 1469–1471.
2. Arends, B. G., Kos, G. P. A., Wobrock, W., Schell, D., Noone, K. J., Fuzzi, S., and Pahl, S. 1992. Comparison of techniques for measurements of fog liquid water content. *Tellus* **44B**, 604–611.
3. Arnott, W. P., Dong, Y. Y., Hallett, J., and Poellot, M. R. 1994. Role of small ice crystals in radiative properties of cirrus: A case study, FIRE II, 22 November 1991. *J. Geophys. Res.* **99**, 1371–1382.
4. Bächmann, K., Haag, I., and Röder, A. 1993. A field study to determine the chemical content of individual raindrops as a function of their size. *Atmos. Environ.* **27A**, 1951–1958.
5. Bächmann, K., Haag, I., and Steigerwald, K. 1995. Determination of transition metals in size-classified rain samples by atomic absorption spectrometry. *Atmos. Environ.* **29**, 175–177.
6. Bächmann, K., Röder, A., and Haag, I. 1992. Determination of anions and cations in individual raindrops. *Atmos. Environ.* **26A**, 1795–1797.
7. Baker, B. A. 1992. Turbulent entrainment and mixing in clouds: a new observational approach. *J. Atmos. Sci.* **49**, 387–404.
8. Baker, M. B., and Latham, J. 1979. The evolution of droplet spectra and the rate of production of embryonic rain drops in small cumulus clouds. *J. Atmos. Sci.* **36**, 1612–1615.

9. Baker, M. B., Latham, J., and Corbin, R. G. 1980. The influence of entrainment on the evolution of cloud droplet spectra: I: A model of inhomogenous mixing. *Q. J. R. Meteorol. Soc.* **106**, 581–598.

10. Baumgardner, D. 1983. An analysis and comparison of five water droplet measuring instruments. *J. Appl. Meteor*, **22**, 891–910.

11. Baumgardner, D., Dye, J. E., Gandrud, B., Rogers, D., Weaver, K., Knollenberg, R. G., Newton, R., and Gallant, R. 1995. The multiangle aerosol spectrometer probe: a new instrument for airborne particle research. *9th AMS Symposium on Meteorological Observations and Instrumentation*, Charlotte, N. C., p. 6.

12. Baumgardner, D., and Spowart, M. 1990. Evaluation of the forward scattering spectrometer probe. Part III: Time response and laser inhomogeneity limitations. *J. Atmos. Oceanic Technol.* **7**, 666–672.

13. Baumgardner, D., Strapp, W., and Dye, J. E. 1985. Evaluation of the forward scattering spectrometer probe. Part II: corrections for coincidence and dead-time losses. *J. Atmos. Oceanic Technol.* **2**, 626–632.

14. Bergeron, T. 1935. On the physics of cloud and precipitation. *5th Assembly of the U.G.G.I.*, Lisbon, pp. 156–178.

15. Berner, A. 1988. The collection of fog droplets by a jet impaction stage. *Sci. Total Environ.* **73**, 217–228.

16. Berner, A., Reischl, G., Enderle, K. H., Jaeschke, W., Fuzzi, S., Orsi, G., and Facchini, M. C. 1988. The liquid water content of a radiation fog measured by an FSSP 100 optical probe and a fog impactor. *Sci. Total Environ.* **77**, 133–140.

17. Beswick, K. M., Hargreaves, K. J., Gallagher, M. W., Choularton, T. W., and Fowler, D. 1991. Size-resolved measurements of cloud droplet deposition velocity to a forest canopy using an eddy correlation technique. *Q. J. R. Meteorol. Soc.* **117**, 623–645.

18. Bigg, E. K. 1977. Some properties of the aerosol at Mauna Loa observatory. *J. Appl. Meteor.* **16**, 262–267.

19. Blyth, A. M. 1993. Entrainment in cumulus clouds. *J. Appl. Meteor.* **32**, 626–641.

20. Borrmann, S., and Jaenicke, R. 1993. Application of microholography for ground-based *in situ* measurements in stratus cloud layers: A case study. *J. Atmos. Oceanic Technol.* **10**, 277–293.

21. Borrmann, S., Jaenicke, R., and Neumann, P. 1993. On spatial distributions and inter-droplet distances measured in stratus clouds with in-line holography. *Atmos. Res.* **29**, 229–245.

22. Bower, K. N., Hill, T. A., Coe, H., and Choularton, T. W. 1991. SO_2 oxidation in an entraining cloud model with explicit microphysics. *Atmos. Environ.* **25A**, 2401–2418.

23. Brenguier, J.-L. 1989. Concidence and dead-time corrections for particle counters. Part II: High concentration measurements with an FSSP. *J. Atmos. Oceanic Technol.* **6**, 585–598.

24. Cederfelt, S.-I., Martinsson, B., Svenningsson, B., Wiedensohler, A., Frank, G., Hansson, H.-C., Swietlicki, E., Wendisch, M., Beswick, K. M., Bower, K. N., Gallagher, M. W., Pahl, S., Maser, R., and Schell, D. 1997. Field validation of the droplet aerosol analizer. *Atmos. Environ.* **31**, 2657–2670.

25. Collett, J. L., Daube, B. C., Munger, J. W., and Hoffmann, M. R. 1990. A comparison of two cloudwater/fogwater collectors: the rotating arm collector and the Caltech Active Strand Cloudwater Collector. *Atmos. Environ*, **24A**, 1685–1692.

26. Collett Jr., J., Oberholzer, B., and Staehelin, J. 1993. Cloud chemistry at MT Rigi, Switzerland: Dependence on drop size and relationship to precipitation chemistry. *Atmos. Environ.* **27A**, 33–42.

27. Cooper, D. W. 1983. Cyclone design: sensitivity, elasticity and error analizes. *Atmos. Environ.* **17**, 485–489.
28. Cooper, W. A., and Lawson, R. P. 1984. Physical interpretation from the HIPLEX-1 Experiment. *J. Appl. Meteor.* **23**, 523–540.
29. Dessens, H. 1947. Brume et noyaux de condensation. *Ann. de Geophys.* **3**, 68–86.
30. Dixon, R. W., and Charlson, R. J. 1994. Development of a new real-time method for measuring S(IV) in cloud water using a counter-flow virtual impactor. *Tellus* **46B**, 193–204.
31. Dlugi, R. 1989. Chemistry and deposition of soot particles in moist air and fog. *Aerosol Sci. Technol.* **10**, 93–105.
32. Dowling, D. R., and Radke, L. F. 1990. A summary of the physical properties of cirrus clouds. *J. Appl. Meteor.* **29**, 970–978.
33. Dye, J. E. 1976. Comparisons of the electrostatic distrometer with impactor slides. *J. Appl. Meteor.* **15**, 783–789.
34. Easter, R. C., and Peters, L. K. 1994. Binary homogeneous nucleation: temperature and relative humidity fluctuations, nonlinearity, and aspects of new particle production in the atmosphere. *J. Appl. Meteor.* **33**, 775–784.
35. Eichel, C., Krämer, M., Schütz, L., and Wurzler, S. 1996. The water soluble fraction of atmospheric aerosol particles and its influence on cloud microphysics. *J. Geophys. Res.* **101D**, 29,299–29,510.
36. Eldridge, R. G. 1969. Mist—the transition from haze to fog. *Bull. American Meteor. Soc.* **50**, 422–426.
37. Enderle, K.-H., and Jaeschke, W. 1991. *Chemie und Mikrophysik des Nebels.* Report 13, Zentrum für Umweltforschung, Universität Frankfurt. 194 pp.
38. Enderle, K. H., and Jaeschke, W. 1987. Problems of fog sampling. *J. Aerosol Sci.* **18**, 793–795.
39. Facchini, M. C., Fuzzi, S., Lind, J. A., Fierlinger- Oberlinninger, H., Kalina, M., Puxbaum, H., Winiwarter, W., Arends, B. G., Wobrock, W., Jaeschke, W., Berner, A., and Kruisz, C. 1992. Phase-partitioning and chemical reactions of low molecular weight organic compounds in fog. *Tellus* **44B**, 533–544.
40. Fletcher, N. H. 1962. *The Physics of Rainclouds.* Cambridge University Press, 386 pp.
41. Forkel, R., Seidl, W., Dlugi, R., and Deigele, E. 1990. A one-dimensional numerical model to simulate formation and balance of sulfate during radiation fog events. *J. Geophys. Res.* **95**, 18501–18515.
42. Fraser, A. B., and Bohren, C. F. 1993. Viewing the vagaries and verities of Virga. *Mon. Wea. Rev.* **121**, 2429–2430.
43. Fuzzi, S. 1994. Clouds in the troposphere. C. Boutron (Ed.), In *Topics in Atmospheric and Interstellar Physics and Chemistry.* Les éditions de physique, Les Ulis, France: pp. 291–308.
44. Fuzzi, S., Cesari, G., Evangelisti, F., Facchini, M. C., and Orsi, G. 1990. An automatic station for fog water collection. *Atmos. Environ.* **24A**, 2609–2614.
45. Fuzzi, S., Facchini, M. C., Orsi, G., and Ferri, D. 1992. Seasonal trend of fog water chemical composition in the Po Valley. *Environ. Pollut.* **75**, 75–80.
46. Gallagher, M. W., Beswick, K. M., and Choularton, T. W. 1992. Measurement and modelling of cloudwater deposition to a snow-covered forest canopy. *Atmos. Environ.* **26A**, 2893–2903.
47. Gallagher, M. W., Choularton, T. W., Morse, A. P., and Fowler, D. 1988. Measurements of cloud droplet deposition at a hill site. *Q. J. R. Meteorol. Soc.* **114**, 1355–1362.
48. Gallagher, M. W., Downer, R. M., Choularton, T. W., Gay, M. J., Stromberg, I., Mill, C. S., Radojevic, M., Tyler, B. J., Bandy, B. J., Penkett, S. A., Davies, T. J.,

Dollard, G. J., and Jones, B. M. R. 1990. Case studies of the oxidation of sulphur dioxide in a hill cap cloud using ground and aircraft based measurements. *J. Geophys. Res.* **95**, 18517–18537.

49. Gerber, H. 1984. Liquid water content of fogs and hazes from visible light scattering. *J. Climate Appl. Meteor.* **23**, 1247–1252.

50. Gerber, H. 1991. Direct measurement of suspended particulate volume concentration and far-infrared extinction coefficient with a laser-diffraction instrument. *Appl. Opt.* **30**, 4824–4831.

51. Gerber, H., Arends, B. G., and Ackerman, A. 1994. New microphysics sensor for aircraft use. *Atmos. Res.* **31**, 235–252.

52. Gieray, R., Lammel, G., Metzig, G., and Wieser, P. 1993. Size dependent single particle and chemical bulk analysis of droplets and interstitial particles in an orographic cloud. *Atmos. Res.* **30**, 263–293.

53. Grabowski, W. W., and Pawlowska, H. 1993. Entrainment and mixing in clouds: the paluch mixing diagram revisited. *J. Appl. Meteor.* **32**, 1767–1773.

54. Graedel, T. E., and Weschler, C. J. 1981. Chemistry within aqueous atmospheric aerosols and raindrops. *Rev. Geophys. Space Phys.* **19**, 505–539.

55. Grunow, J. 1954. Bedeutung und Erfassung des Nebelniederschlags. *UGGI, AIHS Assemblé Gen.* Rome pp. 402–415.

56. Hagen, D. E., Podzimek, J., Heymsfield, A. J., Trueblood, M. B., and Lutrus, C. K. 1994. Potential role of nuclei in cloud element formation at high altitudes. *Atmos. Res.* **31**, 123–135.

57. Hales, J. M. 1982. Precipitation scavenging pathways and mechanisms—a qualitative overview. *Symposium on Acid Forming Emissions and their Ecological Effects*, Alberta, Canada, pp. 3–38.

58. Hallberg, A., Noone, K. J., Ogren, J. A., Svenningsson, I. B., Flossmann, A., Wiedensohler, A., Hansson, H.-C., Heintzenberg, J., Anderson, T., Arends, B., and Maser, R. 1994. Phase partitioning of aerosol particles in clouds at Kleiner Feldberg. *J. Atmos. Chem.* **19**, 107–127.

59. Hallberg, A., Ogren, J. A., Noone, K. J., Heintzenberg, J., Berner, A., Solly, I., Kruisz, C., Reischl, G., Fuzzi, S., Facchini, M. C., Hansson, H.-C., Wiedensohler, A., and Svenningsson, I. B. 1992. Phase partitioning for different aerosol species in fog. *Tellus* **44B**, 545–555.

60. Hallberg, A., Ogren, J. O., Noone, K. J., Okada, K., Heintzenberg, J., and Svenningsson, I. B. 1994. The influence of aerosol particle composition on cloud droplet formation. *J. Atmos. Chem.* **19**, 153–171.

61. Hallett, J. 1976. *Measurements of Size, Concentration and Structure of Atmospheric Particulates by the Airborne Continuous Replicator*. Report Contract AFGL-TR-0149, Air Force Geophys. Lab. 92 pp.

62. Hamill, P., Turco, R. P., and Toon, O. B. 1988. On the growth of nitric and sulfuric acid aerosol particles under stratospheric conditions. *J. Atmos. Chem.* **7**, 287–315.

63. Hansson, H.-C., Johansson, E.-M., and Ekholm, A.-K. 1984. A non-selective preconcentration technique for water analysis by Pixe. *Nucl. Instr. Meth.* **B3**, 158–162.

64. Hansson, H.-C., Martinsson, B. G., Swietlicki, E., Asking, L., Heintzenberg, J., and Ogren, J. A. 1986. PIXE in complex analytical systems for atmospheric chemistry. *Nucl. Instr. Meth.* **22B**, 235–240.

65. Hegg, D. A. 1991. Particle production in clouds. *Geophys. Res. Lett.* **18**, 995–998.

66. Hegg, D. A., and Hobbs, P. V. 1981. Cloud water chemistry and the production of sulfates in clouds. *Atmos. Environ.* **15**, 1597–1604.

67. Heintzenberg, J., Fouquart, Y., Heymsfield, A., Ström, J., and Brogniez, G. 1995. Interactions of radiation and microphysics in cirrus. P. J. Crutzen and V. Ramanathan (Ed.), In *Clouds, Chemistry and Climate*. Springer, Berlin: pp. 29–55.

68. Heintzenberg, J., Ogren, J. A., Noone, K. J., and Gärdneus, L. 1989. The size distribution of submicrometer particles within and about stratocumulus cloud droplets on Mt. Åreskutan, Sweden. *Atmos. Res.* **24**, 89–101.

69. Heintzenberg, J., Okada, K., and Ström, J. 1996. On the composition of non-volatile material in upper tropospheric aerosols and cirrus crystals. *Atm. Res.* **41**, 81–88.

70. Henrici, G., Hofmann, D., Georgii, H.-W., and Groeneveld, K. O. 1991. ppb-PIXE analysis of the trace-element content in precipitation. *J. Atmos. Chem.* **12**, 391–400.

71. Hering, S. V., Blumenthal, D. L., Brewer, R. L., Gertler, A., Hoffmann, M., Kadlecek, J. A., and Pettus, K. 1987. Field intercomparison of five types of fogwater collectors. *Environ. Sci. Technol.* **21**, 654–663.

72. Hering, S. V., and Friedlander, S. K. 1982. Origins of aerosol sulfur size distributions in the Los Angeles basin. *Atmos. Environ.* **16**, 2647–2656.

73. Huebert, B. J., and Baumgardner, D. 1985. A preliminary evaluation of the Mohnen slotted-rod cloud water collector. *Atmos. Environ.* **19**, 843–846.

74. Ikegami, M., Okada, K., Zaizen, Y., and Makino, Y. 1993. Aerosol particles in the middle troposphere over the Northwestern Pacific. *J. Meteor. Soc. Japan* **71**, 517–528.

75. Isaac, G. A., and Daum, P. H. 1987. A winter study of air, cloud and precipitation chemistry in Ontario, Canada. *Atmos. Environ.* **21**, 1587–1600.

76. Ishizaka, Y. 1972. On materials of solid particles contained in snow and rain water: Part 1. *J. Meteor. Soc. Japan* **50**, 362–375.

77. Ishizaka, Y. 1973. On materials of solid particles contained in snow and rain water: Part 2. *J. Meteor. Soc. Japan* **51**, 325–336.

78. Jandik, P., and Bonn, G. 1993. *Capillary Electrophoresis of Small Molecules and Ions*. VCH Publisher, New York: 298 pp.

79. Jensen, E. J., Toon, O. B., Westphal, D. L., Kinne, S., and Heymsfield, A. J. 1994. Microphysical modeling of cirrus 1. Comparison with 1986 FIRE IFO measurements. *J. Geophys. Res.* **99**, 10421–10442.

80. Jensen, J. B., Austin, P. H., Baker, M. B., and Blyth, A. M. 1985. Turbulent mixing, spectral evolution and dynamics in a warm cumulus cloud. *J. Atmos. Sci.* **42**, 173–192.

81. Jensen, J. B., and Baker, M. B. 1989. A simple model of droplet spectral evolution during turbulent mixing. *J. Atmos. Sci.* **46**, 2812–2829.

82. Jervis, R. E., and Landsberger, S. 1983. Trace elements in wet atmospheric deposition: application and comparison of PIXE, INAA, and graphite-furnace AAS techniques. *Intern. J. Environ. Anal. Chem.* **15**, 89–106.

83. Johansson, E. M., and Akselsson, K. R. 1981. A chelating agent-activated carbon-Pixe procedure for sub ppb analysis of trace elements in water. *Nucl. Instr. Meth.* **181**, 221–226.

84. Johansson, T. B., Akselsson, R., and Johansson, S. A. E. 1970. X-ray analysis: elemental trace analysis at the 10^{-12} g level. *Nucl. Instr. Meth.* **84**, 141–143.

85. Junge, C. 1950. Das Wachstum der Kondensationskerne mit der relativen Feuchtigkeit. *Ann. Meteorologie* **3**, 129–135.

86. Junge, C. E. 1963. *Air Chemistry and Radioactivity*. Academic Press, New York and London: 382 pp.

87. Junge, C. E., Chagnon, C. W., and Manson, J. E. 1961. Stratospheric aerosols. *J. Meteorol.* **18**, 81–108.

88. Keily, D. P., and Millen, S. G. 1960. An airborne cloud-drop-size distribution meter. *J. Meteorol.* **17**, 349–356.

89. Kerfoot, O. 1968. Mist precipitation on vegetation. *For. Abstr.* **29**, 8–20.

90. Kerminen, and Wexler, 1995. Enhanced formation and development of sulfate particles due to marine boundary layer circulation. *J. Geophys. Res.* **100D**, 23051–23062.

91. Khalsa, S. J. S. 1993. Direct sampling of entrainment events in a marine stratocumulus layer. *J. Atmos. Sci.* **50**, 1734–1750.

92. King, W. D., Parkin, D. A., and Handsworth, R. J. 1978. A hot-wire liquid water device having fully calculable response characteristics. *J. Appl. Meteor.* **17**, 1809–1813.

93. Klepikova, N. V., and Skhirtlaze, G. Z. 1984. Studies of droplet continental and maritime cumulus clouds: field experiment and model. *9th Int. Cloud Phys. Conf.*, Tallin, pp. 469–471.

94. Knollenberg, R. G. 1970. The optical array: an alternative to scattering or extinction for airborne particle size determination. *J. Appl. Meteorol.* **9**, 86–103.

95. Knollenberg, R. G. 1976. Three new instruments for cloud physics measurements: the 2-D spectrometer, the Forward Scattering Spectrometer Probe, and the Active Aerosol Spectrometer. *International Conference on Cloud Physics*, Boulder, CO, pp. 554–561.

96. Knollenberg, R. G. 1981. Techniques for probing cloud microstructure. P. V. Hobbs and A. Deepak (Ed.), In *Clouds, Their Formation, Optical Properties and Effects*. Academic Press, New York: pp. 15–92.

97. Korolev, A. V. 1994. A study of bimodal droplet size distributions in stratiform clouds. *Atmos. Res.* **32**, 143–170.

98. Krämer, M., Brinkmann, J., Eichel, C., Jaenicke, R., Schell, D., Schüle, M., and Schütz, L. 1995. Field studies on the cloud processing of atmospheric aerosol particles and trace gases. *J. Aerosol Sci.* **26**, S893–S894.

99. Krämer, M., Schütz, L., Svenningsson, B., and Wiedensohler, A. 1991. Number size distribution of insoluble atmospheric aerosol particles in fog/cloud-water. *J. Aerosol Sci.* **22**, S525–S528.

100. Krämer, M., and Schütz, L. 1994. On the collection efficiency of a rotating arm collector and its applicability to cloud-and fogwater sampling. *J. Aerosol Sci.* **25**, 137–148.

101. Landsberger, S., Davies, T. D., Tranter, M., Abrahams, P. W., and Drake, J. J. 1989. The solute and particulate chemistry of background versus a polluted, black snowfall on the Cairngorm mountains, Scotland. *Atmos. Environ.* **23**, 395–401.

102. Landsberger, S., Jervis, R. E., Aufreiter, S., and van Loon, J. C. 1982. The determination of heavy metals (Al, Mn, Ni, Cu, Zn, Cd and Pb) in urban snow using an atomic absorption graphite furnace. *Chemosphere* **2**, 237–247.

103. Lawson, R. P. 1995. Digital holographic measurements of cloud particles. *AMS Conference on Cloud Physics*, Dallas, Texas, 15–20 January, p. 6.

104. Lawson, R. P., and Cormack, R. H. 1995. Theoretical design and preliminary tests of two new particle spectrometers for cloud microphysics research. *Atmos. Res.* **35**, 315–348.

105. Lelieveld, J., and Crutzen, P. J. 1991. The role of clouds in tropospheric photochemistry. *J. Atmos. Chem.* **12**, 229–267.

106. Liesegang, W. 1934. Untersuchungen über die Mengen der in Niederschlägen enthaltenen Verunreinigungen. *Kleine Mitt. Mitglied. Ver. Wasser-Boden-u. Lufthyg.* **10**, 350–355.

107. Lindberg, S. E., and Harriss, R. C. 1983. Water and acid soluble trace metals in atmospheric particles. *J. Geophys. Res.* **88**, 5091–5100.

108. Lovett, G. M. 1984. Rates and mechanisms of cloud water deposition to a subalpine Balsam Fir forest. *Atmos. Environ.* **18**, 361–371.

109. Malyschew, A., Schmidt, H.-J., Weil, K. G., and Hoffmann, P. 1994. Methods for characterization of colloid particles in rain water. *Atmos. Environ.* **28**, 1575–1581.

110. Martin, M., and Heintzenberg, J. 1992. Calibration of an optical counter for liquid-borne particles. *J. Aerosol Sci.* **23**, 373–378.

111. Martinsson, B. 1996. Physical basis for a droplet aerosol analysing method. *J. Aerosol Sci.* **27**, 997–1013.

112. Martinsson, B., Cederfelt, S.-I., Svenningsson, B., Frank, G., Hansson, H.-C., Swietlicki, E., Wiedensohler, A., Wendisch, M., Gallagher, M. W., Colvile, R. N., Beswick, K. M., Choularton, T. W., and Bower, K. N. 1996. Experimental determination of the connection between cloud droplet size and its dry residue size. *Atmos. Environ.* **31**, 2477–2490.

113. Mason, B. J. 1971. *Physics of Clouds*, 2nd ed.. Oxford University Press, London.

114. Mazin, I. P. 1995. Cloud water content in continental clouds of middle latitudes. *Atmos. Res.* **35**, 283–297.

115. McMurry, P. H., and Stolzenburg, M. R. 1989. On the sensitivity of particle size to relative humidity for Los Angeles aerosols. *Atmos. Environ.* **23**, 497–507.

116. Munger, J. W., Collett, J., Daube, B., and Hoffmann, M. R. 1989. Chemical composition of coastal stratus clouds: dependence on droplet size and distance from the coast. *Atmos. Environ.* **23**, 2305–2320.

117. Murakami, M., and Matsuo, T. 1990. Development of hydrometeor video sonde. *J. Atmos. Oceanic Technol.* **7**, 613–620.

118. Murakami, M., Matsuo, T., Nakayama, T., and Tanaka, T. 1987. Development of cloud particle video sonde. *J. Meteor. Soc. Japan* **65**, 803–809.

119. Naruse, H., and Maruyama, H. 1971. On the hygroscopic nuclei in the cloud droplets. *Papers in Meteorology and Geophysics* **22**, 1–21.

120. Naruse, H., and Okada, K. 1989. Change in concentrations of cloud condensation nuclei and aerosol particles during the dissipation of fog. *Papers in Meteorology and Geophysics* **40**, 125–137.

121. Niida, T., Kousaka, Y., and Oda, S. 1990. The measurement of low level impurities in liquid by aerosolizing method. *3rd International Aerosol Conference*, Kyoto, Japan, pp. 871–874.

122. Noone, K. B., and Heintzenberg, J. 1991. On the determination of droplet size distributions with the counterflow virtual impactor. *Atmos. Res.* **26**, 389–406.

123. Noone, K. J., Charlson, R. J., Covert, D. S., Ogren, J. A., and Heintzenberg, J. 1988. Cloud droplets: solute concentration is size dependent. *J. Geophys. Res.* **93D**, 9477–9482.

124. Noone, K. J., Ogren, J. A., and Heintzenberg, J. 1988. An examination of the chemical and physical variability of clouds at a mountain-top site in central Sweden. *Ann. Meteorologie* **25**, 282–284.

125. Noone, K. J., Ogren, J. A., Johansson, K. B., Hallberg, A., Fuzzi, S., and Lind, J. A. 1990. Hydrogen peroxide partitioning in ambient clouds. *EUROTRAC Symposium '90*, Garmisch-Partenkirchen, FRG, pp. 259–260.

126. Ogren, J., and Rodhe, H. 1986. Measurements of the chemical composition of cloudwater at a clean air site in central Scandinavia. *Tellus* **38**, 190–196.

127. Ogren, J. A., and Charlson, R. J. 1992. Implications for models and measurements of chemical inhomogeneities among cloud droplets. *Tellus* **44B**, 208–225.

128. Ogren, J. A., Heintzenberg, J., and Charlson, R. J. 1985. In-situ sampling of clouds with a droplet to aerosol converter. *Geophys. Res. Lett.* **12**, 121–124.

129. Ogren, J. A., Heintzenberg, J., Zuber, A., Hansson, H.- C., Noone, K. J., Covert, D. S., and Charlson, R. J. 1988. Measurements of the short-term variability of aqueous-phase mass- concentrations in cloud droplets. M. H. Unsworth and D. Fowler (Ed.), In *Acid Deposition at High Elevation Sites.* Kluwer, Amsterdam, pp. 125–137.

130. Ogren, J. A., Heintzenberg, J., Zuber, A., Noone, K. J., and Charlson, R. J. 1989. Measurements of the size-dependence of non-volatile aqueous mass concentrations in cloud droplets. *Tellus* **41B**, 24–31.

131. Ogren, J. A., Noone, K. J., Hallberg, A., Heintzenberg, J., Schell, D., Berner, A., Solly, I., Kruisz, C., Reischl, G., Arends, B. G., and Wobrock, W. 1992. Measurements of the size dependence of the concentration of non-volatile material in fog droplets. *Tellus* **44B**, 570–580.

132. Okada, K., Tanaka, T., Naruse, H., and Yoshikawa, T. 1990. Nucleation scavenging of submicrometer aerosol particles. *Tellus* **42B**, 463–480.

133. Ozaki, Y., Utiyama, M., Fukuyama, T., Nakajima, M., and Hayakawa, Y. 1990. Measurement of droplet size distribution by a hot-film anemometer. *Third International Aerosol Conference*, Kyoto, Japan, pp. 663–666.

134. Pahl, S., Winkler, P., Schneider, T., Arends, B., Schell, D., Maser, R., and Wobrock, W. 1994. Deposition of trace substances via cloud interception on a coniferous forest at Kleiner Feldberg. *J. Atmos. Chem.* **19**, 231–252.

135. Pallon, J., and Malmqvist, K. G. 1981. Evaluation of low temperature ashing of biological materials as preconcentration method for PIXE analysis. *Nucl. Instr. Meth.* **181**, 71–75.

136. Paluch, I. 1979. The entrainment mechanism in Colorado cumuli. *J. Atmos. Sci.* **36**, 2467–2478.

137. Paluch, I. R., and Baumgardner, D. G. 1989. Entrainment and fine-scale mixing in continental convective clouds. *J. Atmos. Sci.* **46**, 261–278.

138. Paluch, I. R., and Knight, C. A. 1984. Mixing and the evolution of cloud droplet size spectra in a vigorous continental cumulus. *J. Atmos. Sci.* **41**, 1801–1815.

139. Personne, P., Brenguier, J. L., Pinty, J. P., and Pointin, Y. 1982. Comparative study and calibration of sensors for the measurement of the liquid water content of clouds with small droplets. *J. Appl. Meteor.* **21**, 189–196.

140. Persson, C. 1966. Automatic colorimetric determination of low concentrations of sulphate for measuring sulphur-dioxide in ambient air. *Air & Water Pollut. Int. J.* **10**, 845–852.

141. Peter, T., and Baker, M. 1995. Lifetimes of small ice crystals in the upper troposphere and stratosphere. P. J. Crutzen and V. Ramanathan (Ed.), In *Clouds, Chemistry and Climate.* Springer, Berlin: pp. 57–82.

142. Pilinis, C., Seinfeld, J. H., and Grosjean, D. 1989. Water content of atmospheric aerosols. *Atmos. Environ.* **23**, 1601–1606.

143. Pontikis, C. A., and Hicks, E. M. 1993. The influence of clear air entrainment on the droplet effective radius of warm maritime convective clouds. *J. Atmos. Sci.* **50**, 2889–2900.

144. Pruppacher, H. R., and Klett, J. D. 1978. *Microphysics of Clouds and Precipitation.* Reidel Publishing Co., Dordrecht 714 pp.

145. Ramanathan, V., and Collins, W. 1991. Thermodynamic regulation of ocean warming by cirrus clouds deduced from observations of the 1987 El Niño. *Nature* **351**, 27–32.

146. Rogers, D. C. 1994. Detecting ice nuclei with a continuous-flow diffusion chamber—some exploratory tests of instrument response. *J. Atmos. Oceanic Technol.* **11**, 1042–1047.
147. Rogers, R. R., and Yau, M. K. 1989. *A Short Course in Cloud Physics, 3rd Ed.*. Pergamon Press, Oxford: 293 pp.
148. Rood, M. J., Shaw, M. A., Larson, T. V., and Covert, D. S. 1989. Ubiquitous nature of ambient metastable aerosol. *Nature* **337**, 537–539.
149. Sassen, K. 1978. The continuous-impactor-replicator for laboratory cloud composition analysis. *J. Appl. Meteor.* **17**, 1319–1326.
150. Sassen, K., and Kreuger, S. K. 1993. Toward an empirical definition of Virga: Comments on "Is Virga rain that evaporates before reaching the ground?" *Mon. Wea. Rev.* **121**, 2426–2428.
151. Sassen, K., Liou, K. N., Kinne, S., and Griffin, M. 1985. Highly supercooled cirrus cloud water: confirmation and climatic implications. *Science* **227**, 411–413.
152. Schaefer, V. I. 1941. Method of making replicas of snow flakes, ice crystals and other short-lived substances. *Museum News* **19**, 11–14.
153. Schell, D., Georgii, H.-W., Maser, R., Jaeschke, W., Arends, B. G., Kos, G. P. A., Winkler, P., Schneider, T., Berner, A., and Kruisz, C. 1992. Intercomparison of fog water samplers. *Tellus* **44B**, 612–631.
154. Schemenauer, R. 1986. Acidic deposition to forests: the 1985 Chemistry of High Elevation Fog (CHEF) Project. *Atmosphere-Ocean* **24**, 303–328.
155. Schickel, K.-P. 1975. *Sampling of Droplet Distributions from Water Clouds.* Report DLR-Mitt. 75–25, Institut für Physik der Atmosphäre Oberpfaffenhofen. 177 pp.
156. Schickel, K.-P. 1981. *Methoden zur Messung von Wolkentröpfchen—Eine Bibliographie.* Report DFVLR-Mitt, 81–04, Institut für Physik der Atmosphäre Oberpfaffenhofen. 119 pp.
157. Schickel, K.-P. 1982. Zur Bestimmung repräsentativer Wolkentropfenspektren. *Meteorol. Rdsch.* **36**, 123–126.
158. Scorer, R. S., and Wexler, H. 1963. *A Colour Guide to Clouds.* Pergamon Press, Oxford: 63 pp.
159. Seidl, W. 1994. Initial cloud droplet number concentrations in clean and polluted areas: simulation by a condensation model and comparison with measurements. *Atmos. Res.* **31**, 157–185.
160. Sequeira, R. 1988. On the solubility of some natural minerals in atmospheric precipitation. *Atmos. Environ.* **22**, 369–374.
161. Sheridan, P. J., Brock, C. A., and Wilson, J. C. 1994. Aerosol particles in the upper troposphere and lower stratosphere: elemental composition and morphology of individual particles in northern midlatitudes. *Geophys. Res. Lett.* **21**, 2587–2590.
162. Silverman, B. A., Thomson, B. J., and Ward, J. H. 1964. A laser fog distrometer. *J. Appl. Meteor.* **3**, 792–801.
163. Slanina, J. 1986. Standardized techniques for the collection and analysis of precipitation. W. Jaeschke (Ed.), In *Chemistry of Multiphase Atmospheric Systems.* Springer, Berlin: pp. 91–116.
164. Stössel, R. P., and Prange, A. 1985. Determination of trace elements in rainwater by total-reflection-X-ray fluorescence. *Anal. Chem.* **35**, 2880–2885.
165. Ström, J., and Heintzenberg, J. 1994. Water vapour, condensed water and crystal concentration in orographically influenced cirrus clouds. *J. Atmos. Sci.* **51**, 2368–2383.
166. Ström, J., Schröder, F., Heintzenberg, J., Anderson, T., Strauss, B., and Stingl, J. 1994. Recent measurements in cirrus and contrail using the measurements with the

MISU CVI-payload during ICE'94. *7th Workshop of the European Cloud and Radiation Experiment (EUCREX)*, Villeneuve d'Ascq, France, pp. 83–85.

167. Tanaka, T., Matsuo, T., Okada, K., Ichimura, I., Ichikawa, S., and Tokuda, A. 1989. An airborne video-microscope for measuring cloud particles. *Atmos. Res.* **24**, 71–80.

168. Tang, I. N., and Munkelwitz, H. R. 1993. Composition and temperature dependence of the deliquescence properties of hygroscopic aerosols. *Atmos. Environ.* **27A**, 467–473.

169. Telford, J. W., Keck, T. S., and Chai, S. K. 1984. Entrainment at cloud tops and the droplet spectra. *J. Atmos. Sci.* **41**, 3170–3179.

170. Telford, J. W., Kim, K.-E., Keck, T. S., and Hallett, J. 1993. Entrainment in cumulus clouds. II: Drop size variability *Q. J. R. Meteorol. Soc.* **119**, 631–653.

171. Toon, O. B., Turco, R. P., Jordan, J., Goodman, J., and Ferry, G. 1989. Physical processes in polar stratospheric ice clouds. *J. Geophys. Res.* **94D**, 11359–11380.

172. Turco, R. P., Whitten, R. C., and Toon, O. B. 1982. Stratospheric aerosols: observation and theory. *Rev. Geophys. Space Phys.* **20**, 233–279.

173. Vong, R. J., and Kowalski, A. S. 1995. Eddy correlation measurements of size-dependent cloud droplet turbulent fluxes to complex terrain. *Tellus* **47B**, 331–352.

174. Walcek, C. J., and Brankov, E. 1994. The influence of entrainment-induced variability of cloud microphysics on the chemical composition of cloudwater. *Atmos. Res.* **32**, 215–232.

175. Walters, P. T., Moore, M. J., and Webb, A. H. 1983. A separator for obtaining samples of cloud water in aircraft. *Atmos. Environ.* **17**, 1083–1091.

176. Warneck, P. 1988. *Chemistry of the Natural Atmosphere*. Academic Press, San Diego, 756 pp.

177. Warner, J. 1969. The microstructure of cumulus clouds. Part 1 General features of drop spectrum. *J. Atmos. Sci.* **26**, 1049–1059.

178. Weickmann, H. 1949. *Die Eisphase in der Atmosphäre*. Report Bericht No. 6, Deutscher Wetterdienst in der US-Zone, Bad Kissingen. 54 pp.

179. Welch, R. M., and Wielicki, B. A. 1986. The stratocumulus nature of fog. *J. Appl. Meteor.* **25**, 101–111.

180. Wiedensohler, A., Hansson, H.-C., Orsini, D., Wendisch, M., Wagner, F., Bower, K. N., Choularton, T. W., Wells, M., Parkin, M., Acker, A., Wieprecht, W., Fachini, M. C., Lind, J. A., Fuzzi, S., and Arends, B. G. 1997. Night-time formation of new particles associated with orographic clouds. *Atmos. Environ.* **31**, 2545–2559.

181. Wiedensohler, A., Krämer, M., and Hansson, H.-C. 1991. A new method for measurements of insoluble submicron particles in water. *J. Aerosol Sci.* **22**, S329–S330.

182. Winiwarter, W., Fierlinger, H., Puxbaum, H., Facchini, M. C., Arends, B. G., Fuzzi, S., Schell, D., Kaminski, U., Pahl, S., Schneider, T., Berner, A., Solly, I., and Kruisz, C. 1994. Henry's law and the behavior of weak acids and bases in fog and cloud. *J. Atmos. Chem.* **19**, 173–188.

183. Winiwarter, W., Puxbaum, H., Fuzzi, S., Facchini, M. C., Orsi, G., Beltz, N., Enderle, K., and Jaeschke, W. 1988. Organic acid gas and liquid-phase measurement in Po Valley fall–winter conditions in the presence of fog. *Tellus* **40B**, 348–357.

184. Winkler, P. 1974. Die relative Zusammensetzung des atmosphärischen Aerosols in Stoffgruppen. *Meteorol. Rdsch.* **27**, 129–136.

185. Winkler, P. 1988. The growth of atmospheric aerosol particles with relative humidity. *Physica Scripta* **37**, 223–230.

186. Winters, W., Hogan, A., Mohnen, V., and Barnard, S. 1979. *ASRC Airborne Cloud Water Collection System*. Report 728, State University of New York at Albany, Atmospheric Sciences Research Center. 49 pp.

187. Yamato, M., and Ono, A. 1989. Chemical and physical properties of stratospheric aerosol particles in the vicinity of tropopause folding. *J. Meteor. Soc. Japan* **67**, 147–165.

188. Yangang, L., and Laiguang, Y. 1994. Error analysis of GBPP-100 probe. *Atmos. Res.* **34**, 379–387.

189. Yoshida, H. 1993. Recent research of air-cyclone. *J. Aerosol Res., Japan* **7**, 319–324.

190. Yoshiyama, H., Tamori, I., and Ueda, H. 1990. Development of mist size analyzer with thin nichrome plate. *J. Aerosol Res., Japan* **5**, 44–51.

191. Zheng, Z., and Hong, C. Y. 1994. Electron microscope analizes of insoluble component in acid rain, Guilin City, China. *Atmos. Res.* **32**, 289–296.

192. Zoller, W. H., and Gordon, G. E. 1970. Instrumental neutron activation analysis of atmospheric pollutants utilizing Ge(Li) γ-Ray detectors. *Analyt. Chem.* **42**, 257–265.

16 Influence of Atmospheric Aerosols upon the Global Radiation Balance

H. HORVATH

University of Vienna, Austria

1 INTRODUCTION

The importance of an atmosphere for the temperature of a celestial body can be easily seen by comparing the mean temperature of the earth and the moon. Both receive the same radiation from the sun, one has an atmosphere, the other does not. The average temperature of the earth is about $+13\,°C$, that of the moon varies between $+120$ and $-130\,°C$, thus $-5\,°C$ can be considered an average, which is colder than the earth, although it absorbs more radiation.

Some simple considerations of the fate of the incoming solar radiation and the emitted infrared and reflected visible radiation will show which factors are important for the average temperature. Let us only deal with a long-term global average. In that case the radiation balance has to be zero, i.e. the incoming and outgoing radiation have to be equal. Let us consider a spherical body receiving the solar radiation of $S_0 = 1360$ W m^{-2}. Since the surface area of a sphere is four times its cross-section, on the average a surface element will receive a flux density of one-quarter of this value: $\bar{S} = 340$ W m^{-2}. With no sources for

Atmospheric Particles, Edited by R. M. Harrison and R. Van Grieken.

energy within the planet the same amount has to be emitted/reflected by the body. Schematics of the fluxes considered in the following are shown in Figure 1.

(A) Let us first consider a black body receiving the radiation. The solar flux will be absorbed and emitted as infrared radiation to space. Equating the incoming radiation to the temperature radiation of a black body given by the Stefan Boltzmann law we obtain $340 \ \text{W m}^{-2} = \sigma T_e^4$, giving a mean temperature of $T_e = 278.6K$ or $5.5\,°C$.

(B) No celestial body is a black body in the visible (if it were we could not see it), but reflects a certain amount of light. If it has an albedo of $A = 0.3$ (the global mean albedo has this value, [1]), i.e. a fraction of 30% of the light is reflected, only $(1 - A)$ or 70% of the solar radiation can be emitted as infrared and this is only possible if the surface temperature is lower.

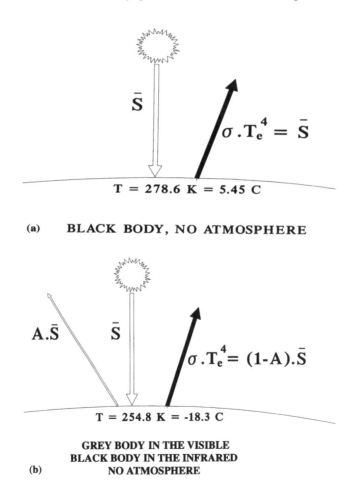

(a)   **BLACK BODY, NO ATMOSPHERE**

GREY BODY IN THE VISIBLE
BLACK BODY IN THE INFRARED
(b)   **NO ATMOSPHERE**

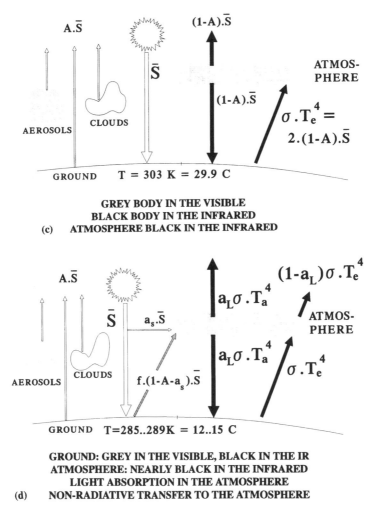

Figure 1. Radiative fluxes of a body receiving solar radiation and emitting infrared radiation: (a) black body; (b) grey body in the visible and black body in the infrared; (c) grey body in the visible and black body in the infrared having a greenhouse atmosphere; (d) model of a realistic atmosphere with light absorption in the visible, a non-radiative energy transfer from the ground to the atmosphere, and some transmission in the infrared

Equating $0.7 \times 340 \, \text{W m}^{-2} = \sigma T_e^4$ we obtain $T_e = 254.8 \text{K}$ or $-18.3°\text{C}$ which is fairly cold. One can immediately see that a change in reflectivity drastically alters the surface temperature. One example of this is the average temperature of the moon, which is below $0°\text{C}$ (however the moon absorbs 93% of the light which is more than the earth does and therefore the moon is warmer than the just obtained value of $-18.3\,°C$).

(C) Let us now consider an atmosphere around the body. It shall contain some substance, which absorbs the infrared radiation emitted by the body. By this absorption the atmosphere heats itself and emits radiation in all directions according to its temperature. Let us assume at first that the atmosphere is a perfect absorber and radiator in the infrared. In that case radiation budget considerations at the outer surface of the atmosphere makes it clear that the average absorbed solar flux density has to be emitted by the atmosphere. But the atmosphere emits this radiation in both directions. Therefore the radiation received by the surface of the earth is twice the absorbed solar radiation. This amount has to be emitted by temperature radiation, which is only possible if the surface temperature is higher. It can thus be obtained as $2 \times 0.7 \times 340\,\mathrm{W\,m^{-2}} = \sigma T_e^4$, giving $T_e = 303\,K$ or $29.9\,°C$. The dramatic increase in temperature due to an absorbing atmosphere is evident. The temperature of the atmosphere is $-18.3°C$.

The atmosphere we have on the earth is not completely absorbing and some other transfers not yet discussed take place as well. Let us consider now as a close to reality atmosphere the following case (a schematic drawing can be found in Figure 1(d). The atmosphere and the earth together have an albedo of A, i.e. of the incoming flux density $\bar{S}$ the amount $A\bar{S}$ is reflected upwards. Some sunlight passing through the atmosphere is absorbed and transferred as heat to the atmosphere, obtained from the absorbance of short-wave radiation $\tilde{a}_s$ as $\tilde{a}_s\bar{S}$. Of the incoming radiation $\bar{S}(1 - A - \tilde{a}_s)$ is absorbed by the surface of the globe. A certain fraction f of the absorbed energy is transported from the ground to the atmosphere by non-radiative processes: the solar energy is transferred both as latent heat via evaporation and condensation of water and by sensible heat through convective transfer from the surface to the atmosphere. The general circulation of the atmosphere and the oceans transports heat from low latitudes to the poles. Circulation is very complex and influenced by many factors. In our model it amounts to $f\bar{S}(1 - A - \tilde{a}_s)$. The other energy is radiated in the infrared from the surface of the earth of temperature T_e to the atmosphere. This is given by the black-body radiation as σT_e^4. The realistic atmosphere which has an absorbance $\tilde{a}_l$ for the infrared (long-wave) radiation absorbs $\tilde{a}_l\sigma T_e^4$ of the radiation from the earth and transmits $(1 - \tilde{a}_l)\sigma T_e^4$ upwards. Finally the atmosphere, having a temperature T_a, emits its grey body infrared radiation $\tilde{a}_l\sigma T_a^4$ both upwards and downwards.

Making a balance of the fluxes for the surface we obtain

$$\bar{S}(1 - A - \tilde{a}_s) + \tilde{a}_l\sigma T_a^4 = \sigma T_e^4 + f\bar{S}(1 - A - \tilde{a}_s)$$

Similarly we obtain for the atmosphere

$$\sigma T_{\mathrm{e}}^4 + f\bar{S}(1 - A - \tilde{a}_{\mathrm{s}}) + \tilde{a}_{\mathrm{s}}\bar{S} = (1 - \tilde{a}_{\mathrm{l}})\sigma T_{\mathrm{e}}^4 + 2\tilde{a}_{\mathrm{l}}\sigma T_{\mathrm{a}}^4$$

After some algebra one obtains

$$\bar{S}\frac{(2 - f)(1 - A - \tilde{a}_{\mathrm{s}}) + \tilde{a}_{\mathrm{s}}}{2 - \tilde{a}_{\mathrm{l}}} = \sigma T_{\mathrm{e}}^4 \qquad (1)$$

With a fraction of $f = 30\%$ non-radiative transfer, $\tilde{a}_{\mathrm{s}} = 1\%$ light absorption of the atmosphere and $\tilde{a}_{\mathrm{l}} = 92$–$97\%$ infrared absorbance of the atmosphere a global mean temperature between 285.0 and 288.4 K or 11.9–15.3 °C is obtained. The corresponding temperatures of the atmosphere are -21.5 and -19.61 °C.

The above equation clearly shows which effects the different variables will have on the mean surface temperature: increasing non-radiative heat transfer decreases the surface temperature, increasing light absorption decreases the surface temperature (but increases the temperature of the atmosphere), increasing infrared absorption increases the temperature both of the surface and the atmosphere.

Finally it should be mentioned that the albedo A and the short-wave absorbance $\tilde{a}_{\mathrm{s}}$ are values for possibly several passes of the light through the atmosphere, thus they are interrelated, since a high albedo of the ground causes a considerable fraction of the sunlight to pass back through the atmosphere, and thus will give a higher integrated light absorption of the atmosphere.

These considerations have clearly shown that the reflectivity of the planet and the infrared absorption in the atmosphere and the non-radiative energy transfer have a great influence on the mean temperature. More infrared absorption by the atmosphere causes an increase in the temperature since the radiation is reradiated to the ground by the atmosphere: the greenhouse effect is well known in the mean time. More reflectivity in the visible will cause a decrease in temperature. Schwartz [2] has coined the name "white-house effect" for this in analogy to the brilliant white houses, which can, for example, be seen in the Mediterranean, where the houses have a cool interior despite a considerable insolation. The reflectivity of the globe is determined by several factors which can all undergo considerable variation: snow or ice reflect radiation well and their coverage of the earth is known to have changed, for example, in the ice ages. Irrigated land has less reflectivity, whereas urban areas have a high reflectivity. But the atmosphere itself reflects sunlight back into space (we will call this upwards in the following), thus reducing the input of solar energy to the globe. The scattering of light by aerosol particles and the reflections by clouds (which need aerosols as condensation nuclei) are the two major effects of the aerosols on the incoming radiation, which will both have a cooling effect. But the black carbon contained in the aerosol can also absorb light, and transfer solar energy to the atmosphere, thus having a heating effect.

The aerosol particles in the atmosphere originate from various sources ranging from chemical reactions (gas-to-particle conversion) of precursor gases to soil erosion in sand-storms. Each process generates its species and size of particles and the atmospheric aerosol consists of a mixture of particles of different chemical composition, shape and size, which come from many sources. On a global scale the yearly produced particle mass amounts to 784–4159 Tg year^{-1}. Natural sources contribute 77–93.5% of the mass and man-made particles 6.5–23% (there is considerable variation between different authors, for an overview see [3], p. 402). In densely populated areas the anthropogenic component obviously dominates. The aerosol shows considerable variation in time and space, since the residence time in the atmosphere is about a week at ground level, which is too short for a homogeneous mixing.

The effect of the aerosol will be discussed in detail below but one (hypothetical) example of its effect will be given here. Let us assume that the aerosol reflects a fraction $R = 5\%$ of the incoming sunlight upwards. We will compare the case with and without aerosol. For an aerosol-free atmosphere let $\bar{S}_0$ be the solar flux density on the ground available for interaction. With an aerosol the solar radiation available will be $\bar{S}_0(1 - R)$. Denoting the global mean temperature with an aerosol-free atmosphere by T_0 and assuming everything else is unchanged when the aerosol is present, the temperature T for an atmosphere with an aerosol is obtained as $T = T_0\sqrt[4]{(1 - R)}$. If we have a temperature of $T_0 = +13\,°C$ without an aerosol, the temperature with the aerosol would be $T = 9.4\,°C$, i.e. the temperature decreases by 3.6 °C. This is not very much, but it should be borne in mind that during the ice age the temperature decreased by only 4–12 °C over land and 4–7 °C over the sea. One can thus see that the aerosol in our atmosphere plays an important role: without any aerosol in the atmosphere we might have a much warmer climate. Any changes in the aerosol could have a dramatic effect on the climate.

Considerable increases in the amount of aerosol particles in the atmosphere occur after large volcanic eruptions, and short-term changes in climate have been reported ever since. The first well-documented event was the eruption of Mt Etna in Sicily in 44 BC during the day when Julius Caesar was murdered. The biographer Plutarch (50–120) reports that the daylight was weak and the sun, the moon and stars were bloody red. [4] The summer following the eruption was extremely cold and the farmers had a bad harvest. The same is also documented at a distance of 10 000 km by Chinese records.

Any change in incoming or outgoing radiation will alter the climatic conditions on the earth. With everything else remaining constant an increase/decrease in surface temperature will take place if more/less radiation reaches the ground. But this change in radiation will invoke other changes which can have both negative or positive feedbacks. Just to mention a few: an increase in radiation can cause the melting of more glaciers and snow fields and the spread of vegetation further north, thus decreasing the reflectivity, which is a positive feedback loop for temperature. On the other hand the molten glacier and snow-water will run off to the ocean and in the case of the North Atlantic it might reduce its salinity such that the ocean currents are altered, which might stop the warm Gulf Stream which very much determines the moderate temperatures in northern Europe. [5] But an increase in temperature could also stimulate the additional production of particles via a mechanism proposed by Charlson *et al.*, [6] which reduces the input of solar energy, and tends to decrease temperature. Thus a very thorough knowledge of the interlacing of meteorology, hydrology, aerosol physics, atmospheric chemistry, biology and others is needed in order to make a prediction of climatic changes caused by variations in aerosol quantity, composition or properties. Although the models have greatly improved over the years at present it seems to be impossible to make any prediction except that a change in climate will occur if the aerosol changes. A discussion of possible effects of the aerosol on the radiation in the atmosphere will be given below. Any temperature changes as given above are only a speculation and are one of the many possible scenarios. In the scientific literature it has thus become a habit to concentrate on a quantity which has not so many unknown variables, radiative forcing. This is the instantaneous increase/decrease of net radiation due to an instantaneous change of a substance of radiative importance in the atmosphere such as aerosols, clouds, greenhouse gases or others. Due to the strong coupling of the ground surface and the lower atmosphere it is reasonable to use the tropopause as the level where radiative forcing is defined. We will only consider radiative forcing in the following.

There is absolutely no doubt that the aerosol is important for the climate issue, as pointed out by Preining [7] in a very concise overview, and the aerosol is included in all climate models today. We will demonstrate its importance for radiation in the atmosphere in the following. With about 20% of the aerosol being man-made, man influences the radiation balance also via the aerosol. With expected increase in aerosol production due to increase in population and energy use an increased impact is to be expected.

2   NOMENCLATURE AND DEFINITIONS

Solar radiation is attenuated, scattered and absorbed by particles and gases in the atmosphere and diffusely reflected by the ground or the clouds. The particles can have various sizes, compositions and properties thus their interaction with radiation will very much depend on these variables. Therefore their properties have to be well described.

A listing of the nomenclature and a few definitions are given in the following, and a summary can be found in Table 1. Unfortunately a variety of symbols and even contradicting definitions for magnitudes used in aerosol and atmospheric optics are in use. In order to avoid confusion the most important

Table 1. Symbols, units and definitions

Quantity	Symbol	Unit	Definition
Number concentration	N	m^{-3}	Number of particulates per volume
Mass concentration	M	gm^{-3}, μgm^{-3}	Mass of particulates per volume
Number size distribution	$\left(\frac{dN}{dr}\right) = n(r)$	m^{-4}	Number of particles per volume and diameter interval dr
Number size distribution	$\left(\frac{dN}{d\log r}\right)$	m^{-3}	Number of particles per volume and logarithmic diameter interval $d\log r$
Mass size distribution	$\left(\frac{dM}{dr}\right) = m(r)$	gm^{-4}, μgm^{-4}	Mass of particles per volume and diameter interval dr
Mass size distribution	$\left(\frac{dM}{d\log r}\right)$	gm^{-3}, μgm^{-3}	Mass of particles per volume and logarithmic diameter interval $d\log r$
Radiant flux	Φ	W	Power of radiation
Radiant flux density	$\frac{d\Phi}{dA} = S$	Wm^{-2}	Flux per area crossing surface element dA
Irradiance	$E = \frac{d\phi}{dA}$	Wm^{-2}	Flux per area incident on surface element dA
Radiative intensity	$I = \frac{d\phi}{d\omega}$	Wsr^{-1}	Light flux emitted per solid angle
Radiance	$L = \frac{d^2\Phi}{(dA\cos\theta)d\omega}$	$Wm^{-2}sr^{-1}$	Emitted flux per projected surface area and solid angle in direction θ
Extinction coefficient attenuation coefficient	$\sigma_e = -\frac{1}{\Phi}\frac{d\Phi}{dx}$	m^{-1}, km^{-1} or Mm^{-1}	Fraction of radiant flux lost from a collimated beam per unit thickness of aerosol
Scattering coefficient	$\sigma_s = -\frac{1}{\Phi}\frac{d\Phi}{dx}$	m^{-1}, km^{-1} or Mm^{-1}	Fraction of radiant flux lost from a collimated beam per unit thickness of aerosol due to scattering
Absorption coefficient	$\sigma_a = -\frac{1}{\Phi}\frac{d\Phi}{dx}$	m^{-1}, km^{-1} or Mm^{-1}	Fraction of radiant flux lost from a collimated beam per unit thickness of aerosol due to absorption
Single scattering albedo	$\tilde{\omega} = \frac{\sigma_s}{\sigma_e}$	Dimension less	Ratio of scattering coefficient and extinction coefficient
Absorption number	$\tilde{\alpha} = \frac{\sigma_a}{\sigma_e}$	Dimension less	Ratio of absorption coefficient and extinction coefficient
Transmittance	$\tau = \frac{\Phi}{\Phi_0}$	Dimension less	Ratio of incident flux Φ_0 and flux Φ transmitted through layer of aerosol
Optical depth	$\delta = -\ln\frac{\Phi}{\Phi_0}$	Dimension less	Negative natural logarithm of ratio of incident flux Φ_0 and flux Φ transmitted through layer of aerosol

Table 1. Contd.

Volume scattering function	$\gamma(\phi) = \frac{d\Phi}{E\,dV\,d\omega}$	$m^{-1}sr^{-1}$	Flux $d\Phi$ scattered in direction ϕ_s per irradiance E, per volume element dV of aerosol, and per solid angle $d\omega$
Scattering phase function	$P(\phi) = 4\pi\frac{\gamma(\phi)}{\sigma_s}$	sr^{-1}	Radiation scattered per unit solid angle divided by average radiation scattered into the unit solid angle
Albedo	A	Dimension less	Fraction of incident radiant flux reflected by a surface

definitions and symbols are given below. Both the CIE (Commission Internationale d'Eclairage) and the IAMAP (the International Association of Meteorology and Atmospheric Physics, Radiation Commission) have issued recommendations, which have been summarized in [8].

An aerosol is characterized by its *mass concentration* M which is the mass of particulate matter per unit volume; for atmospheric aerosols it is usually given in $\mu g\,m^{-3}$. Frequently this is called total suspended particulates (TSP). If a certain species of aerosol particles such as cloud condensation nuclei is considered, the *number concentration* N being the number of particles per unit volume, given for example in m^{-3}, is used. If the total number of particles is given, its value very much depends on the limits of the particle size range considered.

A more complete description of the aerosol is achieved by using the various size distributions. A detailed treatment is given in Chapter 1. [9] In the following we will use the *mass size distribution* $m(r) = \left(\frac{dM}{dr}\right)$ which gives the mass dM of particles per m^3 in the size interval $[r, r + dr]$ by $dM = m(r)dr = \left(\frac{dM}{dr}\right)dr$. Mass size distributions can, for example, be measured by impactors. In a similar way the number size distribution and the volume size distribution are defined. For spherical particles with density ρ the volume and mass size distributions can be obtained by

$$\left(\frac{dV}{dr}\right) = \frac{4\pi}{3}r^3\left(\frac{dN}{dr}\right) \quad \text{and} \quad \left(\frac{dM}{dr}\right) = \frac{4\pi}{3}r^3\rho\left(\frac{dN}{dr}\right)$$

The mass size distributions can also be given for logarithmic size intervals: $\left(\frac{dM}{d\ln r}\right)$. If the radius interval is $[r, r + dr]$ then

$$d\ln r = \ln(r + dr) - \ln r = \frac{1}{r}dr$$

and the mass dM of particles in this interval is

$$dM = \left(\frac{dM}{d\ln r}\right)d\ln r$$

The relation for size distributions for logarithmic and linear intervals is

$$\left(\frac{dM}{d\ln r}\right) = r\left(\frac{dM}{dr}\right).$$

The most frequently used empirical models for the size distribution of the atmospheric aerosol are:

1. The lognormal size distribution is a good representation for the aerosol both in clean and polluted areas. It is given by

$$\left(\frac{dM}{d\ln r}\right) = \frac{1}{\sqrt{2\pi}\ln\sigma_g}\exp\left(\frac{-(\ln r - \ln r_g)^2}{2\ln^2\sigma_g}\right) \tag{2}$$

The geometric standard deviation σ_g gives the width of the size distribution, the geometric mean radius r_g the position of the peak of the curve. The size distribution of the atmospheric aerosol can frequently be represented as the sum of three lognormal size distributions and if the volume or mass size distribution is used, the mean diameters are at 10–50 nm (nucleation mode), 0.2–0.8 μm (accumulation mode) and $\geq$ 3μm (coarse mode), see e.g. [10]. The geometric standard deviation σ_g has values around 2. We will use this characterization in the following. From the point of view of the optics of the atmosphere the nucleation mode is unimportant.
2. Modified gamma functions have been suggested by Deirmendjian [11] for atmospheric aerosols which is represented by $\left(\frac{dN}{dr}\right) = ar^\alpha\exp(-br^\gamma)$ with α, γ and b having the values 1, 0.5 and 8.944 for haze M; 2, 0.5 and 15.1186 for haze L and 2, 1 and 20 for haze H, the constant a is adjusted for the total particle number (10–1000 cm^{-1}). Models for clouds droplets are also given.
3. Junge [12] has suggested a power law size distribution function given by $\left(\frac{dN}{d\log r}\right) = cr^{-\nu}$ with $\nu \approx 3$, which is a good first approximation and is valid for time- and space-averaged aerosol in the diameter interval of 0.1–10 μm. Although a sum of three lognormal distributions and a power law may look quite different a comparison of the two representations when plotting $\left(\frac{dN}{dr}\right)$ makes them appear quite similar [13] and in the range of the optically important diameters of 0.05–2 μm a surprisingly good agreement exists. Therefore both size distributions are used successfully as model size distributions in atmospheric optics.

The *refractive index m* characterizes the material forming a particle optically. For transparent particles such as NaCl or water the refractive index is a real number and is the ratio of the phase velocities in vacuum and in the material. Materials which absorb light have a complex refractive index, an existing imaginary part indicates light absorption. The complex refractive index usually is written as $m = n - ik$ (see e.g. [14]), with n and k positive real numbers. It should be pointed out that some authors (e.g. [15]) use $m = n + ik$ instead. Attention is needed when working with computer codes. Using the wrong sign for the imaginary part can have catastrophic results. If the material has a bulk absorption coefficient of α (i.e. a lamina x of the bulk material transmits light according to $\Phi = \Phi_0 \exp(-\alpha x)$), the imaginary part of the refractive index at wavelength λ is given by $k = \frac{\alpha \lambda}{4\pi}$. Only materials which are extremely black (e.g. having a transmission of $10^{-10\,000}$ at a thickness of 1 mm) have an imaginary part of the refractive index, which is markedly different from zero. Only two substances which form particles of the atmospheric aerosol have this property: black carbon, the main constituent of soot and haematite, a black iron oxide. Whereas the first is a by-product of any combustion and thus present everywhere in the world, the latter might have local importance if haematite is a major constituent of the soil.

For considerations of light or infrared radiation the basic magnitude is the *radiative* (light) *flux* Φ, which is the amount of radiative power transported by the light. Its value is given in *Watts*. A radiative power received by a surface or emitted by a luminous surface is conveniently given as flux received or emitted per unit area, thus it is called the *flux density* or *irradiance E*, which is given in $\mathrm{W\,m^{-2}}$.

The atmospheric aerosol scatters radiation in all directions but not with the same strength. The flux density at a given distance will usually be different in different directions. Therefore it is convenient for point sources to use the *radiative intensity I*, which is the light flux $\mathrm{d}\Phi$ emitted per solid angle $\mathrm{d}\omega$, thus I is given in $\mathrm{W\,sr^{-1}}$. For extended light sources the *radiance*, defined as the flux emitted per unit solid angle and unit area (perpendicular to the direction considered), is used. If ψ is the angle between the normal of the radiating surface and the direction in which the emitted radiation is considered, the radiance is given by

$$L = \frac{\mathrm{d}^2\Phi}{\mathrm{d}\omega\,\mathrm{d}A\cos\psi}$$

its units are $\mathrm{W\,m^{-2}\,sr^{-1}}$.

Solar radiation is attenuated on its way through the atmosphere. For a parallel monochromatic flux the attenuation of initial flux Φ_0 to Φ at distance x is described by the well-known Lambert–Beer law $\Phi = \Phi_0 \exp(-\sigma_e x)$. The constant σ_e is called the *extinction (or attenuation) coefficient*. It is defined as the fraction of radiant flux lost from a collimated beam per unit thickness of aerosol and is given in units of reciprocal length, e.g. in $\mathrm{m^{-1}}, \mathrm{km^{-1}}$ or $\mathrm{Mm^{-1}}$. Chandrasekhar [16] and many others use the expression absorption coefficient for σ_e instead. We will not use this concept, since this expression is reserved for another process.

For spherical particles the extinction coefficient can be obtained by calculating the electromagnetic waves interacting with an obstacle. This can be done in an exact manner by solving the Maxwell equations. This was first done by series expansion (Mie [17]), a solution using special functions has been presented by Debeye [18], newer treatments can be found (e.g. in [19], [20], [15] or [21]). For fractal aggregates, as is the case for soot particles, experimental and theoretical treatments can be found in [22]. For a monodisperse aerosol consisting of N spherical particles per unit volume the extinction coefficient is obtained as $\sigma_e = Nr^2\pi Q_e N$. The *extinction efficiency factor* Q_e is dimensionless and can be understood as the effectiveness with which the particle interacts with the light. If $Q_e = 1$, the particle attenuates the electromagnetic wave incident on the cross-section of the particle. For aerosol particles Q_e can be both smaller than one (for particles considerably smaller than the wavelength of light) or larger than one (for particles comparable to the wavelength of light). The value of Q_e depends on the radius r of the particles and the wavelength λ of the light via the size spherical parameter $x = \frac{2r\pi}{\lambda}$ and the refractive index m, which is a real number for transparent particles and a complex number for light-absorbing materials. The efficiency factor Q_e is directly obtained from Mie theory. Computer codes doing this can be found in books (e.g. [15], [21]) or are available by request from the authors (e.g. [24]). A collection of Mie programs is available on the Internet at "htpp://imperator.cip-iw1. uni-bremen.de/fg01/codes.html".

The extinction coefficient for a polydisperse aerosol consisting of spherical particles is obtained by integration between the smallest size r_1 and the largest size r_2 of the particles:

$$\sigma_e = \pi \int_{r_1}^{r_2} r^2 Q_e \left(\frac{dN}{dr}\right) dr \quad \text{or} \quad \sigma_e = \int_{r_1}^{r_2} \frac{3}{4r} Q_e \left(\frac{dV}{dr}\right) dr \tag{3}$$

For particles which are not homogeneous or spherical only in very special cases can light scattering and extinction be calculated. Examples are thin rods of infinite size, ellipsoidal particles or spherical particles coated by a spherical shell, where the Maxwell equations can be solved (see e.g. [15]). For other shapes one has to rely on experimental work. Here the difficulties arise; the production of aerosol particles of a prescribed shape is not a very easy task. For latex particles it is possible to study the optical properties of doublets or triplets (see e.g. [25]), but for completely irregular particles microwave analogy experiments are the only reliable method of obtaining the scattering properties. [26] For complex particles much larger than the wavelength of light it is also possible to suspend them in a capacitor and induce a rotation and measure the light scattered by multicolour laser light. [27] For the light extinction it can be said that the volume equivalent sphere is the best representation for particles smaller than the wavelength, and the cross-section equivalent sphere for larger particles. [25] The angular dependence of irregularly shaped particles can be off even more than a factor of 2 from the equivalent sphere at some angular ranges, [26] thus adding some insecurity to the optical data, especially of dry particles in the atmosphere. If the particles are wet, they are very close to having a near spherical shape.

The attenuation of light by aerosol particles and gases is due to two processes: (i) *elastic scattering*, being the process by which part of the light from a beam is deflected to other directions, but the light energy is conserved, i.e. some of the photons change their direction; (ii) *absorption*, being the process by which the photons transfer their energy to the molecules of the gas or the particles, increasing its internal energy and causing heating of the particle and eventually of the surrounding medium.

The *scattering coefficient* σ_s is defined as the fraction of radiant flux lost from a collimated beam per unit thickness of aerosol due to scattering (deflection of light into directions different from the directions of the transmitted beam) and is also given in units of reciprocal length. Similar to above we can obtain the following relations: for monodispersions $\sigma_s = r^2 \pi Q_s N$ with Q_s the *scattering efficiency factor* and for a polydispersion $\sigma_s = \pi \int_{r1}^{r2} Q_s n(r) dr$.

The *absorption coefficient* σ_a is defined as the fraction of radiant flux lost from a collimated beam per unit thickness of aerosol due to absorption (transformation of light into other forms of energy) and is also given in units of reciprocal length. Due to a lack of available words Chandrasekhar [16] calls this the true absorption coefficient. In analogy to the procedure given above the light absorption coefficient for a monodispersion of N particles per unit volume is obtained by $\sigma_a = r^2 \pi Q_a N$ with Q_a the *absorption efficiency factor* and for a polydispersion $\sigma_a = \pi \int_{r1}^{r2} r^2 Q_a n(r) dr$. The absorption efficiency factor is also obtainable by application of the above-mentioned computer codes, the relation $Q_e = Q_a + Q_s$ is simply a consequence of $\sigma_e = \sigma_s + \sigma_a$.

Since extinction is due to scattering and absorption, the extinction coefficient is represented by the sum of the scattering coefficient and the absorption coefficient: $\sigma_e = \sigma_s + \sigma_a$. The ratio of the scattering coefficient to the extinction coefficient is called the *single scattering albedo*

$$\tilde{\omega} = \frac{\sigma_s}{\sigma_e} = 1 - \frac{\sigma_a}{\sigma_e}.$$

It is the fraction of scattered light with respect to the total light which interacts with the particles. It can also be understood as the probability of quantum survival. For an aerosol having no light absorption (i.e. consisting only of transparent particles and gases) the single scattering albedo is 1. When using the single scattering albedo $\tilde{\omega}$ it is possible to write the scattering and the absorption coefficient as $\sigma_s = \sigma_e \tilde{\omega}$ and $\sigma_a = \sigma_e(1 - \tilde{\omega})$.

The ratio of the absorption coefficient to the extinction coefficient is called the *absorption number* $\alpha = \frac{\sigma_a}{\sigma_e}$.

As already mentioned, a parallel beam of light flux Φ_0 is attenuated to $\Phi = \Phi_0 \exp(-\sigma_e x)$ after a distance x. We call $\tau = \frac{\Phi}{\Phi_0} = \exp(-\sigma_e x)$ the *transmittance* of the aerosol layer extending the distance x. In case the extinction

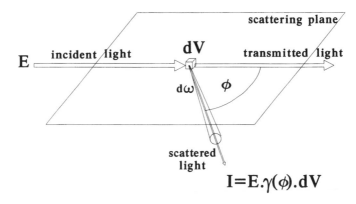

$$I = E.\gamma(\phi).dV$$

Figure 2. Definition of the volume-scattering function. Radiation producing an irradiance E at the location of the volume element dV is scattered by the material contained in the volume element. A certain flux $d\Phi$ of light is scattered into the direction given by the scattering angle ϕ. The intensity of the scattered light $I = \frac{d\Phi}{d\omega}$ is obtained from $I = \frac{d\Phi}{d\omega} = E\gamma(\phi)dV$

coefficient $\sigma_e(x)$ varies with distance x, the transmittance between two locations with coordinates x_1 and x_2 is obtained by $\exp(-\int_{x1}^{x2} \sigma_e(x)dx)$. The argument of the exponential function is called the *optical depth* δ of the aerosol, i.e. $\delta = \int_{x1}^{x2} \sigma_e(x)dx$. It is a dimensionless number. The optical depth δ and the transmittance τ are related by $\tau = e^{-\delta}$ or $\delta = -\ln\tau$.

Sometimes it is useful to introduce the *mass extinction (absorption, scattering) coefficient*, which is the optical coefficient divided by the mass concentration of the aerosol. We will use the symbols $\frac{\sigma_e}{M}$, $\frac{\sigma_a}{M}$, $\frac{\sigma_s}{M}$. They give the effectiveness with which the particles interact with light. They are usually given in units of $m^2\,g^{-1}$.

The *volume-scattering function* $\gamma(\phi)$ is defined in the following way, (see Figure 2): unpolarized light illuminating a small volume element dV of aerosol with an irradiance E is scattered by the material contained in the volume element. Part of the incident flux is scattered into various directions given by the scattering angle ϕ. The scattering angle ϕ is the angle between the transmitted and the scattered beam. Therefore $\phi = 0°$ is forward scatter and $\phi = 180°$ is backscatter. The flux $d\Phi$ scattered by the volume element dV into the solid angle $d\omega$ is given by $d\Phi = E\gamma(\phi)d\omega dV$. Dividing by $d\omega$ we obtain the scattered intensity in direction ϕ as $I = \frac{d\Phi}{d\omega} = E\gamma(\phi)dV$. The volume scattering function has a dimension of $m^{-1}\,sr^{-1}$. The volume scattering function for aerosols can be obtained by any of the aforementioned computer codes. One of the results of the Mie calculations are the functions i_1 and i_2. For an aerosol consisting of spherical particles with size distribution $n(r)$ the volume scattering function for unpolarized illumination and without considering polarization of the scattered light is obtained by

$$\gamma(\phi) = \frac{\lambda^2}{4\pi^2} \int_{r1}^{r2} \frac{i_1 + i_2}{2} \left(\frac{dN}{dr}\right) dr \qquad (4)$$

We will use scattering angles larger than $180°$ below, which has no meaning according to our definition of Figure 2, but we will use the following convention: the scattering angle $(180° + \beta)$ is identical to $(180° - \beta)$ and $\gamma(180° + \beta) = -\gamma(180° - \beta)$.

The light scattered by the volume dV in all directions must be equal to the light lost due to scattering along the distance dx of the volume element. Denoting by dA the cross-section of the volume element, the light flux incident on it is $E dA$ and the totally scattered amount of flux thus is $d\Phi = E dA \sigma_s dx = E \sigma_s dV$. This must be equal to the intensity of the scattered light integrated over all solid angles $d\omega$, i.e. $d\Phi = E dV \int_{4\pi} \gamma(\phi) d\omega$. Thus $\sigma_s = \int_{4\pi} \gamma(\phi) d\omega$. For spherical particles or particles in random orientation, where the scattering has rotational symmetry, we obtain $\sigma_s = 2\pi \int_0^\pi \gamma(\phi) \sin\phi \, d\phi$.

If only the angular dependence of the scattering function is of interest, the volume-scattering function divided by the scattering coefficient which is called the *phase function* $P(\phi) = 4\pi \frac{\gamma(\phi)}{\sigma_s}$ is used. It can be understood as the ratio of *the energy scattered in direction ϕ per unit solid angle to the average energy of all directions scattered per unit solid angle*. The phase function for an isotropically scattering medium is $P(\phi) = 1$. The extinction coefficient σ_e, the single scattering albedo $\tilde{\omega}$, the volume-scattering function $\gamma(\phi)$ and the scattering phase function $P(\phi)$ are related by $P(\phi) = 4\pi \frac{\gamma(\phi)}{\tilde{\omega}\sigma_e}$.

Depending on the size of the particles the volume-scattering function can be quite different. If the particles are smaller than one-tenth of the wavelength Rayleigh scattering can be applied, they scatter light symmetrically in the forward and the backwards region. For particles larger than $0.5\,\mu m$ the forward scattering is several orders of magnitude larger than the backscattering.

The *asymmetry parameter g*, is defined as $g = \int_0^\pi \gamma(\phi) \cos\phi \, \sin\phi \, d\phi / \int_0^\pi \gamma(\phi) \sin\phi \, d\phi$. For isotropic or symmetric scattering (e.g. Rayleigh scattering) the asymmetry parameter is zero, for a purely forward scattering aerosol it is 1. The asymmetry parameter of the cloudless atmosphere [28] ranges from 0.1 (very clean) to 0.75 (polluted); for a cloudy atmosphere the values are between 0.8 and 0.9. For an aerosol consisting of particles having a logarithmic normal distribution with a geometric standard deviation of 1.7 and mean diameters of 0.15, 0.3, 0.5 and 0.7 μm the asymmetry parameter is 0.43, 0.62, 0.71 and 0.75.

Although volume-scattering functions are readily available from Mie computer codes, radiative transfer calculations often use model scattering functions. Simple analytical expressions have proven to be very useful, with the Henyey–Greenstein functions [28] being the favourite of atmospheric optics.

The main characteristic of these scattering functions is the asymmetry parameter g, and the phase function is given by

$$P(\phi) = \frac{1 - g^2}{(1 + g^2 - 2g\cos\phi)^{3/2}} \tag{5}$$

One might question whether this simple phase function is a good representative of the "real" phase function. But for multiple scattering, where this function is used frequently, this is not a point of discussion, since the light field "forgets" details of the scattering function.

The *albedo A* is defined as the fraction of incident light reflected diffusely and/or specularly by a surface. It is a pure number and can be understood as the fraction of radiation which is not absorbed by the surface.

3   OPTICAL PROPERTIES OF THE ATMOSPHERE

More than 99% of the air consists of oxygen and nitrogen and noble gases. They scatter light according to the theory of Rayleigh. The extinction coefficient depends on the inverse fourth power of the wavelength and has a magnitude of $11.62\,\text{Mm}^{-1}$ at a wavelength at 550 nm and standard pressures and temperature. At other temperatures, wavelengths and pressures, σ_s is obtained by

$$\sigma_s = 11.62\,\text{Mm}^{-1}\left(\frac{\lambda}{550\,\text{nm}}\right)^{-4.08}\left(\frac{288.15K}{T}\right)\left(\frac{p}{101.3\,\text{kPa}}\right)$$

and its scattering phase function is

$$P(\phi) = \frac{3}{4\pi}[1 + \cos^2(\phi)]$$

For atmospheric optics it is convenient to define scale height, which is the vertical extent a hypothetical homogeneous substance/aerosol has in order to obtain the same vertical attenuation as the inhomogeneously distributed substance/aerosol. For the air the scale height is obtained by dividing the pressure at sea-level by the product of the density of the air and the gravity, giving a scale height of ≈ 8 km, the optical depth is $\delta = 0.093$, thus the transmittance of the pure air at perpendicular incidence is $\tau = 0.91$ at 550 nm. Due to the symmetry of the scattering, half of the light attenuated by the air, i.e. 4.4%, is scattered upwards (neglecting multiple scattering for the moment).

4   ATMOSPHERIC AEROSOL PARTICLES

One of the trace substances in the atmosphere is the aerosol, amounting to 1 to 100 ppb of the mass of air. The aerosol in the atmosphere exhibits a considerable variation in location, height, time and constitution. Mass concentrations range from $2 \mu g\,m^{-3}$ at polar regions through $10 \mu g\,m^{-3}$ for the background, $30 \mu g\,m^{-3}$ for remote and rural, $170 \mu g\,m^{-3}$ in polluted urban environments to $100\,000 \mu g\,m^{-3}$ in sand-storms (concentrations, mean diameters and standard deviations for the different types of aerosols are given in [3]). Its vertical extent is variable, earlier estimates were 1–2 km; [30] d'Almeida [31] has found a scale height for the Sahara dust of 2.5 km, but recent measurements in southern Europe [32] have shown that the vertical extent of an assumed homogeneous aerosol layer can be 4–8 km. The vertical optical depth is obtained by multiplying the extinction coefficient at ground level by the scale height. Jeske [33] gives a vertical attenuation of 17% at 500 nm, amounting to a vertical optical depth of 0.19 under "average conditions". Esposito et al. [32] have determined vertical optical depths larger than 0.3 in a rural location in Europe.

The main materials which form the aerosol particles in the atmosphere are sulfur compounds (both natural and anthropogenic), nitric compounds (mainly anthropogenic), sea-salt (natural), organic materials (both natural and anthropogenic), soil-derived substances such as quartz or limestone (mainly natural), and combustion products, which among others contain black carbon to a more or less degree (with a few exceptions anthropogenic). All transparent materials (numbers 1 to 7 on Table 2) have similar density and refractive index. Finally we have also to include mixed absorbing particles which will be considered as representative for air pollution. With increasing life time the particles in the atmosphere undergo transformations such as coagulation, incorporation in clouds droplets and re-evaporation or deposition of various chemical compounds on to the surface of the particle and thus the particles can consist of a mixture of several substances.

In general the size of the airborne particles can be represented by a lognormal size distribution. Sulfates, nitrates and organics have mass mean diameters around $0.5 \mu m$, the size of the black carbon particles is smaller, whereas soil-derived particles and sea-salt particles are mainly larger than a few micrometres.

The mass extinction coefficient $\frac{\sigma_e}{M}$, the single scattering albedo, the asymmetry parameter, and the fraction of the attenuated light which is scattered in the backwards direction (given by $2\pi \int_{90°}^{180°} \gamma(\phi)\sin(\phi)d(\phi)/\sigma_e$ for particles consisting of the aforementioned materials and for various particle sizes) are shown in Table 2. One can see that there is a considerable dependence of the efficiency of the interaction with light on the mean size of the particles. For particles consisting of a transparent material the sizes between 0.3 and $0.8 \mu m$ are most efficient in their interaction with light, having mass absorption coefficients in the order of $4\,m^2\,g^{-1}$, substances with a low density have a higher value. Particles with mean diameters < 0.1 or $> 2\mu m$ are less efficient in the interaction with light, with much lower values for particles with mean diameters $<< 0.1$ or $>> 3 \mu m$. There is little difference between the various kinds of

Table 2. Optical parameters of aerosols consisting of various substances and having various sizes. For the computations spherical particles have been assumed, the size distributions are lognormal with mean diameters as given below and a standard deviation of $\sigma_g = 1.8$. The mass extinction coefficient is given in m^2g^{-1}, the single scattering albedo is dimensionless as well as the asymmetry parameter. For perpendicular incidence the fraction of the backscattered (i.e. $90° < \phi < 180°$ light of the totally attenuated light is given. All values are for a wavelength of 550 nm

Substance	$d_g = 0.1\,\mu m$	$0.3\mu m$	$0.5\mu m$	$0.8\mu m$	$3.0\mu m$
No. 1 Ammonium sulfate	Refractive index:		$1.53 - .00i$	Density:	$1769\,kg\,m^{-3}$
Mass extinction coefficient	0.65	3.29	4.03	3.48	$0.78\,m^2g^{-1}$
Single scattering albedo	1.00	1.00	1.00	1.00	1.00
Asymmetry parameter	0.35	0.61	0.66	0.68	0.70
Backscattered fraction	0.258	0.112	0.093	0.095	0.100
No. 2 Ammonium nitrate	Refractive index:		$1.56 - 0.00i$	Density:	$1725\,kg\,m^{-3}$
Mass extinction coefficient	0.74	3.67	4.36	3.65	$0.80\,m^2g^{-1}$
Single scattering albedo	1.00	1.00	1.00	1.00	1.00
Asymmetry parameter	0.35	0.60	0.65	0.66	0.69
Backscattered fraction	0.256	0.117	0.101	0.104	0.104
No. 3 Sulfuric acid	Refractive index:		$1.43 - 0.00i$	Density:	$1841\,kg\,m^{-3}$
Mass extinction coefficient	0.42	2.30	3.10	2.99	$0.77\,m^2\,g^{-1}$
Single scattering albedo	1.00	1.00	1.00	1.00	1.00
Asymmetry parameter	0.34	0.64	0.71	0.74	0.73
Backscattered fraction	0.264	0.098	0.072	0.069	0.085
No. 4 Sodium chloride	Refractive index:		$1.54 - 0.00i$	Density:	$2165\,kg\,m^{-3}$
Mass extinction coefficient.	0.56	2.81	3.39	2.88	$0.64\,m^2\,g^{-1}$
Single scattering albedo	1.00	1.00	1.00	1.00	1.00
Asymmetry parameter	0.35	0.60	0.65	0.67	0.69
Backscattered fraction	0.257	0.114	0.097	0.100	0.102
No. 5 Crustal material	Refractive index:		$1.56 - 0.00i$	Density:	$2000\,kg\,m^{-3}$
Mass extinction coefficient	0.64	3.17	3.77	3.15	$0.69\,m^2\,g^{-1}$
Single scattering albedo	1.00	1.00	1.00	1.00	1.00
Asymmetry parameter	0.35	0.60	0.65	0.66	0.68
Backscattered fraction	0.256	0.117	0.101	0.104	0.104
No. 6 Organic material	Refractive index:		$1.40 - 0.00i$	Density:	$1000\,kg\,m^{-3}$
Mass extinction coefficient	0.66	3.75	5.23	5.23	$1.45\,m^2\,g^{-1}$
Single scattering albedo	1.00	1.00	1.00	1.00	1.00
Asymmetry parameter	0.35	0.65	0.72	0.76	0.75
backscattered fraction	0.266	0.094	0.066	0.061	0.079

Table 2. Contd.

Substance	$d_g = 0.1\,\mu m$	0.3μm	0.5μm	0.8μm	3.0μm
No. 7 "Non-absorbing"	Refractive index:		$1.50 - 0.00i$	Density:	$1800\ kg\,m^{-3}$
Mass extinction coefficient	0.57	2.98	3.75	3.33	$0.77\ m^2\,g^{-1}$
Single scattering albedo	1.00	1.00	1.00	1.00	1.00
Asymmetry parameter	0.35	0.62	0.68	0.69	0.70
Backscattered fraction	0.259	0.107	0.087	0.087	0.096
No. 8 Soot	Refractive index:		$1.50 - 0.47i$	Density:	$1000\ kg\,m^{-3}$
Mass extinction coefficient	9.79	9.68	7.64	5.36	$1.34\ m^2\,g^{-1}$
Single scattering albedo	0.12	0.32	0.39	0.44	0.51
Asymmetry parameter	0.27	0.57	0.70	0.79	0.90
Backscattered fraction	0.038	0.044	0.031	0.023	0.017
No. 9 Mixed absorbing particles, internal mixture of 87% "non-absorbing" and 13% soot	Refractive index:		$1.50 - 0.10i$	Density:	$1630\ kg\,m^{-3}$
Mass extinction coefficient	1.84	3.90	4.14	3.42	$0.84\ m^2\,g^{-1}$
Single scattering albedo	0.31	0.61	0.65	0.64	0.52
Asymmetry parameter	0.33	0.63	0.72	0.78	0.90
Backscattered fraction	0.083	0.060	0.040	0.030	0.015

No. 10 External mixture of 87% "non-absorbing" and 13% soot

Mass extinction coefficient	1.77	3.85	4.25	3.59	$0.85\ m^2\,g^{-1}$
Single scattering albedo	0.27	0.61	0.69	0.70	0.72
Asymmetry parameter	0.33	0.61	0.69	0.70	0.72
Backscattered fraction	0.205	0.125	0.104	0.081	0.087

No. 11 External mixture of 13% soot of diameter 0.1 μm and 87% "non-absorbing" of diameter as given above

Mass extinction coefficient	1.77	3.87	4.53	4.16	$1.94\ m^2\,g^{-1}$
Single scattering albedo	0.37	0.71	0.75	0.73	0.42
Asymmetry parameter	0.33	0.60	0.67	0.67	0.63
Backscattered fraction	0.205	0.103	0.085	0.085	0.086

transparent particles since they have similar density and refractive index, therefore we will consider only one substance, a hypothetical "non-absorbing" substance, given as No. 7 in Table 2, as representative for transparent particles in the following. Obviously the single scattering albedo for the transparent particles is 1.

As representative of strongly light-absorbing particles we have chosen soot. It is the product of any combustion and consists mainly of elemental (black) carbon and usually also some unburnt or partly burnt fuel. It is produced in the flame by condensation and coagulation and usually has a fractal structure. Therefore both its effective refractive index and density are lower than the values for the bulk material. Here commonly used values have been selected. For details see [34]. Soot particles are about a factor of 2 more efficient in their interaction of light compared to the transparent particles. Particles with a mean diameter of 0.1 μm are even more than 10 times more efficient than their transparent counterpart. For sizes < 0.1 μm no decrease in mass extinction coefficient takes place. For particles smaller than 1 μm the single scattering albedo is smaller than 0.5, indicating that the majority of the interaction of the light with the particles is due to absorption.

The atmospheric aerosol contains soot—it can even be found at the South Pole. There are two possibilities for the mixing of soot with the remaining substances of the aerosol: (A) If soot and the other constituents of the aerosol form individual particles we call it an external mixture. This will be the case near a source of black carbon. The optical properties of the externally mixed aerosol are obtained by calculating the (extinction/scattering/absorption coefficient, volume scattering function) for each species of the particles, and adding them. Single scattering albedo and asymmetry parameter are obtained from the added values. (B) If the particles are made of a mixture of elemental carbon and transparent substances, we call it an internal mixture. This can happen when several particles have been incorporated in a cloud droplet which evaporated again resulting in a mixed particle. Also the particles formed from different substances can have coagulated in the atmosphere or the particles are formed during an inefficient combustion process such as biomass burning, where evaporated materials condense, for example on soot particles. For mixed particles it is difficult to calculate the optical properties of the aerosol particles since they have an irregular shape and constitution. It has become a habit to use an "effective refractive index" for mixed particles, which is defined as the (hypothetical) refractive index an assumed spherical particle would have, which has the same optical properties as the mixed particle under consideration. There are a variety of rules how to obtain the refractive index of a particle containing a mixture of substances, but there is no theoretical basis for any of them [35]. The volume additivity of the real and imaginary part of the refractive index is frequently used: the average refractive index is obtained by adding the refractive indices multiplied by the volume fraction. If two substances with volume V_1 and V_2 having refractive indices $m_1 = n_1 - ik_1$ and $m_2 = n_2 - ik_2$ are mixed, the refractive index of the mixture is assumed to be

$$m = \frac{m_1 V_1 + m_2 V_2}{V_1 + V_2} = \frac{n_1 V_1 + n_2 V_2}{V_1 + V_2} - i\frac{k_1 V_1 + k_2 V_2}{V_1 + V_2} \qquad (6)$$

In Table 2, No. 9, we can find values for an aerosol which is assumed to be an internal mixture of a non-absorbing aerosol and black carbon. A commonly used value for the refractive index is 1.5–0.1 *i*, which is considered as a representative for an aerosol in a polluted area. This refractive index is obtained by the volume mixing rule, if the mass of the material forming the aerosol particles contains 87% "non-absorbing material" and 13% soot, which is common, for example in Vienna. Its mass extinction coefficient is higher than for transparent particles. The single scattering albedo is between 0.3 and 0.6.

It is interesting to compare the characteristics of an aerosol consisting of internally mixed particles with the properties of an aerosol having exactly the same amount of substances, but being mixed externally, i.e. a chemical analysis of samples will give exactly the same result for the two aerosols. The last two groups of data have been calculated for this purpose.

Let us first consider the internally mixed aerosol with particles consisting of 13% soot and 87% transparent substance (No. 9 of Table 2) with an aerosol having exactly the same ingredients but with soot and transparent material in separate particles but of the same size as the internally mixed aerosol (No. 10 of Table 2). The mass extinction coefficients are almost equal, i.e. with respect to transmission there will be barely any difference. But the single scattering albedo is definitely lower for the internally mixed particles, i.e. with the same amount of carbon the mixed particles have more light absorption. The internally mixed aerosol has a more asymmetric scattering and together with the lower single scattering albedo the backscattered light is only half or even less of the back-scattered light of the externally mixed aerosol.

Having an externally mixed aerosol with soot particles and transparent particles of the same size is a very unlikely case. Since the soot particles found in most urban environments originate from automotive emissions, they have sizes around 0.1 μm. For this reason the optical properties are given for an aerosol again consisting of 13% soot, but with a fixed diameter of 0.1 μm and non-absorbing particles of sizes between 0.1 and 3 μm (No. 11 in Table 2). The results are not very different from the previous case, except for the 3 μm particles. As before, the amount of backscattered light is less for the internally mixed aerosol.

From this one can conclude that internally and externally mixed aerosols show little difference when only considering transmission, but the internally mixed aerosol particles have more light absorption and less backscattering than an externally mixed aerosol. It should be noted that a chemical analysis of an aerosol sample will give no information on the mixing state. Only simultaneous determination of, for example, size and optical properties will give hints on the mixing state. Aged aerosols have a considerable fraction of internally mixed particles.

We have seen that mixed particles have an increased light absorption and decreased backscatter. For particles of 3 μm or larger minimal quantities of soot can produce considerable light absorption. This is of special importance for cloud droplets: minute quantities of carbon incorporated in the cloud droplets have been found, with sizes of the carbon particles below a diameter of 0.1 μm. Even a small amount of carbon can drastically decrease their albedo (cloud anomaly, [36]).

It is interesting to compare the amount of aerosol mass needed to obtain a given attenuation with the quantity of other substances which are considered important in climatic change. Since the absorption of infrared radiation, which leads to the greenhouse effect, is known to everybody, a comparison with the infrared attenuation of the well-known greenhouse gas CO_2 is performed: CO_2 has absorption at various wavelengths, among others it has a strong broad band absorption at a wavelength of 15 μm. Air containing 340 ppmV of CO_2 (thus a density of 660 mg m^{-3} of carbon dioxide) has a transmission of 0.0001 for a distance of 1 km. [37] Thus the extinction coefficient is 0.0092 m^{-1}. For an aerosol with a mass extinction coefficient of 4.5 m^2 g^{-1} a mass concentration of 2 mg m^{-3} is needed to give the same attenuation. It is astonishing to see how little aerosol material is needed to give the same strong attenuation as the CO_2 in the infrared. The mass is less than one three-hundredth and this might be the reason why the greenhouse effect from increasing carbon dioxide has been discussed for decades, whereas the aerosols, which can give a cooling of the atmosphere, have been neglected for a long time.

Using a scale height of 2–5 km and the mass extinction coefficients for the accumulation and coarse modes, forming the different types of aerosols, the vertical optical depth of the aerosol can be easily obtained. This is left to the reader and instead data derived from a global aerosol climatology (d'Almeida et al., [38] Table 7.3) are shown in Table 3. The table gives a considerable variation in the vertical optical depth. This is typical for all aerosol data, since the aerosol itself is extremely variable as can be deduced from the variability of visibility for example. It is completely normal to observe a visibility of 10 km one day and 50 km the next, which is a variation in the aerosol extinction almost by a factor of 6 (the Rayleigh scattering coefficient of the air has to be

Table 3.   Vertical optical depth for various types of aerosols. A scale height of 2 to 5 km has been assumed, the extinction coefficients were obtained from d'Almeida [38, table 7.3]. A wavelength of 500 nm is used

Type of aerosol	Extinction coefficient (Mm^{-1})	Optical depth
Arctic	13–309	0.026–1.55
Clean continental	19–283	0.038–1.41
Average continental	270–362	0.54–1.81
Polluted urban	204–873	0.4–4.36
Clean maritime	103–145	0.2–0.73
Air (8 km)	11.62	0.092 (at 550 nm)

subtracted). This tremendous variability of the aerosol also creates a problem with the estimates of climatic effects of the aerosol, since it is very difficult to define a "representative" or "average" aerosol.

For clean polar and continental air masses the aerosol vertical attenuation is comparable to that of the air. But values of the optical depth of $\delta \gg 0.5$ are also possible for polar air masses. A global distribution of optical depths using the aforementioned aerosol climatology gives vertical optical depths ranging from 0.015 in the Antarctic, 0.05 in the Arctic, 0.15 over most of the oceans, 0.3–0.6 over Africa, Australia and South America and 0.7–1.52 over Europe (for July, slightly different values for January, see [38] Figures 7.63 and 7.69). In savanna regions the vertical optical depth was found to be ≈ 0.9 as background value and ≈ 1.7 for aged smoke from biomass burning. [39] Let us assume for the following considerations an optical depth of 0.3 to be a "representative value".

We have seen above that aerosols (including clouds) influence the radiative balance by scattering and absorbing part of the incoming or reflected visible radiation and scattering and absorbing outgoing infrared radiation. Therefore, besides the attenuation of visible radiation, it is also important to know the angular scattering and absorption properties as well. Let us consider the scattering properties first. For an aerosol consisting of transparent particles all the light which is attenuated by the particles is scattered, their single scattering albedo therefore is 1. For the soot and the mixed absorbing particles only part of the attenuated light is scattered, with values for the single scattering albedo well below 1.0. For 0.1 μm soot particles the single scattering albedo is only 0.12, but it must be borne in mind that the single scattering albedo is a relative value; it just means that the majority of the light is absorbed. But we also have to consider a peculiarity of light-absorbing particles: for the 0.1 μm soot particles the mass extinction coefficient is $9.79 \, \mathrm{m^2 \, g^{-1}}$, for the "non-absorbing" counterpart it is 0.57. The light-absorbing particles thus have a 17 times larger interaction with light and thus despite a single scattering albedo of only 0.12 the soot particles scatter twice as much light as the non-absorbing particles of 0.1 μm. For particles with diameters between 0.3 and 0.8 μm the soot particles scatter about 60% of the amount scattered by transparent particles. The mixed absorbing particles of 0.1 μm diameter scatter about the same amount as the transparent particles of equal size; the larger ones scatter about 70% although they only contain 10% black carbon.

The material and the particle size very much influence the angular dependence of the scattered light. The scattering phase functions for transparent particles, soot and mixed particles are shown in Figures 3(a) to 3(c). For pure air and also for particles smaller than 50 nm (i.e. nucleation mode particles) the forward and the backscattered amounts are equal (Figure 3(a), solid line) and a slight minimum occurs at $\phi = 90°$. Half of the light which is scattered goes to the back direction.

ammonium sulfate

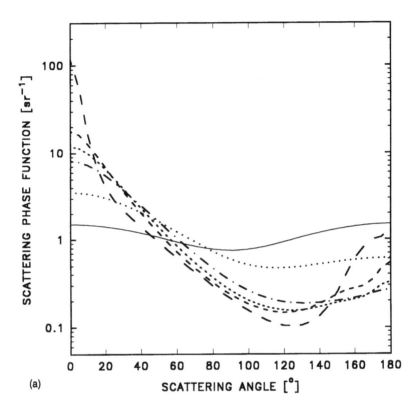

Figure 3.   (a) Scattering phase function for transparent aerosol particles. The optical characteristics for ammonium sulfate and size distributions as in Table 2 have been used. Solid line: pure air; dotted: 0.1 μm; dash-dot: 0.3 μm; short dash: 0.5 μm, medium dash: 0.8 μm; long dash: 3 μm.

Transparent particles, for example of ammonium sulfate (Figure 3(a)), which are larger than a few tenths of a micrometre scatter the majority of the light in the forward direction. For the particles with a mean diameter of 0.1 μm this is not easily visible in the logarithmic representation, but only 26% of the totally scattered light is scattered in the range of $\phi > 90°$, the asymmetry parameter being still small. With increasing particle size the forward scattering becomes more and the backscattering less and the fraction of light scattered in the back direction goes down to $\approx 10\%$. For particles larger than a few micrometres the scattering in a narrow range of the forward direction increases considerably and some increase in the scattering in the back direction is also visible. With increasing particle size the asymmetry parameter increases.

soot

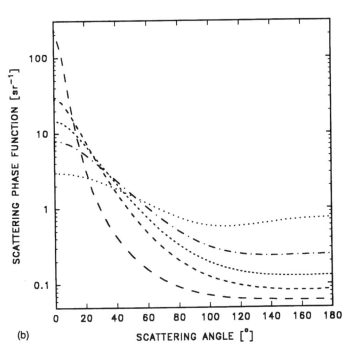

(b)

Figure 3. (b) Scattering phase function for soot aerosol particles. The optical characteristics for soot particles and size distributions as in Table 2 have been used. Dotted line: 0.1 μm; dash-dot: 0.3 μm; short dash: 0.5 μm; medium dash: 0.8 μm; long dash: 3 μm.

For the strongly absorbing soot aerosol (Figure 3(b)) the phase functions for the 0.1 μm particles are similar to the transparent particles. But since the particles mainly absorb light, only a small fraction of the attenuated light is scattered at all and thus also the fraction of light scattered in the backwards direction is small. For larger particle sizes the light scattered in the backwards direction is considerably smaller than for transparent particles, thus the fraction of backscattered light is small and the asymmetry parameter high.

For the mixed absorbing aerosol particles there is astonishingly little difference from the strongly absorbing soot particles (see Figure 2(c)).

For completeness the optical characteristics of cloud drops are given in Table 4 and scattering phase functions are shown in Figure 3(d). The droplets have been assumed to have a lognormal size distribution with a geometric standard deviation of 1.2. As mean sizes the following values have been selected: 5 μm (representative of cumulus clouds), 7 μm (altostratus) and 10 μm (stratus). Due

mixed absorbing aerosol

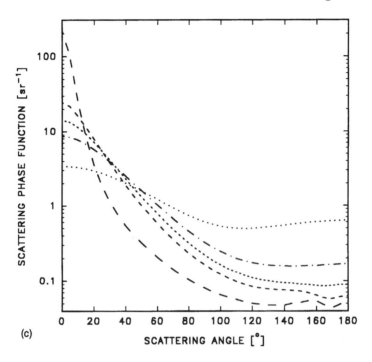

(c)

Figure 3. (c) Scattering phase function for mixed absorbing aerosol particles. The optical characteristics for mixed aerosol particles and size distributions as in Table 2 have been used. Dotted line: 0.1 μm; dash-dot: 0.3 μm; short dash: 0.5 μm; medium dash: 0.8 μm; long dash: 3 μm:

to the larger particle size the mass extinction coefficient is lower than for the usual aerosol particles. But the liquid water content (= mass concentration in our nomenclature) of dense clouds is in the range of 1 g m^{-3}, thus the extinction coefficients for the three types of clouds considered is $\sigma_e = 670\,000$, $470\,000$ and $320\,000$ Mm^{-1}, thus at least 350 times larger than the extinction coefficient of heavy urban pollution. The optical depth of a 500 m thick cloud layer is $\delta = 335$, 235 and 160 respectively and thus no direct radiation ($< 10^{-70}$) is transmitted. The scattering phase function for these droplets is shown in Figure 3(d). Due to the larger particle size the scattering in the narrow forward region is more, the scattering perpendicular to the incoming radiation is less and the value of the asymmetry parameter is closer to 1. The value for the phase function in the back direction is comparable to transparent aerosols. The maximum at 140° is the first indication of the rainbow, which is more pronounced for larger droplets; the increased back-scatter causes the glory. The fraction of backscattered light is less than for

water drops

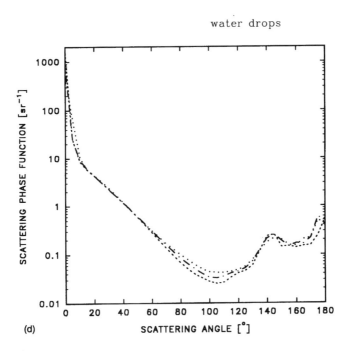

Figure 3. (d) Scattering phase function for cloud drops. The optical characteristics for water droplets and size distributions as in Table 4 have been used. Dotted line: 5 μm; dash-dot: 7 μm; short dash: 10 μm. The scale in the *y*-direction is different from Figures 3(a)–(c)

transparent aerosol particles and comparable to the value for absorbing particles. But it must be stressed that the optical depth of clouds is much larger than for aerosols, therefore a considerable amount of light is scattered backwards by a cloud.

5   REFLECTED AND ABSORBED RADIATION

With the values given in Table 2 it is a simple exercise to calculate the amount of solar radiation reflected upwards by the atmospheric aerosol. Let us consider perpendicular incidence for the moment. With an optical depth δ of the aerosol layer, the attenuation of the sunlight is given by $\exp(-\delta)$, and the fraction of scattered or absorbed light is given by $[1 - \exp(-\delta)]$. Multiplying this by the fraction of backscattered light, as given in Table 2, we obtain the fraction of

Table 4. Optical parameters of cloud droplets. For the computations lognormal size distributions with mean diameters as given below and a standard deviation of $\sigma_g = 1.2$ have been assumed. The mass extinction coefficient is given in m^2g^{-1}, the single scattering albedo is dimensionless as well as the asymmetry parameter. For perpendicular incidence the fraction of the backscattered (i.e. $\phi > 90°$) light of the totally attenuated light is given. All values are for a wavelength of 550 nm

	$d_g = 5\,\mu m$	$7\,\mu m$	$10\,\mu m$
Mass extinction coefficient (m^2g^{-1})	0.67	0.47	0.32
Single scattering albedo	1.00	1.00	1.00
Asymmetry parameter	0.84	0.84	0.85
Fraction of backscattered light	0.047	0.046	0.041

light scattered upwards. In Table 5 values are given for an atmosphere with an optical depth of $\delta = 0.3$ and various particles.

Transparent particles with a (rather small) size of 0.1 μm scatter ≈ 7% of the light upwards for perpendicular incidence on an aerosol layer of $\delta = 0.3$. For the usual size of the accumulation mode aerosol about 3% are scattered back, which definitely cannot be neglected. The particles will reduce the input of solar energy and have a cooling effect. For the strongly absorbing soot aerosol, the backscattered fraction is 1% or less; the mixed absorbing particles also scatter less radiation upwards than the transparent particles.

Table 5 also contains the fraction of sunlight which is absorbed by the aerosol. Obviously the transparent particles do not absorb any solar radiation, but for both the soot and the mixed absorbing particles the amount of absorbed solar energy is larger than the amount scattered upwards. So the particles will

Table 5. Fraction of sunlight scattered upwards and absorbed by the particles. A hypothetical atmosphere with an optical depth of $\delta = 0.3$ and perpendicular incidence and only single scattering have been assumed. For the computations spherical particles have been assumed, the size distributions are lognormal with mean diameters as given below and a standard deviation of $\sigma_g = 1.8$. All values are for a wavelength of 550 nm

Substance		$d_g = 0.1\,\mu m$	$0.3\,\mu m$	$0.5\,\mu m$	$0.8\,\mu m$	$3.0\,\mu m$
Ammonium sulfate	Scattered:	0.067	0.032	0.024	0.027	0.026
	Absorbed:	0.0	0.0	0.0	0.0	0.0
Soot	Scattered:	0.010	0.011	0.008	0.006	0.004
	Absorbed:	0.23	0.18	0.16	0.15	0.13
Mixed absorbing	Scattered:	0.022	0.016	0.01	0.008	0.004
particles	Absorbed:	0.18	0.10	0.09	0.09	0.12

transfer a considerable fraction of the solar radiation as heat to the atmosphere, thus they will have a heating effect, at least for the atmosphere. Both the pure soot aerosol and the mixed aerosol have considerable light absorption which may not be representative of an average of the atmospheric aerosol. But one can already see that the light absorption of the aerosol plays an important role. If the light absorption is sufficiently small, the backscattered radiation will dominate and the aerosol will have a cooling effect; for higher light absorption the absorbed radiation will dominate and a heating effect of the atmosphere will occur.

So far we only have considered perpendicular incidence of the radiation on a layer of aerosol, and the backscattered radiation is obtained by integrating the volume-scattering function between 90 and 180°, or to be more precise, we have to integrate from 90° through 180° to 270°. If the solar radiation has a zenith distance z, all the radiation scattered in directions between $90° - z$ and $270° - z$ goes upwards and thus back to the space. Since the scattering phase function depends very much on the angle, the amount of light scattered upwards depends on the zenith angle. For $z = 0$ the backscattered fraction is given in Table 2, for $z = 90°$ it obviously is $0.5\tilde{\omega}$. But with increasing zenith angle the optical depth of the layer also increases, therefore the backscattered radiation increases with increasing zenith angle. In order to separate the dependence of the backscattered radiation from the variable optical depth, we will consider the backscattered radiation of a volume element of aerosol at solar zenith angle z relative to the backscattered radiation at zenith angle $z = 0°$, i.e. we look at

$$\int_{90°-z}^{270°-z} \gamma(\phi)\sin(\phi)\,d\phi \Big/ \int_{90°}^{270°} \gamma(\phi)\sin(\phi)\,d\phi.$$

An example for the dependence on zenith angle is plotted in Figure 4. The fraction of radiation scattered upwards increases with zenith angle by a factor of 5 for the aerosol considered here.

Besides scattering light the aerosol also absorbs light. The absorbed sunlight heats the atmosphere, whereas the scattered light reduces the input of solar energy, thus cools the atmosphere. Also the fraction of absorbed light very much depends on the aerosol properties, mainly the imaginary part of the refractive index. For an aerosol with a geometric mean diameter of $0.5\,\mu m$ and variable refractive index the fraction of absorbed and scattered light has been calculated and is plotted in Figure 5. As characterization for the light absorption properties the single scattering albedo has been used. For aerosols with very little light absorption, e.g. $\tilde{\omega} = 0.95$, the light absorption transfers less energy to the atmosphere than is scattered at any zenith angle of the sun. For a zenith angle of 30° an aerosol with a single scattering albedo of $\tilde{\omega} \approx 0.9$ will scatter and absorb equal amounts of solar radiation, thus neither have a

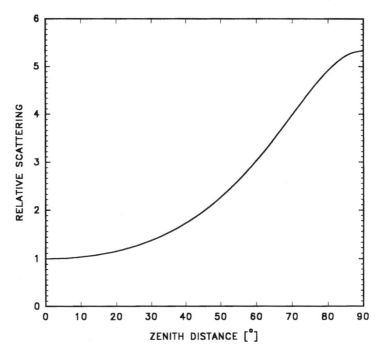

ammonium sulfate, 0.5μm

Figure 4. Variation of upwards scattered light by a volume element of aerosol as a function of the zenith distance. All values are relative to the backscattered light at perpendicular incidence. The aerosol particles consist of ammonium sulfate with $d_g = 0.5\,\mu m$ and $\sigma_g = 1.8$

heating nor a cooling effect. For a zenith angle of 60° the equilibrium will be for an albedo of $\tilde{\omega} \approx 0.8$. One can see that both the aerosol properties (singe scattering albedo) and the zenith angle determine the radiative forcing of an aerosol. One also has to consider the albedo of the underlying surface and multiple scattering as will be discussed below. When considering a global average Shettle and Fenn [40] have found a critical single scattering albedo of $\tilde{\omega} = 0.85$ of the aerosol. If $\tilde{\omega}$ is larger the aerosol mainly cools, if $\tilde{\omega}$ is smaller it predominantly heats the atmosphere.

6 HUMIDITY INFLUENCE

So far we have considered aerosol particles consisting of pure substances such as NaCl or $(NH_4)SO_4$. Many substances which are ingredients of aerosol particles are deliquescent. This means that the substance takes up a certain

$$d_g = 0.5\mu m$$

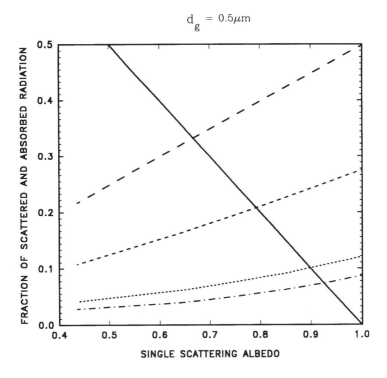

Figure 5. Fraction of absorbed and upwards scattered light of a volume element of aerosol. The particles have a lognormal size distribution with a mean diameter of 0.5 μm, the real part of the refractive index was set to 1.5, the imaginary part was varied between 0.00 and −0.3. The resulting single scattering albedos of the aerosol are plotted in the x-direction. The solid line gives the fraction of the attenuated light which is absorbed, the other curves give the fraction of the light scattered into the hemisphere which goes back to space: dash-dot: zenith angle $z = 0°$ (scattering integrated from $\phi = 90°$ to $\phi = 270°$); short dash: $z = 30°(\phi = 60-240°)$; medium dash: $z = 60°(\phi = 30-210°)$; long dash: $z = 90°(\phi = 0-180°)$

quantity of water when a minimum humidity is exceeded. The amount of water uptake depends on the humidity and the substance. For a given humidity an equilibrium is reached when the saturation vapour pressure over the solution equals the vapour pressure in the air. The relation between the concentration of the solute, usually given as molality (moles of solute per *kg* of solvent) and the water activity (ratio of the equilibrium vapour pressure over the solution and that over pure water, equals the considered relative humidity) have been determined for many substances and are available in the literature (see e.g. [41]). For pure substances the phase transition is abrupt and large hysteresis effects occur. The atmospheric aerosol contains both particles consisting of

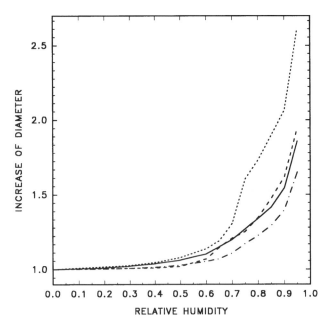

Figure 6. Increase of particle diameter with increasing humidity. Solid line: rural aerosol; dash-dot: urban; medium dash: desert; short dash: maritime

pure substances which show a well-defined phase transition with hysteresis effects and particles which are a mixture of several substances where no exact phase transition can be found. A summary of intense investigations is given in [42]. For the different types of atmospheric aerosols the increase in mass is for example tabulated in [3]. When the humidity increases from < 60 to 80% the rural aerosol takes up about its own mass of water, the urban aerosol about half and the maritime aerosol about twice its mass. The increase in diameter as a function of relative humidity has been calculated and is shown in Figure 6. Up to a humidity of ≈ 50% very little change in particle diameter is visible. At higher humidities the smaller increase is found for the urban aerosol containing the least amount of water-soluble particles, and the largest increase occurs for the maritime aerosol, which obviously contains a large fraction of water-soluble particles. Since the optical properties of an aerosol are roughly proportional to the square of the diameter, an increase in diameter by a factor of 1.4, which is obtained at a humidity of 85% for a rural aerosol, causes a doubling of the extinction and scattering coefficient. Therefore humidity plays an important role in the optical properties of the aerosol.

The dependence on humidity of the extinction coefficient, the absorption coefficient and the asymmetry parameter for the average continental, urban and polluted maritime aerosol models [38] is shown in Figure 7. Extinction and

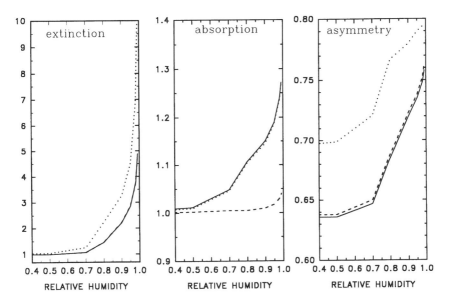

Figure 7. Variation of light extinction, light absorption and asymmetry parameters with relative humidity for model aerosols. Extinction and scattering coefficients are normalized to 1 at 0% relative humidity. Dashed line: average continental aerosol model; solid line: urban aerosol model; dotted line: maritime-polluted aerosol model

absorption coefficients are given relative to the value of the dry particles, i.e. they have a value of 1 at 0% relative humidity. The dramatic increase in the extinction coefficient, especially for the more deliquescent maritime aerosol, at high humidities is clearly visible. Although no light-absorbing substance is added, the absorption coefficient also increases with the uptake of water since in a mixed particle the light absorption of carbon is more efficient. The asymmetry parameter increases with humidity, due to the growth of the particles.

7 MULTIPLE SCATTERING

When passing through a layer of length x of molecules/particles/droplets the attenuation is determined by the optical depth $\delta = \sigma_e x$. For purely absorbing substances a photon which interacts with a molecule/particle disappears. In the case of light scattering the photon is reflected elastically by the molecule/particle, i.e. it just changes direction, but if scattered in the forward direction it still may reach the ground. After being scattered once the photon could be scattered a second time and be scattered upwards. By this secondary scattering more light goes upwards as expected by single scattering. If the optical depth δ

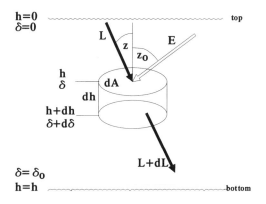

Figure 8. Derivation of the radiative transfer equation for a horizontally homogeneous atmosphere. The change of radiance L passing through volume element dV is considered. The considered direction is given by the zenith angle z and the azimuth φ (not shown). The sunlight has a zenith angle of z_0 and an azimuth φ_0 (not shown) and a flux density E_0 at the top of the atmosphere. The atmosphere extends from $h = 0$ (top) to $h = h_0$ (ground). Its optical depth δ is $\delta = 0$ at the top and $\delta = \delta_0$ at the ground

of the atmosphere is $\delta \ll 1$, the amount of primary scattered light is $\propto \delta$ and the secondary scattered light thus is $\propto \delta^2$ and therefore can be neglected. This might be the case for an extremely clean atmosphere (see e.g. Table 3), continental aerosol with

$$\delta = \delta_{\text{aerosol}} + \delta_{\text{air}} = 0.026 + 0.092 = 0.116$$

but frequently a vertical optical depth of > 0.3 will occur, and for a zenith angle of $60°$ the optical depth of the atmosphere has twice this value. In that case multiple scattering cannot be neglected.

Multiple scattering is treated in radiative transfer theory and only the basic considerations will be summarized here. For details see e.g. [16], [43], [44] or [28].

Let us assume a horizontally homogeneous atmosphere and an unpolarized radiation. All magnitudes used below depend on wavelength but this will not be written explicitly. The radiative transfer equation in principle is the energy conservation for a volume element: a change in radiance is due to (i) loss by attenuation and gains by (ii) scattered sunlight, (iii) scattered diffuse light and (iv) black-body emission. The geometry is shown in Figure 8. The atmosphere extends from $h = 0$ (top) to $h = h_0$ (ground). We will use as dimensionless vertical coordinates the optical depth δ of the atmosphere, with $\delta = 0$ the top and $\delta = \delta_0$ the bottom of the atmosphere. We consider a volume element dV which extends a vertical distance dh, with σ_e the extinction coefficient and the vertical extension can be expressed by $d\delta = \sigma_e \, dh$.

1. Let us now consider the radiance L passing through a volume element at a given height (given by δ) for a direction with an angle z with the vertical (or by $\mu = \cos z$) and an azimuth angle φ, i.e. we consider $L(\delta, \mu, \varphi)$. It changes by $dL(\delta, \mu, \varphi)$. For this layer dh of the atmosphere having the optical depth $d\delta = \sigma_e \, dx$ the loss in radiance is obtained by

$$dL_1(\delta, \mu, \varphi) = -L(\delta, \mu, \varphi)\sigma_e \, dh/\mu = -L(\delta, \mu, \varphi) \, d\delta/\mu$$

In this layer light is scattered which has to be added to the radiance $L(\delta, \mu, \varphi)$.

2. Let us now consider the gains in radiance and first gains due to the scattered sunlight. The solar radiation comes from a direction given by the zenith angle $z_0(\mu_0 = \cos z_0)$ and an azimuth angle φ_0 having a flux density of E_0 at the top of the atmosphere. It is attenuated to $E = E_0 \exp(-\delta/\mu_0)$. The scattering medium contained in the volume element $dV = dA \, dh$, being a cylinder with horizontal base and top surface dA and height dh, scatters the light giving an intensity $dI = E \, dA \, dh \, \gamma(\phi)$, with ϕ being the angle between direction (z_0, φ_0) and (z, φ). The radiance of this scattered light is obtained by $dL_2 = dI/(dA \, \mu)$. With $dh = d\delta/\sigma_e$, $\gamma(\phi) = \frac{1}{4\pi}\sigma_s P(\phi)$ and $\sigma_s/\sigma_e = \tilde{\omega}$, we obtain

$$dL_2(\delta, \mu, \varphi) = \frac{\tilde{\omega}}{4\pi} P(\phi) \, E_0 \exp(-\delta/\mu_0) d\delta/\mu$$

3. We also have to consider the diffuse radiation field $L(\delta, z', \varphi')$ with $\mu' = \cos z'$ which is also scattered by the aerosol in the volume element dV. The radiance $L(\delta, z', \varphi')$ coming from a solid angle $d\Omega' = d\varphi \cos z' dz' = d\varphi d\mu'$ produces an irradiance of $dE = L(\delta, z', \varphi')d\Omega'$ and the radiance d^2L_3 scattered in direction (z, φ) is obtained as above by

$$d^2L_3 = \frac{\tilde{\omega}}{4\pi} L(\delta, \mu', \varphi')P(\phi')d\Omega' d\delta/\mu$$

with ϕ' the angle between direction (z', φ') and (z, φ). Integrating the radiation coming from the full solid angle we obtain the contribution by the scattered diffuse radiation

$$dL_3 = \tilde{\omega}\Big(\int_0^{2\pi} \int_{-1}^{1} P(\phi')L(\delta, z', \varphi')d\varphi' d\mu'\Big)d\delta/\mu$$

4. The source emission which is proportional to the black-body emission $B(T(\delta))$ obtained from Planck's law, with the temperature $T(\delta)$ depending on height (or optical depth), gives a contribution $dL_4 = (1 - \tilde{\omega})B(T(\delta))d\delta/\mu$. Adding the contributions (1)–(4) we obtain

$$\mu\frac{dL(\delta, \mu, \varphi)}{d\delta} = -L(\delta, \mu, \varphi) + \frac{\tilde{\omega}}{4\pi} P(\phi)E_0 \exp\frac{-\delta}{\mu_0}$$
$$+ \frac{\tilde{\omega}}{4\pi}\int_0^{2\pi} \int_{-1}^{1} P(\phi')L(\delta, z', \varphi')d\varphi' d\mu' + (1 - \tilde{\omega})B(T(\delta)) \tag{7}$$

The first term on the right-hand side is the attenuation thus a reduction in radiance, whereas the other three terms add radiance to the considered beam. It is common to call the sum of the last three terms the source term or source function $J(\delta, \mu, \varphi)$ with

$$J(\delta, \mu, \varphi) = \frac{\tilde{\omega}}{4\phi} P(\phi) E_0 \exp \frac{-\delta}{\mu_0}$$
$$+ \frac{\tilde{\omega}}{4\pi} \int_0^{2\pi} \int_{-1}^1 P(\phi') L(\delta, z', \varphi') d\varphi' d\mu' + (1 - \tilde{\omega}) B(T(\delta)) \tag{8}$$

With this the equation of radiative transfer can be written as

$$\mu \frac{dL(\delta, \mu, \varphi)}{d\varphi} = -L(\delta, \mu, \varphi) + J(\delta, \mu, \varphi) \tag{9}$$

This can formally be integrated yielding

$$L(\delta, \mu, \varphi) = L_0 \exp \frac{-\delta}{\mu} + \frac{1}{\mu} \int_0^\delta J(\delta, \mu, \varphi) \exp \frac{\delta - \delta'}{\mu} d\delta' \tag{10}$$

with L_0 the radiance entering at $\delta = 0$. The practical integration of the radiative transfer equation is only possible in very special cases. Many methods have been developed to obtain reasonably good solutions, such as two-stream approximation, doubling and adding or Monte Carlo methods (see e.g. [43], [28] or [43]). Also tables of numerical calculations are available, and nowadays with computing facilities it is possible to obtain the desired solution.

Many published results of multiple scattering calculations use model phase functions, both because the real phase function might not be known and to permitt the computation in a reasonable time frame. A comparison of "real" and model phase functions is shown in Figure 9. For air, non-absorbing particles with mean diameters of 0.1, 0.3 and 3 μm and for cloud droplets, the phase functions obtained by Mie theory and the Henyey–Greenstein phase functions having the same asymmetry parameter are plotted. The Henyey–Greenstein phase function is a reasonably good approximation for the 0.1 and 0.3 μm particles although the increase in backscattering cannot be followed by this simple model phase function. The phase function for air is not well represented at all and for particles of 3 μm and droplets of 7 μm the model neither represents well the peak in forward scattering, the minimum around 100° nor the increase in backscattering. But it has to be mentioned that if considerable multiple scattering occurs, the phase function for single scattering is of less importance. Nevertheless it should be borne in mind that for the cloud-free atmosphere multiple scattering occurs, but the atmosphere is not opaque, therefore some errors in the calculation of the radiation field are introduced by using the model functions.

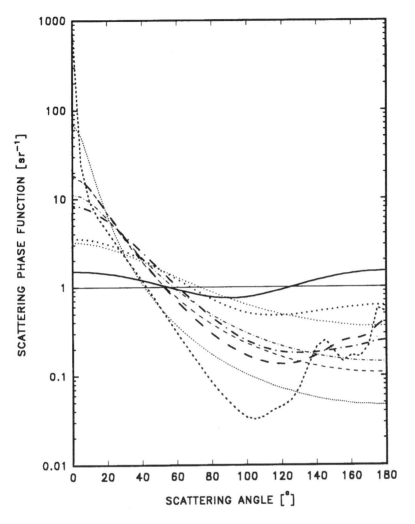

Figure 9. Comparison of model scattering phase functions with phase functions obtained from Mie theory. The Mie phase functions are the thick lines, the corresponding Henyey–Greenstein functions with the same asymmetry parameter are the thin lines. Solid line: pure air; dots: "non-absorbing", 0.1 μm; dash-dot: "non-absorbing", 0.3 μm; long dash: "non-absorbing", 3 μm; short dash: water drops, 7 μm

To see the importance of multiple scattering we will consider one simple example: the reflected and absorbed light by a slab of scattering and absorbing medium. Radiative transfer results are tabulated in [43] for media with a variety of asymmetry parameters of the Henyey–Greenstein scattering phase function, single scattering albedos, and optical depth of the layer. Two types of aerosol both with an asymmetry parameter of 0.75 (a typical value for the atmospheric aerosol) in the slab will be considered: a purely scattering aerosol and an aerosol which is weakly absorbing ($\tilde{\omega} = 0.95$). For a vertical incidence of radiation we will compare the results of the radiative transfer approach with that where no multiple scattering is considered. For single scattering the amount of light reflected upwards is obtained by

$$(1 - \tau)\tilde{\omega}2\pi \int_{90°}^{180°} \frac{1}{4\pi} P(\phi) \sin(\phi)d\phi \qquad (11)$$

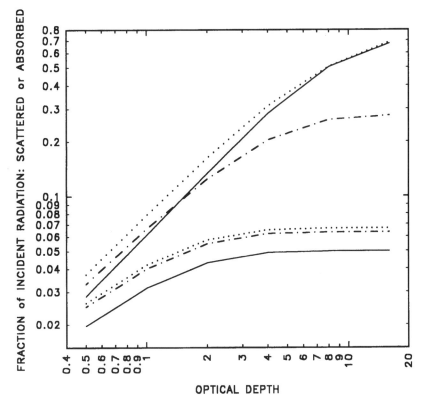

Figure 10. Comparison of the light scattered upwards and the light absorbed by a horizontal slab of scattering medium with vertical incidence and variable optical depth. The Henyey–Greenstein scattering phase function with $g = 0.75$ has been used. The fraction of the incident light scattered or absorbed has been obtained by radiative transfer (upper set of curves) and for comparison the results neglecting multiple scattering (lower set of curves) are shown. The dotted line gives the upwards scattered light by a purely scattering aerosol. The dash-dot line gives the scattered fraction of the incident solar radiation for an aerosol with $\tilde{\omega} = 0.95$ and the solid line gives the absorbed fraction

for Henyey–Greenstein scattering phase functions we obtain

$$\frac{1}{2}\int_{90°}^{180°} P(\phi)\sin(\phi)d\phi = \frac{(1-g^2)}{2g}((1-g)^{-\frac{1}{2}}-(1+g)^{-1}) \qquad (12)$$

The ratio of the radiation scattered upwards and the value obtained for single scattering is plotted in Figure 10. Already for a slab with an optical depth of 0.5 (transmission of the direct sunlight of $\tau = 0.61$) multiple scattering enhances the radiation scattered upwards by 42%. At an optical depth of $\delta = 16$ the light scattered upwards is 10.3 times the value of the result expected by single scattering only. In that case 68.8% of the incident light is reflected. For the slightly absorbing aerosol the reflected light is considerably less, e.g. for $\delta = 16$ it is less than half in comparison to the non-absorbing aerosol although only 5% of the light is absorbed. But 5% is absorbed at each interaction and multiple scattering can add up considerably. Consequently the fraction of absorbed light increases significantly with optical depth.

For the aerosol with $\tilde{\omega} = 0.95$ and single scattering the amount of radiation scattered upwards is 24% more than the amount of light absorbed by the particles. This is independent of the optical depth of the layer of particles as can be seen by the constant vertical distance of the lower dash-dot and solid lines in Figure 10. The radiation reflected back to space is larger than that transferred to the atmosphere, thus the aerosol has a cooling effect. When multiple scattering is also taken into account, the upper two curves show that for an optical thickness of up to $\delta = 1.5$ the backscattered radiation is more than the absorbed; for larger depths the absorption is considerably more. Therefore heating or cooling by a light-absorbing aerosol also depends on the optical thickness. Here we have only considered the simple case of a perpendicular incidence of sunlight. If we consider a Lambertian radiator shining on a plane parallel slab (this is equivalent to the sunlight hitting the atmosphere of our globe, being a spherical shell) the reflected light increases more than the absorbed light with respect to the perpendicular incidence; for an optical depth larger than $\delta = 4$ the absorbed light is more than the light scattered upwards.

Reflections from the ground also lead to multiple passes of the light through the atmosphere which enhances light absorption by the atmosphere. Similarly, as already shown in Figure 10, increased optical depth, also due to a larger zenith angle, enhances absorption. Thus an aerosol can have a positive or negative radiative forcing, depending on the optical depth and surface albedo. Generally, the higher the albedo of the ground, the more the chance that an even slightly absorbing aerosol will favour an energy gain for the earth–atmosphere system (see e.g. [46] or [47]).

8   CLOUDS

Clouds can also be considered as an atmospheric aerosol. Clouds form when the air is supersaturated with water vapour. During cloud formation the supersaturation is usually in the range of less than a few per cent. The cloud droplets have sizes between 5 and 20 μm, the number of droplets varies between 60 and 600 cm^{-3}. The liquid water content of clouds varies from 0.05 to 1.3 gm^{-3} (stratus), 1 gm^{-3} (cumulus) to 4 g m^{-3} (cumulus congestus). Fog has smaller droplet sizes (0.5 to a few tens of micrometres [48] and lower liquid water content. At the low supersaturations in clouds and fogs only heterogeneous condensation occurs, therefore adequate condensation nuclei, i.e. aerosol particles, have to be present. These particles must be able to grow at the low supersaturation and several soluble substances qualify for this, such as $(NH_4)_2SO_4$ or NH_4Cl ([49], Chapter 9). Also a mixture of soluble and insoluble or partly wettable particles can act as cloud condensation nuclei. The number of cloud condensation nuclei very much depends on the supersaturation considered. The number of particles provoking condensation at 0.3 and 3% humidity differ roughly by a factor of 4. Only a small fraction (< 1–10%) of the aerosol particles act as condensation nuclei. Over the continent especially in populated areas of Europe and the USA several hundreds to thousands of cloud condensation nuclei are present, in remote and oceanic regions it may drop to a few tens of particles per cm^3.

In order to form a cloud droplet a cloud condensation nucleus has to surpass its critical radius at the corresponding critical supersaturation, which is given by the maximum of the Köhler curve (see e.g. [49]). The critical radius and supersaturation depend on the material and the dry size of the particle. For example for a 0.1 μm $(NH_4)_2SO_4$ particle the needed supersaturation is 0.053% and it has to grow to a diameter of 1.32 μm. For NaCl the corresponding values are 0.037% and 1.92 μm. If the initial particle is 10 times larger, the corresponding values are 0.0016% and 4.5 μm for $(NH_4)_2SO_4$ and 0.0012% and 6.07 μm for NaCl. During cloud formation the supersaturation is not constant and particles may grow and shrink alternatively. So it will happen that some particles grow and others shrink; in our example at a supersaturation of 0.05% the 0.1 μm $(NH_4)_2SO_4$ will shrink while the NaCl particle will grow. Besides having the "right" supersaturation there must also be sufficient time to grow. For this reason it is difficult to make an exact prediction of the number of particles which are activated. However, particles which require a lower supersaturation will have a greater chance to become cloud droplets. The deliquescent particles which are not activated will still grow in the humid environment of the cloud reaching a radius which is several times the dry radius and thus increase the amount of light scattered by the cloud.

For a given quantity of water vapour it is obvious that the cloud condensation nuclei determine the size of the drops. Let us denote the liquid water content by W, and the average drop radius by $\bar{r}$ (we assume that all particles have this radius), the bulk density of water by ρ_w and the number of drops per volume by N, then $W = (4\pi/3)\bar{r}^3 N \rho_w$. The extinction coefficient of the cloud is obtained by $\sigma_e = N\bar{r}^2\pi Q_e$. For cloud droplets the extinction efficiency Q_e is very close to the asymptotic value of 2. Expressing $\bar{r}$ as $\bar{r} = [3W/(4\pi N \rho_w]^{1/3}$ we obtain the extinction coefficient of a cloud as

$$\sigma_e = N^{\frac{1}{3}} Q_e \pi^{1/3} \rho_w^{-2/3} W^{2/3} \left(\frac{3}{4}\right)^{2/3} \tag{13}$$

The extinction coefficient thus not only increases with the liquid water content but also with the number of cloud droplets, i.e. cloud condensation nuclei. Thus more cloud condensation nuclei mean an optically denser cloud with the same amount of water in the cloud. The extinction coefficients of clouds are considerable, as already mentioned above. The optical depth of cloud layers is much larger than $\delta = 20$, therefore no direct light is transmitted and light reaching the ground through a cloud does this only by multiple scattering. We have already seen in Figure 10 that multiple scattering enhances the scattering of sunlight upwards. This is clearly also the case for clouds, although the forward scattering is dominant (asymmetry parameter of 0.85). The effect of multiple scattering can be seen when looking at very dense thunderstorm clouds from an aeroplane: they are brilliantly white from above although they look quite dark from below. The albedo A of a cloud can be obtained by radiative transfer, and a good approximation is given by [44]: the albedo of a cloud with optical depth δ is

$$A = \frac{(1-g)\delta}{1+(1-g)\cdot\delta} \quad \text{and with } g = 0.85 \quad A \approx \frac{\delta}{\delta + 6.7} \tag{14}$$

The number of cloud condensation nuclei has an influence on the extinction coefficient, thus on the optical depth of the cloud, which consequently influences the albedo. More aerosol particles mean more and smaller cloud droplets and thus an increase in the extinction coefficient and consequently the albedo increases. Smaller cloud droplets have a smaller asymmetry factor, which also slightly increases the albedo. The presence of more aerosol particles which act as cloud condensation nuclei therefore changes the albedo of the cloud with everything else being unchanged, i.e. the same liquid water content, temperature and latent heat release. This has been observed in the form of shiptracks (see e.g. [50] and [51]). They occur when a not very dense cloud cover is formed for example over the Pacific and the emissions by ships create additional cloud condensation nuclei, which change the droplet size and number in the cloud. The effect is visible in satellite images. The effect can best be seen in thin clouds or fog, since a change in optical depth then produces the largest effect (see the above formula). Increases in the albedo of 0.18 have been determined.

One can thus see that aerosol particles have a considerable influence on cloud albedo. A sensitivity analysis [52] shows that the effect of additional particles is largest if the albedo of the cloud is around 0.5, which would mean an optical depth of $\delta = 13$, thus a thin cloud, and low particle numbers. At the maximum one additional particle per cm^3 increases the albedo by 1%. Changing the cloud albedo over oceanic and remote areas of the world by 1% would require about 1000–10 000 t of particles per year. This is a very small amount in comparison to the anthropogenic sulfur emissions which amount to 6×10^6 t year^{-1}; anthropogenic emission can thus easily affect the cloud albedo.

In areas with a considerable contribution of combustion emission (industry, traffic, biomass burning) it is obvious that black carbon particles can be incorporated in cloud droplets. Raindrops containing hundreds of small carbon particles have been collected. Dispersed small light-absorbing particles have a light absorption which is much larger than one would expect according to their mass fraction. With multiple scattering this effect is even more enhanced, and a soot volume fraction of 10^{-5}, corresponding to one soot particle with a diameter of $0.2\,\mu m$ in a $10\,\mu m$ droplet, can decrease the albedo of the cloud from 0.92 to 0.85 and increase the absorptivity to 0.2. [36]. Thus small fractions of soot can have considerable effects. But again many factors play a role: for thin clouds ($\delta < 35$) additional light-absorbing particles make it brighter, whereas the albedo of thick clouds ($\delta > 35$) decreases with increasing number of slightly absorbing particles. The albedo of polluted clouds has also been observed by Kondratyev and Binenko: [53] they compared clouds over a city in the Ukraine with similar clouds over a rural area and the Black Sea. The cloud water contained more ions, black carbon and organic materials over the town. The cloud droplets over the city were smaller and more numerous. The albedo over the rural area was 0.71, decreasing to 0.60 over the urban area. The cloud had a yellowish appearance. The albedo calculated by using the measured carbon content and the method by Chylek et al. [36] agrees with the measured values, which appears to be clear proof of the influence of black carbon on absorption. In contradiction to this, Twomey [51] demonstrates that light absorption of the aerosol does not reduce the reflectance of clouds. This seems to be a contradiction. But a thorough inspection of the two papers makes it appear that both authors are right, since they deal with different subjects. Kondratyev deals with fairly thick clouds where the soot particles are incorporated into the droplets, thus light absorption by the droplet is larger than by the soot particles themselves and multiple scattering in the cloud enhances the absorption further, whereas Twomey considers thin clouds and does not account for incorporation of the soot particles into the cloud droplets.

The magnitude of the light absorption by cloud seems as yet to be an unresolved issue. Cess et al. [54] report satellite and ground-based measurements of solar radiation and conclude that a significant absorption by the clouds has been found, which is $25\,\mathrm{W\,m^{-2}}$ larger than the prediction by models for a cloudy atmosphere.

9 THE ATMOSPHERE–EARTH SYSTEM

So far we have mainly considered the atmospheric aerosol and its interaction with solar radiation. Scattering by the aerosol causes less radiation to the surface, and absorption by the aerosol causes the atmosphere to be heated, but at the same time less radiation reaches the ground which will lead to cooling of the ground. If the ground has a high albedo, the reflected radiation will again be absorbed by the atmosphere, increasing the heating effect further. Due to the strong coupling of the atmosphere and the ground surface it is reasonable to consider them together, and discuss effects of the aerosol on the whole system.

Let us first look at a simplification which gives an easy insight in the processes involved. We will discuss the case of an atmosphere–earth system with and without an additional aerosol. Without the additional aerosol the albedo for the sunlight be A, thus of the incident flux density $\bar{S}$ the part $A\bar{S}$ is reflected upwards and $(1-A)\bar{S}$ is absorbed. Let us now add some aerosol to the atmosphere. The aerosol will reflect a fraction $\tilde{r}$ of the solar radiation upwards (single or multiple reflections from the ground are included in this number, see below) and absorb a fraction $\tilde{a}$ of the incident radiation (again multiple passes are included in this number). With the additional aerosol the atmosphere–earth system reflects $\bar{S}\tilde{r}+(1-\tilde{r}-\tilde{a})\bar{S}$. The reflected part can be written as $\bar{S}A(1-\tilde{r}-\tilde{a}+\frac{\tilde{r}}{A})$. If we compare this to the reflected amount of the atmosphere without the additional aerosol which was $\bar{S}A$ one can immediately see that we have to consider the expression in parentheses

$$\left(1-\tilde{r}-\tilde{a}+\frac{\tilde{r}}{A}\right) \tag{15}$$

If this expression is >1 the atmosphere–earth system reflects more radiation with the additional aerosol thus it suffers an energy loss. If it is <1 it gains energy, the radiative forcing is positive and heating might occur.

Let us first consider simple cases and we exclude an albedo of $A=1$ or $A=0$ of the earth. If we have only a scattering aerosol $\tilde{r}\neq0$ and $\tilde{a}=0$. Thus $(1-\tilde{r}-\tilde{a}+\frac{\tilde{r}}{A})>1$ and we have negative forcing in any case. Similarly for a purely absorbing substance we would have positive forcing, no matter what the albedo.

If we have a scattering and absorbing aerosol, a positive forcing and heating of the atmosphere earth system will occur, if $\frac{\tilde{r}}{A}-\tilde{r}-\tilde{a}<0$. Let us assume a strongly absorbing aerosol with $\tilde{r}=\tilde{a}$. Then the above equation will be fulfilled if $A>0.5$, i.e. the surface has to reflect well, as would be the case over snow or arid land, and positive forcing will occur. On the other hand, for the same aerosol negative forcing will occur if $\frac{\tilde{r}}{A} - \tilde{r} - \tilde{a} > 0$. This is the case for $A < 0.5$, e.g. over the sea. Thus it is clear that energy gain due to an additional aerosol depends not only on the absorption and scattering properties of the aerosol, but on the albedo of the surface as well.

The scattering and absorption properties of the aerosol $\tilde{r}$ and $\tilde{a}$ which we have used before depend on the albedo of the ground, since the sunlight will traverse the aerosol many times, and the flux density depends on the albedo. The above results are quantitatively right, but if we want to be more precise we have to consider the fate of the radiation reaching the ground being absorbed, scattered, etc. in more detail. We will follow a method used by Twomey. [50]

The geometry is shown in Figure 11. The additional aerosol is symbolized by the detached layer. The flux density $\bar{S}$ passes through the aerosol layer of optical density δ, thus the direct solar radiation is reduced to $\bar{S}\exp(-\delta)$, with $\exp(-\delta) = \tau$. This flux density plus the forward scattered light reaches the ground. If f is the fraction of the light which is scattered downwards and $(1 - f)$ is scattered upwards, then the total flux reaching the ground is given by $\bar{S}(\tilde{\omega}f(1 - \tau) + \tau)$. We denote this by $t\bar{S}$, and call it the transmitted light. The light which is scattered upwards by the aerosol (before reaching the ground) is given by $\bar{S}\tilde{\omega}(1 - \tau)(1 - f)$ which we will denote by $r\bar{S}$ and call the reflected light. The flux density leaving the ground after reflection is $At\bar{S}$. Of this $At^2\bar{S}$ leaves the atmosphere upwards and $Atr\bar{S}$ is scattered back to the ground. After a second reflection from the ground $rA^2t\bar{S}$ goes to the aerosol with $rA^2t^2\bar{S}$ going upwards and so on. The total sunlight going upwards after one or several reflections or scattering by the aerosol is given by

$$\bar{S}(r + t^2A(1 + Ar + A^2r^2 + \cdots)) = \bar{S}\left(r + \frac{t^2A}{1 - Ar}\right) \qquad (16)$$

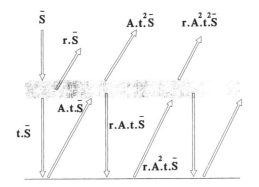

Figure 11. Energy balance of the atmosphere–earth system. To the existing system additional aerosol is added, symbolized by the detached layer. Fluxes scattered and reflected are shown as discussed in the text

Without additional aerosol the light reflected upwards is $A\bar{S}$. For the additional aerosol to cause a positive radiative forcing (net heating) of the atmosphere–earth system the radiation reflected/scattered upwards has to be smaller than without the additional aerosol, therefore

$$\left(r+\frac{t^2 A}{1-Ar}\right) < A \tag{17}$$

is the condition for a net heating.

With the smaller than sign replaced by an equals sign it is possible to separate cases where heating or cooling due to the additional aerosol occurs. This is shown in Figure 12 for two different albedos of the surface of the globe. We consider aerosols with different reflective and transmittive properties and various optical depths. In the r, t plane the conditions for no additional radiative forcing are given as a solid line. To the left the forcing is positive, to the right it is negative. A comparison of the curves for two different albedos shows clearly that the albedo of the surface has an eminent importance. For several types of aerosols the values of r and t are plotted for various optical depths. The aerosol with $\tilde{\omega} = 0.99$ represents an unpolluted aerosol, $\tilde{\omega} = 0.95$ could be a continental aerosol, $\tilde{\omega} = 0.90$ is polluted urban aerosol and $\tilde{\omega} = 0.40$ would be pure carbon, but could also be a value found for aged savanna smoke plumes: Liousse et al. [55] report a mass absorption coefficient of $\approx 25\,m^2\,g^{-1}$ for the black carbon in the aged smoke plume containing on the average 22% black carbon. With a mass scattering coefficient of $4\,m^2\,g^{-1}$ for the total aerosol a single scattering albedo of $\tilde{\omega} = 0.42$ is obtained. For the determination of r and t the tables of multiple scattering of van de Hulst [43] for a Lambertian radiator have been used.

The importance of the albedo is clearly evident from Figure 12. A slightly absorbing aerosol with a single scattering albedo of $\tilde{\omega} = 0.99$ is neutral or causes a slightly positive forcing for an albedo of $A = 0.75$ (snow with some vegetation in between) and strong negative forcing for an albedo A of the ground of $A = 0.1$ (ocean). An urban pollution ($\tilde{\omega} = 0.90$) will definitely cause a net heating over a bright ground and net cooling over dark ground. Only for black carbon or the savanna smoke plume is the forcing always positive.

10   WHITE-HOUSE EFFECT VS GREENHOUSE EFFECT

From the previous considerations we know that the atmospheric aerosol reflects some solar radiation back into space by scattering and that the aerosol also transfers a small fraction of the solar energy to the atmosphere by light absorption. These effects can be directly by the aerosol particles themselves or via clouds, the drops of which form by condensation on aerosol particles.

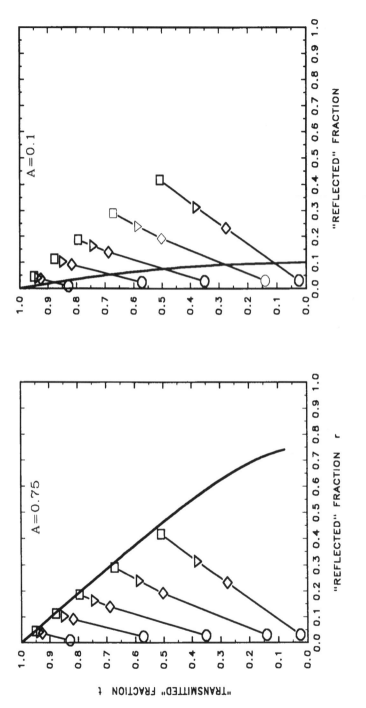

Figure 12. Conditions for an additional aerosol to cause a positive or negative radiative forcing of the atmosphere–earth system. The solid line gives the transmission t and the reflection r an aerosol has to have in order to cause a zero radiative forcing. To the left of the curve the forcing is positive (heating) and to the right it is negative (cooling). The points represent different aerosols: square: $\bar{\omega} = 0.99$; triangle: $\bar{\omega} = 0.95$; diamond: $\bar{\omega} = 0.90$; circles: $\bar{\omega} = 0.40$. Different optical depths are considered and aerosols with the same optical depth but different absorption properties are connected by a line. From the top to the bottom the densities are 0.2, 0.5, 1.0, 2.0 and 4.0. The albedo of the ground is given as a parameter

Reflection and absorption depend on the quantity, size and nature of the particles, the light-scattering and the light- absorbing properties of the particles and their state of mixing, the relative humidity and the growth of particles with humidity, the vertical aerosol load, the surface albedo, the zenith angle of the solar radiation, the number of cloud condensation nuclei, the water vapour available for condensation, the type of clouds formed and others. The source strength of the particles, the lifetime of the particles and their transformation have to be known in order to estimate the effect of the aerosols on radiative forcing. The atmospheric aerosol is highly variably both spatially and temporarily. Of considerable interest obviously is the contribution of man-made aerosols or which changes in radiative forcing might occur if one source is eliminated or another added. It is clear that a complete information on the aerosol is not available and that the missing information has to be obtained by models and assumptions. Thus a considerable uncertainty in radiative forcing does exist. But it is clear that the aerosol particles have an effect. With increasing effort in research on aerosol–climate interaction a better understanding of the mechanisms involved has been achieved. As learning usually occurs with progress of time, estimates on radiative forcing may contradict each other as new information becomes available, but this is completely natural for science in progress.

For completeness it has to be added that the aerosol particles also emit and absorb infrared radiation and in analogy to the greenhouse effects by the gases a greenhouse effect by aerosols is credible. But this effect is small as can be seen by a simple consideration: the global mean optical depth at 550 nm is estimated as $\delta_{550\,nm} = 0.125$, [56] the extinction coefficient of aerosols in the IR ($10\,\mu m$) is about one-tenth of the value at 550 nm [38], the average optical depth in the IR will be $\delta_{10\,\mu m} = 0.0125$, i.e. the transmission is 98.8%. On the other hand, the transmission in the IR due to the greenhouse gases is somewhere around 5%, thus the aerosols are of marginal influence. But even if there were an influence, most of the aerosols are in the lower troposphere with similar temperatures as the surface, thus any IR emissions are similar to those of the ground.

In the previous sections we have discussed the effect of the aerosol on radiation mainly in exaggerated examples. Let us now look at some real data. An early study [57] using a hemispherically symmetric and latitude-dependent aerosol resulted in a radiative forcing due to aerosols between -0.7 and $-1.0\,W\,m^{-2}$. Later Charlson et al. [58] estimated the direct forcing due to sulfates as $-1.02\,W\,m^{-2}$ on a global average and $-1.57\,W\,m^{-2}$ in the more industrialized northern hemisphere with two-thirds contribution by anthropogenic aerosols. Two years later Kiehl and Brieglib [59] halved the values, by considering the whole visible spectral range another mass extinction coefficient has been chosen and another asymmetry parameter has been assumed appropriate. This shows clearly that assumptions, which appear reasonable in both cases, considerably influence the results. Since it is impossible to have information on the aerosol at any volume element of our atmosphere we have to rely on assumptions and it is by all means possible that future work will give different results again. Nevertheless both works have shown that an increased negative forcing does exist in densely populated areas. Data on the extinction and absorption properties of urban and rural aerosols in Europe [60, 61] indicate that a negative forcing due to scattering could be compensated by light absorption due to black carbon.

Estimation of direct forcing by thin aerosols is most simple, because it is proportional to the aerosol loading, since multiple scattering can be neglected. A study concerning the anthropogenic contribution to the forcing due to sulfate emissions, [62] conversion to aerosol particles, the residence time, the mass extinction coefficient (which requires a size distribution of the aerosol formed by oxidation of SO_2 including humidity growth), the albedo of the underlying surface, the cloud cover and the transmission of the atmosphere above the aerosol layer. The result is $-1.1\,\mathrm{W\,m^{-2}}$ with an accuracy of a factor of 2.4.

Besides sulfur emissions, which are one major source for anthropogenic particulates in industrialized regions, biomass burning can contribute vast amounts of particles especially in the regions around the equator. Biomass burning produces particles which also contain soot, but due to the incompleteness of burning the particles also contain considerable amounts of organic material. The light absorption by the black carbon contained in the particles from biomass burning seem to have no positive forcing effect [39]. The estimated radiative forcing is $-0.6\,\mathrm{W\,m^{-2}}$ with an uncertainty of a factor of 2. [63]

In urban environments nitrates give a considerable contribution to the light extinction budget; in the densely populated regions of Europe extinction due to nitrates is frequently larger than the contribution due to sulfates, especially nowadays with efficient sulfur removal. The effect on a global scale might not be very big, since nitrates are volatile and upon dilution may only exist in the gas phase. Organic material is also a major contributor to the extinction budget but unfortunately very little is known about this component of the atmospheric aerosol.

The residence time for aerosols in the atmosphere of about 1 week is very short compared to the residence time for the greenhouse gases of several decades. It is puzzling that one week of aerosol emissions almost compensate for decades of CO_2 emissions. If we stopped combustion of fossil fuels now the greenhouse effect will still be there for decades whereas the white-house effect will be gone within a week. Radiative forcing by aerosols is already like walking on insecure terrain, but this speculation does not have one single piece of safe ground. The many positive and negative feedback mechanisms and the occasionally chaotic climate need to be mentioned here.

Every experimentalist is happy if there is a chance to test a theory. In principle a test is simple: alter one of the parameters and read the results. For the atmosphere on a global scale this is difficult, since, for example, the global sulfur emissions are of the order of 10^{14} g year^{-1} and doubling this amount for an extended period is beyond our reach. The volcanic eruptions have qualitatively demonstrated the effect of increased aerosol load. Except for recent eruptions no aerosol data are available, besides reports mentioning a faint sun which was visible with the naked eye without glare e.g. after the Krakatau eruption, or a sky whiter than usual. But colder summers have been documented after large volcanic eruptions. The eruption of Mount Pinatubo might have caused some global cooling [64] and although the aerosol was mainly in the stratosphere it might still support the climate forcing by anthropogenic aerosols. Industrialized regions are especially suited to investigate aerosol effects. Unfortunately a significant difference between mean temperatures in industrial and comparable remote areas has not yet been found. Apparently the global circulation manages a very efficient heat flow. The sulfur emissions on the northern and southern hemisphere differ considerably. Due to the short residence time the two hemispheres are essentially decoupled for aerosols. Therefore one might expect a difference in radiative forcing north and south of the equator. Significant temperature differences have not been observed. Using a global circulation model [65] the response to different forcing at the two hemispheres has been calculated. The resulting differences are smaller than from the estimates using the difference in radiative forcing. Therefore there seems to be a substantial net heat flow from the southern to the northern hemisphere.

A different approach has been selected [66] by separating day and night temperatures: the white-house effect only is effective during the day, compensating for the greenhouse effect more or less. At night, with almost no interaction of the aerosol with the infrared radiation we have the "pure" greenhouse effect. Therefore the nights become hotter as compared to several decades ago and less warming as expected by the greenhouse effect is found during the daytime. Again it is an indication, but the effect is small and it is difficult to separate it from other effects. Also the IPCC [67] noted an increase in night-time temperature over land.

Very little is known about the white-house effect due to changes in clouds caused by anthropogenic aerosols. Most authors agree that the aerosols have an effect, but have difficulty quantifying it. The problems have already been mentioned and include questions like which particles act as cloud condensation nuclei, what is the thickness of the cloud and—in polluted regions—what is the effect of black carbon. Yet there is experimental evidence of an increased albedo by additional aerosol particles: the cloud reflectivity observed from space increases in the vicinity of shiptracks which can be at least partly an effect of sulfur emitted with the exhaust gases. [50]

Estimated radiative forcings over the industrial period on the global average have been given by the IPCC (1994). [67] The radiative forcing of the long-lived anthropogenic infrared active greenhouse gases (CO_2, CH_4 N_2O, halocarbons) is approximately $+2.5 \pm 0.4 \, \text{Wm}^{-2}$. The anthropogenically caused decrease of ozone in the stratosphere contributes $-0.1 \pm 0.05\,\mathrm{Wm^{-2}}$, the increased tropospheric ozone contributes $+0.4 \pm 0.2\,\mathrm{W\,m^{-2}}$, the direct effect of the tropospheric aerosol is $-0.9 \pm 0.6\,\mathrm{Wm^{-2}}$ and the indirect effect due to changes in clouds (density, number, reflectivity) caused by the anthropogenic aerosol are difficult to estimate and therefore no value but only a range between 0 and $-1.5\,\mathrm{Wm^{-2}}$ is given. Summing this one can see that the white-house effect by the aerosols can largely compensate for the greenhouse effect.

11 CONCLUSION

The author hopes, that the reader realizes that the aerosol has an influence on climate and that climate models can only exist if a variety of assumptions are made, since we do not have sufficient data on the aerosol. So the assumptions are at least to some degree arbitrary. Let me just list some of the variables which are needed for a model: aerosol concentration, vertical and horizontal distribution, single scattering albedo, scattering phase function, cloud-forming properties, cloud thickness, ground albedo. All these variables are needed in three dimensions. When making climate predictions we also need the feedback mechanisms such as change in circulation due to change in radiative forcing, change in ground albedo, additional aerosol sources, change in aerosol properties. It is clear that none of these variables is known accurately.

With this in mind it is worth while finishing this contribution with the most important findings given by the IPCC working group (1995) [68], which has been very carefully analysing all available data and gives a different and more cautious statement:

It is evident that greenhouse gases, caused by human activities, mainly fossil fuel use, land-use change and agriculture, increase in the atmosphere. The direct radiative forcing of the long lived greenhouse gases is estimated as 2.45 $\mathrm{Wm^{-2}}$. Tropospheric aerosols which mainly result from fossil fuel combustion and biomass burning have a negative direct forcing, the IPCC working group estimates it as $-0.5\,\mathrm{Wm^{-2}}$ as a global average. The possibility of additional negative indirect forcing due to clouds of similar magnitude is considered. Since aerosols have a short lifetime compared to the greenhouse gases the forcing can be focused to particular regions, where the forcing can offset the positive forcing due to greenhouse gases. The white-house effect rapidly adjusts to increases or decreases in emissions.

The formulations dealing with the atmospheric aerosol just given are very cautious and from the previous considerations we know why: the database for the atmospheric aerosol, especially the natural aerosol, is very poor. We have to rely on many assumptions and models. In any case the outcome of a model depends on the assumptions made explicitly or implicitly. Finally it has to be mentioned that we have concentrated on radiative forcing. Any secondary effects, such as change in circulation, albedo, cloud cover, etc., are difficult to estimate and much variability in the outcome of a model again depends on the assumption. Therefore, although contributions considering the climate issue may look like hard science, they are not since they use assumptions at its base which are speculative. [69]

Nevertheless, this clearly shows that there is no doubt about the negative forcing due to atmospheric aerosol particles, but there are still many unknown variables. Thus research on all aspects of atmospheric aerosol particles will be a big challenge in the future.

REFERENCES

1. Ramanathan V., Cess R.D., Harrison E.F., Minnis P., Barkstron B.R., Ahmad E., Hartman D. Cloud-radiative forcing and climate: Results from the earth radiation budget experiment. *Science* **243**, 57–63 (1989)
2. Schwartz S.E. The whitehouse effect: Shortwave radiative forcing of climate by anthropogenic aerosols. *J. Aerosol Science* **27**, 359–382 (1996)
3. Jaenicke R. Aerosol physics and chemistry, Chapter 9 of Landolt Börnstein, *Numerical Data and Functional Relationships in Science and Technology*, Vol. 4 *Meteorology*, Subvolume b. *Physical and Chemical Properties of the Air*. (K.H. Helwange and O. Madelung eds.) pp. 402–457, Springer, Berlin (1988)
4. *Universum, das neue Natur & Umwelt, Vulkane ändern das Klima*. Südwest Verlag, Munich, Germany, pp. 72–77 (1983)
5. Broecker W.S. Chaotic climate, *Scientific American*, November 1995, 44–50 (1995)
6. Charlson R.J., Lovelock J.E., Andreae M.O. and Warren S.G. Oceanic phytoplankton, atmospheric sulfur, cloud albedo and climate. *Nature* **326**, 655–661 (1987)
7. Preining O. Aerosol and climate—an overview. *Atmospheric Environment* **25A** 2443–2444 (1991)
8. Horvath H. Remarks and suggestions on nomenclature and symbols in atmospheric optics. *Atmospheric Environment* **28**, 757–759 (1994)
9. Jaenicke R. (1996) Atmospheric aerosol size distribution. Chapter 1 in *Environmental Particles* Vol IV, *Atmospheric Particles*, R.M.-Harrison, Editor (1996)
10. Whitby K.T. The physical characteristics of sulfur aerosol. *Atmospheric Environment* **12**, 135–159 (1978)
11. Deirmendjian D. (1969) *Electromagnetic Scattering of Polydispersions*. Elsevier, New York (1969)
12. Junge C. *Air Chemistry and Radioactivity*. Academic Press, New York (1963)
13. Willeke K. and Whitby K.T. Atmospheric aerosols: Size distribution interpretation. *J. Air Pollution Association* **25**, 529–534 (1975)
14. Born M. and Wolf E. *Principles of Optics*. Pergamon, Oxford (1975)
15. Bohren C.F., Huffman D.R. *Absorption and Scattering of Light by Small Particles*, Wiley Interscience, New York (1983)
16. Chandrasekhar S. (1960) *Radiative Transfer*. Dover, New York (1960)
17. Mie G. Beiträge zur Optik trüber Medien, speziell kolloidaler Metallösungen. *Ann. Phys.* **25**, 377–455 (1908)
18. Debeye P. Der Lichtdruck auf Kugeln von beliebigem Material. *Ann. Phys.* **30**, 57–137 (1909)

19. Van de Hulst H.C. *Light Scattering by Small Particles*. John Wiley (1957), reprinted by Dover, New York (1981)
20. Kerker M. (1969) *The Scattering of Light by Small Particles*. Academic Press, New York (1969)
21. Barber P.W., and S.C. Hill *Light Scattering by Particles: Computational Methods*. World Scientific, Singapore (1990)
22. Berry M.V. and Percival I.C., Optics of fractal clusters such as smoke. *Optica Acta* **33**, 577–591 (1986)
23. Colbeck I., Hardman E.J. and Harrison R.M. (1989) Optical and dynamical properties of fractal clusters such as carbonaceous smoke. *J. Aerosol Sci.* **20**, 765–774 (1989)
24. Reist P.C. and Wilson W. (1989) An interactive model for predicting scattering functions of polydisperse aerosols. Paper presented at the International Conference on Aerosols and Background Pollutions, Galway, Ireland, 13–15 June 1989
25. Gebhart J. and Anselm A. (1988) Effect of particles shape on the response of single particle optical counters. In: *Optical Particles Sizing, Theory and Practice*. G. Gouesbet and G. Gréhan Eds. Plenum, New York, pp. 393–409 (1988)
26. Zerull R.H. Scattering measurements of dielectric and absorbing particles. *Contr. Atm. Physics* **49**, 168–188 (1976)
27. Killinger R.T. and Zerull R.H. Effects of shape and orientation to be considered for optical particle sizing. In: *Optical Particles Sizing, Theory and Practice*. G. Gouesbet and G. Gréhan Eds. Plenum, New York, pp. 419–429 (1988)
28. Zege E.P., Ivanov A.P., Katsev I.L. Image transfer through a scattering medium. Springer Berlin (1991)
29. Henyey L.G., Greenstein, J.L. Diffuse radation in the galaxy *Astrophys. J.* **93**, 70–83 (1941)
30. McCartney E.J. *Optics of the Atmosphere*. John Wiley, New York, 152 pp. (1976)
31. D'Almeida G.A. Der Staubtransport aus der Sahara. PhD. Thesis. Mainz (1983)
32. Esposito F., Horvath H., Romano F., and Serio C. Daily variation of the vertical aerosol optical depth and the horizontal extinction coefficient of the atmospheric aerosol at a rural location in Southern Europe. *Journal of Geophysical Research* accepted for publication (1996)
33. Jeske N. (1988) Meteorological optics and radiometeorology. Chapter 7 of Landolt Börnstein, *Numerical Data and Functional Relationships in Science and Technology*, Vol. 4 *Meteorology*, Subvolume b. *Physical and Chemical Properties of the Air*. (K.H. Helwange and O. Madelung eds.) Springer, Berlin (1988)
34. Horvath H. Atmospheric light absorption, a review. *Atmospheric Environment*, **27A**, 293–317 (1993)
35. Chylek P. and Srivastava V. (1983) Dielectric constant of a composite inhomogeneous medium. *Phys. Rev.* **B27**, 5098–5106 (1983)
36. Chylek P., Ramaswamy V, and Cheng R.J. (1984) Effect of graphitic carbon on the albedo of clouds. *J. Atmospheric Sciences* **41**, 3076–3084 (1984)
37. Bakan S. and Hintzpeter H. (1988) Atmospheric radiation. Landolt Börnstein, *Numerical Data and Functional Relationships in Science and Technology*, Vol. V/4b *Meteorology*, Subvolume b. *Physical and Chemical Properties of the Air*. (K.H. Helwange and O. Madelung eds.) pp. 110–186 Springer, Berlin (1988)
38. D'Almeida G.A., Koepke P., Shettle E.P. *Atmospheric Aerosol. Global Climatology and Radiative Characteristics*. A. Deepak Publishing, Hampton, Virginia, USA (1991)
39. Liousse C., Penner J.E., Chung C. Walton J.J., Molenkamp C.R., Eddleman H., Cachier H. (1995) Modelling carbonaceous aerosols. Paper ES32 presented at

the GAIM Science Conference, Garmisch-Partenkirchen, Germany 25–29 September

40. Shettle E.P. Fenn R.W. (1979) *Models for the Aerosols of the Lower Troposphere and the Effects of Humidity Variations on Their Optical Properties.* Report AFCRL TR 79 0214, Environmental research paper No. 675, Air Force Cambridge Lab., Hambscom A.F.B. NTIS, ADA 085951, 94 pp. (1979)

41. Robinson R.A. and Stokes R.H. (1959) *Electrolyte Solutions.* Butterworth, London (1959)

42. Hänel G. The properties of atmospheric aerosol particles as a function of the relative humidity at thermodynamic equilibrium with the surrounding moist air. *Advances in Geophysics,* 19, Academic Press, New York, USA, pp. 73–188. (1976)

43. Van de Hulst H.C. *Multiple Light Scattering.* Academic Press, New York (1980)

44. Meador W.E. and Weaver W.R. Two-stream approximations to radiative transfer in planetary atmospheres: a unified description of existing methods and a new improvement. *J. Atmos. Sci.* 37, 630–643 (1980)

45. Rozé C., Maheu B., Gréhan G., Ménard J. Evaluation of the sighting distance in a foggy atmosphere by a Monte Carlo simulation. *Atmospheric Environment* 28, 769–775 (1994)

46. Grassl H., Newiger N. Changes of local planetary albedo by aerosol particles. In: *Atmospheric Pollution 1982,* Studies in Environmental Sciences 20, pp. 313–320 edited by M. Benarie, Elsevier, Amsterdam, The Netherlands (1982)

47. Twomey S. *Atmospheric Aerosols.* Developments in Atmospheric Sciences 7, Elsevier, Amsterdam, The Netherlands (1979)

48. Lillequist G.H. and Cehak K. *Allgemeine Meteorologie,* Vieweg & Sohn, Braunschweig, Germany 119 pp. (1984)

49. Pruppacher H.R. and Klett J.D. *Microphysics of Clouds and Precipitation.* D. Reidel, Norwell, Mass. USA (1980)

50. Twomey, S.A., Piepgrass M. and Wolf T.L. An asessment of the impact of pollution on global cloud albedos *Tellus* 36B, 356–366 (1984)

51. Twomey, S., Gall R., Leuthold M. Pollution and cloud reflectance. *Boundary-Layer Meteorology* 41, 335–348 (1987)

52. Hobbs P.V. International Geophysics Series, Vol. 54. Aerosol cloud interactions. Chapter 2 of *Aerosol Cloud Climate Interactions.* Edited by P.V. Hobbs. Academic Press, San Diego, USA, pp. 33–73 (1993)

53. Kondratyev K. Ya. and Binenko V.I. (1987) Optical properties of dirty clouds. *Boundary-Layer Meteorology,* 41, 349–354 (1987)

54. Cess R.D., Zhang M.H., Minnis P., Corsetti L., Dutton E.G., Forgan B.W., Garber D.P., Gates W.L., Hack J.J., Harrison E.F., Jing X., Kiehl J.T., Long C.N., Morcrette J.J., Potter G.L., Ramanthan V., Subasilar, B., Whitlock C.H., Young D.F., Zhou Y. Absorption of clouds: observation versus models. *Science* 267, 496–499 (1995)

55. Liousse C., Devaux C., Dulac F., Cachier H. Aging of savanna biomass burning aerosols; consequences on their optical properties. *J. Atmospheric Chemistry* 22, 1–17 (1995)

56. Toon O.B. and Pollack J.B. A global average model of atmospheric aerosols for radiative forcing calculations. *J. Appl. Meteorol.* 15, 225–246 (1976)

57. Coakley J.A. Jr., Cess R.D., Yurevich R.D. The effect of tropospheric aerosols on the earth's radiation budget: a parameterization for climate models. *J. Atmos. Sci.* 40, 116–138 (1983)

58. Charlson R.J., Langner J., Rohde H., Leovy C.B., Warren S.G. (1991) Perturbation of the northern hemsphere radiative balance by backscattering from anthropogenic sulfate aerosols. *Tellus* 43AB, 152 (1991)

59. Kiehl J. T. and Briegleb B. P. The relative role of sulfate aerosols and greenhouse gases in climate forcing. *Science* **260**, 311–314 (1993)

60. Horvath H. Spectral extinction coefficients of rural aerosol in southern Italy—a case study of cause and effect of variability of atmospheric aerosol. *J. Aerosol Sci.* **27**, 437–453 (1996)

61. Horvath H., Kasahara M., Pesava P. The size distribution and composition of the atmospheric aerosol at a rural and a nearby urban location. *J. Aerosol Sci.* **27**, 417–435 (1996)

62. Penner J.E., Charlson R.J., Laulainen N., Leifer R., Novakov T., Ogren J., Radke L.F. Schwartz S.E., Travis L. Quantifying and minimizing uncertainty of climate forcing by anthropogenic aerosols. *Bull. Amer. Meteorol. Soc.* **75**, 375–400 (1994)

63. Penner J.E., Dickinson R.E., O'Neill C.A. Effects of aerosol from biomass burning on the global radiation budget. *Science* **256**, 1432–1433 (1992)

64. Dutton E.G. and Christy J.R. Solar radiative forcing at selected locations and evidence for global local cooling following the eruptions of El Chicónand Pinatubo. *Geophys. Res. Letters* **19**, 2313–2316 (1992)

65. Cox S. J., Wang W. Ch., and Schwartz S. E. Climate response to radiative forcings by sulfate aeosols and green house gases. *Geophysical Research Letters* **22**, 2509–2512 (1995)

66. Karl T.R., Johnes P.D., Knight R.W., Kukla G., Plummer N., Razuvayev V., Gallo K.P., Lindseay J., Charlson R.J., Peterson T.C. A new perspective on recent global warming: Asymmetric trends of daily maximum and minimum temperature. *Bull. Amer. Met. Soc.* **74**, 1007–1023 (1993)

67. IPCC (Intergovernmental Panel on Climate Change) Climate change 1994. Radiative forcing of climate change and evaluation of the IPCC IS92 emission scenarios. Cambridge University Press, pp. 194–195 (1995) and IPCC Radiative forcing of climate change. *The 1994 Report of the Scientific Assessment Working Group of IPCC. Summary for Policymakers.* World Meteorological Office, United Nations Environmental Programme (1994)

68. IPCC (Intergovernmental Panel on Climate Change) *Working Group I. Summary for Policymakers.* World Meteorological Office, United Nations Environmental Programme (1995)

69. Preining O. Global warming: greenhouse gases versus aerosols. *The Science of the Total Environment* **126**, 199–204 (1992)

Index

Abietane diterpenoids 303
Abietic acid 303
Absorbance spectra 156
Absorbed radiation 569–72
Absorption coefficient 555, 575
Absorption efficiency factor 555
Absorption number 555
Absorption process 555
Accommodation coefficient 207
Acid ammonium nitrate–sulfate solutions 221
Acid ammonium nitrate–sulfate system 222
Adiabatic cloud water content (LWC$_{ad}$) 517
Aerodynamic conditions 32
Aerodynamic diameter 36, 42
Aerodynamic transport. *See* Transport
Aerometric balance studies 443–4
Aerosol analysis
  AAS 115
  AES 118
  ASV 113–14
  GFAAS 115
  ICP-MS 119–20
  NAA 111
  PIXE 108
  XRF 104–6
Aerosol kinetics 204–12
Aerosol samples, INAA 111
Aerosol tunnel sampler 52–3
Aerosol–vapour equilibrium 205–8
Aerosols 558–9
  optical parameters 560–1
  with both aqueous and solid phases 218–25
AES
  applications 118
  fly-ash particles 186
Aethalometer 308, 309, 333
Air quality standards, PAHs 269–70
Airborne particles, structural heterogeneity within 173–202
Aitken particle size range 6
Albedo 555, 558, 561, 565, 571, 580, 584, 586, 589

Alternaria 355
Ambient aerosols 69, 75
  cascade impactors for 71
Ambrosia artemisiifolia 353, 358
Ammonia 125–6, 129, 217
  as precursor to secondary aerosol particles 378–9
Ammonia–nitric acid–ammonium nitrate system 151
Ammonium 130, 134
Ammonium acid sulfate speciation 155
Ammonium bisulfate 156
Ammonium chloride 135
Ammonium nitrate 135, 207, 211, 216–17
Ammonium sulfate 159, 566
Andersen cascade impactor 49–50
Anodic stripping voltammetry (ASV)
  analysis 112–14
  applications 113–14
Antarctic continent 194
Anthropogenic aerosols 180–91, 305
Anthropogenic emissions 98
Anthropogenic particles 476
Anthropogenic sources 98–100
  trace metals 391–3, 414
Aqueous phase particles 214–17
Artificially generated aerosols 195–7
Aspiration 40–1
Aspiration efficiency 40
Asymmetry parameters 557, 558, 579–80
Atmosphere, optical properties 558
Atmosphere–earth system 585–7
  positive or negative radiative forcing 588
Atmospheric aerosol particles 558–69
  basic physical properties 31–42
  nucleation and growth 477–81
  size 41
  sources 1, 30
Atmospheric aerosol sampling 29–94
  aerosol tunnel sampler 52–3
  Black Smoke and SO$_2$ Directive 56
  components of instruments 44
  deposited particles 76–83
  efficiency 49, 67

Index compiled by Geoffrey C. Jones